Lecture Notes in Computer Science 9142

Commenced Publication in 1973
Founding and Former Series Editors:
Gerhard Goos, Juris Hartmanis, and Jan van Leeuwen

More information about this series at http://www.springer.com/series/7407

Ying Tan · Yuhui Shi
Fernando Buarque · Alexander Gelbukh
Swagatam Das · Andries Engelbrecht (Eds.)

Advances in Swarm and Computational Intelligence

6th International Conference, ICSI 2015
held in conjunction with the Second BRICS
Congress, CCI 2015
Beijing, China, June 25–28, 2015
Proceedings, Part III

 Springer

Editors
Ying Tan
Peking University
Beijing
China

Yuhui Shi
Xi'an Jiaotong-Liverpool University
Suzhou
China

Fernando Buarque
Universidade de Pernambuco
Recife
Brazil

Alexander Gelbukh
Instituto Politécnico Nacional
Mexico
Mexico

Swagatam Das
Indian Statistical Institute
Kolkata
India

Andries Engelbrecht
University of Pretoria
Pretoria
South Africa

ISSN 0302-9743   ISSN 1611-3349   (electronic)
Lecture Notes in Computer Science
ISBN 978-3-319-20468-0   ISBN 978-3-319-20469-7   (eBook)
DOI 10.1007/978-3-319-20469-7

Library of Congress Control Number: 2015941139

LNCS Sublibrary: SL1 – Theoretical Computer Science and General Issues

Springer Cham Heidelberg New York Dordrecht London

Printed on acid-free paper

Springer International Publishing AG Switzerland is part of Springer Science+Business Media
(www.springer.com)

Preface

This book and its companion volumes, LNCS vols. 9140, 9141, and 9142, constitute the proceedings of the 6th International Conference on Swarm Intelligence in conjunction with the Second BRICS Congress on Computational Intelligence (ICSI-CCI 2015) held during June 25–28, 2015, in Beijing, China.

The theme of ICSI-CCI 2015 was "Serving Our Society and Life with Intelligence." With the advent of big data analysis and intelligent computing techniques, we are facing new challenges to make the information transparent and understandable efficiently. ICSI-CCI 2015 provided an excellent opportunity for academics and practitioners to present and discuss the latest scientific results and methods as well as the innovative ideas and advantages in theories, technologies, and applications in both swarm intelligence and computational intelligence. The technical program covered all aspects of swarm intelligence, neural networks, evolutionary computation, and fuzzy systems applied to all fields of computer vision, signal processing, machine learning, data mining, robotics, scheduling, game theory, DB, parallel realization, etc.

The 6th International Conference on Swarm Intelligence (ICSI 2015) was the sixth international gathering for researchers working on all aspects of swarm intelligence, following successful and fruitful events in Hefei (ICSI 2014), Harbin (ICSI 2013), Shenzhen (ICSI 2012), Chongqing (ICSI 2011), and Beijing (ICSI 2010), which provided a high-level academic forum for the participants to disseminate their new research findings and discuss emerging areas of research. It also created a stimulating environment for the participants to interact and exchange information on future challenges and opportunities in the field of swarm intelligence research. The Second BRICS Congress on Computational Intelligence (BRICS-CCI 2015) was the second gathering for BRICS researchers who are interested in computational intelligence after the successful Recife event (BRICS-CCI 2013) in Brazil. These two prestigious conferences were held jointly in Beijing this year so as to share common mutual ideas, promote transverse fusion, and stimulate innovation.

Beijing is the capital of China and is now one of the largest cities in the world. As the cultural, educational, and high-tech center of the nation, Beijing possesses many world-class conference facilities, communication infrastructures, and hotels, and has successfully hosted many important international conferences and events such as the 2008 Beijing Olympic Games and the 2014 Asia-Pacific Economic Cooperation (APEC), among others. In addition, Beijing has rich cultural and historical attractions such as the Great Wall, the Forbidden City, the Summer Palace, and the Temple of Heaven. The participants of ICSI-CCI 2015 had the opportunity to enjoy Peking operas, beautiful landscapes, and the hospitality of the Chinese people, Chinese cuisine, and a modern China.

ICSI-CCI 2015 received 294 submissions from about 816 authors in 52 countries and regions (Algeria, Argentina, Australia, Austria, Bangladesh, Belgium, Brazil, Brunei Darussalam, Canada, Chile, China, Christmas Island, Croatia, Czech Republic,

Egypt, Finland, France, Georgia, Germany, Greece, Hong Kong, India, Ireland, Islamic Republic of Iran, Iraq, Italy, Japan, Republic of Korea, Macao, Malaysia, Mexico, Myanmar, New Zealand, Nigeria, Pakistan, Poland, Romania, Russian Federation, Saudi Arabia, Serbia, Singapore, South Africa, Spain, Sweden, Switzerland, Chinese Taiwan, Thailand, Tunisia, Turkey, UK, USA, Vietnam) across six continents (Asia, Europe, North America, South America, Africa, and Oceania). Each submission was reviewed by at least two reviewers, and on average 2.7 reviewers. Based on rigorous reviews by the Program Committee members and reviewers, 161 high-quality papers were selected for publication in this proceedings volume with an acceptance rate of 54.76 %. The papers are organized in 28 cohesive sections covering all major topics of swarm intelligence and computational intelligence research and development.

As organizers of ICSI-CCI 2015, we would like to express our sincere thanks to Peking University and Xian Jiaotong-Liverpool University for their sponsorship, as well as to the IEEE Computational Intelligence Society, World Federation on Soft Computing, and International Neural Network Society for their technical co-sponsorship. We appreciate the Natural Science Foundation of China and Beijing Xinhui Hi-tech Company for its financial and logistic support. We would also like to thank the members of the Advisory Committee for their guidance, the members of the international Program Committee and additional reviewers for reviewing the papers, and the members of the Publications Committee for checking the accepted papers in a short period of time. Particularly, we are grateful to Springer for publishing the proceedings in their prestigious series of *Lecture Notes in Computer Science*. Moreover, we wish to express our heartfelt appreciation to the plenary speakers, session chairs, and student helpers. In addition, there are still many more colleagues, associates, friends, and supporters who helped us in immeasurable ways; we express our sincere gratitude to them all. Last but not the least, we would like to thank all the speakers, authors, and participants for their great contributions that made ICSI-CCI 2015 successful and all the hard work worthwhile.

April 2015

Ying Tan
Yuhui Shi
Fernando Buarque
Alexander Gelbukh
Swagatam Das
Andries Engelbrecht

Organization

Honorary Chairs

Xingui He   Peking University, China
Xin Yao   University of Birmingham, UK

Joint General Chair

Ying Tan   Peking University, China

Joint General Co-chairs

Fernando Buarque   University of Pernambuco, Brazil
Alexander Gelbukh   Instituto Politécnico Nacional, Mexico, and Sholokhov
     Moscow State University for the Humanities, Russia
Swagatam Das   Indian Statistical Institute, India
Andries Engelbrecht   University of Pretoria, South Africa

Advisory Committee Chairs

Jun Wang   Chinese University of Hong Kong, HKSAR, China
Derong Liu   University of Chicago, USA, and Institute
     of Automation, Chinese Academy of Science, China

General Program Committee Chair

Yuhui Shi   Xi'an Jiaotong-Liverpool University, China

PC Track Chairs

Shi Cheng   Nottingham University Ningbo, China
Andreas Janecek   University of Vienna, Austria
Antonio de Padua Braga   Federal University of Minas Gerais, Brazil
Zhigang Zeng   Huazhong University of Science and Technology,
     China
Wenjian Luo   University of Science and Technology of China, China

Technical Committee Co-chairs

Kalyanmoy Deb   Indian Institute of Technology, India
Martin Middendorf   University of Leipzig, Germany

Yaochu Jin   University of Surrey, UK
Xiaodong Li   RMIT University, Australia
Gary G. Yen   Oklahoma University, USA
Rachid Chelouah   EISTI, Cergy, France
Kunqing Xie   Peking University, China

Special Sessions Chairs

Benlian Xu   Changshu Institute of Technology, China
Yujun Zheng   Zhejing University of Technology, China
Carmelo Bastos   University of Pernambuco, Brazil

Publications Co-chairs

Radu-Emil Precup   Polytechnic University of Timisoara, Romania
Tom Hankes   Radboud University, Netherlands

Competition Session Chair

Jane Liang   Zhengzhou University, China

Tutorial/Symposia Sessions Chairs

Jose Alfredo Ferreira Costa Federal University of Rio Grande do Norte, Brazil
Jianhua Liu   Fujian University of Technology, China
Xing Bo   University of Limpopo, South Africa
Chao Zhang   Peking University, China

Publicity Co-chairs

Yew-Soon Ong   Nanyang Technological University, Singapore
Hussein Abbass   The University of New South Wales, ADFA, Australia
Carlos A. Coello Coello   CINVESTAV-IPN, Mexico
Eugene Semenkin   Siberian Aerospace University, Russia
Pramod Kumar Singh   Indian Institute of Information Technology and
                               Management, India
Komla Folly   University of Cape Town, South Africa
Haibin Duan   Beihang University, China

Finance and Registration Chairs

Chao Deng   Peking University, China
Suicheng Gu   Google Corporation, USA

Conference Secretariat

Weiwei Hu   Peking University, China

ICSI 2015 Program Committee

Hafaifa Ahmed	University of Djelfa, Algeria
Peter Andras	Keele University, UK
Esther Andrés	INTA, Spain
Yukun Bao	Huazhong University of Science and Technology, China
Helio Barbosa	LNCC, Laboratório Nacional de Computação Científica, Brazil
Christian Blum	IKERBASQUE, Basque Foundation for Science, Spain
Salim Bouzerdoum	University of Wollongong, Australia
David Camacho	Universidad Autonoma de Madrid, Spain
Bin Cao	Tsinghua University, China
Kit Yan Chan	DEBII, Australia
Rachid Chelouah	EISTI, Cergy-Pontoise, France
Mu-Song Chen	Da-Yeh University, Taiwan
Walter Chen	National Taipei University of Technology, Taiwan
Shi Cheng	The University of Nottingham Ningbo, China
Chaohua Dai	Southwest Jiaotong University, China
Prithviraj Dasgupta	University of Nebraska, Omaha, USA
Mingcong Deng	Tokyo University of Agriculture and Technology, Japan
Yongsheng Ding	Donghua University, China
Yongsheng Dong	Henan University of Science and Technology, China
Madalina Drugan	Vrije Universiteit Brussel, Belgium
Mark Embrechts	RPI, USA
Andries Engelbrecht	University of Pretoria, South Africa
Zhun Fan	Technical University of Denmark, Denmark
Jianwu Fang	Xi'an Institute of Optics and Precision Mechanics of CAS, China
Carmelo-Bastos Filho	University of Pernambuco, Brazil
Shangce Gao	University of Toyama, Japan
Ying Gao	Guangzhou University, China
Suicheng Gu	University of Pittsburgh, USA
Ping Guo	Beijing Normal University, China
Fei Han	Jiangsu University, China
Guang-Bin Huang	Nanyang Technological University, Singapore
Amir Hussain	University of Stirling, UK
Changan Jiang	RIKEN-TRI Collaboration Center for Human- Interactive Robot Research, Japan
Liu Jianhua	Fujian University of Technology, China
Colin Johnson	University of Kent, UK

Chen Junfeng	Hohai University, China
Liangjun Ke	Xian Jiaotong University, China
Farrukh Khan	FAST-NUCES Islamabad, Pakistan
Thanatchai Kulworawanichpong	Suranaree University of Technology, Thailand
Germano Lambert-Torres	Itajuba Federal University, Brazil
Xiujuan Lei	Shaanxi Normal University, China
Bin Li	University of Science and Technology of China, China
Xuelong Li	Xi'an Institute of Optics and Precision Mechanics of Chinese Academy of Sciences, China
Jane-J. Liang	Zhengzhou University, China
Andrei Lihu	Polytechnic University of Timisoara, Romania
Bin Liu	Nanjing University of Post and Telecommunications, China
Ju Liu	Shandong University, China
Wenlian Lu	Fudan University, China
Wenjian Luo	University of Science and Technology of China, China
Chengying Mao	Jiangxi University of Finance and Economics, China
Bernd Meyer	Monash University, Australia
Martin Middendorf	University of Leipzig, Germany
Hongwei Mo	Harbin University of Engineering, China
Jonathan Mwaura	University of Pretoria, South Africa
Yan Pei	The University of Aizu, Japan
Radu-Emil Precup	Polytechnic University of Timisoara, Romania
Kai Qin	RMIT University, Australia
Quande Qin	Shenzhen University, China
Robert Reynolds	Wayne State University, USA
Guangchen Ruan	Indiana University, USA
Indrajit Saha	University of Warsaw, Poland
Dr. Sudip Kumar Sahana	BITMESRA, India
Yuhui Shi	Xi'an Jiaotong-Liverpool University, China
Zhongzhi Shi	Institute of Computing Technology, Chinese Academy of Sciences, China
Xiang Su	OULU, France
Ying Tan	Peking University, China
T.O. Ting	Xi'an Jiaotong-Liverpool University, China
Mario Ventresca	Purdue University, USA
Dujuan Wang	Dalian University of Technology, China
Guoyin Wang	Chongqing University of Posts and Telecommunications, China
Jiahai Wang	Sun Yat-sen University, China
Lei Wang	Tongji University, China
Ling Wang	Tsinghua University, China
Lipo Wang	Nanyang Technological University, Singapore
Qi Wang	Xi'an Institute of Optics and Precision Mechanics of CAS, China

Zhenzhen Wang	Jinling Institute of Technology, China
Fang Wei	Southern Yangtze University, China
Ka-Chun Wong	University of Toronto, Canada
Zhou Wu	City University of Hong Kong, HKSAR, China
Shunren Xia	Zhejiang University, China
Bo Xing	University of Limpopo, South Africa
Benlian Xu	Changshu Institute of Technology, China
Rui Xu	Hohai University, China
Bing Xue	Victoria University of Wellington, New Zealand
Wu Yali	Xi'an University of Technology, China
Yingjie Yang	De Montfort University, UK
Guo Yi-Nan	China University of Mining and Technology, China
Peng-Yeng Yin	National Chi Nan University, Taiwan
Ling Yu	Jinan University, China
Zhi-Hui Zhan	Sun Yat-sen University, China
Defu Zhang	Xiamen University, China
Jie Zhang	Newcastle University, UK
Jun Zhang	Waseda University, Japan
Junqi Zhang	Tongji University, China
Lifeng Zhang	Renmin University, China
Mengjie Zhang	Victoria University of Wellington, New Zealand
Qieshi Zhang	Waseda University, Japan
Yong Zhang	China University of Mining and Technology, China
Wenming Zheng	Southeast University, China
Yujun Zheng	Zhejiang University of Technology, China
Zhongyang Zheng	Peking University, China
Guokang Zhu	Chinese Academy of Sciences, China
Zexuan Zhu	Shenzhen University, China
Xingquan Zuo	Beijing University of Posts and Telecommunications, China

BRICS CCI 2015 Program Committee

Hussein Abbass	The University of New South Wales, Australia
Mohd Helmy Abd Wahab	Universiti Tun Hussein Onn Malaysia
Aluizio Araujo	Federal University of Pernambuco, Brazil
Rosangela Ballini	State University of Campinas, Brazil
Gang Bao	Huazhong University of Science and Technology, China
Guilherme Barreto	Federal University of Ceará, Brazil
Carmelo J.A. Bastos Filho	University of Pernambuco, Brazil
Antonio De Padua Braga	Federal University of Minas Gerais, Brazil
Felipe Campelo	Federal University of Minas Gerais, Brazil
Cristiano Castro	Universidade Federal de Lavras, Brazil
Wei-Neng Chen	Sun Yat-Sen University, China
Shi Cheng	The University of Nottingham Ningbo, China

Ding Wang	Institute of Automation, Chinese Academy of Sciences, China
Qinglai Wei	Northeastern University, China
Benlian Xu	Changshu Institute of Technology, China
Takashi Yoneyama	Aeronautics Institute of Technology (ITA), Brazil
Yang Yu	Nanjing University, China
Xiao-Jun Zeng	University of Manchester, UK
Zhigang Zeng	Huazhong University of Science and Technology, China
Zhi-Hui Zhan	Sun Yat-sen University, China
Mengjie Zhang	Victoria University of Wellington, New Zealand
Yong Zhang	China University of Mining and Technology, China
Liang Zhao	University of São Paulo, Brazil
Zexuan Zhu	Shenzhen University, China
Xingquan Zuo	Beijing University of Posts and Telecommunications, China

Additional Reviewers for ICSI 2015

Bello Orgaz, Gema
Cai, Xinye
Chan, Tak Ming
Cheng, Shi
Deanney, Dan
Devin, Florent
Ding, Ke
Ding, Ming
Dong, Xianguang
Fan, Zhun
Geng, Na
Gonzalez-Pardo, Antonio
Han, Fang
Helbig, Marde
Jiang, Yunzhi
Jiang, Ziheng
Jordan, Tobias
Jun, Bo
Junfeng, Chen
Keyu, Yan
Li, Jinlong
Li, Junzhi
Li, Wenye
Li, Xin
Li, Yanjun
Li, Yuanlong

Lin, Ying
Liu, Jing
Liu, Zhenbao
Lu, Bingbing
Manolessou, Marietta
Marshall, Linda
Menéndez, Héctor
Oliveira, Sergio Campello
Peng, Chengbin
Qin, Quande
Ramírez-Atencia, Cristian
Ren, Xiaodong
Rodríguez Fernández, Víctor
Senoussi, Houcine
Shang, Ke
Shen, Zhe-Ping
Sze-To, Antonio
Tao, Fazhan
Wang, Aihui
Wen, Shengjun
Wu, Yanfeng
Xia, Changhong
Xu, Biao
Xue, Yu
Yan, Jingwei
Yang, Chun

Yaqi, Wu
Yassa, Sonia
Yu, Chao
Yu, Weijie
Yuan, Bo

Zhang, Jianhua
Zhang, Tong
Zhao, Minru
Zhao, Yunlong
Zong, Yuan

Additional Reviewers for BRICS-CCI 2015

Amaral, Jorge
Bertini, João
Cheng, Shi
Ding, Ke
Dong, Xianguang
Forero, Leonardo
Ge, Jing
Hu, Weiwei
Jayne, Chrisina
Lang, Liu
Li, Junzhi
Lin, Ying

Luo, Wenjian
Ma, Hongwen
Marques Da Silva, Alisson
Mi, Guyue
Panpan, Wang
Rativa Millan, Diego Jose
Xun, Li
Yan, Pengfei
Yang, Qiang
Yu, Chao
Zheng, Shaoqiu

Contents – Part III

Neural Networks

Evolutionary and Genetic Algorithms

Fuzzy Systems

Simulation

Image and Texture Analysis

Dimension Reduction

System Optimization

Other Applications

Segmentation and Detection System

Machine Translation

Virtual Management and Disaster Analysis

Other Applications

Neural Networks

The Effective Neural Network Implementation of the Secret Sharing Scheme with the Use of Matrix Projections on FPGA

Nikolay Ivanovich Chervyakov, Mikhail Grigorevich Babenko(✉),
Nikolay Nikolaevich Kucherov, and Anastasiia Igorevna Garianina

North-Caucasus Federal University NCFU Stavropol, Stavropol, Russia
k-fmf-primath@stavsu.ru, mgbabenko@ncfu.ru,
{nik.bekesh,xrizoberil_09}@mail.ru

Abstract. In this paper neural network implementation of the modified secret sharing scheme based on a matrix projection is offered. Transition from a finite simple Galois field to a complex field allows to reduce by 16 times memory size, necessary for storage of the precalculated constants. Implementation of the modified secret sharing scheme based on a matrix projection with use of the neural network of a finite ring for execution of modular arithmetical addition and multiplication operations in a finite field allows to save on average 30% of the device area and increases the speed of scheme's work on average by 17%.

Keywords: Secret Sharing Scheme · Matrix projections · Neural network of a finite ring

1 Introduction

Secret sharing schemes (SSS) have found wide application in creation of group protocols in systems of information security. Firstly SSS were proposed by Shamir [19] and Blakely [4] independently from each other. One of the researches directions is creation of multiple secret sharing schemes. In work [2] the multiple secret sharing scheme based on matrix projections which is used for creation of copyright of images protection systems by means of water signs [17], graphic secret sharing schemes [1,16,18], the pro-active secret sharing schemes [3,18], secret sharing schemes on general linear groups [9] is offered. In the work [10] parallel hardware implementation of the graphic secret sharing scheme based on a method of matrix projections is offered, in work it is shown that the hardware implementation of the offered algorithms on FPGA on average is 300 times faster, than implementation of the same algorithms on CPU. We will note that in case of the graphic secret sharing scheme implementation in work [10] for generation an accidental matrix they use the pseudorandom number generator constructed on shift registers with the linear back coupling for which implementation the sequential scheme with digit carry is used. It is offered to use pseudorandom

© Springer International Publishing Switzerland 2015
Y. Tan et al. (Eds.): ICSI-CCI 2015, Part III, LNCS 9142, pp. 3–10, 2015.
DOI: 10.1007/978-3-319-20469-7_1

number generator based on inversed congruent sequences and implemented in residue number system (RNS) from work [12], in which the unit of residue to binary conversion is improved by means of the method from work [6] application and its parallel implementation with use of neural network of a finite ring [8].

In the next section we define modification of the secret sharing scheme constructed on matrix projections over a complex field and determinate effective arithmetic operations implementation over the complex field. In Sect. 3 the effective neural networking implementation of SSS is realized.

2   Modification of the Secret Sharing Scheme Constructed on Matrix Projections over a Complex Field

2.1   Matrix Projections over a Complex Field

We will propose defeninition of a matrix projection over a complex field by analogy with operation of a matrix projection from work [2], having replaced operation of a matrix transposing over the valid field with complex conjugation of a matrix for the purpose of more simple matrix projection computation.

Definition 1. *Let A be a complex matrix of the size $m \times k$ and the rank k $(m \geq k > 0)$, and*

$$S = A\left(A^*A\right)^{-1}A^*, \tag{1}$$

where $(\cdot)^*$ - complex conjugate matrix. We will call a matrix S of the size $m \times m$ a matrix A projection, and we will designate this operation: $S = proj\,(A)$

We will calculate vectors v_i using k linearly independent vectors x_i of the size $k \times 1$, by the formula: $v_i = Ax_i$,

where $1 \leq i \leq k$, thus we will obtain vectors v_i of the size $m \times 1$ and we will put them in a vector $B = [v_1, v_2, \ldots, v_k]$.

Theorem 1. *For each complex matrix A of the size $m \times k$ and the rank k $(m \geq k > 0)$ and matrix $B = [v_1, v_2, \ldots, v_k]$, where $v_i = Ax_i$, for $i = 1, 2, \ldots, k$ and k linearly independent vectors x_i of the size $k \times 1$, the equality $proj\,(A) = proj\,(B)$ is satisfied.*

Proof. As $B = [v_1, v_2, \ldots, v_k]$, where $v_i = Ax_i$, vector B can be represented as: $B = [v_1, v_2, \ldots, v_k] = [Ax_1, Ax_2, \ldots, Ax_k] = A[x_1, x_2, \ldots, x_k] = AX$, where X is matrix $X = [x_1, x_2, \ldots, x_k]$, that consists of k linearly independent vectors x_i. Therefore, the rank of a square matrix X of the size $k \times k$ is k. We will calculate the projection of the matrix B.

$$proj\,(B) = B\left(B^*B\right)^{-1}B^* = AX\left(X^*A^*AX\right)^{-1}(AX))^* \tag{2}$$

As complex matrixes X and X^* of the size $k \times k$ and the rank k, are square there are reciprocal matrixes X^{-1} and $(X^*)^{-1}$, therefore the formula (2) can be rewritten as: $proj\,(B) = AX\left(X^*A^*AX\right)^{-1}(AX))^* = AXX^{-1}\left(A^*A\right)^{-1}(X^*)^{-1}X^*A^* = AE\left(A^*A\right)^{-1}EA^* = A\left(A^*A\right)^{-1}A^* = proj\,(A)$.

The matrix S has the following properties:

1. $SA = A$. As $SA = A(A^*A)^{-1}A^*A = EA = A$.
2. S - idempotent matrix, that is $S^2 = S$.
   As $S^2 = A(A^*A)^{-1}A^*A(A^*A)^{-1}A^* = AE(A^*A)^{-1}A^* = S$.
3. $\operatorname{tr}(S) = \operatorname{rank}S = k$, where $\operatorname{tr}(S) = \sum_{i=1}^{m} s_{i,i}$. As matrix S is idempotent matrix.

2.2 The Modulo p Arithmetic Operations Implementation over the Complex Field, Where p Is Prime Number $p = 4k + 1$

For effective computation S under the prime module p, where $p = 4k + 1$, it is possible to use approach from works [13–15] which allows to reduce number of elementary operations of multiplication and addition for execution of basic arithmetical operations with complex numbers, such as addition and multiplication under the module and modular inverse.

Let $y = a_y + b_y i$ be a complex number, where $a_y, b_y \in N \cup (0)$. We will represent it with a pair $A_y = |a_y + jb_y|_p$ and $B_y = |a_y - jb_y|_p$, where $j^2 \equiv 1 \bmod p$ and $j \in F_p^*$, then the complex number y under module p can be represented as $y \bmod p \to (A_y, B_y)$.

Modulo p arithmetic operations with complex numbers $y = a_y + b_y i$ and $z = a_z + b_z i$, are calculated as follows:

1. Addition: $(y + z) \bmod p \to \left(|A_y + A_z|_p, |B_y + B_z|_p\right)$.
2. Multiplication: $(y \times z) \bmod p \to \left(|A_y \times A_z|_p, |B_y \times B_z|_p\right)$.
3. Complex conjugation: $\bar{y} \bmod p \to (B_y, A_y)$.
4. Complex number y inversion under module p:

   $$y^{-1} \bmod p \to \left(|B_y R_y|_p, |A_y R_y|_p\right), \text{ where } R = \left|\left(|A_y B_y|_p\right)^{-1}\right|_p$$

Thus, the use of complex number representation under the prime module $p = 4k + 1$ allows to reduce number of multiplication operations with integers in case of multiplication of complex numbers execution. Operation of a complex number conjugation doesn't require any arithmetical actions. Arithmetical operation of finding the inverse over a finite field requires three modular multiplication more than computation of modular inverse over a simple field.

2.3 Secret Sharing Scheme

Let the secret be represented as a square matrix I of the size $m \times m$, we need to share it among n participants so as each k participants could recover a confidential matrix I, and each $k - 1$ participants, having their parts integrated, could learn nothing about the secret.

Phases of secret sharing:

1. We generate a prime number p of l bit length so that $p \equiv 1 \bmod 4$.
2. We calculate j, where $j \equiv -1 \bmod p$.
3. We construct a complex matrix A of the size $m \times k$ and the rank k, where $m > 2(k-1) - 1$.
4. We generate n random vectors x_i of the size $k \times 1$ so as each k vectors are linearly independent.
5. We calculate vectors $v_i = (A \times x_i) \bmod p$ for $i = \overline{1,n}$.
6. We calculate $S = \text{proj}(A)$
7. We calculate $R = I \oplus \text{ComplexToInt}(S)$, where operation assigns number $\left(A_{S_{i,j}} << l\right) + B_{S_{i,j}}$ to each complex matrix element $S_{i,j} \bmod p \rightarrow \left(A_{S_{i,j}}, B_{S_{i,j}}\right)$.
8. We delete matrixes I, x_i and S.
9. We display matrix R and prime number p.
10. We give to each i participant of the scheme his part of a secret v_i, where $i = \overline{1,n}$.

Phase of secret recovering:

1. We construct matrix B, using k parts: $B = [v_1, v_2, \ldots, v_k]$.
2. We calculate matrix B projections: $S = \text{proj}(B)$.
3. We check $\text{tr}(S) = k$.
4. We calculate matrix $I = R \oplus \text{ComplexToInt}(S)$.

3   The Effective Neural Networking Implementation of SSS

3.1   Neural Network of a Finite Ring

The Neural Network (NN) is a high-parallel dynamic system with topology of directed graph which can output information by means of response of its status to input influences. In NN processing elements and directed graphs are called as nodes [8, 11]. The structure of data processing algorithm, provided in residue number system, also as well as NN structure have natural parallelism that allows to use NN as the formal device to describe the algorithm.

From this point of view algorithms of modular computation correspond to algorithms of computation by means of Basic processing Elements (artificial neurons). Artificial neural networks and the main modular structures represent the connective devices received by serial connection among themselves of Basic Elements. Neural and modular educations layerwise will be defined if the algorithm of connection of Basic Elements is set.

We will consider the general approach of application of NN to computation in finite rings and to formation of the NN model of a finite ring (NNFR). Thus neurons are arithmetical elements which have characteristics of the operator on the module, but not the ordinary non-linear functions of activation applied when training NN. The analysis of arithmetic of a finite ring showed that the computational model based on the iterative mechanism of reducing by the module is the main operation in case of modular data handling.

For effective implementation of the secret sharing scheme on matrix projections it is necessary to expedite addition and multiplication operations under the module. Computation in a finite ring (field) can be defined as follows.

NN architecture. The general interpretation of NN architecture is a mass parallel interdependent network of simple elements and its hierarchical arrangement. The NN structure somewhat models biological nervous system.

Power of NN is its ability to use the initial knowledge base for the solution of the existing problem. All neurons work is competitive, and direct computation is influenced by the knowledge ciphered in communications of a network [5, 8, 21, 22].

Interaction of neurons is considered in three-level hierarchy of the network consisting of layers [8, 20, 23]:

1. Display of parameter (this layer contains the residual connected to the weighed value of each computing discharge).
2. Display of bit computation (defines the function of a finite ring applied to each computing discharge).
3. Display of operation of a finite ring (defines the main operations which are used for implementation of arithmetics of a finite ring).

Well consider the structure of NNFR from work [7]. Based on a computational model of a finite ring the main operator in which is the operator of extraction of separate discharges of binary representation of the converted number, multi-layer subnets can be constructed. The structure of the subnet is shown in figure 1, where synaptic weights are $z_i = \left|2^i\right|_p$, $i = 0, 1, \ldots, n - 1$.

1. Assembly (it is used for collection of the inputs belonging to one binary place of input sources). The result of binary to residue conversion operations, multiplication, addition calculated by means of NNFR is function of the amount of the weighed input discharges.
2. The computing. The result of computation is determined by the positive logic. The end result of NNFR will have the steady form.

3.2   The Modeling

The primary reason for implementing secret image sharing in the FPGA hardware is to decrease execution time. Execution time is of particular concern when dealing with large images that are required to be processed in real-time.

The decrease in execution time is primarily due to the parallel computations and the intrinsic pipelined architecture of the fine-grained FPGA. FPGA hardware in the pipelined architecture allows the simultaneous computation of multiple data sets. Furthermore, the high speed I/O capabilities of the hardware

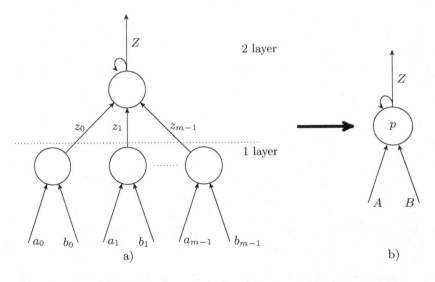

Fig. 1. Structure of a subnet (a) and its symbolic designation (b)

platform, such as Ethernet and USB ports, facilitate high speed data communication to the Virtex 6 XC6VLX760. The FPGA hardware pipeline has an observed initial delay of 124 clock cycles.

The secret sharing schemes are being modeled in case of $k = 4$ for images of the size 128×128, 256×256, 512×512, 768×768, 1024×1024. In case of modeling scheme from work [10] we do all the computations under the module $p = 251$, while in proposed scheme $p = 17$. As $251 < 17^2$, the schemes has the same level of security. According to work [10] the array of constants of multiplicative inverses consisting of 250 8-bit words is stored in memory while in case of implementation of the offered scheme the array of 16 5-bit words and an array of synoptic weights that consists of 9 5-bit words are stored. Consequently memory size necessary for storage precalculated constants in work [10] is 16 times more than memory size necessary for storage precalculated constants of the proposed scheme. For finding a reciprocal matrix we use parallel architecture from work [10] where all modular operations are realized with use of a neural network of a finite ring. Simulation is carried out on a board of Virtex 6 XC6VLX760.

From results of simulation it is possible to draw(figure 2) a conclusion that neural network implementation of the secret sharing scheme allows to save on matrix projections on average 30% of the device area and to increase the speed of schema work on average for 17%.

4   Conclusion

In this paper the neural networking implementation of the modified secret sharing scheme based on matrix projection is proposed. Transition from a finite simple Galois field to a complex field allows to reduce by 16 times memory size

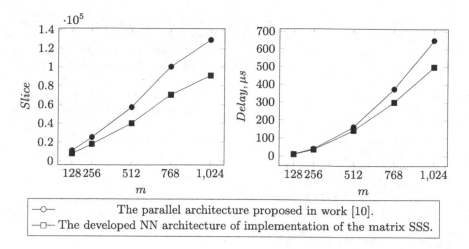

Fig. 2. Technical parameters of implementation of parallel architecture of secret sharing schemes on FPGA

necessary for storing precalculated constants. Implementation of the modified secret sharing scheme of the neural network of a finite ring based on a matrix projection with use of modular arithmetic addition and multiplication in a finite field allows to save on average 30% of the device area and to increase the speed of schema work on average for 17%.

Acknowledgments. Current work was performed as a part of the State Assignment of Ministry of Education and Science (Russia) No. 2563.

References

1. Bai, L.: A reliable (k, n) image secret sharing scheme. In: Autonomic and Secure Computing 2nd IEEE International Symposium on Dependable, pp. 31–36 (2006)
2. Bai, L.: A strong ramp secret sharing scheme using matrix projection. In: International Symposium on World of Wireless, Mobile and Multimedia Networks, WoW-MoM 2006 (2006)
3. Bai, L., Zou, X.: A Proactive Secret Sharing Scheme in matrix projection method. Int. J. Security and Networks 4(4), 201–209 (2009)
4. Blakley, G.: Safeguarding crypographic keys. In: Proceedings of the AFIPS 1979 National Computer Conference, vol. 48, pp. 313-317 (1997)
5. Chervyakov, N.I.: The conveyor neural network of a finite ring. Patent RU 2317584 from (30.05.2008)
6. Chervyakov, N.I., Babenko, M.G., Lyakhov, P.A., Lavrinenko, I.N.: An Approximate Method for Comparing Modular Numbers and its Application to the Division of Numbers in Residue Number Systems. Cybernetics and Systems Analysis 50(6), 977–984 (2014)
7. Chervyakov, N.I., Galkina, V.A., Strekalov, U.A., Lavrinenko, S.V.: Neural network of a finite ring. Patent RU 2279132 from (07.08.2003)

8. Chervyakov, N.I., Sakhnyuk, P.A., Shaposhnikov, V.A., Makokha, A.N.: Neurocomputers in residual classes. M.: Radiotechnique, p. 272 (2003)
9. Dong, X., Zhang, Y.: A multi-secret sharing scheme based on general linear groups. In: International Conference on Information Science and Technology (ICIST), pp. 480–483 (2013)
10. Esposito, R., Mountney, J., Bai, L., Silage, D.: Parallel architecture implementation of a reliable (k, n) image sharing scheme. In: 14th IEEE International Conference on Parallel and Distributed Systems, pp. 27–34 (2008)
11. Galushkin, A.I: The theory of neural networks. M.: INGNR, p. 416 (2000)
12. Gayoso, C.A., Gonzalez, C., Arnone, L., Rabini, M., Castineira Moreira, J.: Pseudorandom number generator based on the residue number system and its FPGA implementation. In: 7th Argentine School of Micro-Nanoelectronics, Technology and Applications (EAMTA), pp. 9–14 (2013)
13. Jenkins, W.K., Krogmeier, J.V.: The Design of dual-mode complex signal processors based on Quadratic Modular Number Codes. IEEE Transactions on Circuits and Systems **34**, 354–364 (1987)
14. Jullien, G.A., Krishnan, R., Miller, W.C.: Complex digital Signal Processing over ftnite rings. IEEE Transactions on Circuits and Systems **34**, 365–377 (1987)
15. Krishnan, R., Jullien, G.A., Miller, W.C.: Complex digital signal processing using quadratic Residue Number Systems. IEEE Transactions on ASSP **34**, 166–177 (1986)
16. Patil, S., Deshmukh, P.: Enhancing Security in Secret Sharing with Embedding of Shares in Cover Images. International Journal of Advanced Research in Computer and Communication Engineering **3**(5), 6685–6688 (2014)
17. Patil, S., Deshmukh, P.: Verifiable image secret sharing in matrix projection using watermarking. In: International Conference on Circuits, Systems, Communication and Information Technology Applications (CSCITA), pp. 225–229 (2014)
18. Patil, S., Rana, N., Patel, D., Hodge, P.: Extended Proactive Secret Sharing using Matrix Projection Method. International Journal of Scientific & Engineering Research **4**(6), 2024–2029 (2013)
19. Shamir, A.: How to share a secret. Communications of the ACM **22**(11), 612–613 (1979)
20. Zhang, D., Zhang, D.: Parallel VLSI Neural System Designs, vol. 257. Springer-Verlag, Berlin (1998)
21. Zhang, C.N., Yun, D.Y.: Parallel designs for Chinese remainder conversion. In: IEEE 16-th International Conference on Parallel Processing - ICPP, pp. 557–559 (1987)
22. Zang, D., Jullien, G.A., Miller, W.C.: A neural-like approach to finite ring computation. IEEE Transactions on Circuits and Systems **37**(8), 1048–1052 (1990)
23. Zhang, D., Jullien, G.A., Miller, W.C.: VLSI implementations of neural-like networks for finite ring computations. In: Proceedings of the 32nd Midwest Symposium on Circuits and Systems, vol. 1, pp. 485–488 (1989)

A ROP Optimization Approach
Based on Improved BP Neural Network PSO

Jinan Duan[1,2(✉)], Jinhai Zhao[1,2], Li Xiao[1,2], Chuanshu Yang[1,2], and Changsheng Li[1,2]

[1] Sinopec Research Institute of Petroleum Engineering, Beijing 100029, China
{duanjn.sripe,Zhaojh,xiaoli.sripe,yangcs,lics}@sinopec.com
[2] China University of Petroleum, Beijing 100029, China

Abstract. Effective optimization of ROP (Rate of Penetration) is a crucial part of successful well drilling process. Due to the penetration complexities and the formation heterogeneity, traditional way such as ROP equations and regression analysis are confined by their limitations in the drilling prediction. Intelligent methods like ANN and PSO become powerful tools to obtain the optimized parameters with the accumulation of the geology data and drilling logs. This paper presents a ROP optimization approach based on improved BP neural network and PSO algorithm. The main idea is, first, to build prediction model of the target well from well logs using BP neural network, and then obtain the optimized well operating parameters by applying PSO algorithm. During the modelling process, the traditional BP training algorithm is improved by introducing momentum factor. Penalty function is also introduced for the constraints fulfillment. We collect and analyze the well log of the No.104 well in Yuanba, China. The experiment results show that the proposed approach is able to effectively utilize the engineering data to provide effective ROP prediction and optimize well drilling parameters.

Keywords: ROP optimization · BP neural network · PSO optimization · ROP modelling · Intelligent well drilling

1   Introduction

The ROP in petroleum engineering usually refers to the speed of moving forward of the drilling tools during the drilling process. It is an important technical indicator to the oil/gas development process. With the deepening of oil/gas exploration and development, successful well drilling has become increasingly difficult. In many cases, oil and gas reservoirs lie deeply underground. It is common that the depths range from 5,000 to 10,000 meters underground. Due to the complex formation properties, so-called non-productive time takes over a very high proportion in the total drilling time[1]. All these factors ultimately lead to longer drilling cycles, and a low overall ROP, which seriously slow down the progress of the exploration and development. How to obtain the best operating parameters of the optimized ROP remains a problem of the industry.

© Springer International Publishing Switzerland 2015
Y. Tan et al. (Eds.): ICSI-CCI 2015, Part III, LNCS 9142, pp. 11–18, 2015.
DOI: 10.1007/978-3-319-20469-7_2

An important job to effectively optimize ROP is modelling and predicting. Unfortunately, most of the existing ROP predicting methods all seriously rely on a lot of physical experiment sor human experiences. There are limitations for the on-site practices using these methods, so it is necessary to find a convenient and relatively accurate ROP prediction method. In recent years, the applications of artificial intelligence methods in the petrol engineering have been gradually evolving. In this paper we proposed a ROP optimization approach based on the improved BP neural network and PSO. The approach is mainly to build a prediction model of target well through the improved BP neural network from well logs by introducing momentum factor, and then use PSO to find the optimized operating parameters. It is showed in the results that the proposed approach is able to obtain better operating parameters which effectively optimize ROP.

2   ROP Modelling and Predicting

2.1   Traditional Modelling Approaches

The researchers have found several major factors affecting ROP during the last decades, including weight of bit, rotation rate, drilling pressure and other parameters. They also summarized the basic relation between ROP and these various factors by establishing a series ROP equations, which are usually regarded as the traditional approaches to ROP prediction[2]. The Bourgoyne and Young Equation[3] is a widely used ROP equation:

$$R = \frac{KC_pC_H(W-W_0)N^b}{1+CH}$$
$$\frac{dH}{dt} = e^{(a_1 + \Sigma \, a_j x_j)} \tag{1}$$

Where R is generally used for penetration rate; N is speed; W is weight of bits; W_0 is the threshold bit weight; S_d is lithology parameter; H is teeth wear; K, b, C are constants; C_p is pressure difference coefficient; C_H is hydraulics coefficient. Most of the practical ROP prediction applications are the improvements or modifications upon it[4].

However, the disadvantages of the equations are obvious. First, there are usually too many undetermined coefficients in the equations, and the common way to determine the values is by directly assigning according human experiences or using the multiple regression, which are both not precise enough[5]. In practice, the prediction error is commonly more than 30%, and ROP equations are regarded as useless[6]. Thus, ROP equations are abandoned by many engineers[7]. Second, in order to get relatively precise prediction result, the ROP equations require a huge amount of work to do, including all kinds of condition simulations and physical experiments, which all bring a lot of cost of money and human resources[8]. At last, maybe the most frustrating one, the ROP equations lack a good adaptability, which directly leads to the recalibration to the equations for the next application. Engineers have to repeat the experiments each time in different areas with different geology conditions[9].

2.2   Proposed ROP Prediction Approach

Due to the disadvantages of the traditional methods, we turned to artificial intelligence methods to try to get better prediction result with less cost. BP neural network is one of most widely used neural network in many engineering fields. The basic idea of BP training algorithm is to propagate the errors back forward by changing the weights with an additional adjustment through all layers[10]. In this paper we introduce the momentum factor to improve basic BP neural network, and its core propagation method is as follows:

$$W_{jk}(t+1) = W_{jk}(t) + a_j \times \Delta_k + a \times \left(W_{jk}(t) - W_{jk}(t-1)\right). \qquad (2)$$

Where W_{jk} is the weight of links between layer j and k; t is number of training times; a_j is output of nodes in layer j; Δ_k is the adjustment of layer k. Based on the improved BP neural network, we finally build a standard BP network with three layers z upon these data. As Fig.1 shows, the input layer has 6 nodes, covering pressure difference, mud density, weight of bits, rotation rate, formation drillability and delivery capacity; only node in the output layer is ROP.

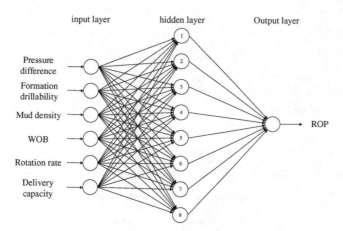

Fig. 1. The ROP prediction BP network

The training algorithm is based on the improved training algorithm mentioned in the last section. The momentum factor a is determined according to the training error by the function showed as follow:

$$\alpha = \begin{cases} 0, & E(t) > 1.05 \times E(t-1) \\ 0.95, & E(t) < E(t-1) \\ \alpha & else \end{cases} \qquad (3)$$

Where, E is sum of error squares，t is the number of training times.

2.3   Prediction Modelling

We have collected 408 samples in total from 104# well logof Yuanba in China. Each sample represented the geology condition and working parameters for a certain meter underground. And by using normalization technologies, we finally made the samples ready for the prediction modelling process. All of samples are loaded to the software, and are split into two groups, two third as the training samples, and the rest as the generalizing samples. We train the network with the basic and improved BP algorithms for hundreds of times separately. The results shows that the improved algorithms greatly enhanced the network performance. As Fig. 2 shows, the improved BP training method make network converge faster, and the improved network is able to jump out of the local optimum.

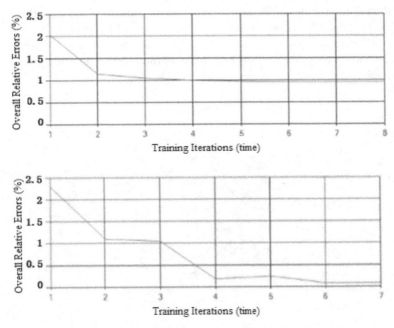

Fig. 2. Typical training curves of basic BP and improved BP

Similarly, the improved BP network naturally has a better generalization performance. Fig. 3 shows a typical generalization results of the two BP networks.

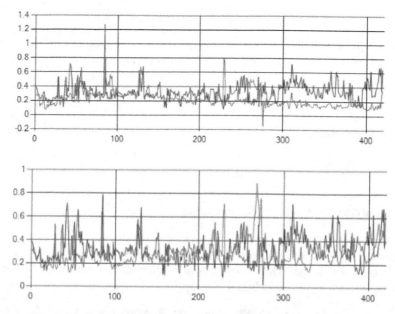

Fig. 3. Typical generalization results of basic BP and improved BP

The results show that with the improved BP neural network, we get a ROP prediction model whose error is below 15%, which is much more precise than the traditional way.

3   ROP Optimization by Using PSO

3.1   Optimizing Parameters and Target Function

After obtaining a satisfying ROP prediction model, the next job is to try to make model output reach the maximum value. There are a few number of parameters that can be adjusted, however, some of them are usually not allowed to change because they are often coupled with other performance aspects of well drilling such as wellbore stability. Inappropriate adjustment to theses parameters may cause drilling accidents, which will eventually decrease the ROP. On the other hand, those parameters that can be changed also have value boundaries.

In this case, we have to decide which parameters to optimize.According to the drilling practice, we have chosen two major operating parameters: the WOB and the rotation rate. The two parameters are allowed and easily to adjust in the practice, and they are usually adjusted manually in many real world well drilling process. The manually adjustment mainly rely on human experiences. Considering the value boundaries, we introduces the penalty function to deliberately fulfill the constraints of the problem, which will confine the particles travelling within the boundaries as possible as they can, and we finally establish the target function of the ROP optimization:

$$fitness = n + 9 \times p(w) + 8 \times p(r). \tag{4}$$

where, n represent the BP network output, and is determined by the following function, in which N_{output} is the real output of the network:

$$n = \begin{cases} 10, & N_{output} \leq 0 \\ N_{output}, & else \end{cases}. \tag{5}$$

And w is the WOB, r is the rotation rates, p is a value boundaries penalty function:

$$p(x) = \begin{cases} 1, & x \geq 2 \; or \; x \leq 0 \\ 0, & else \end{cases}. \tag{6}$$

3.2   Applying PSO to ROP Optimization

The PSO algorithm is a powerful optimized solution searching tool. We have implemented a PSO searching software using .NET Framework 4.5 and C# 5.0, which support us to create particle swarm, adjust searching preferences and control the searching process, as shown in Fig. 4.

Fig. 4. ROP optimization software

In this case study, we create a swarm optimizer that is composed of 20 particles, and the velocity position updating method is as follow:

$$V_{i+1} = w \times V_i + 2 \times \text{rand}() \times (pBest - X_i) + 2 \times \text{rand}() \times (gBest - X_i). \tag{7}$$

$$X_{i+1} = X_i + V_i. \tag{8}$$

where, w is the linearly decreasing inertiaweight, which is determined by the following function for each time searching:

$$w(t) = \frac{(w_{ini} - w_{end})(T - t)}{T} + w_{end}. \tag{9}$$

Typically, we set w_{ini} to 0.9, and w_{end} to 0.4. By applying above swarm optimizer, we load the ROP model and samples in the last chapter, and finally obtain an optimized ROP curve.

3.3   Discussions

As the Fig. 5 shows, the optimized parameters directly lead to a better ROP than the original output. The average relative increasement is about 25%. The operating parameters sequence are able to be saved for the future applications such as drilling simulation or control. The experiment result has showed that PSO is an effective way to find the best input parameters in the ROP optimizing problem. The result is exciting, and is able to avoid the disadvantages of manual adjustment.

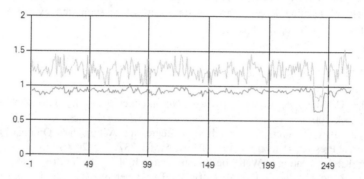

Fig. 5. Comparison between PSO optimized ROP and the original BP output

Interestingly, even if our predicting model are still not 100% accurate, as shown in Fig.6, the proposed approach can still obtain a better ROP than the real process in the most of time.

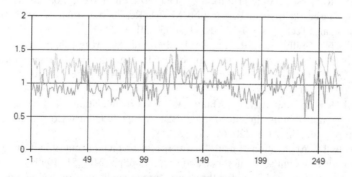

Fig. 6. Comparison between PSO optimized ROP and real-world data comparison

4   Conclusions

In this paper, we have presented a ROP optimization approach which makes full use of the self-learning ability of the ANN and the rapid solution searching ability of the PSO. The approach successfully retrieves the ROP prediction model from the well logs, and also obtains the best operating parameters through PSO algorithms. Case study results shows that our approach is reliable and stable. The optimizing result are also exciting, the approach are proved to be able to increase the ROP effectively. The work in this paper indicates that we have proposed a simple, effective and efficient method to optimize the ROP for the modern petroleum industry.

Acknowledgements. The work presented in this paper is part of the project "Research and Development of Drilling Risk Control System", which is supported by the Science and Research Program of Sinopec Group in 2015.

References

1. Rhodes, C.: Intelligent Planning Reduces Non-productive Drilling Times. Oilfeld Review **13**(2), 1–4 (2001)
2. Bourgoyne, A.T, Young, F.S. A Multiple Regression Approach to Optimal Drilling and Abnormal Pressure Detection. SPEJ, Trans. AIME, **257**, 371 (2000)
3. Wang, K.: Application of Well Log Information in ROP Prediction. **16**(3), 889-900 (2007)
4. Lin, Y., Zong, Y., Zheng, L., Shi, T., Li, R.: The Developments of ROP Prediction for Oil Drilling. Petroleum Drilling Techniques **32**(1), 10–13 (2004)
5. Bahari, A., Seyed, B.: Trust-region approach to find constants of bourgoyne and young penetration rate model in khangiran iranian gas field. In: SPE Latin American and Caribean Petroleum Engineering Conference, SPE 107520, Buenos Aires, vol. 26, pp. 20-22 (2007)
6. Irawan, S., Rahman, A.M.A., Tunio, S.Q.: Optimization of Weight on Bit During Drilling Operation Based on Rate of Penetration Model. Research Journal of Applied Sciences, Engineering and Technology **4**(12), 1690–1695 (2012)
7. Marana, A.N., Guilherme, I.R., Papa, J.P., Ferreira, M., Miura, K., Torres, F.A.C.: An intelligent system to detect drilling problems through drilled cuttings return analysis. In: IADC/SPE Drilling Conference and Exhibition, IADC/SPE 128916, New Orleans, pp. 103-110 (2010)
8. Fruhwirth, R.K., Thonhauser, G., Mathis, W.: Hybrid simulation using neural networks to predict drilling hydraulics in real time. In: SPE Annual Technical Conference and Exhibition, San Antonio, pp. 24-27 (2006)
9. Dashevskly, D.: Application of neural networks for predictive control in drilling dynamics. In: SPE Annual Technical Conference and Exhibition, pp. 248-251 (1999)
10. Karri, V.: RBF neural network for thrust and torque predictions in drilling operations. In: Third International Conference Computational Intelligence and Multimedia Applications(ICCIMA 1999), pp. 55-59 (1999)

Structure Determination of a Generalized ADALINE Neural Network for Application in System Identification of Linear Systems

Wenle Zhang[✉]

Department of Engineering Technology, University of Arkansas at Little Rock,
Little Rock, AR 72204, USA
wxzhang@ualr.edu

Abstract. This paper presents a structure determination method of a GADALINE based neural network used for linear system identification and parameter estimation. In GADALINE linear system identification, the past input data are used as its input and the past output data are also used as its input in the form of feedback because in such a linear system, the current system output is dependent on past outputs and on both the current and past inputs. The structure determination is then to determine how many past inputs should be included as its input and how many past output should be fed-back as its input also. The measured data set can then be used to train the GADALINE and during training, the performance error can be used to determine the network structure in our method just as the Final Prediction Error used in Akaike's criterion. One advantage of the method is its simplicity. Simulation results show that the proposed method provides satisfactory performance.

Keywords: System identification · Structure determination · Neural network · ADALINE

1 Introduction

System identification of dynamic systems is to establish the mathematical model that best describe actual model structure and parameters of a dynamic system based on the set of the measured input and output data of the system. There exist many well established system identification methods, such as least squares method, maximum likelihood method and instrumental variable method (L. Ljung 1999, Söderström and Stoica 1989). Recently, neural network techniques (Haykin 1999, Mehrotra 1997) have been increasingly studied for system identification.

Most neural networks used for system identification are in the category called recurrent (feedback) neural networks, such as, Multi-Layer Perceptron (Rumelhart 1986) with feedback and Hopfield neural network (Hopfield, 1982). And most neural network methods studied for system identification are mainly for nonlinear systems based on MLP (Sjöberg et al. 1994, Narendra and Parthasarathy 1990, Qin et al. 1992). Other types of neural networks are also seen, Hopfield (Chu et al. 1990), Radial Basis Function (Valverde 1999). A generalized ADALINE (GADALINE)

© Springer International Publishing Switzerland 2015
Y. Tan et al. (Eds.): ICSI-CCI 2015, Part III, LNCS 9142, pp. 19–26, 2015.
DOI: 10.1007/978-3-319-20469-7_3

based neural network method for linear systems was given by the author (Zhang 2007). However, not much work has been done on neural network structure determination for effective system identification. The structure of a GADALINE neural network for identification of a single input single output (SISO) linear system is simply the number of inputs to the ADALINE, which consists of precisely how many past inputs and how many past outputs taken as its inputs (fed-back). In this way, the structure determination of the neural network can be similarly done as the model order estimation of a SISO system. Model order estimation have been well studied, e.g., determinant ratio of convariance matrix method by (Woodside 1971), Final Prediction Error criterion by (Akaike 1974a) and Instrumental Product Moment test method by (Wellstead 1978).

In this work, the author will present the structure determination method of a GADALINE based neural network for linear system identification. The structure determination is to determine how many past inputs should be included as its input and how many past output should be fed-back as its input also. The measured data set can then be used to train the GADALINE and during training, the performance error can be used to determine these numbers in our method just as the Final Prediction Error used in Akaike's criterion for traditional model order estimation. One advantage of the proposed method is that it is very simple. Simulation results showed that the proposed method is effective. The rest of paper is organized as: section II describes the linear system order and parameter estimation problem, section III gives the GADALINE based order and parameter estimation method, section IV presents the simulation results and finally the conclusion section.

2   System Identification of Linear Systems

Consider a stable and observable discrete time linear SISO system, the input and output data can be measured over an enough long period of time and arranged in the form of a input-output pair-wise data set of size N: $\{u(kT), y(kT)\}$, $k = 1, 2, \ldots, N$, T is the sampling period, a constant, thus the data set is often simply written as $\{u(k), y(k)\}$. Then the system can be modeled by the following polynomial difference equation,

$$y(k) + a_1 y(k-1) + a_2 y(k-2) + \ldots + a_{na} y(k-n_a)$$
$$= b_1 u(k-1) + b_2 u(k-1) + \ldots + b_{nb} u(k-n_b) + e(k) \quad (1)$$

or

$$P(z^{-1}) y(k) = Q(z^{-1}) u(k) + e(k)$$
$$P(z^{-1}) = 1 + a_1 z^{-1} + a_2 z^{-2} + \ldots + a_{na} z^{-na}, \text{ and}$$
$$Q(z^{-1}) = b_1 z^{-1} + b_2 z^{-2} + \ldots + b_{nb} z^{-nb} \quad (2)$$

Where e is the model estimation error, z^{-1} is the unit delay operator, n_a and n_b are system structure parameters, $n_b \leq n_a$, n_a is the order of the system; a_i and b_j are system parameters, $i = 1, 2, \ldots, n_a$ and $j = 1, 2, \ldots, n_b$. If a_i and b_j are constants, the system is said to be time invariant, otherwise time varying.

Then the system identification is to determine the order of both polynomial $P(z^{-1})$ and $Q(z^{-1})$, that is, n_a and n_b, and to estimate coefficients a_i and b_j from the input and output data set according to a certain criterion, such as minimizing an error function, which is a measure of how close the model is to the actual system. Fig. 1 shows a general system identification architecture, $y_u(k)$ is the deterministic output, $n(k)$ is the noise (white) and $e(k)$ is the error.

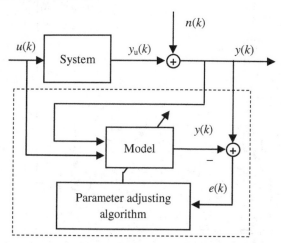

Fig. 1. System identification for ARX model

3   Structure Determination of the GADALINE

The proposed method is how the model in the dashed box in Fig. 1 can be built by a generalized ADALINE whose number of inputs is $m = n_b + n_a$ and whose inputs correspond to delayed inputs and delayed outputs. So, input weights correspond to coefficients a_i and b_j. By adjusting the order n_b and n_a and the GADALINE training performance can then be evaluated and the proper n_b and n_a values can be selected.

The GADALINE for our system identification method is configured as shown in Fig. 2, where the bias input is removed, and the input is passed through a series of unit delay operators to become a number of inputs and the output is fed-back to the input, also passing through a series of unit delay operators to become a number of inputs. If we still keep the notation for inputs as $u_1 \sim u_m$, then the output y can be found by,

$$y = \Sigma_1^m w_i u_i = \boldsymbol{u}^T \boldsymbol{w} \tag{3}$$

where $\boldsymbol{u}$ is the input vector and $\boldsymbol{w}$ is the weight vector,

$$\boldsymbol{u} = [u_1\, u_2\, \dots\, u_m]^T = [u(k-1)\dots u(k-n_b)\, -y(k-1)\dots -y(k-n_a)]^T \tag{4}$$

$$\boldsymbol{w} = [w_1\, w_2\, \dots\, w_m]^T = [b_1\, b_2\, \dots\, b_{nb}\, a_1\, a_2\, \dots a_{na}]^T \tag{5}$$

3.1   GADALINE Training

The GADALINE training algorithm is based on the Least Mean Square algorithm (LMS) generalized by introducing a momentum term in the weight adjustment equation. Training is performed by minimizing the cost function,

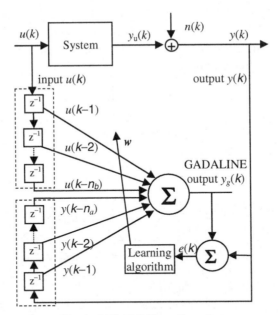

Fig. 2. GADALINE configuration

$$E(w) = 1/N \, \Sigma_1^{\,N} \, e^2(k) \tag{6}$$

where k is the iteration index and $e(k)$ is given by,

$$e(k) = d(k) - y(k) \tag{7}$$

where the desired output $d(k)$ at each iteration k is constant. Then the generalized LMS weight adjustment is

$$\Delta w(t) = \eta \, e(t)u(t) + \alpha \Delta w(t-1) \tag{8}$$

where the learning–rate is usually in the range of $0 < \eta < 1$, and α is the momentum weight factor – a small non-negative number in the range of $1 > \alpha \geq 0$.

GADALINE training is performed by presenting the training data set – the sequence of input-output pairs. During training, each input pair from the training data set is presented to the GADALINE and it computes its output y. This value and the corresponding target value from the training data set are used to generate the error and the error is then used to adjust all the weights.

3.2   Structure Determination

Here structure determination is to determine the two values (termed orders) k_a and k_b for n_a and n_b respectively. Ideally, if some a priori information about the system to be identified is available, meaning a guess can be made about the proper value for n_a and n_b, then the estimation can be tested for k_a and k_b from the range $n_a \pm C_1$ and $n_b \pm C_1$, here C_1 defines the range for each order and it is a small positive integer, such as 2 or 3. If no such information is available, then the test should start from 1 and increment to C_2 (a positive integer which is a guessed upper bound for the order parameter, such as 7). This test is performed during the GADALINE training by examining the accumulated training error in a way similar to the Final Prediction Error method in traditional model order estimation method. The system polynomial coefficients are then estimated by the weights in the trained GADALINE as in the mapping described in (4). The FPE for each pair k_a and k_b can be established as,

$$FPE(k_a, k_b) = 1/N \sum_1^N e^2(k) \qquad (9)$$

According to Akaike, once $k_a = n_a$ and $k_b = n_b$ which means the proper value for n_a and n_b are found, this value should be minimum or practically, there should be no further significant reduction for this value. Then based on the model order parsimony principle, the order pair should be determined as $(k_a, k_b) = \min(FPE(k_a, k_b))$.

This method of Structure Determination of the GADALINE can be summarized as an algorithm, called SDG algorithm, in the following:

   i)   Choose a proper value for η, such as 0.3 and for α, such as 0.9. Initialize weight vector w with small random numbers such as of uniform $(-0.3, +0.3)$. Then choose a proper C_2 for the problem, say 7.
   ii) Let $k_a = 1, k_b = 1$
   iii) Train GADALINE for $k = 1$ to N
   iv) Calculate $FPE(k_a, k_b) = 1/N \sum_1^N e^2(k)$
   v)   Increment k_b until k_a, this makes sure $n_b \leq n_a$ in (2)
   vi) Increment k_a until C_2
   vii) Let $(n_a, n_b) = \min(FPE(k_a, k_b))$
   viii) Let $[b_1\ b_2\ ...\ b_{nb}] = [w_1\ w_2...w_{nb}]$
   ix) Let $[a_1\ a_2\ ...a_{na}] = [w_{nb+1}\ w_{nb+2}...w_{na+nb}]$

3.3   Implement the SDG Algorithm

The SDG algorithm is computational simple and can be easily implemented in any modern programming language. Here it is implemented in Mathworks MATLAB. The algorithm has similar drawback to other neural network based methods. That is, trial and error is often performed on selecting proper values for the learning rate η and momentum α for solving different problems.

3.4   Procedure to Run SDG Algorithm

In order to obtain effective result, proper experiment should be designed first, then input/output data be collected and SDG algorithm can be applied.

1) Experiment design: As in any traditional system identification, the input signal should be selected such that it is "rich" enough to excite all modes of the system, this is called persistent exciting. For linear systems, input signal should excite all frequencies and amplitude is not so important. Common used persistent exciting signals that can be tried are: random signal, multi-sine signal and random binary sequence signal (RBS).

2) Data collection: Measured input-output $u_S(k)$ and $y_S(k)$ are used to set up $u(k)$.

3) Run SDG algorithm: Apply the algorithm determine network structure order n_a and n_b and estimate the parameters as given in $w(N)$.

4   Simulation Results

Simulation is performed on 3 example systems that are often seen in the literature.

Example 1. First consider a second order linear SISO system described by the following difference equation,

$$y(k) - 1.5\ y(k-1) + 0.7\ y(k-2) = 1.0\ u(k-1) + 0.5\ u(k-2)$$

The system is to be simulated with an RBS input signal switching between -1 and 1 with equal probability for a total number of samples N=500. The FPE matrix is obtained as in the following for $\eta=0.02$ and $\alpha=0.9$. The row number of FPE corresponds to n_a and column number corresponds to n_b. FPE(2,2) = 0.0, then choose order and parameters are, na = 2, nb = 2

```
[b1 b2 a1 a2]= [1.0000 0.5000 -1.5000 0.7000]
```

$$FPE_{4,4} = \begin{bmatrix} 2.3818 & \infty & \infty & \infty \\ 0.2131 & 0.0000 & \infty & \infty \\ 0.0511 & 0.0137 & 0.0000 & \infty \\ 1.8656 & 0.0979 & 0.0407 & 0.1848 \end{bmatrix}$$

Here estimated parameters are almost the same as the true values. Note that FPE(3,3) is also 0.0, but the order is bigger. The training trajectories for both cases are shown in Fig. 3. The later is considered as over-fitting.

Example 2. Now consider a third order linear SISO system described by the following difference equation,

$$y(k) - 1.5\ y(k-1) + 1.0\ y(k-2) - 0.5\ y(k-3) = 0.5\ u(k-1)$$

The system is to be simulated with an RBS input signal with N=1000. The obtained FPE matrix is given below for $\eta=0.04$ and $\alpha=0.9$. Notice FPE(3,1) = 0.0001, then choose order and parameters as, na = 3, nb = 1

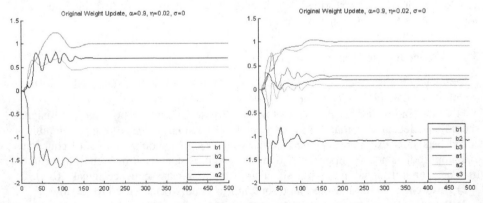

Fig. 3. Parameter trajectories for example 1

```
[b1 a1 a2 a3]  =[0.5000 -1.4998 0.9997 -0.4999]
```

$$\begin{pmatrix} 0.1801 & \infty & \infty & \infty & \infty \\ 0.2513 & 0.3239 & \infty & \infty & \infty \\ 0.0001 & 0.0158 & 0.0270 & \infty & \infty \\ 0.0031 & 0.0000 & 0.0005 & 0.0008 & \infty \\ 0.0818 & 0.0014 & 0.0000 & 0.0128 & 0.0288 \end{pmatrix}$$

The parameter trajectories are shown in Fig. 4. In this case, the estimated parameters are very close to the system's true values. Note that FPE(4,2) and FPE(5,3) are both slight lower than FPE(3,1) but the difference is not as significant. Then parsimony principle is applied.

Example 3. Now consider another third order linear SISO system (Ljung 1999) described by the following difference equation,

$$y(k) - 1.2\,y(k{-}1) - 0.15\,y(k{-}2) + 0.35\,y(k{-}3) = 1.0\,u(k{-}1) + 0.5\,u(k{-}2)$$

The system is to be simulated with an RBS input signal with N=1000. The obtained FPE matrix is given below for η=0.015 and α=0.9. The minimum FPE(3,2) = 0.0083, then choose order and parameters as, na = 3, nb = 2

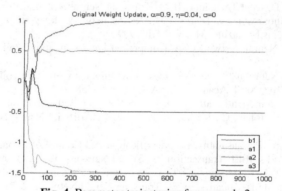

Fig. 4. Parameter trajectories for example 2

```
[b1 b2 a1 a2 a3]=[0.9976 0.5276 -1.1668 -0.1983 0.3650]
```

Note that the parameters estimated are slightly off. Also as n_a and n_b got higher, the FPE became extra large indicating training is about to diverge. This is an indication that the input is under exciting. The original author used multi-sinusoidal input signal.

$$\begin{pmatrix} 3.0272 & \infty & \infty & \infty & \infty \\ 0.0329 & 0.1574 & \infty & \infty & \infty \\ 0.3338 & 0.0083 & 0.0180 & \infty & \infty \\ 0.5836 & 0.0901 & 0.0370 & 0.0419 & \infty \\ 48.5774 & 70.8158 & 79.0960 & 110.7572 & 122.4904 \end{pmatrix}$$

5   Conclusions

Presented in this work is the author's first effort on the GADALINE based neural network structure determination method for linear system identification. The neural network's final prediction error is calculated during training of the network. The two structure order values are then determined based on the minimum FPE and polynomial coefficients are estimated at the same time as order obtained. The FPE value is evaluated as in a traditional model order selection criterion given by Akaike such that the order pair (n_a, n_b) is chosen where FPE becomes significantly small for the first time. The proposed algorithm is computational very simple due to the fact it is an ADALINE network based. Simulation results showed the method has satisfactory performance.

References

1. Akaike, H.: A New Look at the Statistical model identification. IEEE Trans. On Automatic Control **19**, 716–723 (1974)
2. Atencia, M., Sandoval, G.: Gray Box Identification with Hopfield Neural Networks. Revista Investigacion Operacional **25**(1), 54–60 (2004)
3. Bhama, S., Singh, H.: Single Layer Neural Network for Linear System Identification Using Gradient Descent Technique. IEEE Trans. on Neural Networks **4**(5), 884–888 (1993)
4. Chu, S.R., Shoureshi, R., Tenorio, M.: Neural networks for system identification. IEEE Control Systems Magazine, 31–34, April 1990
5. Haykin, S.: Neural Networks, A Comprehensive Foundation, 2nd edn. Prentice Hall (1999)
6. Hopfield, J.: Neural networks and physical systems with emergent collective computational abilities. Proc. Nat'l. Acad. Sci. USA **79**, 2554–2558 (1982)
7. Ljung, L.: System Identification - Theory for the User, 2nd edn. Prentice-Hall (1999)
8. Mehrotra, K., Mohan, C., Ranka, S.: Elements of Artificial Neural Networks. MIT press (1997)
9. Narendra, K.S., Parthasarathy, K.: Identification and Control of Dynamical Systems using Neural Networks. IEEE Transactions on Neural Networks **1**, 1–27 (1990)
10. Qin, S.Z., Su, H.T., McAvoy, T.J.: Comparison of four neural net learning methods for dynamic system identification. IEEE Trans. on Neural Networks **2**, 52–262 (1992)
11. Rumelhart, D.E., Hinton, G.E., Williams, R.J., Learning internal representations by error propagation. In: Parallel Distributed Processing: Explorations in the Microstructure of Cognition, vol. I. MIT Press, Cambridge (1986b)
12. Sjöberg, J., Hjalmerson, H., Ljung, L.: Neural Networks in System Identification. In: Preprints 10th IFAC symposium on SYSID, vol. 2, Copenhagen, Denmark, pp. 49–71 (1994)
13. Söderström, T., Stoica, P.: System Identification. Prentice Hall, Englewood Cliffs (1989)
14. Valverde, R.: Dynamic systems identification using RBF neural networks. Universidad Carlos III de Madrid. Technical report (1999)
15. Wellstead, P.E.: An Instrumental Product Moment Test for Model Order Estimation. Automatica **14**, 89–91 (1978)
16. Widrow, B., Lehr, M.A.: 30 years of Adaptive Neural Networks: Perceptron, Madaline, and Backpropagation. Proc. IEEE **78**(9), 1415–1442 (1990)
17. Woodside, C.M.: Estimation of the Order of Linear Systems. Automatica **7**, 727–733 (1971)
18. Zhang, W.: System Identification Based on a Generalized ADALINE Neural Network. Proc. 2007 ACC, New York City, pp. 4792–4797 (2007)

Evolutionary and Genetic Algorithms

A Self-configuring Metaheuristic for Control
of Multi-strategy Evolutionary Search

Evgenii Sopov[(✉)]

Siberian State Aerospace University, Krasnoyarsk, Russia
evgenysopov@gmail.com

Abstract. There exists a great variety of evolutionary algorithms (EAs) that represent different search strategies for many classes of optimization problems. Real-world problems may combine several optimization features that are not known beforehand, thus there is no information about what EA to choose and which EA settings to apply. This study presents a novel metaheuristic for designing a multi-strategy genetic algorithm (GA) based on a hybrid of the island model, cooperative and competitive coevolution schemes. The approach controls interactions of GAs and leads to the self-configuring solving of problems with a priori unknown structure. Two examples of implementations of the approach for multi-objective and non-stationary optimization are discussed. The results of numerical experiments for benchmark problems from CEC competitions are presented. The proposed approach has demonstrated efficiency comparable with other well-studied techniques. And it does not require the participation of the human-expert, because it operates in an automated, self-configuring way.

Keywords: Metaheuristics · Self-configuration · Genetic algorithm · Multi-objective optimization · Non-stationary optimization

1   Introduction

Evolutionary algorithms were introduced as search and adaptation heuristics and are widely used in solving many complex optimization problems [1, 2]. The performance of EAs mostly depends on the search strategy represented by the structure, operations and parameters of the algorithm.

The set of EA search strategies can be classified with respect to classes of optimization problems. Each type of problem can be solved in a straightforward way by using a general EA scheme. However many researchers prefer to apply modified EAs that are specially designed for a certain problem and are based on additional heuristics, which take into account the specific features of the problem.

Many real-world optimization problems are usually complex and not well-studied, so they are viewed as black-box optimization problems. In such cases, the researcher usually has no possibility of choosing the proper search strategy and fine tuning the parameters of the algorithm. Thus, there exists a problem of designing self-configuring techniques that can operate with many search strategies and adaptively control their usage.

© Springer International Publishing Switzerland 2015
Y. Tan et al. (Eds.): ICSI-CCI 2015, Part III, LNCS 9142, pp. 29–37, 2015.
DOI: 10.1007/978-3-319-20469-7_4

All EAs in this study are assumed to be genetic algorithms (GAs). There exists a great variety of modifications of GAs covering many classes of optimization problems.

The rest of the paper is organized as follows. Section 2 describes the related work. Section 3 describes the proposed approach. Section 4 and 5 demonstrate the performance of approach with respect to multi-objective and non-stationary optimization, and the results of numerical experiments are discussed. In the conclusion the results and further research are discussed.

2   Related Work

In the field of EAs, the term "self-configuration" has several meanings [3]:

1. Deterministic Parameter Control using some deterministic rules without taking into account any feedback from the evolutionary search. For example, the time-dependent change of the mutation rates [1, 4].
2. Self-control or self-adaptation of the control parameters of the EA. Usually optimization of the control parameters is viewed as a part of the optimization objective. Values of controlled parameters are included in the population or in the chromosome of individuals [5].
3. Meta-heuristics for the EA control that combine the EA with some other adaptive search techniques. Usually the approach is based on the external control procedure. Good results are obtained using the fuzzy-logic technique [6, 7].
4. Self-configuration via the automated and adaptive design of an efficient combination of the EA components from the predefined set.

The fourth way is more appropriate for the given problem of the multi-strategy EA design. The predefined set of EA search strategies can contain different versions (modifications) of the EA for the given class of optimization problems. The automated selection, combination and interaction of search strategies should provide the dynamic adaptation to the problem landscape over each stage of the optimization process.

3   Designing a Self-Configuring Metaheuristic for Multi-strategy Evolutionary Search Control

In the field of statistics and machine learning, ensemble methods are used to improve decision making. This concept can be used in the field of EA. There are at least two well-studied approaches to the interaction of several EAs: the coevolutionary approach and the island model.

The island model was introduced as a parallel version of an EA. The coevolution algorithm is an evolutionary algorithm in which fitness evaluation is based on interactions between individuals [8]. There are two types of interaction: competitive and cooperative [9, 10].

Coevolution can also be applied to perform the self-configuring [11]. In [12] the self-configuring coevolutionary algorithm in a form of a hybrid of the island model, competitive and cooperative coevolution was proposed. The similar approach, named Population-based Algorithm Portfolios (PAP), was proposed in [13].

The same idea can be applied to the problem of the self-configuring control of many evolutionary search strategies. We can denote a general scheme of the self-configuring EA in the form of "Self*EA" or "Self*GA", where * refers to a certain class of optimization problems. The main idea is as follows (the algorithm scheme is presented in Fig. 1):

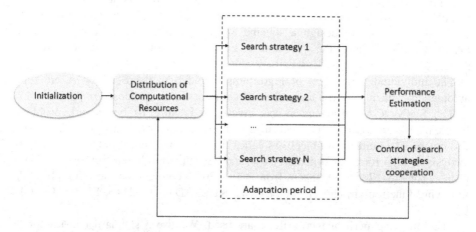

Fig. 1. Self*EA and Self*GA scheme

The total population is divided into disjoint subpopulations of equal size. The portion of the population is called the computational resource. Each subpopulation corresponds to certain search strategy and evolves independently (corresponds to the island model). After a certain period, which is called the adaptation period, the performance of individual algorithms is estimated and the computational resource is redistributed (corresponds to the competitive coevolution). Finally, random migrations of the best solutions are presented to equate start positions of EAs for the run with the next period (corresponds to the cooperative coevolution).

Two examples of implementation of the approach for multi-objective and non-stationary optimization problems will now be considered.

4   Self-Configuring Multi-Objective GA

EAs are highly suitable for multi-objective (MO) optimization and are flexible and robust in finding an approximation of the Pareto set. At the same time, many MOPs are still a challenge for EA-based techniques. There exist various MO search strategies, which are implemented in certain different EAs [15]. We can combine different MO search strategies in the form of an ensemble and design a Self*GA that can be named SelfMOGA.

At the first step, we need to define the set of search strategies included in the SelfMOGA. We use five basic and well-studied techniques: VEGA, FFGA, NPGA, NSGA-II, SPEA-II.

The key point of any coevolutionary scheme is the performance evaluation of a single algorithm. The following criteria, combined into two groups, are proposed. The first group includes the static criteria (the performance is measured over the current adaptation period). *Criterion 1* is the percentage of non-dominated solutions. *Criterion 2* is the uniformity (dispersion) of non-dominated solutions. The second group contains the dynamic criteria (the performance is measured in a comparison with previous adaptation periods). *Criterion 3* is the improvement of non-dominated solutions. The solutions of the previous and current adaptation period are compared.

Finally, we need to design a scheme of the redistribution of resources. In the SelfMOGA, we apply the probability sampling scheme "substitution by rank selection".

The individual parameters of algorithms included in the SelfMOGA are tuned automatically during the run using the self-configuration as proposed in [14].

The SelfMOGA performance was investigated using 19 benchmark problems. Problems 1-6 are various 2D test functions based on quadric, Rastrigin, Rosenbrock and others. The number of objectives varies from 2 to 4. These problems are good for analysis of algorithm performance as we can visualize the Pareto set and the Pareto front. Problems 7-19 are taken from CEC 2009 competition on MO [16]. The following function are chosen: FON, POL, KUR, ZDT1-4, ZDT6, F2, F5, F8, UP4, UP7.

The following performance criteria are used. We use 3 standard metrics for the problems with the analytically defined Pareto set. They are *generational distance (GD), maximum spread (MS) and spacing (S)*. For other problems, the following criteria based on the non-dominated ranking are used: *Criterion 1 (C1)* – a minimal rank in the population of the single algorithm (after non-dominated sorting of all solutions from all algorithms). *Criterion 2 (C2)* - a proportion of solutions with the minimal rank. *Criterion 3 (C3)* – dispersion of distances in the criterion space for solutions with the minimal rank.

Numerical results for some problems are presented in Tables 1 and 2. All results are the averages of 100 independent runs. We have estimated the performance for each of 5 single algorithms independently. The column "Average" corresponds to the average of 5 single algorithms. The last column corresponds to the SelfMOGA (the ensemble of 5 search techniques).

The experiments show that there is no single search strategy that shows the best results for all problems. In real-world problems, the researcher usually cannot choose the proper technique beforehand. Thus the average value demonstrates the average performance of the randomly chosen technique. The SelfMOGA can yield to the best algorithm in some cases, but always outperforms the average value.

Table 1. Numerical results for problems with the analytically defined Pareto set

Metric	Algorithm						
	SPEA-II	VEGA	FFGA	NPGA	NSGA-II	Average	SelfMOGA
Problem: F2							
GD	0,849	1,468	3,334	1,699	**0,263**	1,522	0,46
MS	**0,750**	0,768	2,928	0,730	0,850	1,205	0,832
S	1,289	0,446	0,096	0,400	0,525	0,551	**1,483**
Problem: UP4							
GD	0,421	0,441	0,616	0,616	0,426	0,504	0,266
MS	**0,928**	0,933	1,127	0,905	**0,928**	0,964	0,973
S	0,368	**0,612**	0,218	0,279	0,340	0,363	0,464
Problem: UP7							
GD	1,149	1,881	4,683	2,274	**0,305**	2,058	0,629
MS	0,719	0,757	1,571	**0,713**	0,785	0,909	0,792
S	1,459	0,444	0,092	0,358	0,556	0,582	**1,735**

Table 2. Numerical results for problems with the algorithmically defined Pareto set

Metric	Algorithm						
	SPEA-II	VEGA	FFGA	NPGA	NSGA-II	Average	SelfMOGA
Problem: KUR							
C1	1	1	21	2	1	5	1
C2	0,190	0,158	0,620	0,101	0,344	0,283	**0,358**
C3	0,691	0,699	0,395	0,676	**0,743**	0,641	0,741
Problem: ZDT1							
C1	3	9	26	10	2	10	1
C2	0,158	0,115	0,318	0,297	0,578	0,293	**0,993**
C3	0,239	0,521	0,758	0,455	0,206	0,436	**0,159**
Problem: ZDT6							
C1	4	21	48	12	1	17	1
C2	0,210	0,235	0,627	0,230	0,162	0,293	**0,948**
C3	0,458	0,703	1,100	0,655	**0,380**	0,659	0,311

Some real-world applications of the SelfMOGA can be found in [17], where the SelfMOGA was successfully applied for the problem of designing the convolutional neural networks for context image analysis in the Dialog HCI. In [18] – for the problem of designing network intrusion detectors. In [19] – for designing the SVM with feature selection for text categorization problem.

5   Self-Configuring GA for Non-stationary Optimization

Non-stationary optimization problems are also called dynamic optimization problems (DOP) or changing (non-stationary, dynamic) environment optimization. A good survey on DOP methods is proposed in [20]. There exist many types of environmental changes. There also exist a great variety of DOP techniques to deal with the certain type of changes.

We introduce a novel approach based on Self*GA concept. It is called SelfDOPGA. We will discuss the SelfDOPGA in detail. The first step defines the search strategies. In this study, we will use the following list of 5 basic and well-studied DOP techniques: Self-adaptive technique, Restarting optimization, Local adaptation, Diversity increasing technique, The explicit memory.

In the SelfDOPGA the adaptation period is defined by the period between changes in the environment. The detection of changes can be performed by re-evaluating the fitness of the current best solutions (called detectors). After each adaptation period is completed, we need to estimate the performance of single algorithms. The performance measure is the offline error.

The redistribution scheme is random migrations. We do not use the standard scheme "the best displaces the worst" to prevent the convergence and the decrease of the population diversity.

The performance of the SelfDOPGA was investigated at solving the following benchmark problems of non-stationary optimization. Two standard non-stationary optimization problems: the moving peaks benchmark (MPB) and the dynamic Rastrigin function. The MPB uses "Scenario 2" with the number of peaks equal to 1, 5, 10 and 20. The dimension for the Rastrigin problem is equal to 2, 5 and 10 [21]. Two problems from the CEC'2009 competition on dynamic optimization: Rotation peak function (F1), Composition of sphere's function (F2) and Composition of Rastrigin's function (F3). The types of changes for F1-F3 are denoted as T1-T6 [22].

The results of numerical experiments for the MPB and the dynamic Rastrigin function are presented in Table 3. The efficiency criteria for all functions are mean average error over 100 independent runs. We have estimated the performance for each of 5 used single DOP techniques independently. The column "Average" corresponds to the average of 5 single algorithms. The last column corresponds to the SelfDOPGA (the ensemble of 5 DOP techniques).

Table 3. Performance for the MPB and the Dynamic Rastrigin Problem (DRP)

Dimen-sion	Algorithm						
	Self-adaptive GA	Re-starting	Local adaptation	Diversity increasing	The explicit memory	Average	SelfDOP GA
			Problem: MPB				
1	0,5	1,25	0,96	**0,4**	1,98	1,018	0,82
5	**3,27**	5,03	4,85	3,93	7,63	4,942	4,56
10	9,96	10,87	10,84	9,01	16,25	11,385	**7,12**
20	8,72	9,35	11,5	7,97	17,82	11,072	**7,33**
			Problem: DRP				
2	**0**	**0**	**0**	0,7	1,3	0,4	**0**
5	3,56	4,02	6,74	6,87	8,15	5,868	**2,83**
10	14,53	15,01	21,23	21,8	34,36	21,386	**11,56**

As we can see, the SelfDOPGA shows better performance than the average performance of individual techniques. The most significant difference is observed in problems with higher dimensionality.

Some results of numerical experiments for F1-F3 functions are shown in Table 4. The results of the SelfDOPGA runs are compared with the performance of algorithms from the CEC'09 competition [23-25]. The algorithms are a self-adaptive differential evolution algorithm (jDE), a clustering particle swarm optimizer (CPSO), Evolutionary Programming with Ensemble of Explicit Memories (EP with EEM), a standard particle swarm optimizer (PSO) and a standard genetic algorithm (GA). All values in Table 4 are "mean±STD" values over 20 independent runs (according to CEC'09 competition rules). Here the "mean" value is the absolute function error value after each period of the change occurs. The "STD" value is standard deviation.

Table 4. Performance for the CEC'09 problems

Type of changes	Algorithm						SelfDOP GA
	jDE	CPSO	EP with EEM	GA	PSA	Average	
Problem: F1							
T1	0,028 ±0,44	0,035 ±0,42	5,71 ±9,67	5,609 ±9,35	5,669 ±7,73	3,41 ±5,52	2,21 ±7,03
T3	2,999 ±7,12	4,131 ±8,99	10,87 ±13,4	13,13 ±13,8	11,73 ±13,5	8,57 ±11,36	4,27 ±7,56
T6	1,145 ±5,72	1,056 ±4,80	13,12 ±14,9	29,25 ±25,6	32,01 ±25,6	15,31 ±15,32	2,03 ±4,29
Problem: F2							
T1	0,96 ±3,08	1,247 ±4,17	6,214 ±9,62	33,05 ±53,7	45,79 ±59,3	17,45 ±25,97	1,64 ±3,87
T3	50,1 ±124	10,27 ±33,4	4,988 ±8,24	128,5 ±188	135,8 ±185	65,93 ±107,72	7,87 ±13,5
T6	3,36 ±12,9	3,651 ±6,92	3,478 ±7,59	43,25 ±69,8	73,34 ±99,9	25,41 ±39,42	3,55 ±7,34
Problem: F3							
T1	11,39 ±58,11	137,5 ±221,6	151,98 ±190,71	158,1 ±264,5	553,6 ±298,1	202,51 ±206,60	150,6 ±233,1
T3	572,10 ±386,09	765,9 ±235,8	136,67 ±183,86	573,9 ±399,8	827,1 ±212,6	575,13 ±283,63	512,61 ±257,9
T6	243,27 ±384,98	753,0 ±361,7	107,54 ±158,36	491,7 ±464,3	803,5 ±375,0	479,80 ±348,86	236,8 ±342,5

In a case of the F1-F3 problems, the SelfDOPGA outperforms both the GA and PSO algorithms. It also outperforms EP in some cases, but yields to jDE and CPSO. It should be noted that the SelfDOPGA outperforms the average value and the value of the SelDOPGA performance is closer to the best value than to the average.

CPSO, jDE and EP are specially designed techniques. Numerical experiments show that their performance can decrease for some types of changes. The SelfDOPGA demonstrates sufficiently good performance within the whole range of types of changes.

6   Conclusions

In this study, a novel self-configuring meta-heuristic is proposed. Different evolutionary search strategies are interacted in a form of ensemble. The control of the interactions and cooperation of strategies is based on a hybrid of the island model, cooperative and competitive coevolution schemes.

We have demonstrated the performance of the Self*GA with respect to multi-objective (SelfMOGA) and non-stationary (SelfDOPGA) optimization problems. Results show that the Self*GA outperforms the average performance of single algorithms. Thus the Self*GA performs better than a randomly chosen technique. The results for benchmark problems from CEC competitions are comparable with widely used techniques. The main advantage of the Self*GA is that it operates in an automated, self-configuring way.

In further works, the Self*GA will be applied to other classes of optimization problems.

Acknowledgments. The research was supported by President's of the Russian Federation grant (MK-3285.2015.9). The author expresses his gratitude to Mr. Ashley Whitfield for his efforts to improve the text of this article.

References

1. Holland, J.: Adaptation In Natural and Artificial Systems. University of Michigan Press (1975)
2. Goldberg, D.: Genetic Algorithms in Search, Optimization and Machine Learning. Reading. Addison-Wesley, MA (1989)
3. Schaefer, R., Cotta, C., Kołodziej, J.: Parallel problem solving from nature. In: Proc. PPSN XI 11th International Conference, Kraków, Poland (2010)
4. Back, T.: Self-adaptation in genetic algorithms. In: Proceedings of 1st European Conference on Articial Life (1992)
5. Eiben, A.E., Hintering, R., Michalewicz, Z.: Parameter control in evolutionary algorithms. IEEE Transactions on Evolutionary Computation **3**(2) (1999)
6. Lee, M., Takagi, H.: Dynamic control of genetic algorithms using fuzzy logic techniques. In: Proceedings of the Fifth International Conference on Genetic Algorithms (1993)
7. Liu, J., Lampinen, J.: A fuzzy adaptive differential evolution algorithm. Soft Comput.: Fusion Found., Methodologies Applicat. **9**(6) (2005)
8. Ficici, S.G.: Solution Concepts in Coevolutionary Algorithms. A Doctor of Philosophy Dissertation, Brandeis University (2004)
9. Mühlenbein, H.: Strategy adaptation by competing subpopulations. In: Davidor, Y., Männer, R., Schwefel, H.-P. (eds.) PPSN 1994. LNCS, vol. 866, pp. 199–208. Springer, Heidelberg (1994)
10. Potter, M.A.: A cooperative coevolutionary approach to function optimization. In: Davidor, Y., Männer, R., Schwefel, H.-P. (eds.) PPSN 1994. LNCS, vol. 866. Springer, Heidelberg (1994)
11. Mallipeddi, R., Suganthan, P.N.: Ensemble differential evolution algorithm for CEC2011 problems. In: IEEE Congress on Evolutionary Computation, New Orleans, USA (2011)
12. Sergienko, R.B., Semenkin, E.S.: Competitive cooperation for strategy adaptation in coevolutionary genetic algorithm for constrained optimization. In: Proc. of 2010 IEEE Congress on Evolutionary Computation (2010)
13. Peng, F., Tang, K., Chen, G., Yao, X.: Population-based algorithm portfolios for numerical optimization. IEEE Trans. Evol. Comput. **14**(5) (2010)

14. Semenkin, E., Semenkina, M.: Self-configuring genetic algorithm with modified uniform crossover operator. In: Tan, Y., Shi, Y., Ji, Z. (eds.) ICSI 2012, Part I. LNCS, vol. 7331, pp. 414–421. Springer, Heidelberg (2012)
15. Zhoua, A., Qub, B.-Y., Lic, H., Zhaob, S.-Zh., Suganthanb, P.N., Zhangd, Q.: Multiobjective evolutionary algorithms: A survey of the state of the art. Swarm and Evolutionary Computation 1(1) (2011)
16. Zhang, Q., Zhou, A., Zhao, Sh., Suganthan, P.N., Liu, W., Tiwari, S.: Multiobjective optimization test instances for the CEC 2009 special session and competition. In: IEEE Congress on Evolutionary Computation, IEEE CEC 2009, Norway (2009)
17. Sopov, E., Ivanov, I.: Design efficient technologies for context image analysis in dialog HCI using self-configuring novelty search genetic algorithm. In: Proc. of the 11th International Conference on Informatics in Control, Automation and Robotics (ICINCO 2014), Vienna, Austria (2014)
18. Sopov, E., Panfilov, I.: Intrusion detectors design with self-configuring multi-objective genetic algorithm. In: Proc. of 2014 International Conference on Network Security and Communication Engineering (NSCE2014), Hong Kong (2014)
19. Sopov, E., Panfilov, I.: Self-tuning SVM with feature selection for text categorization problem. In: Proc. of International Conference on Computer Science and Artificial Intelligence (ICCSAI2014), Wuhan, China (2014)
20. Nguyena, T.T., Yang, S., Branke, J.: Evolutionary dynamic optimization: A survey of the state of the art. Swarm and Evolutionary Computation 6 (2012)
21. Morrison, R.W., De Jong, K.A.: A test problem generator for non-stationary environments. In: Proc. of the 1999 Congr. on Evol. Comput. (1999)
22. Li, C., Yang, S., Nguyen, T.T., Yu, E.L., Yao, X., et al.: Benchmark Generator for CEC2009 Competition on Dynamic Optimization. Technical Report 2008, Department of Computer Science, University of Leicester, UK (2008)
23. Brest, J., Zamuda, A., Boskovic, B., Maucec, M.S., Zumer, V.: Dynamic optimization using self-adaptive differential evolution. In: Proc. of IEEE Congr. Evol. Comput. (2009)
24. Li, C., Yang, S.: A clustering particle swarm optimizer for dynamic optimization. In: Proc. of the Congr. on Evol. Comput. (2009)
25. Yu, E.L., Suganthan, P.N.: Evolutionary programming with ensemble of external memories for dynamic optimization. In: Proc. of IEEE Congr. Evol. Comput. (2009)

Fast Genetic Algorithm for Pick-up Path Optimization in the Large Warehouse System

Yongjie Ma$^{(\boxtimes)}$, Zhi Li, and Wenxia Yun

College of Physics and Electronic Engineering, Northwest Normal University,
Lanzhou, Gansu Province, China
myjmyj@163.com, {colorblind,oyfmydear}@yeah.net

Abstract. The order-picking processes of fixed shelves in a warehouse system require high speed and efficiency. In this study, a novel fast genetic algorithm was proposed for constrained multi-objective optimization problems. The handling of constraint conditions were distributed to the initial population generation and each genetic process. Combine the constraint conditions and objectives, a new partial-order relation was introduced for comparison of individuals. This novel algorithm was used to optimize the stacker picking path in an Automated Storage/Retrieve System (AS/RS) of a large airport. The simulation results indicates that the proposed algorithm reduces the computational complexity of time and space greatly, and meets the needs of practical engineering of AS/RS optimization control.

Keywords:: Fast genetic algorithm · Automated storage/retrieve system · Multi-objective optimization · Stacker · Pick-up path

1 Introduction

A large Automated Storage Retrieval System (AS/RS) is usually a discrete, random, dynamic, multi-factor and multi-objective complex system. It is difficult to optimize using traditional ways. However, the intelligent methods would be efficient [1].

The scheduling of large warehouse includes location assignment, job scheduling, picking path optimization. This paper is limited to the pick-up path optimization problem. Path optimization is a typical combinatorial optimization problem. [2].

The path optimization problem of the large warehouse system can be regarded as constrained multi-objective optimization problem (MOP), and the solving of multi-objective optimization problem is still a difficult issue. Although some evolutionary algorithms were used to solve the problem, the Genetic Algorithm (GA) has been the one of the most effective way to solve the MOP[3-5].

2 The Optimization Objective and Constraints Conditions of Large Warehouse

This paper only considers the certain numbers of operations and types in pick-up path optimization.

© Springer International Publishing Switzerland 2015
Y. Tan et al. (Eds.): ICSI-CCI 2015, Part III, LNCS 9142, pp. 38–44, 2015.
DOI: 10.1007/978-3-319-20469-7_5

2.1   The Optimization Objectives of Large Warehouse

The arrangement of shelves and initial storage state of a warehouse for food are shown in Fig.1 and 2. The storage location can be expressed as a six-tuple (i,j,k,s,w,p), where (i,j,k) is the location coordinates, s is the storage time, w is the weight, and p is the types.

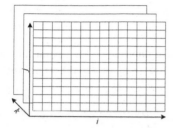
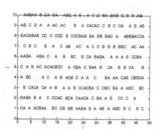

Fig. 1. The arrangement of shelves for food   **Fig. 2.** The initial storage state

Assuming shelf direction is z axis, horizontal direction is x axis, and vertical direction is y axis. The stacker travels in the horizontal and vertical directions, with velocities of Vx and Vy. The travel time of stacker can be expressed as the follows:

$$\min T = \sum_{k=1}^{K} \sum_{i_1,i_2=0}^{I} \sum_{j_1,j_2=0}^{J} c^k_{(i_1,j_1)} \cdot t_{(i_1,j_1)-(i_2,j_2)} + 2t_0 = \sum_{k=1}^{K} \sum_{i_1,i_2=0}^{I} \sum_{j_1,j_2=0}^{J} c^k_{(i_1,j_1)} \cdot \max \left[\frac{L \cdot |i_2 - i_1|}{v_x}, \frac{H \cdot |j_2 - j_1|}{v_y} \right] + 2t_0 \quad (1)$$

where $t_{(i1,j1)-(i2,j2)}$ is the stacker travel time from location (i_1,j_1) to (i_2,j_2), t_0 is the time between I/O location and aisle exit, I and J are the maximal number of storage locations in x and y directions, K is picking times, and L and H are the length and height of storage shelves, respectively.

For goods such as food should be chosen the earlier as far as possible:

$$\max \ S = \sum_{k=1}^{K} \sum_{i=0}^{I} \sum_{j=0}^{J} c^k_{(i,j)} \cdot s_{(i,j)} \quad (2)$$

where $s(i,j)$ is storage time, $C^k_{(i,j)}$ represents whether the good is chosen:

$$c^k_{(i,j)} = \begin{cases} 1, & \textit{if good of } (i, j) \textit{ is chosen} \\ 0, & \textit{otherwise} \end{cases}, \ k = 1,\dots ,K; \ i = 1, \dots ,I; \ j = 1, \cdots ,J \quad (3)$$

2.2   The Constraint Conditions of Large Warehouse

The storage/retrieval operations must meet the capacity constraints of the stacker :

$$\sum_{k=1}^{K} \sum_{i=0}^{I} \sum_{j=0}^{J} c^k_{(i,j)} \cdot w_{(i,j)} \leq Q \quad (4)$$

Storing goods at (i1,j1) and (i2,j2), the order should meets from-near-to-far strategy:

$$i_1 \leq i_2, j_1 \leq j_2 \tag{5}$$

Retrieving goods at (i3,j3) and (i4,j4), the orders should meet from-far-to-near strategy:

$$i_3 \geq i_4, j_3 \geq j_4 \tag{6}$$

The path may be not the optimal. However, it can reduce the operation time. That is, two picking paths of different endpoint composed of (i1,j1), (i2,j2)and (i3,j3), (i4,j4) , the crossing point (i,j) does not exist.

$$\min(i_1,i_2,i_3,i_4) < i < \max(i_1,i_2,i_3,i_4) \quad and \quad \min(j_1,j_2,j_3,j_4) < j < \max(j_1,j_2,j_3,j_4), \tag{7}$$
$$\forall i_1 \neq i_3 \ and \ i_1 \neq i_4 \ or \ j_1 \neq j_3 \ and \ j_1 \neq j_4$$

Prior to storing goods, one should ensure that the storage locations are empty; when retrieving the goods, one should ensure that there are goods prepared for retrieval at the storage locations:

$$P_{(i,\,j)} = \begin{cases} X, & output, X \in \{A,B,\cdots\} \\ 0, & input \end{cases} \tag{8}$$

where $P_{(i,j)}$ is the goods of location (i,j), A and B are types of goods.

3   Fast Genetic Algorithm Design

The proposed algorithm adopts symbol encoding. The job types include single, dual and quadruple command cycle. A chromosome represents a pick-up path expressed as $x=\{(0,0),(i_1j_1)(i_2j_2),\ldots(i_nj_n)\}, 1 \leq n \leq K$, where K is the number of command. This study uses vector $X=(x_1;\ldots;x_m)^T$ as an individual. D is the search space, $D_0=\{X|X \in D\}$ is the searched space, $D_1=\{X_i|i=1,2,\ldots,pop\}$ is set of population, pop is population size, D_2 is subset of crossover, and D_3 is subset of mutation, t is iterations, and G_{max} is maximum number of iterations. When starting iteration, setting $D_0=\Phi, D_2=\Phi, D_3=\Phi$.

3.1   Initial Population

The storage location space is divided into three sub-space of A, B, C, D, then empty storage location is $D_\Phi=(D_A \cup D_B \cup D_C)$.   According to the constraint conditions (10), the   feasible   solution   space $D_A=\{p(i,j)|p(i,j)=A\}$, $D_B=\{p(i,j)|p(i,j)=B\}$, $D_C=\{p(i,j)|p(i,j)=C\}$, are generated in three sub-space, $i=1,\ldots,I, j=1,\ldots,J$.

(1) In D_A, D_B and D_C, pop, $X_{Ai}(0)(i=1,2,\ldots,pop)$, $X_{Bi}(0)(i=1,2,\ldots,pop)$, $X_{Ci}(0)$ $(i=1,2,\ldots,pop)$ with uniform distribution are randomly generated respectively.

(2) According to the commands, randomly selecting the individual from $X_{Ai}(0)$, $X_{Bi}(0)$, $X_{Ci}(0)$, and forming *pop* chromosomes $X_l(0)=\{(0,0)(i_{l1}j_{l1}),(i_{l2}j_{l2}),\ldots,(i_{lK}j_{lK})$ $|l=1,2,\ldots,pop\}$, K is the number of command.

3.2   Individual Ranking

Given the problems (1) as the minimization problems, the smaller fitness set is, the better results are.

If $s_k=max\{g_k(X),0\}(k=1,2,\ldots,r)$, for the vector with n-dimensional $f(X)=(f_1(X)$, $f_2(X),\ldots,f_n(X))$ in the solution space Rn, the function is defined as follow

$$Better(X_1,X_2) = \begin{cases} true, & if\ s_k(X_1) < s_k(X_2) \\ true, & if\ f_i(X_1) < f_i(X_2)\ and\ s_k(X_1) \leq s_k(X_2) \\ false, & otherwise \end{cases} \quad (9)$$

where $i=1,2,\cdots,n$; and $k=1,2,\cdots,r$.

If *Better=true*, then X_1 is better than X_2.

3.3   Crossover Operator

Single-point crossover is adopted, because each node of chromosome contains constraint conditions. Assuming that $R_1, R_2 \in X_l(0)$, Pc is crossover probability, a random number $\lambda_1 \in (0,1)$ is generated. If $\lambda_1 < Pc$, then

(1) Generate a random number $\lambda_2 \in [1, K]$. K is the number of nodes on a chromosome.

(2) Single-point crossover is carried out in R_1, R_2 and generating two new chromosomes.

Since λ_2 is a random value, there may exist as offspring atavism phenomena. For the new individuals Ci,

(1) If $C_i \in D_0$, then discard;   (2) If $C_i \notin D_0$, then $D_2=D_2 \cup \{C_i\}$.

Finally, $N_2(N_2 \leq pop)$ individuals are obtained in subset D_2.

3.4   Mutation Operator

Since the chromosome contains K nodes, K nodes mutation strategy is applied, Pm is the mutation probability, a random number $\lambda_3 \in (0,1)$ is generated. If $\lambda_3 < Pm$, then

(1) Generate a random number $\lambda_4 \in [1, K]$, the mutation of λ_4 nodes are implemented in the chromosomes, j is the times of mutation, $j=0$.

(2) Generate a random number $\lambda_{5i} \in (0,1)$, $i=1,2,\ldots,\lambda_4$. If $\lambda_5 \leq 0.5$, the first node of chromosome will mutate, $j=j+1$; otherwise, go to step (3).

(3) Continually generate a random number, $\lambda_{5i} \in (0,1)$, $i=1,2,\ldots,\lambda_4$. If $\lambda_5 > 0.5$, the second node of chromosome will mutate, $j=j+1$; if $j=\lambda_4$, go to step (4), otherwise, go to step (3).

(4) Generate a new chromosome, and end mutation.

For the new individuals Z_i:

(1) If $Z_i \in D_0$, then discard;   (2) If $Z_i \notin D_0$, then $D_3 = D_3 \cup \{Z_i\}$.

Finally, $N_3(N_3 \leq pop)$ individuals are obtained in D_3.

3.5   Selection Operator

To avoid degradation of the offspring, the $(\mu + \lambda)$ evolution strategy is adopted to maintain the elite, with $\lambda/\mu = 7$. The parent populations are combined $D_0 \cup D_1$, crossover subset D_2, mutation subset D_3, and

$$D_4 = D_0 \cup D_1 \cup D_2 \cup D_3 \qquad (10)$$

The temporary population D_4 is obtained:

$$D_4 = \{X_i \mid i = 1, 2, \cdots, pop, pop + N_2 + N_3\} \qquad (11)$$

In D_4, the individual Rank is calculated according to Eq.(9). The former μ are chosen according to Eq.(12).

$$D_1 = \{X_i(t+1) \mid (i = 1, 2, \cdots, \mu; t = 1, 2, \cdots, G_{\max}; X_i(t+1) \notin [L, U])\} \qquad (12)$$

The λ are randomly generated from D_4, and these two parts are considered as a feasible solution set D_1.

3.6   Stopping Rule

If the algorithm reaches its maximum G_{max} or $D_2 \cup D_3 = \Phi$, then the algorithm stops. The D_1 is the set as the global optimal solution.

3.7   Constraint Processing

The constraint conditions (4), (8) can be handled when generating initial population, the constraint conditions (5), (6) can be handled by adjusting I/O order after the algorithm stops. Then only the constraint condition (7) needs to be considered in the genetic process.

4   Experiments and Examples

The three types of goods are raw materials (A), auxiliary materials (B), and machine offerings (C). According to the problem mentioned above (population number $pop=100$, $G_{max}=50$, $Pc=0.7$, and $Pm=0.2$). The dynamic storage location assignment and stacker travel time optimization are obtained of Single Command Cycle, Dual Command Cycle, and Quadruple Command Cycle.

4.1   Single Command Cycle

In a single command cycle either storage or retrieval is performed between two visits. The simulation results are shown in 3.

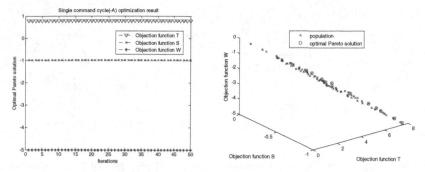

Fig. 3. Single Command Cycle (-A)

4.2   Dual Command Cycle

In a dual command cycle, the stacker performs storage and retrieval. The simulation results are shown in Fig.4.

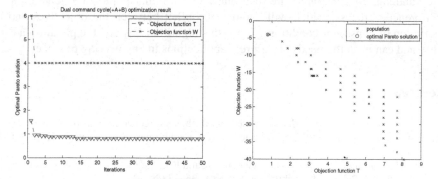

Fig. 4. Dual Command Cycle (+A +B)

Results presented in the following table are based on the simulation analyses of travel times.

Table 1. The optimization result of airport warehouse pick-up

command	types	The optimal Pareto solution (i,j)=goods	The optimal picking time/min	The worst picking time/min	Average of picking time/min	Probability of success
single command cycle	+A	(1,3)=0	0.7720	0.7720	0.7720	100%
	-A	(10,33)=A	0.7720	0.7720	0.7720	100%

Table 1. (*Continued*)

dual command cycle	+A+B	(1,3)=0,(1,5)=0	0.7720	0.7937	0.7859	95%
	-A-B	(1,23)=A,(1,24)=B	0.7828	0.7937	0.7844	95%
	+A-B	(2,5)=0,(7,19)=B	0.7828	0.8262	0.7900	95%
quadruple command cycle	+A+B+C+C	(2,5)=0,(4,10)=0, (5,9)=0,(4,8)=0	1.7065	1.9488	1.7485	91%
	-A-B-C-C	(6,1)=A,(5,5)=B, (3,6)=C,(1,1)=C	1.6060	3.3370	2.1639	88%
	+C-A-B-C	(1,3)=0,(1,2)=A, (2,1)=B,(1,1)=C	1.5548	1.7173	1.6213	89%
	+A+B-C-C	(7,5)=0,(8,8)=0, (9,4)=C,(1,1)=C	1.5737	2.3810	1.7350	90%
	+A+B+C-A	(7,8)=0,(10,7)=0, (9,10)=0,(6,7)=A	1.9557	2.4785	2.0423	90%

Node: "+" indicates storage, "-" indicates retrieval.

Results presented in Fig.3-4 and Table 1 are given on basis of performed simulation analyses for travel time and pick-up path.

5   Conclusions

Results showed that the proposed algorithm could quickly find the optimal solution. The handlings of constraints are distributed to the initial population generation and each genetic process, which will avoid the conflict between the objectives and the restriction of constraint conditions. Thus, this algorithm has good application adaptability and can meet the needs of different applications in engineering practice.

References

1. Lu, H.M., Yen, G.: Rank-density-based multi-objective genetic algorithm and benchmark test function study. J. IEEE Transactions on Evolutionary Computation **7**(4), 325–342 (2003)
2. Sun, H.: Multiple People Picking Assignment and Routing Optimization Based on Genetic Algorithm. J. Science & Technology Vision **1**, 26–27 (2014)
3. Shen, C.P., Wu, Y.H., Zhou, C.: The study of orders structure and the adaptation of picking system. J. Chinese Journal of Mechanical Engineering **5**, 820–828 (2011)
4. Yao, C.L., Zhang, G.J., Zhang, B.J.: Multi-depots distribution problem study based on rich network road model. In: The 25th Chinese Control and Decision Conference, vol. 2, pp. 876–881 (2011)
5. Ma, Y.J., Jiang, Z.Y., Yang, Z.M.: Dynamic location assignment of AS /RS based on genetic algorithm. J. Journal of Southwest Jiao Tong University **43**(3), 415–421 (2008)

A Randomized Multi-start Genetic Algorithm for the One-Commodity Pickup-and-Delivery Traveling Salesman Problem

Hiba Yahyaoui, Takwa Tlili[✉], and Saoussen Krichen

LARODEC, Institut Supérieur de Gestion Tunis, Université de Tunis,
Tunis, Tunisia
takwa.tlili@gmail.com

Abstract. The One-Commodity Pickup-and-Delivery Traveling Salesman Problem (1-PDTSP) is a generalization of the standard travelling salesman problem. 1-PDTSP is to design an optimal tour that minimizes the overall travelled distance through the depot and a set of customers. Each customer requires either a pickup service or a delivery service. We propose a Randomized Multi-start Genetic Algorithm (RM-GA) to solve the 1-PDTSP. Experimental investigations show that the proposed algorithm is competitive against state-of-the-art methods.

Keywords: One-commodity pickup-and-delivery TSP · Randomized multi-start optimization · Genetic algorithm

1 Introduction

The traveling salesman problem (TSP) is one of the most challenging NP-hard optimization problem. TSP has been widely studied in the fields of combinatorial mathematics [6], operational research [7] and artificial intelligence [1] due to its theoretical and practical significance. A more general extension of the TSP, called the One-commodity Pickup-and-Delivery TSP (1-PDTSP), is the focus of this paper. This variant was firstly evoked by Hernández-Pérez & Salazar-González, 2004 [3]. The 1-PDTSP variant is about seeking the shortest possible route that visits each customer exactly once and returns to the origin customer. A vehicle must either pick up or deliver known amounts of a single commodity to a set of customers. The collected goods from the pickup customers can be supplied to the delivery customers.

Due to its NP-hardness, the 1-PDTSP requires efficient approximate approaches to find the near optimal solutions in a reasonable amount of time. Several meta-heuristics have been proposed for the 1-PDTSP in the last 10 years. Hernández-Pérez et al. [2] combined the GRASP and VND as a hybrid approach for solving the 1-PDTSP. In GRASP/VND, the local search phase of GRASP has been replaced by a VND with two neighborhood structures (2-opt and 3-opt operators). The results showed that the GRASP/VND for 1-PDTSP

© Springer International Publishing Switzerland 2015
Y. Tan et al. (Eds.): ICSI-CCI 2015, Part III, LNCS 9142, pp. 45–49, 2015.
DOI: 10.1007/978-3-319-20469-7_6

is robust in terms of quality and computational effort. In Zhao et al. [8], Genetic
Algorithm (GA) has been applied for the first time to 1-PDTSP. To construct
the offsprings, a new pheromone-based crossover operator is used in the proposed GA. Experimentations showed that GA outperformed almost all larger
instances as well as new best solutions are reached. More recently, Mladenović
et al. [4] developed a general variable neighborhood (GVND) that improved the
best known solutions in all benchmark instances. GVND is proved to be slightly
better than GRASP/VND.

The goal of this research is to present an effective solution approach to deal
with a TSP with pickup and delivery. The experimental results shows that the
developed genetic algorithm is effective and very competitive compared to the
other state-of-the-art methods. The remainder of this paper is organized as follows. In section 2, our GA meta-heuristic is described by given some pseudo-
codes, followed by computational results in Section 3. Conclusions are presented
in Section 4.

2   Randomized Multi-Start Genetic Algorithm for 1-PDTSP

Genetic algorithm (GA) is a global search meta-heuristic belonging to the class
of evolutionary algorithms, it is based on the principles of natural biological
evaluation. The basic concept consists in starting with an initial population of
random individuals. Each individual is a single possible solution to the problem
under consideration and the population is the search space of the solutions.
In each generation, the fitness of each individual is evaluated, and a the best
individuals are then selected to be in the next population. Some other operators
are considered in GA such as the crossover and mutation which are used where
the individuals in the current population generate the offsprings. The selection,
crossover and mutation is looped until a termination criterion is reached which
is evaluated for each member of this population.

To improve the efficiency of GA for 1-PDTSP, in this paper, we proposed a
randomized multi-start GA (RM-GA), which combines the advantages of GA in
the search convergence and the randomized multi-start hill climbing strategy in
escaping local optimality. The pseudo-code of RM-GA is depicted in algorithm 1.

3   Experimental Results

In this experiments, a benchmark dataset is used for evaluating the proposed
RM-GA for 1-PDTSP. This tested instances have been obtained with a random
generator similar to the Mosheiov [5] benchmark. The solution approach was
built in Java and ran on a personal computer with 2.4 GHz $Intel^{®}$ $Core^{TM}$
processor, 4 GB RAM and Windows 7 as an operating system. This section
experimentally evaluates the performance of the proposed RM-GA compared to
Hybrid GRASP/VND approach of Hernández-Pérez et al. [2]. The GRASP/VND

Algorithm 1. Randomized Multi-start Genetic algorithm for the 1-PDTSP

Input: $cmd \leftarrow list\ of\ customers$
      $C \leftarrow capacity$
1: Find the initial solution x
2: $x* \leftarrow x$
3: **while** terminate criterion t **do**
4:     $p \leftarrow$ Construct random population $(x*)$
5:     **for** $i < p_{size}$ **do**
6:         Select A_i as father individual and B_i as mother individual from p
7:         Crossover $x' \leftarrow A_i$ and B_i
8:         **if** x' is feasible and improves the best solution $x*$ **then**
9:            $x* \leftarrow x'$
10:       **end if**
11:     **end for**
12:     Mutate $x*$
13: **end while**
Output: Best tour

Table 1. Results on the benchmark instances; $n = 20$ and $Q \in \{10, 15, 20\}$

	$Q = 10$			$Q = 15$			$Q = 20$	
Inst	H-best	RM-GA	Inst	H-best	RM-GA	Inst	H-best	RM-GA
A	4963	**4963**	A	-	4085	A	3816	**3816**
B	4976	**4976**	B	-	4309	B	4224	**4224**
C	6333	**6333**	C	-	5121	C	4492	**4492**
D	6280	**6280**	D	-	5470	D	4706	**4706**
E	6415	**6415**	E	-	5658	E	4673	**4673**
F	4805	**4805**	F	-	4352	F	4118	**4118**
G	5119	**5119**	G	-	4538	G	4369	**4369**
H	5594	**5594**	H	-	4564	H	4159	**4159**
I	5130	**5130**	I	-	4117	I	4116	**4116**
J	4410	**4410**	J	-	3977	J	3700	3703

results are directly extracted from the original paper. Tables 1, 2 and 3 report the minimum total cost obtained for our proposed RM-GA with comparison to the state-of-the-art algorithm. In all tables, the columns n, $Inst$, q and H-best represent respectively, number of customers, name of instance, capacity of vehicle and GRASP/VND solution.

For the first subset of instances, shown in table 1, we reached the best solution found in Hernández-Pérez et al. [2] for almost all instances where $Q = 10$ and $Q = 20$. Table 2 show that we found the H-best in 67% of cases. From table 3, we can conclude that our approach performs better for the large scaled instances. In table 3, we reached H-best in 90% of instances.

Table 2. Results on the benchmark instances; $n = 20$ and $Q \in \{25, 30, 35\}$

	$Q = 25$			$Q = 30$			$Q = 35$	
Inst	H-best	RM-GA	Inst	H-best	RM-GA	Inst	H-best	RM-GA
A	3816	**3816**	A	3816	**3816**	A	3816	**3816**
B	4224	**4224**	B	3942	3942	B	3942	3942
C	4492	**4492**	C	3989	3990	C	3897	3900
D	4706	**4706**	D	4112	4113	D	3743	**3743**
E	4673	**4673**	E	4381	44390	E	4299	**4299**
F	4118	**4118**	F	4118	**4118**	F	4118	**4118**
G	4369	4370	G	4248	**4248**	G	4248	**4248**
H	4159	4160	H	4007	**4007**	H	4007	**4007**
I	4116	4116	I	4026	**4026**	I	4026	**4026**
J	3700	3703	J	3678	**3678**	J	3678	**3678**

Table 3. Results on the benchmark instances; $n = 20$ and $Q \in \{40, 45, 1000\}$

	$Q = 40$			$Q = 45$			$Q = 1000$	
Inst	H-best	RM-GA	Inst	H-best	RM-GA	Inst	H-best	RM-GA
A	3816	**3816**	A	3816	**3816**	A	3816	**3816**
B	3942	**3942**	B	3942	**3942**	B	3942	**3942**
C	3897	**3897**	C	3897	**3897**	C	3897	**3897**
D	3743	3744	D	3743	3744	D	3743	3744
E	4299	**4299**	E	4299	**4299**	E	4299	**4299**
F	4118	**4118**	F	4118	**4118**	F	4118	**4118**
G	4248	**4248**	G	4248	**4248**	G	4248	**4248**
H	4007	**4007**	H	4007	**4007**	H	4007	**4007**
I	4026	**4026**	I	4026	**4026**	I	4026	**4026**
J	3678	**3678**	J	3678	**3678**	J	3678	**3678**

4 Conclusion

In this paper, we addressed the 1-PDTSP which concerns with finding a feasible vehicle route with the optimal traveling cost, such that each customer requires either a pickup service or a delivery service. To handle the 1-PDTSP, we proposed a Randomized Multi-start Genetic Algorithm (RM-GA). The experimental results showed that RM-GA is competitive compared to the other existing method.

References

1. Brassai, S.T., Iantovics, B., Enachescu, C.: Artificial intelligence in the path planning optimization of mobile agent navigation. Procedia Economics and Finance **3**, 243–250 (2012)
2. Hernandez-Perez, H., Rodrguez-Martn, I., Salazar-Gonzalez, J.J.: A hybrid grasp/vnd heuristic for the one-commodity pickup-and-delivery traveling salesman problem. Computers & Operations Research **36**, 1639–1645 (2009)
3. Hernandez-Perez, H., Salazar-Gonzalez, J.J.: Heuristics for the one-commodity pickup and delivery traveling salesman problem. Transportation Science **38**, 245–255 (2004)
4. Mladenovic, N., Urosevic, D., Hanafi, S., Ilic, A.: A general variable neighborhood search for the one-commodity pickup-and-delivery travelling salesman problem. European Journal of Operational Research **220**, 270–285 (2012)

5. Mosheiov, G.: The travelling salesman problem with pick-up and delivery. European Journal of Operational Research **79**, 299–310 (1994)
6. Rodriguez, A., Ruiz, R.: The effect of the asymmetry of road transportation networks on the traveling salesman problem. Computers & Operations Research **9**, 1566–1576 (2012)
7. Wang, Y.: The hybrid genetic algorithm with two local optimization strategies for traveling salesman problem. Computers & Industrial Engineering **70**, 124–133 (2014)
8. Zhao, F., Li, S., Sun, J., Mei, D.: Genetic algorithm for the one-commodity pickup-and-delivery traveling salesman problem. Computers & Industrial Engineering **56**, 1642–1648 (2009)

The Optimum Design of High-Rise Building Structure Based on the Strength and Stiffness of Genetic Algorithm

Kaike Sun[✉], Shoucai Li, Shihong Chu, Shuhai Zheng, and Shuai Guo

Shandong Academy of Building Research, Jinan 250031, China
jiurenkaihei@sina.cn

Abstract. The high-rise building structure have characteristics that multi-operating mode, multivariate, multivariate, multiple constraints and multi-objective, and the characteristics of complex discrete have a great influence on structural strength and stiffness. So this paper built a comprehensive and practical objective function for optimization design of high-rise building structure. This function made the design not only meet the requirement of structure safety and good performance, but also should be make structure is reasonable and material consumption as low as possible to achieve optimal allocation of resources. On this basis, it used genetic algorithm by strict derivation to code variables, and then evolved until the objective function converged to complete the optimization design of strength and stiffness for high-rise building strength. The work put forward a method of quantitative mathematical description for structure optimization design, and has brought considerable economic benefits.

Keywords: Objective Function · High-rise Buildings · Genetic Algorithm · Optimization design

1   Introduction

In recent years, with the rapid development of economy and technology in China, the urban construction process have put forward higher request for building structure design, because building structure gradually develop to top, diversity and innovative, also make more and more tall buildings beyond the national relevant specification. However the traditional architectural design method usually depend on experience to assumes a design scheme, then analyzed structure strength and stiffness by use of engineering mechanics to check and examine whether it meet the specification requirements, this design method need to repeat "hypothesis-check-modified" process many times, and also Spend a lot of manpower and time. So the optimal design of high-rise building structural strength and stiffness is more important to achieve a kind of cost resources optimize configuration, sequentially, it would bring huge economic benefits to the society.

© Springer International Publishing Switzerland 2015
Y. Tan et al. (Eds.): ICSI-CCI 2015, Part III, LNCS 9142, pp. 50–57, 2015.
DOI: 10.1007/978-3-319-20469-7_7

2   The Optimal Design of High-Rise Building Structural Strength and Stiffness

2.1   The Concept of Optimum Design

With the deepening research of building structure, the optimization design is not limited to simple single target, there are more and more uncertain concepts, such as, in the design process of building structural strength and stiffness, on the one hand, the air pollution and rainwater acidification will weaken the strength of the building materials, he earthquake and typhoon will affect the stiffness of structures; on the other hand, the good corrosion resistance and seismic performance of materials lead to the increase of cost. So in the subjective point of view, the optimal design of building structural strength and stiffness is to ensure that it can cost the most reasonable budget to be able to resistance adverse factors and reach the normal performance requirements.

2.2   The Mathematical Model of Optimization Design

Building structure strength and stiffness are relation to material properties, component section, construction scale and so on, the material properties are work ability, strength, deformation and durability of concrete, there have decisive influence on building strength and stiffness, but to a certain extent, it subject to the cost. Above all, the essence of optimization problem is cost optimization based on that both strength and stiffness of building structure meet the design requirements of domestic and foreign relevant regulations under considering the influence of controllable and predictable factors , such as material, structure and environment.

In the final analysis, that is a mathematical model to solve the optimal solution of objective function under conditions of multiple variables and multiple constraints.

Design variables: Material properties (E_i, σ_i, η_i)

Section size (b_i, h_i)

Objective function:

$$\min w = \sum_{i=1}^{n} (w_i E_i \varepsilon_i \eta_i b_i h_i) \lambda \xi \chi \qquad (2\text{-}1)$$

Constraint functions:

$$\begin{cases} \delta(k) < u(k) & (k=1,2,3,,,,t) & a \\ \sigma(j) < \bar{\sigma}(j) & (j=1,2,3,,,,t) & b \\ b_i^{min} < b_i < b_i^{max} & (i=1,2,3,,,,n) & c \\ h_i^{min} < h_i < h_i^{max} & (i=1,2,3,,,,n) & d \end{cases} \qquad (2\text{-}2)$$

In equation(2-1),"w" is the total weight or total cost of building structures, "wi" is weight and cost comprehensive coefficient of the "i" component in the project, "Ei" is elastic modulus of building materials, "λ" is material deformation coefficient, "ζ" is material durability coefficient, "χ" is load coefficient, and "bi, hi " are section size.

In equation (2-2a), "δ(k)" is displacement of "k" floor under loading, and u(k) is allowable displacement of "k" floor; σ(j) and σ(j) are stress value and allowable stress of material for any unit in the building structure under the action of load; the "bimin, bimax, himin, himax" define the maximum and minimum section size of beam and column.

$$\delta(k) = \sum_{\substack{i=j}}^{n} \int_{0}^{L_x} \left(\frac{F_x f_x}{EA_x} + \frac{F_y f_y}{GA_y} + \frac{F_z f_z}{EA_z} + \frac{M_y m_y}{GI_y} + \frac{M_x m_x}{EI_x} + \frac{M_z m_z}{GI_z} \right) dx \qquad (2\text{-}3)$$

Because the torsion effect in building structural unit is very small, so it can be equivalent to the following form:

$$I_{ix} = \beta b_i^3 h_i \qquad (2\text{-}4)$$

After substituting the above equation into (2-3), the "k" floor's displacement can be organized into the following expression:

$$\delta(k) = \sum_{i=j}^{n} \int_{0}^{L_x} \left(\frac{F_x f_x}{Eb_i h_i} + \frac{F_y f_y}{G\frac{5}{6}b_i h_i} + \frac{F_z f_z}{E\frac{5}{6}b_i h_i} + \frac{12M_y m_y}{Gb_i h_i^3} + \frac{M_x m_x}{E\beta b_i^3 h_i} + \frac{12M_z m_z}{Gb_i^3 h_i} \right) dx \qquad (2\text{-}5)$$

$$= \sum_{i=1}^{n} \left(\frac{E_{0ij}}{b_i h_i} + \frac{E_{1ij}}{b_i h_i^3} + \frac{E_{2ij}}{b_i^3 h_i} \right)$$

$$E_{0ij} = \int_{0}^{L_x} \left(\frac{F_x f_x}{E} + \frac{F_y f_y}{G\frac{5}{6}} + \frac{F_z f_z}{E\frac{5}{6}} \right) dx \qquad (2\text{-}6)$$

$$E_{1ij} = \int_{0}^{L_x} \left(\frac{12M_z m_z}{G} \right) dx \qquad (2\text{-}7)$$

$$E_{2ij} = \int_{0}^{L_x} \left(\frac{M_x m_x}{E\beta} + \frac{12M_z m_z}{G} \right) dx \qquad (2\text{-}8)$$

The roles of internal stress of building structure unit depend on the components' load conditions, because the principle of balance, it can obtain internal stress level of any unit, and its display expression as follows:

$$\frac{\partial \sigma_{x_i}}{\partial x} + \frac{\partial \tau_{y_i x_i}}{\partial y} + \frac{\partial \tau_{z_i x_i}}{\partial z} + F_{xi} + f_{xi} + G_{xi} = 0$$

$$\frac{\partial \tau_{x_i y_i}}{\partial x} + \frac{\partial \sigma_{y_i}}{\partial y} + \frac{\partial \tau_{z_i y_i}}{\partial z} + F_{yi} + f_{yi} + G_{y_i} = 0 \qquad (2\text{-}9)$$

$$\frac{\partial \tau_{x_i z_i}}{\partial x} + \frac{\partial \tau_{y_i z_i}}{\partial y} + \frac{\partial \sigma_{z_i}}{\partial z} + F_{zi} + f_{zi} + G_{zi} = 0$$

Among, "F" is the external load, and "G" is gravity load.

The expression shows that the strength and stiffness of structures constraints depends on external load, unit internal force, the size of the interface and the characteristics of the material, then, it can get optimization explicit mathematical model of strength and stiffness constraints after simplifying the above formulas.

$$\left\{ \begin{array}{l} \delta(k) = \sum_{i=1}^{n} (\dfrac{E_{0ij}}{b_i h_i} + \dfrac{E_{1ij}}{b_i h_i^3} + \dfrac{E_{2ij}}{b_i^3 h_i}) \le u(k) \\ \sigma(i) = \sqrt{\sigma x_i^2 + \sigma y_i^2 + \sigma z_i^2} \le \bar{\sigma} \end{array} \right\} \qquad (2\text{-}10)$$

3　The Optimization Design of a High-Rise Residential Based on Genetic Algorithm

Genetic algorithm is a calculation model of simulating Darwin the evolution natural selection and Genetic mechanism of biological evolution process, and also is a kind of searching optimal solution by simulating the natural evolution process. The genetic algorithm is starting from a population of potential solution based on representative problems, whereas a population includes a certain number of individuals after genes code. Each individual is actually chromosomes with the characteristics of the entity, as the main carrier of genetic material, chromosome is a collection of multiple genes, their internal performance is a combination of genes, it determines the external performance of the shape of individual.

Therefore, it is necessary to firstly compete genetic code that is from phenotype to genotype. This process will result that the population of natural evolution are more adapted to the environment than the epigenetic population, so the last individual in the population after decoding can be used as an approximate optimal solution.

The basic operation processes of genetic algorithm are as follows:

A) Initialization: Set the evolution algebra counter t = 0, and set the maximum evolution algebra as T, randomly generated M individuals as the initial group P(0).
B) Individual evaluation: Calculate the fitness of the individuals in the group P (t).
C) Choose operation: The purpose of choosing is making the optimum individual directly heritage to the next generation or through matching cross to produce new individual, and then genetic to future generations. Select operation is established on the basis of evaluating the fitness of individuals in the group.D) Crossover operation: The crossover operator is key of genetic algorithm.
E) Mutation operation: Change a group of certain genes' value on individuals .Group P (t) after selection,The most common crossover operator is one-point crossover. Specific operation: random set a intersection in the individual string, when crossing, the structure of two individuals before or after the point are swapping and generating two new individual. The following gives an example of single point crossover:

Table 1. The example of single point crossover

Individual A	Individual B
1001×111	0011×000
to	to
1 0 0 1 0 0 0	0 1 1 1 1 1
New individual	New individual

Now, it is commissioned by an owner that we should optimize the design of a 30 layer frame shear wall structure(as shown in figure 1)by the application of the above model and genetic algorithm,1-5 layers of the building structure are commodity service outlets, and every layer's high is 4 meters;while,6-30 stores are residential buildings, the height is 3.3 meters. As this structure, the constant load and active load of bottomed shop front room's floor were 4.5 KN/m2 and 2.5 KN/m2,the roofing load has little effect, so that it can be ignored; the constant load and active load of high-rise flats' floor were 4.0 KN/m2 and 2.0 KN/m2, and as well as roof were 6.5 KN/m2 and 0.5KN/m2.Construction sites of structure is top, and the earthquake level's magnitude is 7,the peak acceleration is 0.10g,there are 15 formations to consider, the basic wind pressure rating is 0.7 kN/m, the surface coarse degree is "C". The original design section of column section is 400~ 1100mm, most of the beam section are 300*750 mm. The concrete strength of 1 ~ 4 standard layer are respectively C40, C35 and C3O, the strength of the reinforcement fy = 300 kn/m2.beam allow The reinforcement ratio of beam allowed is 0.8~ 1.8, and the axial compression ratio of column allowed is 0.5 ~0.7, the maximum inter layer displacement angle is 1/800,the comprehensive unit price of concrete, steel and template respectively: Cc = $450 / m3, Cs = 6000 yuan/t, Cf = 36 yuan/m2.

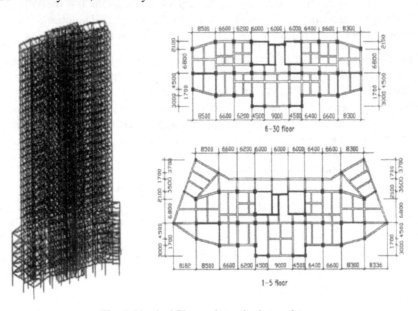

Fig. 1. Vertical Plan and standard story layer

3.1   The Optimized Design Variables and Original Design Domain

In the optimization process, there were four standard layer in the entire high-rise building, and columns and beams are grouping design. As a standard layer,1~5 layers had 147 root beams and 46 columns; then the rest of the residential buildings were divided into three standard layer, so every layer had 120 root beams and 34 columns, the position of the wall and cross section remains unchanged, beam's width is constant, that is b=250mm,In the case of considering square column, there are a total of 507 columns and 148 beams, both are design variables, according to the relevant criteria, the beam's high hmin = 300, hmax = 650, Δ h = 50 mm; The column in hc min = 200 mm,and hc min =200mm ,hc max=1000mm and Δh=50mm.

3.2   The Application of Genetic Algorithm

In this example, according to the engineering structure, there were 16 section height to be chosen by the beams ,so the length of chromosome coding was 4;as well as 32 section height to choose by columns , and the length of chromosome coding was 5. There were 3 individuals in the population, and the crossover probability and mutation probability were PM = 0.8 and PC = 0.1. The crossover operation and mutation operation of genetic algorithm were shown in figure 2 and figure 3.

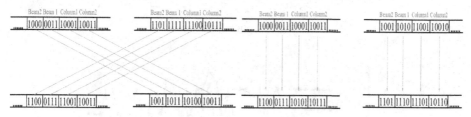

Fig. 2. Crossover operation   **Fig. 3.** Mutation operation

4   The Optimization Results

There were a total of 152 design variable, which included column and beam in the standard layers, and we regarded the within all of the design variables as individuals, and we have exerted a certain operation according to their fitness to the environment (fitness evaluation), so that it can realize the evolutionary process of evolution, and complete the optimization design. In the figure 4 and figure 5, we can obtained the optimization process of construction cost which is objective function of structure optimization, and the angle of inter layer lateral displacement. In the process of intensity optimization, it was not tending towards stability until iterative had been six times, ultimately, the total cost was 15.81 million Yuan, but ,the angle of lateral displacement of the structure floor exceed allowable values, so it had to turn to stiffness optimization , after being four iterations, the result converges to 17.61 million Yuan; but, it also hadn't some local components meet the requirements of strength, so it also needed to return to the optimization of intensity, finally after the optimization of two strength designs and two stiffness designs were completed, we can obtain the most reasonable optimal design scheme of structure under the conditions of considering all constraints.

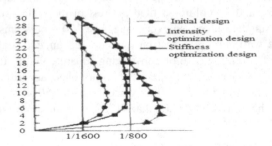

Fig. 4. Inter-story drift in y direction under earthquake

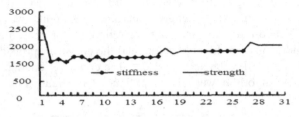

Fig. 5. Optimal history

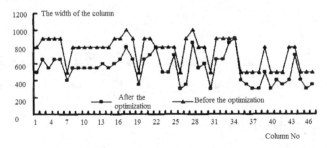

Fig. 6. Distribution of column depth of the initial and final deign

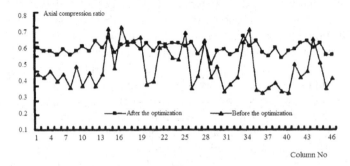

Fig. 7. Distribution of column axial compress ratio of the initial and final deign

From the above images, we can see that the section size of column have reduced a lot, and compared with the original design, the axial compression ratio is greater, and also it was misdistribution, while after the optimization, the axial compression ratio was distributed between 0.5 and 0.6,the index of axial compression ratio was becoming more and more reasonable. In the end, we can obtain the most reasonable optimal design scheme of structure under the conditions of considering all constraints in the case of that the optimization of two strength designs and two stiffness designs were completed, it were these optimization designs that the project's total cost had reduced 2,650,000 Yuan from 20,670,000 to 18,020,000 Yuan, it made up about 12.8 percent of all cost, so that it brought huge economic benefits.

5   Conclusion

In this paper, the author roundly built a comprehensive and comply optimization design model with all parameters and constraints, which was also meeting the ergonomics design, and it was aim at high-rise building structure. So that the author had a in-depth discussion about the optimization problems of strength and stiffness, and it puts forward a simple and practical calculation method to solve constrained problems, and this method was based on genetic algorithm which had the characteristic of adaptive design domain policy and can avoid using penalty function effectively, most important of all, it can effectively solve the problem of low efficiency when analyzing the large complex structures, while, all of those were based on that the objective function was project total cost. The paper combined the characteristics of building structure and china's structure design codes currently, the author carried out a case to optimize a 30 building structure, the example analysis shown that the optimization of structural strength and stiffness based on the genetic algorithm can produce a reasonable, reliable and economic structure design, as well as bring huge economic benefits to the society.

Acknowledgements. The authors thank Prof. Zhang xiao and Prof. Hao Shangfu for the valuable discussion and recommendation.

References

1. Sahaba, M.G., Ashourb, A.F.: ToropovcV.V. Cost optimization of reinforced concrete flat slab buildings. Engineering Structures **27**(3), 313–322 (2005)
2. Kurpati, A., Azarm, S., Wu, J.: Constraint handling improvements for multiobjective genetic algorithms. Structural and Multidisciplinary Optimization **23**, 204–213 (2002)
3. Soibelman, L., Pena-Mora, F.: Distributed Multi-Reasoning Mechanism to Support Conceptual Structural Design. Journal of structural Engineering **126**(6), 733–742 (2000)
4. Gong, Y., Xu, L., Grierson, D.E.: Performance-based desisensitivity analysis of steel moment frames under earthquake loading. J. Numer. Meth. Engng. **63**, 1229–1249 (2005)
5. Park, H.S., Kwon, J.H.: Optimal drift design model for multi-story buildings subjected to dynamic Lateral forces. Structural Design of Tall Building **12**, 317–333 (2003)

Fuzzy Systems

Fuzzy Concepts in Formal Context

Luodan Meng and Keyun Qin[✉]

College of Mathematics, Southwest Jiaotong University,
Chengdu 610031, Sichuan, China
1586128477@qq.com, keyunqin@263.net

Abstract. Formal concept analysis (FCA) provides a theoretical framework for learning hierarchies of knowledge clusters. This paper is devoted to the study of the fuzzy concept in FCA. We propose a fuzzy relation on the universe to characterize the similarity of the objects. Based on fuzzy rough set model, we present a kind of approximation operators to characterize the fuzzy concept and its accuracy degree in FCA. The basic properties of these operators are investigated.

Keywords: Formal concept analysis · Formal context · Fuzzy concept · Accuracy degree

1 Introduction

Formal concept analysis (FCA) was independently introduced by Wille in the 1980's [1]. FCA deals with relational information structures (formal contexts) and provides a theoretical framework for learning hierarchies of knowledge clusters called formal concepts. As an efficient tool of data analysis and knowledge processing, FCA has been applied in many fields, such as knowledge engineering, data mining, information searches, and software engineering [1,2]. It has become increasingly popular among various methods of conceptual data analysis and knowledge representation. Most of the researches on FCA concentrate on such topics as: construction of the concept lattice [3,4], pruning of the concept lattice [5,6], acquisition of rules [7], relationship between the concept lattice and rough set [8–11], and applications [12,13]. The combination of FCA and fuzzy set theory is another important issue.

In FCA, formal concept is a key notion. It is defined by an (set of objects, set of attributes) pair. From extent point of view, these sets of objects are (exact) concepts. Yang and Qin[14] proposed the notion of uncertain concept in a formal context. Based on covering rough set approach, the uncertain concept was characterized by approximation operators. In this approach, each subset of the universe may be a concept, but with different accuracy degrees. Qin and Meng [15] conduct a further study on the uncertainty in a formal context. It is pointed out that some covering approximation operators are not suitable to characterize uncertain concept in a formal context. Furthermore, some new approximation operators are employed to study the accuracy degree of uncertain concept.

Y. Tan et al. (Eds.): ICSI-CCI 2015, Part III, LNCS 9142, pp. 61–67, 2015.
DOI: 10.1007/978-3-319-20469-7_8

This paper is devoted to the discussion of fuzzy concepts in a formal context. Based on the formal (exact) concepts, a fuzzy similarity relation between objects is presented. The fuzzy concept and its accuracy degree are characterized by fuzzy rough approximation operators. The paper is organized as follows: In Section 2, we recall some notions and properties of FCA. In Section 3, we propose a fuzzy similarity relation associated with a formal context. Its basic properties are analyzed. Furthermore, we present an approach to characterizing the fuzzy concepts by using fuzzy approximation operators. The paper is completed with some concluding remarks.

2   Preliminaries

Formal Concept Analysis [1] provides a theoretical framework for learning hierarchies of knowledge clusters called formal concepts. A basic notion in FCA is the formal context. Given a set G of objects and a set M of attributes (also called properties), a formal context consists of a triple $k = \{G, M, I\}$ where I specifies (Boolean) relationships between objects of G and attributes of M, i.e.,$I \subseteq G \times M$. Usually, formal contexts are given under the form of a table that formalizes these relationships. A table entry indicates whether an object has the attribute (this is denoted by 1), or not (it is often indicated by 0).

Let$k = (G, M, I)$ be a formal context. The set-valued operators $\uparrow: P(G) \to P(M)$ and $\downarrow: P(M) \to P(G)$ are defined as [1]:for each $A \in P(G)$ and $B \in P(M)$,

$$A^{\uparrow} = \{m \in M; \forall a \in A((a, m) \in I)\} \tag{1}$$

$$B^{\downarrow} = \{g \in G; \forall b \in B((g, b)) \in I)\} \tag{2}$$

A formal concept of k is defined as a pair (A, B) with $A^{\uparrow} = B, B^{\downarrow} = A$. A is called the extent of the formal concept (A, B), whereas B is called the intent. The main problem in formal concept analysis is that of extracting formal concepts from object/attribute relations. The set of all formal concepts equipped with a partial order (denoted by $\leq$) defined as:$(X_1, Y_1) \leq (X_2, Y_2)$ if and only if $X_1 \subseteq X_2$ or equivalently, $Y_1 \subseteq Y_2$)), forms a complete lattice, called the concept lattice of k and denoted by $L(k)$. Its structure is given by the following theorem.

Theorem 1. *[1] The concept lattice $L(k)$ is a complete lattice in which infimum and supremum are given by:*

$$\wedge_{j \in J} (X_j, Y_j) = (\cap_{j \in J} X_j, (\cup_{j \in J} Y_j)^{\downarrow\uparrow}) \tag{3}$$

$$\vee_{j \in J} (X_j, Y_j) = ((\cup_{j \in J} X_j)^{\uparrow\downarrow}, \cap_{j \in J} Y_j) \tag{4}$$

Example 1. [16] We consider the following formal context $k = \{G, M, I\}$, where $G = \{x_1, x_2, x_3, x_4, x_5, x_6, x_7, x_8\}$, $M = \{a, b, c, d, e, f, g, h, i\}$. and I is represented in the following table.

Table 1. The formal context $\kappa = (G, M, I)$

	a	b	c	d	e	f	g	h	i
x_1	1	1	0	0	0	0	0	1	0
x_2	1	1	0	0	0	0	1	1	0
x_3	1	1	1	0	0	0	1	1	0
x_4	1	0	1	0	0	0	1	1	1
x_5	1	1	0	1	0	1	0	0	0
x_6	1	1	1	1	0	1	0	0	0
x_7	1	0	1	1	1	0	0	0	0
x_8	1	0	1	1	0	1	0	0	0

The formal concepts in this formal context are listed as follows:
$(\emptyset, \{a, b, c, d, e, f, g, h, i\}), (\{x_3\}, \{a, b, c, g, h\}), (\{x_4\}, \{a, c, g, h, i\}),$
$(\{x_6\}, \{a, b, c, d, f\}), (\{x_7\}, \{a, c, d, e\}), (\{x_2, x_3\}, \{a, b, g, h\}),$
$(\{x_3, x_4\}, \{a, c, g, h\}), (\{x_3, x_6\}, \{a, b, c\}), (\{x_5, x_6\}, \{a, b, d, f\}), (\{x_6, x_8\},$
$\{a, c, d, f\}),$
$(\{x_1, x_2, x_3\}, \{a, b, g\}), (\{x_2, x_3, x_4\}, \{a, g, h\}), (\{x_5, x_6, x_8\}, \{a, d, f\}),$
$(\{x_6, x_7, x_8\}, \{a, c, d\}), (\{x_1, x_2, x_3, x_4\}, \{a, g\}),$
$(\{x_5, x_6, x_7, x_8\}, \{a, d\}), (\{x_1, x_2, x_3, x_5, x_6\}, \{a, b\}),$
$(\{x_3, x_4, x_6, x_7, x_8\}, \{a, c\}), (\{x_1, x_2, x_3, x_4, x_5, x_6, x_7, x_8\}, \{a\}).$

3   Uncertain Concepts in a Formal Context

Let $k = \{G, M, I\}$ be a formal context. Every formal concept can be described from extent point of view and $L(k) = \{(X^{\uparrow\downarrow}, X^{\uparrow}); X \subseteq G\}$. In what follows, we denote $L(k') = \{X^{\uparrow\downarrow}; X \subseteq G\}$. $L(k')$ can be looked upon the set of all formal concepts with respect to k. Let $C \in L(k')$ and $x, y \in G$. If $x \in C$ and $y \in C$, then x and y have all the attributes from $C^{\uparrow}$. In this case, we may think that x and y are similar with respect to the concept C. On the contrary, if $x \in C, y \notin C$ or $x \notin C, y \in C$, x and y may be looked upon discernible with respect to C . Based on this observation, we define a similarity relation R on the universe as follows: for any $x, y \in G$,

$$R(x, y) = \frac{|\{C \in L(k'); \{x, y\} \subseteq C\}|}{|\{C \in L(k); \{x, y\} \cap C \neq \emptyset\}|} \tag{5}$$

Intuitively speaking, $R(x, y)$ represents similarity degree of objects x and y. Furthermore, for each $C \in L(k')$, we know that $\{x, y\} \subseteq C$ if and only if $\{x, y\}^{\uparrow} \downarrow \subseteq C$. Thus, $R(x, y)$ can be equivalently represented as

$$R(x, y) = \frac{|\{C \in L(k'); \{x, y\}^{\uparrow\downarrow} \subseteq C\}|}{|\{C \in L(k'); \{x\}^{\uparrow\downarrow} \subseteq C \vee \{y\}^{\uparrow\downarrow} \subseteq C\}|} \tag{6}$$

The theory of fuzzy sets initiated by Zadeh [17] provides an appropriate framework for representing and processing vague concepts by allowing partial memberships. Let U be a nonempty set, called universe. A fuzzy subset of U is defined by a membership function $\mu : U \to [0,1]$. For any $x \in U$, the membership value $\mu(x)$ essentially specifies the degree to which x belongs to the fuzzy subset μ. There are many different definitions for fuzzy subset operations. With the min-max system proposed by Zadeh [17], fuzzy set intersection, union and complement are defined as follows:

$$(\mu \cap \nu)(x) = \mu(x) \wedge \nu(x) \tag{7}$$

$$(\mu \cup \nu)(x) = \mu(x) \vee \nu(x) \tag{8}$$

$$\mu^c(x) = 1 - \mu(x) \tag{9}$$

where μ and ν are fuzzy subsets of U and $x \in U$. We denote by $F(U)$ the set of all fuzzy subsets of U. The following theorem is trivial.

Theorem 2. *Let $\kappa = \{G, M, I\}$ be a formal context.*
(1) $R(x,y)$ is a fuzzy relation on G, i.e., $R : G \times G \to [0,1]$.
(2) $R(x,y)$ is reflexive, i.e., $R(x,x) = 1$ for each $x \in G$.
(3) $R(x,y)$ is symmetric, i.e., $R(x,y) = R(y,x)$ for each $x,y \in G$.

Dubois and Prade [18] first introduced the concept of fuzzy rough sets by combining fuzzy set and rough set. For a fuzzy relation R on a universe U, (U, R) is called a fuzzy approximation space. The fuzzy rough approximations of a fuzzy set are constructed. Based on fuzzy rough set model, we propose the following definition.

Definition 1. *Let $\kappa = (G, M, I)$ be a formal context. For any $\mu \in F(G)$,the lower approximation $\underline{R}(\mu)$ and upper approximation $\overline{R}(\mu)$ of μ are fuzzy subsets of G, and defined as: for each $x \in G$,*

$$\underline{R}(\mu)(x) = \wedge_{u \in U}((1 - R(x,u)) \vee \mu(u)) \tag{10}$$

$$\overline{R}(\mu)(x) = \vee_{u \in U}(R(x,u) \wedge \mu(u)) \tag{11}$$

In this definition, if $X \subseteq G$ is a subset of G, then
$\underline{R}(X)(x) = \wedge_{u \in U}((1 - R(x,u)) \vee X(u)) = \wedge_{u \in \sim X}(1 - R(x,u))$,
$\overline{R}(X)(x) = \vee_{u \in U}(R(x)(u) \wedge X(u)) = \vee_{u \in X} R(x,u)$.

Example 2. We consider the the formal context $\kappa = (G, M, I)$ given in Example 1. By routine computation, we have
$\{C \in L(\kappa'); \{x_2, x_3\} \cap C \neq \emptyset\}$
$= \{\{x_3\}, \{x_2, x_3\}, \{x_3, x_4\}, \{x_3, x_6\}, \{x_1, x_2, x_3\}, \{x_2, x_3, x_4\}, \{x_1, x_2, x_3, x_4\},$
$\{x_3, x_4, x_6, x_7, x_8\}, \{x_1, x_2, x_3, x_5, x_6\}, \{x_1, x_2, x_3, x_4, x_5, x_6, x_7, x_8\}\}$,

$$R(x_2, x_3) = \frac{|\{C \in L(\kappa'); \{x_2, x_3\} \subseteq C\}|}{|\{C \in L(\kappa'); \{x_2, x_3\} \cap C \neq \emptyset\}|} = \frac{6}{10} = \frac{3}{5}.$$

$R(x,y)$ for any $x,y \in G$ can be calculated similarly. It is represented in the following table:

Table 2. The similarity relation

	x_1	x_2	x_3	x_4	x_5	x_6	x_7	x_8
x_1	1	$\frac{2}{3}$	$\frac{2}{5}$	$\frac{1}{4}$	$\frac{2}{7}$	$\frac{1}{6}$	$\frac{1}{8}$	$\frac{1}{9}$
x_2	$\frac{2}{3}$	1	$\frac{3}{5}$	$\frac{1}{3}$	$\frac{2}{9}$	$\frac{1}{7}$	$\frac{1}{10}$	$\frac{1}{11}$
x_3	$\frac{2}{5}$	$\frac{3}{5}$	1	$\frac{5}{11}$	$\frac{2}{13}$	$\frac{1}{4}$	$\frac{2}{13}$	$\frac{1}{7}$
x_4	$\frac{1}{4}$	$\frac{1}{3}$	$\frac{5}{11}$	1	$\frac{1}{10}$	$\frac{1}{7}$	$\frac{2}{9}$	$\frac{1}{5}$
x_5	$\frac{2}{7}$	$\frac{2}{9}$	$\frac{2}{13}$	$\frac{1}{10}$	1	$\frac{1}{2}$	$\frac{1}{4}$	$\frac{3}{8}$
x_6	$\frac{1}{6}$	$\frac{1}{7}$	$\frac{1}{4}$	$\frac{1}{7}$	$\frac{1}{2}$	1	$\frac{4}{11}$	$\frac{3}{5}$
x_7	$\frac{1}{8}$	$\frac{1}{10}$	$\frac{2}{13}$	$\frac{2}{9}$	$\frac{1}{4}$	$\frac{4}{11}$	1	$\frac{4}{7}$
x_8	$\frac{1}{9}$	$\frac{1}{11}$	$\frac{1}{7}$	$\frac{1}{5}$	$\frac{3}{8}$	$\frac{3}{5}$	$\frac{4}{7}$	1

(1) Let $\mu \in F(G)$ be given by:
$\mu = \frac{0.1}{x_1} + \frac{0}{x_2} + \frac{0.6}{x_3} + \frac{0.9}{x_4} + \frac{0}{x_5} + \frac{0.8}{x_6} + \frac{0}{x_7} + \frac{0.3}{x_8}$.

The lower approximation $\underline{R}(\mu)$ and upper approximation $\overline{R}(\mu)$ of μ can be calculated as follows:

$\underline{R}(\mu)(x_1) = \wedge_{u \in U}((1 - R(x_1, u)) \vee \mu(u)) = 0.1$.
$\overline{R}(\mu)(x_1) = \vee_{u \in U}(R(x_1, u)) \wedge \mu(u)) = 0.4$.
Similarly, we have $\underline{R}(\mu)(x_2) = 0, \underline{R}(\mu)(x_3) = 0.4, \underline{R}(\mu)(x_4) = 0.6$,
$\underline{R}(\mu)(x_5) = 0, \underline{R}(\mu)(x_6) = 0.4, \underline{R}(\mu)(x_7) = 0, \underline{R}(\mu)(x_8) = 0.3$.
$\overline{R}(\mu)(x_2) = 0.6, \overline{R}(\mu)(x_3) = 0.6, \overline{R}(\mu)(x_4) = 0.9, \overline{R}(\mu)(x_5) = 0.5$,
$\overline{R}(\mu)(x_6) = 0.8, \overline{R}(\mu)(x_7) = \frac{4}{11}, \overline{R}(\mu)(x_8) = 0.6$.

(2) Let $X \subseteq G$ is given by $X = \{x_2, x_3, x_6\}$. It follows that
$\underline{R}(X)(x_2) = \wedge_{u \in \sim X}(1 - R(x_2, u)) = (1 - R(x_2, x_1)) \wedge (1 - R(x_2, x_4))$
$\wedge (1 - R(x_2, x_4)) \wedge (1 - R(x_2, x_5)) \wedge (1 - R(x_2, x_7)) \wedge (1 - R(x_2, x_8))$
$= (1 - \frac{2}{3}) \wedge (1 - \frac{1}{3}) \wedge (1 - \frac{2}{9}) \wedge (1 - \frac{1}{10}) \wedge (1 - \frac{1}{11}) = \frac{1}{3}$.
$\overline{R}(X)(x_1) = \vee_{u \in X}(R(x_1, u)) = R(x_1, x_2) \vee R(x_1, x_3) \vee R(x_1, x_6) = \frac{2}{3}$.
Similarly, $\underline{R}(X)(x_1) = 0, \underline{R}(X)(x_3) = \frac{6}{11}, \underline{R}(X)(x_4) = 0, \underline{R}(X)(x_5) = 0$,
$\underline{R}(X)(x_6) = 0.4, \underline{R}(X)(x_7) = 0, \underline{R}(X)(x_8) = 0$.
$\overline{R}(X)(x_2) = 1, \overline{R}(X)(x_3) = 1, \overline{R}(X)(x_4) = \frac{5}{11}, \overline{R}(X)(x_5) = 0.5$,
$\overline{R}(X)(x_6) = 1, \overline{R}(X)(x_7) = \frac{4}{11}, \overline{R}(X)(x_8) = 0.6$.

Theorem 3. *Let $\kappa = (G, M, I)$ be a formal context, $\mu, \nu \in F(G)$.*
(1) $\underline{R}(\mu^c) = (\overline{R}(\mu))^c, \overline{R}(\mu^c) = (\underline{R}(\mu))^c$.
(2) $\underline{R}(G) = G, \overline{R}(\emptyset) = \emptyset$.
(3) $\underline{R}(\mu \cap \nu) = \underline{R}(\mu) \cap \underline{R}(\nu), \overline{R}(\mu \cup \nu) = \overline{R}(\mu) \cup \overline{R}(\nu)$.
(4) $\underline{R}(\mu) \subseteq \mu \subseteq \overline{R}(\mu)$.
(5) If $\mu \subseteq \nu$, then $\underline{R}(\mu) \subseteq \underline{R}(\nu), \overline{R}(\mu) \subseteq \overline{R}(\nu)$.

Definition 2. *Let $\kappa = \{G, M, I\}$ be a formal context. For any $\mu \in F(G)$, the accuracy degree $Ad(\mu)$ of μ is defined as $Ad(\mu) = \frac{|\underline{R}(\mu)|}{|\overline{R}(\mu)|}$.*

Clearly, this definition is generalization of the accuracy degree for classical subset in rough set theory.

Example 3. For $\mu \in F(G)$ and $X \subseteq G$ in Example 2, we have
$|\underline{R}(\mu)| = \sum_{i=1}^{8} \underline{R}(\mu)(x_i) = 0.1 + 0 + 0.4 + 0.6 + 0 + 0.4 + 0 + 0.3 = 1.8$.
Similarly, $|\overline{R}(\mu)| = \sum_{i=1}^{8} \overline{R}(\mu)(x_i) = 4.76$, $|\underline{R}(X)| = \sum_{i=1}^{8} \underline{R}(X)(x_i) = 1.28$,
$|\overline{R}(X)| = \sum_{i=1}^{8} \overline{R}(X)(x_i) = 5.58$. Thus we have $Ad(\mu) = \frac{|\underline{R}(u)|}{|\overline{R}(u)|} = \frac{1.8}{4.76} = 0.38$,
$Ad(X) = \frac{|\underline{R}(X)|}{|\overline{R}(X)|} = \frac{1.28}{5.58} = 0.23$.

Based on generalized fuzzy rough approximation operators, we propose another kind of approximations.

Definition 3. *Let $\kappa = (G, M, I)$ be a formal context. If $X \subseteq G$ is a subset of G the lower approximation $\underline{apr}(X)$ and upper approximation $\overline{apr}(X)$ of X are fuzzy subsets of G, and defined as*
$$\underline{apr}(X)(x) = \frac{1}{|X|} \sum_{y \in X} R(x, y),$$
$$\overline{apr}(X)(x) = 1 - \frac{1}{|\sim X|} \sum_{y \in \sim X} R(x, y).$$

Clear, we have

Corollary 1. *Let $\kappa = \{G, M, I\}$ be a formal context. If $X \subseteq G$ is a subset of G, then $\underline{apr}(X^c) = (\overline{apr}(X))^c$, $\overline{apr}(X^c) = (\underline{apr}(X))^c$.*

Example 4. We consider $X \subseteq G$ in example 2 given by $X = \{x_2, x_3, x_6\}$. It follows that
$$\underline{apr}(X)(x_1) = \frac{1}{|X|} \sum_{y \in X} R(x_1, y) = \frac{1}{3}(\frac{2}{3} + \frac{2}{5} + \frac{1}{6}) = 0.41.$$
$$\overline{apr}(X)(x_1) = 1 - \frac{1}{(|\sim X|)} \sum_{y \in \sim X} R(x, y) = 1 - \frac{1}{5}(1 + \frac{1}{4} + \frac{2}{7} + \frac{1}{8} + \frac{1}{9}) = 0.65.$$
Similarly, $\underline{apr}(X)(x_2) = 0.58$, $\underline{apr}(X)(x_3) = 0.62$, $\underline{apr}(X)(x_4) = 0.31$, $\underline{apr}(X)(x_5) = 0.29$, $\underline{apr}(X)(x_6) = 0.46$, $\underline{apr}(X)(x_7) = 0.21$, $\underline{apr}(X)(x_8) = 0.28$. $\overline{apr}(X)(x_2) = 0.72$, $\overline{apr}(X)(x_3) = 0.7$, $\overline{apr}(X)(x_4) = 0.65$, $\overline{apr}(X)(x_5) = 0.6$, $\overline{apr}(X)(x_6) = 0.69$, $\overline{apr}(X)(x_7) = 0.57$, $\overline{apr}(X)(x_8) = 0.55$.
Thus $|\underline{apr}(X)| = \sum_{i=1}^{8} \underline{apr}(X)(x_i) = 3.16$, $|\overline{apr}(X)| = \sum_{i=1}^{8} \overline{apr}(X)(x_i) = 5.13$, and $Ad(X) = \frac{|\underline{apr}(X)|}{|\overline{apr}(X)|} = \frac{3.16}{5.13} = 0.62$.

Remark: There are different approaches to characterizing the similarity relation of objects with respect to a formal context. For example, it can be defined by: $R(x, y) = \frac{|C \in L(\kappa'); \{x,y\} \subseteq C \vee \{x,y\} \cap C = \emptyset|}{|L(\kappa')|}$. We can conduct relative study based on this similarity relation.

4   Concluding Remarks

Fuzzy set theory, soft set theory and rough set theory are all mathematical tools for dealing with uncertainties. This paper is devoted to the discussion of the combinations of fuzzy set, rough set and soft set. The notions of similarity measures induced by soft set and soft fuzzy set are presented. Based on these similarity measures, some new soft fuzzy rough set models are proposed and their properties are surveyed.

In further research, the axiomatization of the approximation operators is an important and interesting issue to be addressed.

Acknowledgments. This work has been supported by the National Natural Science Foundation of China (Grant No. 61473239, 61175055, 61175044), the Fundamental Research Funds for the Central Universities of China (Grant No. 2682014ZT28) and the Open Research Fund of Key Laboratory of Xihua University (szjj2014-052).

References

1. Wille, R.: Restructuring lattice theory: an approach based on hierarchies of concepts. In: Rival, I. (ed.) Ordered Sets, pp. 445–470. Reidel, Dordrecht-Boston (1982)
2. Gediga, G., Wille, R.: Formal Concept Analysis-Mathematic Foundations. Springer, Berlin (1999)
3. Hu, K.Y., Lu, C.Y., Shi, C.Y.: Advances in concept lattice and its application. Journal of Tsinghua University (Science and Technology) 40(9), 77–81 (2000)
4. Ho, T.B.: An approach to concept formation based on formal concept analysis. IEICE Trans. Information and Systems **E782D**(5), 553–559 (1995)
5. Oosthuizen, G.D.: The application of concept lattice to machine learning. Technical Report, University of Pretoria, South Africa (1996)
6. Zhang, W.X., Wei, L., Qi, J.J.: Attribute reduction theory and approach to concept lattice. Science in China, Ser. F Information Sciences 6(48), 713–726 (2005)
7. Godin, R.: Incremental concept formation algorithm based on Galois (concept) lattices. Computational Intelligence **11**(2), 246–267 (1995)
8. Yao, Y.Y.: Concept lattices in rough set theory. In: Dick, S., Kurgan, L., Pedrycz, W., et al. (eds.) Proceedings of 2004 Annual Meeting of the North American Fuzzy Information Processing Society (NAFIPS 2004), pp. 796–801. IEEE Catalog Number: 04TH8736, June 27–30, 2004
9. Oosthuizen, G.D.: Rough sets and concept lattices. In: Ziarko, W.P. (ed.) Rough Sets, and Fuzzy Sets and Knowledge Discovery (RSKD93), pp. 24–31. Springer-Verlag, London (1994)
10. Zhang, X.H., Dai, J.H., Yu, Y.C.: On the union and intersection operations of rough sets based on various approximation spaces. Information Sciences **292**, 214–229 (2015)
11. Zhang, X.H., Zhou, B., Li, P.: A general frame for intuitionistic fuzzy rough sets. Information Sciences **216**, 34–49 (2012)
12. Tonella, P.: Using a concept lattice of decomposition slices for program understanding and impact analysis. IEEE Transactions on Software Engineering **29**(6), 495–509 (2003)
13. Grigoriev, P.A., Yevtushenko, S.A.: Elements of an agile discovery environment. In: Grieser, G., Tanaka, Y., Yamamoto, A. (eds.) DS 2003. LNCS (LNAI), vol. 2843, pp. 311–319. Springer, Heidelberg (2003)
14. Yang, J.L., Qin, K.Y.: Uncertain concepts in a formal context. In: ICAMMS (2014)
15. Qin, K.Y., Meng, L.D.: Analysis on uncertain concepts of formal context. In: 2014 International Conference on Control Engineering and Automation (ICCEA 2014), pp. 315–319 (2015)
16. Ganter, B., Wille, R.: Formal Concept Analysis. Springer-Verlag, Heidelberg (1999)
17. Zadeh, L.A.: Fuzzy sets. Information and Control **8**, 338–353 (1965)
18. Dubois, D., Prade, H.: Rough fuzzy set and fuzzy rough sets. International Journal of General Systems **17**, 191–209 (1990)

Fuzzy Clustering-Based Quantitative Association Rules Mining in Multidimensional Data Set

Jining Jia$^{(\boxtimes)}$, Yongzai Lu, Jian Chu, and Hongye Su

State Key Laboratory of Industrial Control Technology,
Institute of Cyber-Systems and Control, Zhejiang University, Hangzhou, China
jiajining@163.com, y.lu@ieee.org

Abstract. In order to solve the problem of mining quantitative association rules, an algorithm named Fuzzy Pattern Fusion based on Competitive Agglomeration (FPF-CA) is developed in this paper. The proposed algorithm is based on the superior functionalities of Fuzzy Pattern Fusion (FPF) for mining quantitative association rules and Competitive Agglomeration (CA) for finding the optimal number of clusters. The popular data set of UCI machine learning repository is used to demonstrate the feasibility of the FPF-CA algorithm. The simulation experiment results show that the proposed algorithm can efficiently mine quantitative association rules according to the actual data distribution.

Keywords: Algorithms · Data mining · Quantitative association rules · Fuzzy clustering

1 Introduction

With the development of huge data collection and storage technologies, scientific, engineering data can be amassed at much higher speeds and lower costs. The data has multidimensional quantitative attributes (e.g., temperature, pressure, flow rate). Quantitative associations may exist in the data. In [1], quantitative association rule is defined as a rule describes associations between quantitative attributes.

Some proposed algorithms and methodologies for mining quantitative association rules can be classified into two basic categories. In the first category, static discretization of quantitative attributes is used to mine rules. For example, in [2], quantitative attributes were discretized to treat as categorical attributes by partitioning the data and combining adjacent partitions as necessary. Interest measure must be used to identify the rules, which can avoid generating too many similar rules. In [3], [4], [5], algorithms were proposed to generate fuzzy association rules by dividing quantitative attributes into various predefined linguistic values. However, there is still a problem that how to determine the suitable numbers of linguistic values. In the second category, clustering-based techniques are employed to find association rules at relatively dense clusters of quantitative attributes. Clustering methods can be divided into hard clustering and fuzzy clustering. Based on these clustering methods (e.g., HCM, FCM, improved FCM), mining algorithms were proposed in [6], [7], [8], which can partition

© Springer International Publishing Switzerland 2015
Y. Tan et al. (Eds.): ICSI-CCI 2015, Part III, LNCS 9142, pp. 68–75, 2015.
DOI: 10.1007/978-3-319-20469-7_9

the attributes and extract rules. Compared with those algorithms based on static discretization, the fuzzy clustering-based algorithms are competitive which can effectively reflect the actual data distribution. One kind of fuzzy clustering algorithm, Competitive Agglomeration (CA) algorithm has been proposed in [9], which can produce the partition with the "optimal" number of clusters.

In this paper, an algorithm named Fuzzy Pattern Fusion based on Competitive Agglomeration (FPF-CA) is developed. The algorithm scans a database only once and partitions each quantitative attribute domain into a sequence of clusters, then, generates fuzzy partition matrix and cluster centers. Patterns generated by fusing the clusters of some attributes are examined to see if such a fusion passes the minimum support threshold. If it does, fusion will proceed until the longest pattern is generated. Quantitative association rules will be generated from these long patterns which pass the minimum confidence threshold.

The rest of the paper is organized as follows. In section 2, problem formulation for mining quantitative association rules and some basic concepts are introduced. In Section 3, FPF-CA algorithm is proposed. In Section 4, two experiments are presented to demonstrate the feasibility of the proposed algorithm. Finally, in section 5, the conclusions are drawn and future works are introduced.

2   Problem Formulation

Let $D = \{\mathbf{x}_1, \ldots, \mathbf{x}_j, \ldots, \mathbf{x}_n\}$, $\mathbf{x}_j = (x_{1j}, \ldots, x_{ij}, \ldots, x_{mj})^T$ $(1 \leq i \leq m, 1 \leq j \leq n)$ be a data set consisting of m instances with n quantitative attributes $a_1, \ldots, a_n$.

In order to discover the interesting knowledge in D, the data of n quantitative attributes are partitioned into $c_1, \ldots, c_n$ fuzzy clusters by using a fuzzy clustering algorithm. Let $a_1(1), \ldots, a_1(c_1), \ldots, a_n(1), \ldots, a_n(c_n)$ be the fuzzy attributes. For the jth quantitative attribute a_j, we find c_j fuzzy clusters, cluster centers $v_j = \{v_1, \ldots, v_{c_j}\}^T$, fuzzy partition matrix $U_j = (u_i(a_j(t)))_{m \times c_j}$ $(u_i(a_j(t)) \in [0, 1], 1 \leq i \leq m, 1 \leq t \leq c_j)$, where $u_i(a_j(t))$ is the membership degree of x_{ij} in the tth fuzzy cluster of the jth quantitative attribute. Let all the fuzzy partition matrix be merged into a new matrix $U = \{U_1, \ldots, U_j, \ldots, U_n\}$.

Suppose that the attribute a_1 is the rule consequent. The other attributes form the rule antecedent. For each fuzzy attribute of a_1, we fuse it with the fuzzy attribute of another quantitative attribute. We then see such a fusion whether pass the minimum support threshold *minsup*. The fusion can be expressed as fuzzy attribute set $F = \{y_1, \ldots, y_p\}(2 \leq p \leq n)$. Refer to [10], let *fuzzysup(F)* in (1) be the fuzzy support of F, where $u_i(y_j)$ is the membership degree of the corresponding data of the ith instance in the y_jth fuzzy attribute.

$$fuzzysup(F) = \frac{\sum_{i=1}^{m} \prod_{j=1}^{p} u_i(y_j)}{m}, (1 \leq i \leq m, 1 \leq j \leq p) \tag{1}$$

If *fuzzysup(F)* ≥ *minsup*, *F* is a fuzzy frequent pattern, fusion will proceed until all the fuzzy frequent patterns $P = \{P_1, ..., P_k\}$ ($2 \leq k \leq n$) are generated. We search for the longest frequent pattern P_l ($P_l \in P_i, i \leq k$), let the rules "($P_l - a_1(t_1)) \rightarrow a_1(t_1)$" generated from P_l pass the minimum fuzzy confidence threshold *minconf*. The quantitative association rules are shown in (2).

$$a_e(t_e) \wedge ... \wedge a_j(t_j) \wedge ... \wedge a_f(t_f) \rightarrow a_1(t_1), (2 \leq e < j < f \leq n, 1 \leq t_j \leq c_j) \qquad (2)$$

3   FPF-CA Algorithm

Main
Input: $D = \{x_j | j=1,..., n\}$, $x_j = (x_{1j},..., x_{mj})^T$, a data set; *minsup*, minimum support threshold; *minconf*, minimum confidence threshold; η_0, τ, c_{max}, parameters of CA.
Output: *U*, fuzzy partition matrix; *V*, matrix of cluster center; *C*, matrix of cluster number; *P*, set of frequent patterns; *S*, fuzzy support matrix of frequent patterns; *R*, set of quantitative association rules; *SC*, matrix of rule support and confidence.

```
U ← Ø;  V ← Ø;  C ← Ø;  P ← Ø;  S ← Ø;
R ← Ø;  SC ←Ø;
[m, n] = size(D);
c₀ = 0;
for j = 1 to n
    [Uⱼ, Vⱼ, c] ← CA(ε₁, m, η₀, τ, Cₘₐₓ, xⱼ);
    U ← U∪Uⱼ;  V ← V∪Vⱼ;  C ←C∪(c₀ + 1, c₀ + c)ᵀ;
    for i = 1 to c
        P₁(c₀+i,1)←(j×100+i);
        S₁(c₀+i,1) = sum(U(:, c₀+i))/ m;
    c₀ = c₀ + c;
kₘₐₓ = 0;
for k = 2 to n
    if Pₖ₋₁= Ø
        break;
    else
        kₘₐₓ = kₘₐₓ + 1;
        [Pₖ, Sₖ] ← FPF(minsup, U, m, n, C, P₁, Pₖ₋₁);
        P ←P∪Pₖ;  S ←S∪Sₖ;
if kₘₐₓ ≥ 2
    for l = C(1, 1) to C(2, 1)
        count = 0;
        for k = kₘₐₓ to 2
            [mₖ, nₖ] = size(Pₖ);
            for i = 1 to mₖ
                if mod(Pₖ(i, 1),100) = l
                    fuzzyconf = Sₖ(i, 2)/ Sₖ(i, 1);
```

```
            if fuzzyconf ≥ minconf
                count = count + 1;
                R₁ (count, :) = Pₖ(i);
                SC₁(count) = (Sₖ(i, 2), fuzzyconf);
        if count ≥ 1
            break;
    R ← R∪R₁; SC ←SC∪SC₁;
return R, SC, V
```

CA(ε_1, m, η_0, τ, c_{max}, $\mathbf{x}_j$)

Input: ε_1, η_0, τ, c_{max}, parameters of CA; m, number of instances; $\mathbf{x}_j$, data of the jth attribute.

Output: U_j, fuzzy partition matrix of the jth attribute; V_j, cluster center matrix of the jth attribute; c, cluster number of the jth attribute;

```
ε₂; k = 0;
U_c ← FCM(c_max, x_j);
for i = 1 to c_max
    N_i (i) = sum(U_c (i, :));
    v_i (i) = (U_c(i, :)· U_c(i, :))·x_j/ (U_c(i, :)· U_c(i, :));
while(1)
    [c_c, ~] = size(U_c);
    sum₁ = 0;
    for i = 1 to c_c
        for j = 1 to m
            sum₁ = sum₁ + U_c (i, j)· U_c (i, j)d ²(x_j, v_i );
    sum₂ = sum(N_i· N_i);
    α_k = η₀ exp(-k / τ) (sum₁ / sum₂);
    for j = 1 to m
        N_j(j) = sum(1/d ²(x_j, v_i ))· N_i/sum(1/d ²(x_j, v_i ));
    for i = 1 to c_c
        for j = 1 to m
            U^FCM(i, j)=(1/d ²(x_j, v_i ))/sum(1/d ²(x_j, v_i ));
            U^Bias(i, j)=α_k (N_i(i) - N_j(j))/ d ²(x_j, v_i );
    U_c = U^FCM + U^Bias;
    for i = 1 to c_c
        N_i(i) = sum(U_c (i, :));
    class = N_i < ε₁;
    n_c = sum(class);
    flag = zeros(1, n_c);
    q = 1;
    for i = 1 to c_c
        if class(i) = 1
            flag(q) = i;
```

```
                q = q + 1;
        U_c (flag, :) = [ ]; N_i (flag) = [ ];
        [c, ~] = size(U_c);
        for i = 1 to c
            v*_i (i) = (U_c (i, :)· U_c (i, :))· x_j/
                       (U_c (i, :)· U_c (i, :));
        if  size(v_i) = size(v*_i)
            c_h = sum(abs(v_i - v*_i));
        v_i = v*_i;
        if  c_c = c & c_h ≤ ε_2
            break;
        k = k + 1;
V_j = sort(v_i);
seq = zeros(1, c);
for i = 1 to c
    seq(i) = find(v_i = V_j);
U_j = zeros(c, m);
for i = 1 to c
    U_j (i,:) = U_c (seq(i),:);
return U_j ^T, V_j , c;
```

FPF(*minsup*, *U*, *m*, *n*, *C*, P_1, P_{k-1})
Input: *minsup*, minimum support threshold; *U*, fuzzy partition matrix; *m*, *n*, number of instances, attributes; *C*, matrix of cluster number; P_1, P_{k-1}, set of frequent 1, *k*-1 patterns;
Output: P_k, set of frequent *k* patterns; S_k, support of frequent *k* patterns;

```
[m_{k-1}, n_{k-1}] = size(P_{k-1});
count = 0;
for i = 1 to m_{k-1}
    if fix(P_{k-1} (i, 1)/100) = 1
        U_{i1} = U(:, mod(P_{k-1} (i, 1),100));
        if n_{k-1} ≥ 2
            U_{i2} = ones(m, 1);
            for j = 2 to n_{k-1}
                U_{i2} = U_{i2} (:).*U(:,fix(P_{k-1} (i, j)/100)+
                    mod(P_{k-1} (i, j),100)-1);
        else
                U_{i2} = U_{i1};
        for r = fix(P_{k-1} (i, n_{k-1})/100) + 1 to n
            for l = C (1, r) to C (2, r)
                U_{i3} = U_{i2} (:).*U(:, l);
                U_{i4} = U_{i3} (:).*U_{i1} (:);
                fuzzysup_1 = sum(U_{i3})/ m;
                fuzzysup_2 = sum(U_{i4})/ m;
```

```
        if fuzzysup₂ ≥ minsup
           count = count +1;
           Pₖ (count, :) = Pₖ₋₁ (i, :) ∪ P₁ (1);
           Sₖ (count, 1) = fuzzysup₁;
           Sₖ (count, 2) = fuzzysup₂;
if count = 0
   Pₖ ← ∅; Sₖ ← ∅
return Pₖ, Sₖ
```

4   Experiments

Iris data set is a famous database. The source is in [11]. The data set contains 150 instances with 4 quantitative attributes. There are 3 classes of 50 instances each in the data set, where each class refers to a type of iris plant. Let's add a quantitative attribute a_1 to form a new data set (Table 1), where the 1st attribute is class identifier (1-Iris Setosa, 2-Iris Versicolour, 3-Iris Virginica).

Table 1. Iris Data Set

Instance	Class (a_1)	Sepal length/cm (a_2)	Sepal width/cm (a_3)	Petal length/cm (a_4)	Petal width/cm (a_5)
i_1	1	5.1	3.5	1.4	0.2
i_2	1	4.9	3	1.4	0.2
...	...	...	...	...	...
i_{51}	2	7	3.2	4.7	1.4
i_{52}	2	6.4	3.2	4.5	1.5
...	...	...	...	...	...
i_{101}	3	6.3	3.3	6	2.5
i_{102}	3	5.8	2.7	5.1	1.9
...	...	...	...	...	...

In the new data set, the data of each quantitative attribute are partitioned into 2-3 fuzzy clusters according to the actual data distribution by using the FPF-CA algorithm. Table 2 shows the cluster centers. Table 3 shows the fuzzy partition matrix.

Table 2. Cluster Centers of the Iris Data Set

v_1	v_2	v_3	v_4	v_5
1.00	4.94	2.95	1.48	0.24
2.00	5.95	3.85	4.39	1.36
3.00	6.97		5.74	2.11

Table 3. Fuzzy Partition Matrix of the Iris Data Set

Membership degree	Class			...	Petal Width		
	$a_1(1)$	$a_1(2)$	$a_1(3)$	...	$a_5(1)$	$a_5(2)$	$a_5(3)$
u_1	1.000	0.000	0.000	...	0.997	0.003	0.000
u_2	1.000	0.000	0.000	...	0.997	0.003	0.000
...	...	...	...	...	...	...	...
u_{51}	0.000	1.000	0.000	...	0.000	1.000	0.000
u_{52}	0.000	1.000	0.000	...	0.011	0.952	0.037
...	...	...	...	...	...	...	...
u_{101}	0.000	0.000	1.000	...	0.027	0.106	0.867
u_{102}	0.000	0.000	1.000	...	0.015	0.145	0.840
...	...	...	...	...	...	...	...

Quantitative association rules are mined by using the fuzzy partition matrix in the Core i5-2400 3.10GHz/4G PC. The parameters of CA are shown in Table 4. If the minimum support threshold is 0.1 and the minimum confidence threshold is 0.9, data mining results are shown in Table 5. It costs 1.14 seconds.

Table 4. Parameters of CA

Quantitative Attributes	τ	ε	η_0	c_{max}
a_1	10	15	0	5
a_2	10	15	2	20
a_3	10	15	3.5	20
a_4-a_5	10	15	3	20

Table 5. Quantitative Association Rules

Quantitative Association Rules	Fuzzy Support	Fuzzy Confidence
$a_2(1) \wedge a_3(1) \wedge a_4(1) \wedge a_5(1) \rightarrow a_1(1)$	0.2143	0.9961
$a_2(2) \wedge a_3(1) \wedge a_4(2) \wedge a_5(2) \rightarrow a_1(2)$	0.1880	0.9167
$a_2(3) \wedge a_3(1) \wedge a_4(3) \wedge a_5(3) \rightarrow a_1(3)$	0.1152	0.9920

The quantitative association rules contain the knowledge for classification. For example, the 2nd rule illustrates that if the sepal length is about 5.95 cm, the sepal width is about 2.95 cm, the petal length is about 4.39 cm, the petal width is about 1.36 cm, the class identifier may be Iris Versicolour. The fuzzy support of this rule is 0.1880. The fuzzy confidence of this rule is 0.9167.

5   Conclusions and Future Works

In this paper, we have developed the FPF-CA algorithm to solve the problem of mining quantitative association rules in multidimensional data set. The proposed algorithm is based on the superior functionalities of FPF for mining quantitative association rules and CA for finding the optimal number of clusters. The simulation experiment results show that the FPF-CA algorithm can efficiently mine quantitative association rules according to the actual data distribution.

Our future work is concerned with applying the FPF-CA algorithm to complex engineering problems. We focus on the operation optimization of thermal power plants.

Acknowledgments. The authors gratefully acknowledge Dr. Chenlong Yu for his friendly cooperation. The authors thank the anonymous referees for their many useful comments and constructive suggestions.

References

1. Han, J.W., Kamber, M., Pei, J.: Data Mining: Concepts and Techniques, 3rd edn, pp. 281–291. China Machine Press, Beijing (2012). English ed
2. Srikant, R., Agrawal, R.: Mining quantitative association rules in large relational talbes. In: Proc. ACM SIGMOD. International Conference on Management of Data, pp. 1–12 (1996)
3. Hu, Y.C., Chen, R.S., Tzeng, G.H.: Discovering Fuzzy Association Rules Using Fuzzy Partition Methods. Knowledge-Based Systems. **16**, 137–147 (2003)
4. Delgado, M., Marín, N., Sánchez, D., etc.: Fuzzy association rules: general model and applications. In: IEEE Transactions on Fuzzy Systems, vol. 11, pp. 214–225 (2003)
5. Hong, T.P., Chiang, M.J., Wang, S.L.: Mining from quantitative data with linguistic minimum supports and confidences. In: IEEE International Conference on Fuzzy Systems, pp. 494–499 (2002)
6. Watanabe, T., Takahashi, H.: A Quantitative association rule mining algorithm based on clustering algorithm. In: Proc. IEEE International Conference on Systems, Man and Cybernetics (SMC 2006), vol. 3, pp. 2652–2657 (2006)
7. Mangalampalli, A., Pudi, V.: FPrep: Fuzzy clustering driven efficient automated pre-processing for fuzzy association rule mining. In: Proc. IEEE International Conference on Fuzzy Systems, pp. 1–8 (2010)
8. Pi, D., Qin, X.-L., Yuan, P.: A modified fuzzy c-means algorithm for association rules clustering. In: Huang, D.-S., Li, K., Irwin, G.W. (eds.) ICIC 2006. LNCS (LNAI), vol. 4114, pp. 1093–1103. Springer, Heidelberg (2006)
9. Frigui, H., Krishnapuram, R.: Clustering by Competitive Agglomeration. Pattern Recognition **30**, 1109–1119 (1997)
10. Dubois, D., Hüllermeier, E., Prade, H.: A Systematic Approach to the Assessment of Fuzzy Association Rules. Data Min Knowl Disc **13**, 167–192 (2006)
11. Fisher, R.A.: The use of multiple measurements in taxonomic problems. Annual Eugenics **7**, 179–188 (1936)

The Application of Fuzzy Pattern Fusion Based on Competitive Agglomeration in Coal-Fired Boiler Operation Optimization

Jining Jia[✉], Yongzai Lu, Jian Chu, and Hongye Su

State Key Laboratory of Industrial Control Technology,
Institute of Cyber-Systems and Control, Zhejiang University, Hangzhou, China
jiajining@163.com, y.lu@ieee.org

Abstract. In order to solve the coal-fired boiler operation optimization problem with multiple main controllable parameters, a fuzzy pattern fusion based on competitive agglomeration (FPF-CA) algorithm is developed and applied in this paper. A simplified mathematical model for coal-fired boiler systems is applied in terms of both historical operational and thermal efficiency data under different load conditions. The FPF-CA algorithm can be applied to perform information fusion in terms of combining fuzzy clusters of some quantitative attributes with generated quantitative association rules. The historical data collected from a 130/2.82-M circulating fluidized bed (CFB) boiler being installed in a production scale thermal power plant is used to specifically analyze, and the simulation experiment results show the application of FPF-CA algorithm is a decision oriented intelligent technology and may provide an efficient results for coal-fired boiler operation optimization in thermal power plant.

Keywords: FPF-CA · Coal fired boilers · Optimization · Data fusion · Quantitative association rules · Fuzzy clustering

1 Introduction

Coal-fired circulating fluidized bed (CFB) boilers are widely used in thermal power plants, with the increase in the size of power plants, the capacity and the number of CFB also increase. But in the aspect of management is more concerned with the safety of boiler operation than its thermal performance. Coal-fired CFB boiler is a multivariable, strong-coupling, time-varying and nonlinear system, the thermal performance of which is strongly influenced by the change of load, coal type and operator inputs. When coal type remains unchanged for a period of time, the influence of operator inputs can be analyzed to optimize the operating procedures on the premise of adapting to changes in load. The heart of the coal-fired boiler operation optimization problem is the determination of the optimization values for the main controllable parameters [1]. Under different load conditions, several mining on operation parameters achieve a set of rules to provide a framework for operation optimization.

© Springer International Publishing Switzerland 2015
Y. Tan et al. (Eds.): ICSI-CCI 2015, Part III, LNCS 9142, pp. 76–83, 2015.
DOI: 10.1007/978-3-319-20469-7_10

The rest of the paper is organized as follows. In section 2, fuzzy clustering-based quantitative association rule algorithms for boiler operation optimization are introduced. In Section 3, a fuzzy pattern fusion based on competitive agglomeration (FPF-CA) algorithm proposed in authors' earlier research is introduced simply. In Section 4, based on real production data in a thermal power plant, an example of the application of FPF-CA is presented to demonstrate the feasibility of the algorithm. Finally, in section 5, the conclusions are drawn and future works are introduced.

2   Fuzzy Clustering-Based Quantitative Association Rule Algorithms for Boiler Operation Optimization

Large amounts of engineering data of CFB boiler are accumulated in thermal power plants databases with multiple quantitative attributes. By analyzing the historical operation data and the corresponding thermal efficiency data, fuzzy clustering-based quantitative association rule algorithms are usually employed to obtain the optimal target values of the main controllable parameters. In the process of partitioning quantitative attributes, fuzzy clustering algorithms are more competitive than those static discretization algorithms, which can reflect the actual operation data distribution. Based on fuzzy clustering algorithms, data mining algorithms are designed to partition every quantitative attribute of operation data sets into several fuzzy attributes and to generate quantitative association rules from fuzzy frequent patterns.

For example, in [2], the optimal setpoints of fuel and air parameters are obtained by using clustering algorithm and fuzzy association rule method. In [3], under typical load conditions, the optimal curve of excess air coefficient can be obtained by analyzing the optimal values determined by association rule mining. In [4], the optimal knowledge base is established by using fuzzy association rule mining approach, which suit for online using and timely updating.

Some modified algorithms of clustering or association rule are proposed. A dynamic evolutionary clustering algorithm is proposed in [5] to search for the optimal cluster number. By introducing the fuzzy sets theory, a method is proposed in [6] which translates the quantitative association rule problem into the boolean association rule problem. In [7], a FCM-based language association rule method is applied in finding how load, excessive air coefficient and flue gas temperature affect boiler efficiency.

Based on the analysis of these algorithms, the major functions of fuzzy clustering-based quantitative association rule are to construct the partition of the quantitative attributes meeting the actual data distribution and to take full advantage of the information of membership degree.

3   Fuzzy Pattern Fusion Based on Competitive Agglomeration

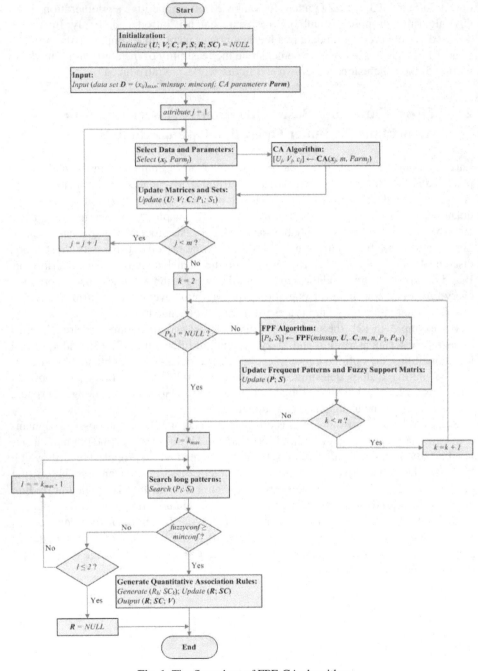

Fig. 1. The flow chart of FPF-CA algorithm

In authors' earlier research, a fuzzy pattern fusion based on competitive agglomeration (FPF-CA) algorithm was proposed. The flow chart of the algorithm is shown in Fig.1.

Competitive agglomeration (CA) algorithm was proposed in [8], which can generate the fuzzy partition of data with the "optimal" number of clusters. Based on the CA algorithm, data set D is analyzed to generate fuzzy partition matrix U, cluster center matrix V and cluster number matrix C. Using these matrices, fuzzy attributes of different quantitative attributes are fused by employing fuzzy pattern fusion (FPF) algorithm to find the fuzzy frequent patterns set P. At the same time, fuzzy support set S is generated. Finally, quantitative association rules set R is generated from P. Outputs also contain rules support and confidence set SC and the corresponding cluster centers.

4   Application

The FPF-CA algorithm developed was applied to analyze the database built up from a 130/3.82-M circulating fluidized bed (CFB) boiler being installed in a production scale thermal power plant. This database contains a large number of quantitative attributes which represent the operation parameters.

The key operation parameters are air-coal ratio r_{ac}, secondary air ratio r_{sa}, bed temperature T_b, oxygen concentration C_{O2}, bed pressure drop $\varDelta P_b$ and flue gas temperature T_{fg}, etc. There is a correlation between these key operation parameters and boiler thermal efficiency. The thermal efficiency of CFB boiler η can be calculated by using a simplified mathematical model proposed in [9] which utilizes data of a dozen operation parameters. These operation parameters include main steam flow D_{ms}, main steam pressure P_{ms}, main steam temperature T_{ms}, feed water pressure P_{fw}, feed water temperature T_{fw}, bed temperature T_b, primary air flow F_{pa}, primary air duct pressure P_{pa}, primary air temperature T_{pa}, secondary air flow F_{sa}, secondary air duct pressure P_{sa}, secondary air temperature T_{sa}, induced draft flow F_{id}, induced draft pressure P_{id}, induced draft temperature T_{id}, oxygen concentration C_{O2}. The simplified mathematical model of boiler thermal efficiency is shown in (1).

$$\eta = (Q_{ms} - Q_{fw}) / Q_r = [D_{ms} (h_{ms} - h_{fw})] / N_{CO2} / (-H) \qquad (1)$$

In Equation (1), D_{ms} is stored in data set D. h_{ms} is steam enthalpy. h_{fw} is feed water enthalpy. h_{ms} and h_{fw} can be calculated by the method based on the IAPWS-IF97, which is the International Association for the Properties of Water and Steam (IAPWS) Industrial Formulation 1997 presented in [10]. The calculation of h_{ms} and h_{fw} needs the data of operation parameters P_{ms}, T_{ms}, P_{fw} and T_{fw}. N_{CO2} is the amount of CO_2 generation, which is approximately equal to the amount of O_2 consumption. The calculation of N_{CO2} is shown in (2), (3), (4), (5), (6), which needs the data of operation parameters F_{pa}, P_{pa}, T_{pa}, F_{sa}, P_{sa}, T_{sa}, F_{id}, P_{id}, T_{id} and C_{O2}. H is the reaction heat in reaction system of carbon reacting with oxygen to yield carbon dioxide. The calculation of H is shown in (7), which needs the data of operation parameters T_b.

$$N_{CO2} \approx N_{O2} \approx F_{pa} / (V_m)_{pa} + F_{sa} / (V_m)_{sa} - F_{id} / (V_m)_{id} \tag{2}$$

$$V_m = RT / P + RT_c (B^0 + \omega B^1) / p_c \tag{3}$$

$$T_r = T / T_c \tag{4}$$

$$B^0 = 0.083 - 0.422 / T_r^{1.6} \tag{5}$$

$$B^1 = 0.139 - 0.172 / T_r^{4.2} \tag{6}$$

$$H = -405.52549 + 0.03709T_b + 8.17397 \times 10^{-6}T_b^2 \tag{7}$$

A data set in Table 1 contains 332 instances with 8 quantitative attributes was exported from the aforementioned database. The time frame within the data set was from 00:00 2009-10-19 to 23:00 2009-11-01. The interval between each instance is 1 hour. Some empty data points had been removed. The 1st and 2nd key quantitative attributes are boiler thermal efficiency η and main steam flow D_{ms}. Fig. 2 shows the distribution of 1st and 2nd quantitative attributes along instances.

Table 1. The Data Set of 130/3.82-M CFB Boiler

| Instance | $\eta/\%$ | $D_{ms}/t/h$ | T_b/K | r_{ac} | $r_{sa}/\%$ | $C_{O2}/\%$ | $\Delta P_b/kPa$ | T_{fg}/K |
	(a_1)	(a_2)	(a_3)	(a_4)	(a_5)	(a_6)	(a_7)	(a_8)
i_1	84.63	93.73	1166.65	5.78	37	2.20	9.49	395.05
i_2	86.74	100.93	1159.55	6.36	33	2.68	8.84	394.95
...	...	...	...	...	...	...	...	...
i_{332}	85.86	112.99	1208.95	5.57	45	1.10	9.10	403.95

The calculation results of h_{ms} and h_{fw} using the IAPWS-IF97 are shown in Table 2.

Table 2. Calculation of the Properties of Water and Steam Using the IAPWS-IF97

| Instance | h_{ms} | h_{fw} | P_{ms} | T_{ms} | P_{fw} | T_{fw} |
	(kJ/kg)	(kJ/kg)	(MPa)	(K)	(MPa)	(K)
i_1	3318.83	664.46	3.565	715.21	5.29	429.94
i_2	3319.75	679.08	3.595	715.8	5.42	433.31
...	...	...	...	...	...	
i_{332}	3317.62	662.48	3.84	716.38	5.51	429.45

The data of each quantitative attribute are partitioned into 2-4 fuzzy clusters reflecting the actual data distribution by using the FPF-CA algorithm. The cluster centers and the fuzzy partition matrix are shown in Table 3, Table 4.

Table 3. Cluster Centers of the Data Set of 130/3.82-M CFB Boiler

$v_1/\%$	$v_2/t/h$	v_3/K	v_4	$v_5/\%$	$v_6/\%$	v_7/kPa	v_8/K
83.69	101.61	1164.61	5.83	34.4	1.76	8.80	396.54
89.68	123.50	1189.69	6.50	54.9	3.01	8.99	402.91
		1208.95				9.20	410.01
							416.50

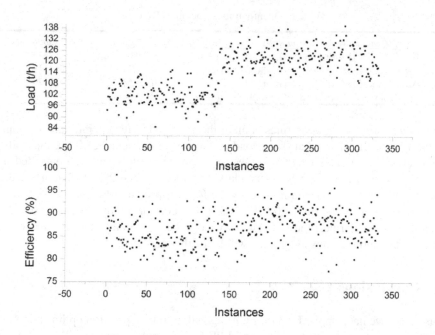

Fig. 2. Distribution of efficiency and load attributes along instances

Table 4. Fuzzy Partition Matrix of the Data Set of 130/3.82-M CFB Boiler

Membership	η		...	T_{fk}			
degree	$a_1(1)$	$a_1(2)$	...	$a_8(1)$	$a_8(2)$	$a_8(3)$	$a_8(4)$
u_1	0.9396	0.0604	...	0.9445	0.0406	0.0107	0.0042
u_2	0.4311	0.5689	...	0.9367	0.0460	0.0123	0.0050
...	...	...	...		...	...	...
u_{332}	0.7235	0.2765	...	0.0372	0.8852	0.0655	0.0121

Quantitative association rules are mined by using the fuzzy partition matrix in the Core i7-M620 2.67GHz/4G PC. The parameters of CA are shown in Table 5. If the minimum support threshold is 0.05 and the minimum confidence threshold is 0.7, data mining results are shown in Table 6. It costs 2.47 seconds.

Table 5. Parameters of CA

Quantitative Attributes	τ	ε	η_0	c_{max}
a_1, a_3, a_5, a_7, a_8	10	33	2	20
a_2, a_6	10	33	3	20
a_4	10	33	1	20

Table 6. Quantitative Association Rules

Quantitative Association Rules	Fuzzy Support	Fuzzy Confidence
$a_2(1) \wedge a_3(1) \wedge a_4(1) \wedge a_5(1) \wedge a_6(1) \wedge a_7(2) \to a_1(1)$	0.0746	0.7245
$a_2(1) \wedge a_3(1) \wedge a_4(1) \wedge a_5(1) \wedge a_6(1) \wedge a_8(1) \to a_1(1)$	0.1004	0.7118
$a_2(2) \wedge a_3(3) \wedge a_4(1) \wedge a_5(2) \wedge a_6(1) \wedge a_7(2) \to a_1(2)$	0.0786	0.7326
$a_2(2) \wedge a_4(1) \wedge a_5(2) \wedge a_6(1) \wedge a_7(2) \wedge a_8(3) \to a_1(3)$	0.0696	0.7241

The quantitative association rules contain the knowledge for operation. For example, the 3rd rule illustrates that if the load of the boiler is about 123.5 t/h, the boiler thermal efficiency is about 89.68%, the operation parameters can be controlled as follows: air-coal ratio r_{ac}=5.83, secondary air ratio r_{sa}=54.9%, bed temperature T_b=1208.95K, oxygen concentration C_{O2}=1.76%, bed pressure drop $\triangle P_b$ =8.99kPa. The fuzzy support of this rule is 0.0786. The fuzzy confidence of this rule is 0.7326. By integrating these rules, the optimal knowledge base is established. By analyzing more data, the optimal knowledge base will be further enriched.

5　　Conclusions

In this paper, we have applied the FPF-CA algorithm to the operation optimization for a production scale coal-fired boiler. The FPF-CA algorithm translates the operation optimization of multiple main controllable parameters problem into the quantitative association rule problem. Based on the superior functionalities of Competitive Agglomeration (CA) for finding the optimal number of clusters and Fuzzy Pattern Fusion (FPF) for mining quantitative association rules, the optimal knowledge base including several types of rules can be efficiently established to guide the operation in thermal power plant. The application of FPF-CA is a decision oriented intelligent technology and provides an efficient approach for coal-fired boiler operation optimization in thermal power plant.

Acknowledgments. The authors gratefully acknowledge Dr. Liang Chen for his friendly help. The authors thank the anonymous referees for their many useful comments and constructive suggestions.

References

1. Gu, J.J., Sun, Q.L., Gao, D.M.: The application of association rules in boiler operation optimization based on organizational evolutionary. In: Asia-Pacific Power and Energy Engineering Conference, pp. 1–4. IEEE Press(2009)
2. Yang, T.T., Liu, J.Z., Zeng, D.L., et al.: Application of data mining in boiler combustion optimization. In: 2nd International Conference on Computer and Automation Engineering, pp. 225–228. IEEE Press (2010)
3. Li, J.Q., Niu, C.L., Gu, J.J., et al.: The determination of optimal excess air coefficient based on data mining in power plant. In: 5th International Conference on Fuzzy Systems and Knowledge Discovery, pp. 384–388. IEEE Press (2008)

4. Zhao, W.J., Liu, C.: The optimizing for boiler combustion based on fuzzy association rules. In: 2011 International Conference of Soft Computing and Pattern Recognition, pp. 306–311. IEEE Press (2011)

5. Jiang, W.J.: A novel dynamic clustering algorithm and its application in fuzzy modeling for thermal processes. In: International Conference on Machine Learning and Cybernetics, pp. 1221–1226. IEEE Press (2006)

6. Yu, X.N., Niu, C.L., Li, J.Q.: The application of fuzzy data mining in coal-fired boiler combustion system. In: IEEE Region 10 Annual International Conference, TENCON, pp. 1–5. IEEE Press (2007)

7. Yu, X.N., Niu, C.L., Li, J.Q.: The enactment of optimal target value in power plant operation based on language association rule. In: IEEE Region 10 Annual International Conference, TENCON, pp. 1–4. IEEE Press (2007)

8. Frigui, H., Krishnapuram, R.: Clustering by Competitive Agglomeration. Pattern Recognition. **30**, 1109–1119 (1997)

9. Song, B., Jiang, Q.Y., Cao, Z.K.: On-line Calculation and Analysis of Heat Efficiency of Circulating Fluidized Bed Boiler Unit. Chemical Industry and Engineering Progress. **27**, 616–620 (2008)

10. Wagner, W., Cooper, J.R., Dittmann, A., et al.: The IAPWS Industrial Formulation 1997 for the Thermodynamic Properties of Water and Steam. Journal of Engineering for Gas Turbines and Power **122**, 150–184 (2000)

Forecasting Algorithms

Technical Indicators for Forex Forecasting:
A Preliminary Study

Yoke Leng Yong[✉], David C.L. Ngo, and Yunli Lee

Department of Computing and Information Systems,
Sunway University, Bandar Sunway, Malaysia
14071856@imail.sunway.edu.my,
{dngo,yunlil}@sunway.edu.my

Abstract. Traders and economists are often at odds with regards to the approach taken towards Forex financial market forecasting. Methods originating from the Artificial Intelligence (AI) area of study have been used extensively throughout the years in predicting the trading pattern as it is deemed to be robust enough to handle the uncertainty associated with Forex trading time series data. Herein this paper, the effects of different input types, in particular: close price as well as various technical indicators derived from the close price are investigated to determine its effects on the Forex trend predicted by an intelligent machine learning module.

Keywords: Forex forecasting · Technical analysis · Linear regression line · Artificial Neural Network · Dynamic Time Warping

1 Introduction

Financial markets encompasses a huge variety of markets such as: Stock Markets, Commodity Markets, Money Markets, Derivatives Markets, Futures Markets, Insurance Markets, and Foreign Exchange Markets (Forex). However, most traders gravitate towards the Stock Market and Foreign Exchange. With an average daily trading value of 3.2 trillion [1], it is a fast paced and dynamic market where traders often watch every little movement like a hawk. With the advancement of research in the application of Artificial Intelligence in computational finance, it has been slowly been adopted in to assist traders in making informed decisions while trading. Numerous publications have attempted to construct an accurate model for Forex time series prediction. Most of these works focus on time series prediction with various AI models, or some hybrid combinations as well as statistical techniques, such as moving average. While the introduction of AI models in Forex prediction have increased dramatically as it has been adopted by the traders, these methods inevitably have their own limitations at different stages of its prediction pipeline. Herein, the investigation focuses on finding the most efficient input parameter to produce a more accurate prediction of the Forex time series data from an intelligent machine learning module such as implemented by Tiong, Ngo and Lee [14], [15] which will be further discussed in Section 3.

© Springer International Publishing Switzerland 2015
Y. Tan et al. (Eds.): ICSI-CCI 2015, Part III, LNCS 9142, pp. 87–97, 2015.
DOI: 10.1007/978-3-319-20469-7_11

2   Forex Trading

2.1   Background

Market analysis has been extremely crucial in the attempt to forecast the future trading patterns to prevent loss and generate profit. However, Forex forecasting is not a straight forward and easy task to accomplish as it is constantly affected by various external factors such as: political stability, economic events, terms of trade, economic performance and etc. All the aforementioned factors contribute to establishing the complex and volatile nature of Forex prediction and the challenge presented has made it one of the hottest topics in research community. There are currently 5 major schools of thoughts which contribute to different analysis method applied to the Forex time series data, namely: Random Walk Analysis, Efficient Market Hypothesis, Technical Analysis, Fundamental Analysis, and Sentiment Analysis as shown in Fig 1.

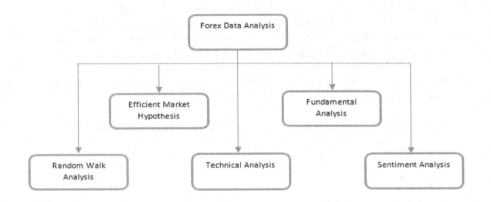

Fig. 1. Forex Analysis Methods

The Random Walk Analysis was popularized by the Burton Malkiel in his book, "A Random Walk Down Wall Street" [2]. The theory works under the premises that the Forex price is defined by the random walk theory whereby the fluctuation of exchange price is random and unpredictable given prior values. It was the same foundation on which Eugene Fama build his theory of Efficient Market Hypothesis [3] which takes a slightly different approach states that the effect of all the known information are already reflected on the price of the stocks/currency. On the contrary, technical analyst / chartist believe that the analysis of historical Forex data reveals a pattern that is constantly repeating itself which can be beneficial for forecasting [4]–[6]. Fundamental Analysis and Sentiment Analysis endeavors to explain the fluctuating price from the angle of cause and effect. While Fundamental Analysis views the fluctuation of currency price with relation to the supply and demand of the currency due to its popularity, social, economic and political factors; Sentiment Analysis concentrates on the trader's own gut feeling or personal opinion of how the market would perform.

There are currently two different approaches in performing technical analysis for the Forex time series data, namely: soft computing and hard computing. Hard computing refers to the more traditional approach to computing which often require the analytical model to be precisely stated and extensive computation time. Soft computing on the other hand, offers the flexibility of dealing with uncertainty and imprecisions. Mochón, Quintana, Sáez and Isasi [7] gave a good overview on the use of soft computing techniques applied in the finance world. Along with the research conducted in the recent years [8], [9], soft computing techniques such as machine learning, neural network, evolutionary computing,   and support vector machines have been gaining popularity within the Forex market research community compared to hard computing techniques such as: MARS and CART as mentioned by Abraham [10].

2.2   Technical Analysis

While there are numerous external factors affecting the Forex trading daily which can be classified into one of the categories in Fig 1, time series analysis has become an important component for Technical Analysis as it provides the necessary mechanism to analyze Forex historical data. Various researchers have performed a lot of in depth research in the various technical indicators used by analyst. Neely and Weller provides an extremely good introduction and background on technical analysis. Taking things a step further, technical indicator can also be gathered from the daily trading values could reveal a lot when analyzed thoroughly.

From the analysis of historical data, traders often form their own heuristic trading rule where a preconceived fluctuating pattern is detected. Among the well-known technical indicators are: Moving Averages and Moving Average Convergence Divergence (MACD). Recent studies such as undertaken by Canelas, Neves and Horta [11], [12] have also started looking into other feature representation which provides a more elegant solution whereby a more compact feature representation for prediction (SAX) can be obtained with faster computation which is important when dealing with a huge dataset. It has also later been expanded to include multi-dimensional information financial time series information [13].

2.3   Technical Indicators

Dating back to the 1700s [6], technical indicators have evolved through the years. The various technical indicators available for the analyst to use such as: Moving Averages, Bollinger Band, Elliot Wave Analysis, and Relative Strength Index (RSI) ranges from the simplistic to the more complicated method in analyzing the Forex time series data. However, the question still remains on how efficient and profitable the technical indicators really are. Schulmester [5] endeavors to answer this question in his paper "Components of the Profitability of Technical Analysis".

Moving average is one of the most often used technical indicators used by Forex traders. It is a lagging indicator which is extremely important to confirm that a pattern is occurring or about to occur.   Among the well-known technical indicators are: Simple Moving Average (SMA), Exponential Moving Average (EMA), and Moving Average Convergence Divergence (MACD) where the formulae to obtain it are as denoted below in Equation (1) – (5).

Referring to Equation (1) and (2), it can be observed that traders would often use the most familiar moving average such as SMA 10, 30, 50, 100 and 200. What makes the aforesaid value of SMA so special? As it happens, the reason the makes it a good practice to pay attention to the movements of the SMA values mentioned as most traders react to these SMAs. The more traders anticipating movements and bounce of a particular SMA, it affects the trading pattern and the anticipated movement would likely happen.

$$SMA_n(P,m) = \frac{\sum_{i=n-(m-1)}^{n} P_i}{m} \tag{1}$$

where m denotes the time period window defined, n is the nth data of the exchange price and P_i is the applied price (closing price).

$$EMA_n(P,m) = (P_n - EMA_{n-1}) * \alpha + EMA_{n-1} \tag{2}$$

$$= (1 - \alpha) * EMA_{n-1} + \alpha * P_n \tag{3}$$

where P_i denotes the current exchange price, n is the nth data of the exchange price and EMA_n is the EMA value of the nth price index. α is the decay factor which can be calculated as: $2/(m+1)$ whereby m is the time period window defined.

$$MACD = EMA(12) - EMA(26) \tag{4}$$

$$SIGNAL = EMA(9) \tag{5}$$

3   Proposed Research

3.1   Algorithm

Utilizing the algorithm previously proposed by Tiong, Ngo and Lee [14], [15], the effects of different input data on the prediction results generated are investigated. The proposed prediction algorithm focuses on prediction of two of the main Forex trend observed by traders; uptrend and downtrend as depicted in Fig 2. Going back to the basis of technical analysis adhered to by most of the analyst, the trend patterns are predicted to reoccur and the prediction of trend development is carried out throughout 2 main stages: Data Analysis Stage as well as the Training and Learning Stage.

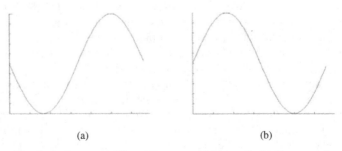

(a)   (b)

Fig. 2. Archetypes of Trend Patterns: (a) Uptrend; (b) Downtrend

In order to obtain the clustering results, the input data provided will have to go through the process of feature selection, pattern analysis and segmentation as shown in Fig 3. As the input data used is essentially a streaming input of Forex time series data, it needs to be pre-processed and analyzed in detail. The raw Forex time series data are segmented with the aid of Linear regression Line (LRL) prior to feature extraction. The features that have been selected to represent the pattern trends for clustering are as listed in Table 1. Further illustration of the feature point (A - E) noted in Table 1 with reference to its precise location in the graph is depicted in Fig 4. With the features selected, Linear Regression Line (LRL) will be employed to identify the trend of the input data. After the segmentation has been performed on the data, K-means clustering is used to cluster the data.

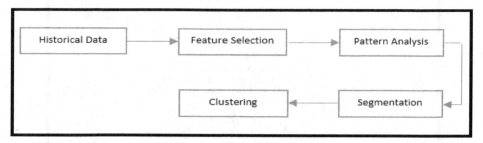

Fig. 3. Data Analysis Stage from [14]

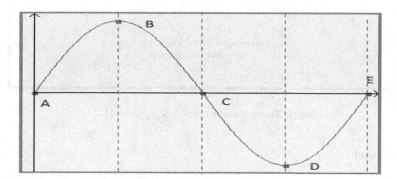

Fig. 4. Feature Selection Description

Table 1. Feature Selection

Feature	Description	Feature Point
FEATURE 1	Distance between starting point and first turning point	A – B
FEATURE 2	Distance between first turning point and changing point	B – C
FEATURE 3	The area between starting point and changing point	Area under curve between A – C
FEATURE 4	Distance between changing point and second turning point	C – D
FEATURE 5	Distance between second turning point and ending point	D – E
FEATURE 6	The area between changing point and end point	Area under curve between C – E

AI algorithms have always played a crucial role with the research focusing on the prediction of financial market. In depth studies have been conducted into the implementation of Artificial Neural Network in forecasting the Forex market trend [16]–[20]. As previous research proved to produce reasonable results, ANN has been implemented to train and learn data and DTW to predict data with the algorithm as shown in Fig 5. More specifically, the Multilayer Perceptron Neural Network has been utilized in the training and learning the patterns obtained from the previous stage.

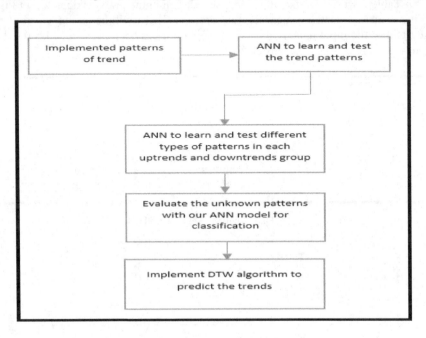

Fig. 5. Training and Learning Stage from [14]

3.2   Dataset

In order to investigate the effects of the different input data on the prediction model, different data inputs were tested on the system. The dataset used are the historical data downloaded from HistData website [21] for EUR/USD and AUD/USD. The time period chosen for testing purposes are from: 2 Jan 2012 to 29 June 2012. As the downloaded dataset is in minutes, it is pre-processed to have the time interval of 30 minutes. The original research involves using the raw close price as input. Herein, the performance is compared against the input of SMA10, SMA30, SMA50, SMA100, EMA10, EMA30, EMA50 and EMA100. The full description of the data used for testing are summarized as shown in Table 2. The full dataset are the segregated in the ratio of 7:3 where 70% of the data are used for training and 30% are used for testing.

Table 2. Dataset Details

Currency dataset	EUR/USD and AUD/USD data (Date/Time, open, high low and close)
Total Time Frame	2 Jan 2012 to 29 Jun 2012
Training	2 Jan 2012 to 4 May 2012
Testing	6 May 2012 to 29 Jun 2012

4   Results

The Forex trend predicted by the DTW algorithm as previously discussed in Section 3.1 provides us with a new time series to be compared against the original segmented Forex time series data. Table 3 clearly shows an example of comparison between the two different time series for 5 segments. The data exhibited in Table 3 is derived from AUD/USD original time series data where the actual/original data (indicated by the dashed line) can be seen to be compared against the partial (indicated by the solid line) and predicted data (indicated by the dotted line). As the results obtained are in the form of time series data, the Mean Absolute Error (MAE) measurement have been chosen in the analyses of the prediction results obtained. Table 4 below denotes the MAE results gathered for both the AUD/USD and EUR/USD currency. It can be observed that using the close price alone produces results with the highest MAE compared to SMA10, SMA30, SMA 50, SMA100, EMA10, EMA30, EMA50 and EMA100.

Table 3. Comparison of the original and forecasted time series

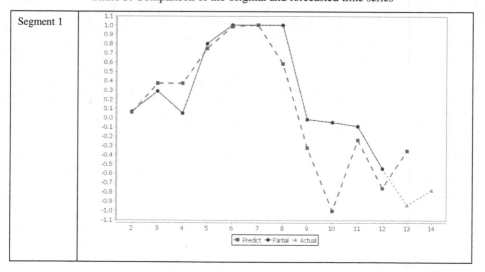

Table 3. (*Continued*)

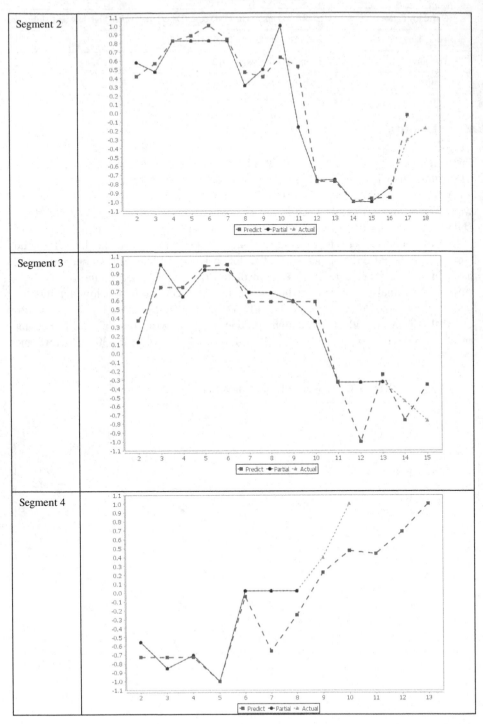

Table 3. (*Continued*)

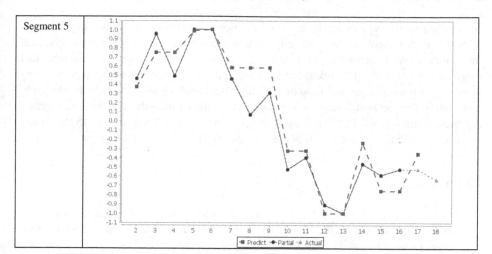

| Segment 5 | |

Comparing both the moving averages, it can be observed that while the faster moving average allows for faster response to the Forex price fluctuation, it is ultimately the slower moving average which encapsulates more data and normalizes noises more effectively and produces a smaller MAE result up to a certain extend. This is due to the fact that the data are more intensively normalized with the slow moving average and therefore affects the prediction result as can be seen from Table 3.

Table 4. MAE Result for Currency AUD/USD and EUR/USD

| Currency | Dataset | | | | | | | | |
	Close Price	*SMA 10*	*SMA 30*	*SMA 50*	*SMA 100*	*EMA 10*	*EMA 30*	*EMA 50*	*EMA 100*
AUD/USD	0.1073	0.0484	0.0336	0.0217	0.0367	0.0538	0.0360	0.0360	0.0363
EUR/USD	0.0953	0.0617	0.0390	0.0337	0.0295	0.0603	0.0341	0.0413	0.0361

Investigations on the effects of the different patterns extracted for classification have also been carried out. This is due to the fact that depending on the Forex price fluctuation, the patterns observed during the time frame used for prediction might contain a combination of uptrends and downtrend patterns or even none at all if the market is slow moving. While the K- Means classifier employed in clustering have shown great versatility, partial patterns proves to be a challenge to be classified as there are not enough data provided when the partial pattern provided is less than 1/3 of the trend as seen in Table 4.

Table 5. K- Means classification Results

Length of partial pattern used	Percentage of correctly classified partial pattern
1/2	100%
1/3	100%
1/4	49.75%

5   Conclusion

In conclusion, the type of input data used for Forex price forecasting is a crucial component which cannot be taken lightly. Our experiments have proven that technical indicators as implemented by most technical analyst can and should be incorporated more in the research of Forex price forecasting. This signifies that by incorporating more of the trading rules and fundamental technical analysis as performed by the technical analyst in the initial stage of the forecasting algorithm will contribute in increasing the accuracy of the Forex price prediction. Further research looking into the fusion of different technical indicators is also a possibility that could be carried out.

References

1. Dunis, C.L., Laws, J., Schilling, U.: Currency trading in volatile markets: Did neural networks outperform for the EUR/USD during the financial crisis 2007–2009? J. Deriv. Hedge Funds. **18**(1), 2–41 (2012)
2. Malkiel, B.: A Random Walk Down Wall Street. W.W. Norton & Company Inc., New York (1973)
3. Fama, E.F.: Efficient Capital Markets: A Review of Theory and Empirical Work. J. Finance. **25**(2), 383–417 (1970)
4. Abdullah, M.H.L.B., Ganapathy, V.: Neural network ensemble for financial trend prediction. In: TENCON 2000, pp. 157–161 (2000)
5. Schulmeister, S.: Components of the Profitability of Technical Currency Trading. Appl. Financ. Econ. **18**(11), 917–930 (2008)
6. Neely, C.J., Weller, P.A.: Technical Analysis in the Foreign Exchange Market. Federal Reserve Bank of St. Louis Working Paper Series 2011-001 (2011)
7. Mochón, A., Quintana, D., Sáez, Y., Isasi, P.: Soft Computing Techniques Applied to Finance. Appl. Intell. **29**(2), 111–115 (2008)
8. Vanstone, B., Tan, C.: A Survey of the Application of Soft Computing to Investment and Financial Trading. Inf. Technol. Pap. **13** (2003)
9. De Brito, R.F.B., Oliveira, A.L.I.: Sliding window-based analysis of multiple foreign exchange trading systems by using soft computing techniques. In: IEEE International Joint Conference on Neural Networks (IJCNN), pp. 4251–4258 (2014)
10. Abraham, A.: Analysis of hybrid soft and hard computing techniques for forex monitoring systems. In: Proceedings of the 2002 IEEE International Conference on Fuzzy Systems, pp. 1616–1622 (2002)
11. Canelas, A., Neves, R., Horta, N.: A new SAX-GA methodology applied to investment strategies optimization. In: Proceedings of The Fourteenth International Conference on Genetic And Evolutionary Computation Conference, pp. 1055–1062 (2012)
12. Canelas, A., Neves, R., Horta, N.: A SAX-GA Approach to Evolve Investment Strategies on Financial Markets Based on Pattern Discovery Techniques. Expert Syst. Appl. **40**(5), 1579–1590 (2013)
13. Canelas, A., Neves, R., Horta, N.: Multi-dimensional pattern discovery in financial time series using SAX-GA With extended robustness. In: Proceeding of The Fifteenth Annual Conference Companion on Genetic and Evolutionary Computation Conference Companion, pp. 179–180 (2013)

14. Tiong, L.C.O., Ngo, D.C.L. Ngo, Lee, Y.: Forex trading prediction using linear regression line, artificial neural network and dynamic time warping algorithms. In: Proc. Fourth Int. Conf. Comput. Informatics, pp. 71–77 (2013)
15. Tiong, L.C.O., Ngo, D.C.L. Ngo, Lee, Y.: Forex prediction using support vector machine and dynamic time warping. In: Proceeding of Thirteenth International Conference on Electronics, Information and Communications (ICEIC), pp. 141–142 (2014)
16. Choi, J.H., Lee, M.K., Rhee, M.W.: Trading S&P 500 stock index futures using a neural network. In: Proceedings of the Third Annual International Conference on Artificial Intelligence Applications on Wall Street, pp. 63–72 (1995)
17. Quah, T.S., Srinivasan, B.: Improving Returns on Stock Investment Through Neural Network Selection. Expert Syst. Appl. **17**(4), 295–301 (1999)
18. Yao, J., Tan, C.L.: A Case Study on using Neural Networks to Perform Technical Forecasting of Forex. Neurocomputing **34**(1), 79–98 (2000)
19. Emam, A.: Optimal Artificial Neural Network Topology for Foreign Exchange Forecasting. In: Proceedings of the 46th Annual Southeast Regional Conference on XX, pp. 63–68 (2008)
20. Sher, G.I.: Forex trading using geometry sensitive neural networks. In: Proceedings of The Fourteenth International Conference on Genetic and Evolutionary Computation Conference Companion, pp. 1533–1534 (2012)
21. HistData.com. http://www.histdata.com/download-free-forex-data

Short Term Load Forecasting Based on Hybrid ANN and PSO

Ellen Banda[✉] and Komla A. Folly[✉]

Department of Electrical Engineering, University of Cape Town, Private Bag,
Rondebosch 7701, Cape Town, South Africa
ellen.banda@gmail.com, Komla.Folly@uct.ac.za

Abstract. Short term load forecasting (STLF) is the prediction of electrical load
for a period that ranges from one hour to a week. The main objectives of the
(STLF) are to predict future load for the generation scheduling at power stations; assess the security of the power system as well as for timely dispatching
of electrical power. The traditional load forecasting tools utilize time series
models which extrapolate historical load data to predict the future loads. These
tools assume a static load series and retain normal distribution characteristics.
Due to their inability to adapt to changing environments and load characteristics, they often lead to large forecasting errors. In an effort to reduce the
forecasting error, hybrid artificial neural network (ANN) and particle swarm
optimization (PSO) is used in this paper.It is shown that the hybridization of
ANN and PSO gives better resultscompared to the standard ANN with back
propagation.

Keywords: Artificial neural network · Particle swarm optimization · Back
propagation · Short term load forecasting

1 Introduction

Short term load forecasting plays an important role in the power system in terms of
generation commitment as well as timely dispatcher information [1, 2, 3, 4]. It is an
essential component of any energy management system and is very important for
power system security studies such as contingency analysis and load management
[1, 2, 3]. Since the availability of electricity plays a vital role in the economic development of a country, it is imperative that future load demand is accurately predicted
[5]. There have been a number of traditionaltechniques that have been applied to the
problem of load forecasting such aslinear regression, time series, etc. as described in
[5]. Traditional load forecasting tools utilise time series models which extrapolated
historical load data to predict the future loads. These tools assumed a static load series
and retained normal distribution characteristics. Due to their inability to adapt to
changing environments and load characteristics, large forecasting errors result when a
deviation between historical load data and present conditions occurs [6]. In recent
years, Computational Intelligence (CI) techniques such as Expert Systems [5, 7, 8],
Fuzzy logic [9, 10, 11], Artificial Neural Network (ANN) [12, 13, 14] etc. have been

© Springer International Publishing Switzerland 2015
Y. Tan et al. (Eds.): ICSI-CCI 2015, Part III, LNCS 9142, pp. 98–106, 2015.
DOI: 10.1007/978-3-319-20469-7_12

proposed for short term load forecasting. Hybrid approaches whereby two or more CI techniques are combined to bring about a load forecasting tool are also avenues that have been explored [15]. For example, Fuzzy Logic can be combined with ANNs [9, 16], PSO with ANN [17, 18, 19, 20], PSO with Fuzzy logic [21], Expert systems [10, 11] or there can be a combination of GA with ANNs [12, 15], etc. This paper presents a hybrid ANN and Particle Swarm Optimization (PSO) method for short term load forecasting. It is shown that the hybridization of ANN and PSO gives better results compared to the standard ANN with back propagation.

2 Artificial Neural Networks and Back Propagation

Artificial neural networks (ANNs) are a type of computational intelligence inspired by the way the biological systems of humans such as the brain process information [22]. The human brain is made up of neurons which are interconnected by dendrites and collects information via this connection. ANNs are made up of a number of simple and highly interconnected processing elements called neurons [22]. All the neurons in the brain work in unison to make sure that all information received is processed as efficiently and accurately as possible. The artificial neurons try to simulate this kind of behaviour displayed by the real neurons in the brain. ANNs learn by example and are configured for particular classes of problems or applications through a learning system [4].

There are a number of topologies that are utilized in ANNs such as feed-forward networks, recurrent networks, etc. The commonly used ANNs are the feed-forward ANNswhere the input signals are propagatedfrom input layer via a hidden layer to the outputlayer[4].Recurrent Neural Networks (RNN) orfeedback ANNshave signals traveling in both directions within the network [23, 24].

Figure 1 illustrates a feed-forward neural network which is the most commonly used neural network architecture for short term load forecasting. It comprises of an input vector which would generally contain inputs made up of historical load data, historical and forecasted weather parameters, day types as well as other load affecting factors [14]. It also contains a hidden layer and then an output layer, usually one output is sufficient; however this can be configured as required.

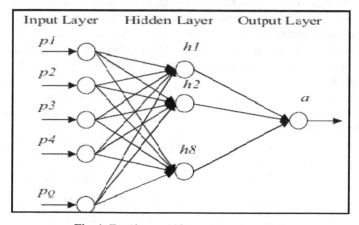

Fig. 1. Feedforward Network topology [14]

In general, a training algorithm such as back propagation is used in ANNs. Back propagation is basically a training method which finds the difference between the actual output and the reference or target output. The weights are then adjusted according to the errors and then propagated back into the system until the error is minimized [22]. Back propagation algorithm is excellent in its ability to accommodate load affecting variables however; the main drawback with this training algorithm is that the training process can become very cumbersome and time consuming [25, 26]. Convergence also becomes a problem. Different proposals for dealing with the convergence problem are presented in [25] where the back propagation algorithm with a momentum factor is said to help the neural network to converge much faster.

Because back propagation is based on the gradient descent, which is local search algorithm, the solution obtained with this method may not be optimal. Recently, training methods based on Evolutionary Algorithms such as Genetic Algorithm (GA) and PSO have been proposed [12, 18]. GA and PSO training algorithms may have a better convergence and search capability [27] than back propagation.

3   Particle Swarm Optimization

Evolutionary Algorithms are search functions which operate on a population and take their inspiration from natural selection and survival of the fittest in the biological world [28]. Particle Swarm Optimization (PSO) is an evolutionary algorithm based on a model of social interaction between independent particles that use social knowledge (also called swarm intelligence i.e. the experience accumulated during the evolution) in order to find the global maximum or minimum of a function [29]. PSO technique was first proposed by Eberhart and Kennedy [30]. It is a population based search procedure where individuals (particles) change their position with time [18]. The model has a set of n particles each representing a dimension of solution space. These particles move in the solution space in order to obtain the optimal solution by changing its position based on the influence by its nearest neighbour. The particles position changes based on the velocity it has over a certain number of iterations.

Whilst the particles search for the best position, they influence each other and thereafter converge to an optimal solution. This algorithm has been used to train neural networks to solve a number of problems such as described by Kennedy et al [28]. The basic principle is described as follows [17]:

1. A particle i is associated with a current position in the search space w_i , a current velocity v_i and a personal best position p_i. A swarm s consists of particles i.

2. The personal best position p_i corresponds to the particles position in the solution space where particle i presents the smallest error as determined by an objective function f.

3. The global best position p_g represents the position with the lowest error amongst all the p_i's

The personal and global best positions are updated by equations (1) and (2)

$$p_i(t + 1) = \begin{cases} p_i(t), & if\ f(p_i(t)) \leq f(w_i(t + 1)) \\ w_i(t + 1), & if\ f(p_i(t)) > f(w_i(t + 1)) \end{cases}. \tag{1}$$

$$p_g \in \{p_0(t), p_1(t), ..., p_s(t)\} \text{ and } p_g = min\{f(p_0(t)), f(p_1(t)), ... f(p_s(t))\}. \quad (2)$$

Each particles velocity and position is updated using equations (3) and (5) as follows:

$$v_{i,j}(t+1) = \omega v_{i,j}(t) + c_1 r_{1,i}(t) \left(p_{i,j}(t) - w_{i,j}(t) \right) + c_2 r_{2,i}(t)(p_{g(i,j)}(t) - w_{i,j}(t)).(3)$$

In equation (3), r_1 and r_2 are random values between 0 and 1 and are used to affect the stochastic nature of the algorithm. $w_{i,j}$, $p_{i,j}$ and $v_{i,j}$ are the current position, current personal best position and velocity of the jth dimension of the ith particle, respectively. The acceleration coefficients, c_1 and c_2, control how far a particle can move in a single iteration. These are typically set to the value of 2 however they can be varied between the range of 0 to4 [30, 31].The inertia weight ω is used to control the convergence of the PSO and is calculated as in (4).

$$\omega = \omega_{max} - \frac{\omega_{max} - \omega_{min}}{maxit} * iter . \quad (4)$$

Equation (4) has ω_{max} and ω_{min} which signify the maximum and minimum inertia. The maximum number of iterations is denoted by maxit and the current iteration value is represented by iter.The new velocity is then added to the current position of the particle as follows in order to get its next position:

$$w_i(t+1) = w_i(t) + v_i(t+1) . \quad (5)$$

The fitness of the ith particle is measured by the optimisation function f which is configured according tothe problem that needs to be solved. The particle with the minimum error is chosen as the best particle.

In the PSO method, the objective is to obtain the particle with the lowest mean square error (MSE) from the neural network.   MSE is calculated as follows:

$$f(w_i) = \frac{1}{N}\sum_{k=1}^{N} \left[\frac{1}{O}\sum_{l=1}^{O}\{T_{kl} - P_{kl}(w_i)\}^2 \right]. \quad (6)$$

where, f is the fitness value, T_{kl} is the target output value, P_{kl} is the predicted output value based on the position vector W_i, N is the number of training set samples, O is the number of output neurons. The particle with the lowest fitness obtained in equation (6) is then utilized in the ANN to forecast the next day's half hourly load.

The procedure is discussed in the next section.

4   Hybrid ANN and PSO

4.1   Data Analysis

The data set used for this analysis was obtained from a real distribution network. Weather data for the area was obtained from the South African Weather Services (SAWBS). The data sets for both weather and historical load were obtained for the period from 2009 – 2011. Data from the years 2009 to 2010 was used to train and validate the neural networks. Selected days in the year 2011 were then used to test the performance.The data was normalized between 0 and 1.

A correlation study to determine the weather variables that play a significant role in influencing the loading of the substation was conducted. The following variables were obtained from South African weather services: hourly humidity and temperature values as well as monthly rainfall measurements. The results show that temperature has a stronger correlation to the load (0.67) than humidity (0.56) [32].

4.2 ANN Input

The ANN networks take the following as inputs:
- Previous day half hourly load data
- Forecast and previous day type
- Forecast and previous day minimum and maximum temperature
- Forecast and previous day minimum and maximum humidity

A total number of 58 neurons were required to form the input layer; each neuron corresponds to an input variable. The output layer consisted of 48 neurons which correspond to the half hour in a day. The number of hidden layer neurons was determined by trial and error whereby the numbers of neurons were increased iteratively from 5 to 100 in steps of 5. The network with the best performance in terms of mean square error was then utilised to forecast the next day's half hourly load.

4.3 PSO Parameter

ThePSO is used to alter the weights of the ANN such that the resulting MSE for the training data is reduced. This process runs until a stop criterion is met which in this case is until the maximum number of iterations has been reached. The particle position in the search space of the PSO corresponds to the weights of the ANN. The PSO variables that were used in this study are: coefficient (c1)=2, coefficient (c2)= 2, inertia (w) $\in$ [0.41.0], the number of particles (S) =20, maximum iterations=100, constriction factor (K)=0.729, r1=r2= random (0, 1), Vmax$\in$[-0.41.0].

The fitness function f in (1) corresponds to the MSE of the ANN network.Each particle represents a possible solution of the weights.

5   Simulation Results

In this paper, only forecasting for weekdays was conducted. Weekends, deemed as Saturday and Sunday, were removed during the training in order for the ANN to learn the weekday dependencies. This separation was introduced because of the differencesbetween weekday and weekend load profiles. Twenty(20) independent runs were performed.Figures 2 to 4 show the predicted loads and the actual load for 3 days in the month of March 2011.It can be seen that the overall performance of the ANN is vastly improved when PSO is combined with ANN algorithm.PSO-ANN has less error and much closer to the actual load profile than ANN.The mean absolute performance error (MAPE) for the selected forecast days is shown in table 1 together with the standard deviation. It can be seen that PSO-ANN has the smallestMAPE and the smallest standard deviation.

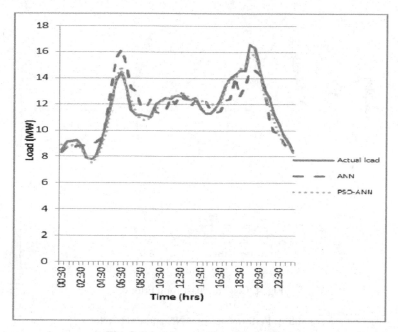

Fig. 2. Load forecast for a Wednesday

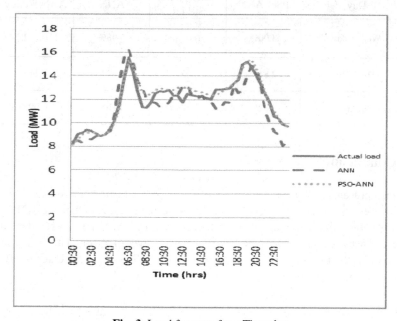

Fig. 3. Load forecast for a Thursday

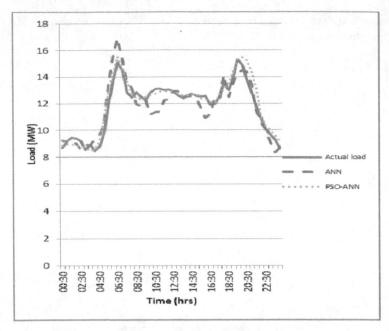

Fig. 4. Load forecast for a Friday

Table 1. Performance of ANN and PSO-ANN

Day	PSO-ANN	PSO-ANN Stddev	ANN	ANN Stddev
Wednesday	3.00%	0.072	5.98%	0.172
Thursday	2.84%	0.035	5.82%	0.115
Friday	3.32%	0.041	4.56%	0.093

6   Conclusions

This paper presents a hybrid ANN and PSO for short term load forecasting. The simulation results show that hybrid PSO-ANN performs better than the ANN using back propagation. The mean absolute performance error (MAPE) and standard devia-tionalso confirmed that PSO-ANN has the lowest overall error. This validates the hypothesis that the hybridization of computational intelligence techniques is effective in reducing forecasting errors.

Acknowledgement. This work is based on the research supported in part by the National Research Foundation of South Africa, UID 83977 and UID 85503.

References

1. Galiana, F.D., Gross, G.: Short term load forecasting. In: Proc. of the IEEE, vol. 75, no. 12 (1987)
2. Papalexopoulos, A.D., Hesterberg, T.C.: A regression based approach to short term load forecasting. IEEE Trans. on Power Systems, **5**(4) (1990)
3. Charytoniuk, W., Chen, M.: Very Short term load forecasting using Artificial Neural Networks. IEEE Transactions on Power Systems, **15**(1) (2000)
4. Banda, E.: Short term load forecasting using Artificial Intelligence techniques. Department of Electrical Engineering, University of Cape Town, Student Thesis (2006)
5. Moghram, I., Rahman, S.: Analysis and evaluation of five short term load forecasting techniques. IEEE Transactions on Power Systems, **4**(4) (1989)
6. Farahat, M.A., Talaat, M.: A new approach for short-term load forecasting using curve fitting prediction optimized by genetic algorithms. In: Proc. of the 14th International Middle East Power Systems Conference, MEPCON 2010, Cairo University (2010)
7. McDonald, J.R., Asar, A., Rattray, W.: Experience with artificial neural network models for short term load forecasting in electrical power systems: a proposed application of expert networks. In: Third International Conference on Artificial Neural Networks, pp. 123–127 (1993)
8. Liang, C.C., Ho, K.L., Lai, T.S., Hau, Y.Y.: Short term Load Forecasting of Taiwan Power System Using A Knowledge-Based Expert System. IEEE Transactions on Power Systems **5**(4), 1214–1221 (1990)
9. Rothe, J.P., Wadhwani, W.K., Wadhwani, S.: Hybrid and integrated approach to short term load forecasting. International Journal of Engineering Science and Technology **2**(12), 7127–7132 (2010)
10. Dash, P.K., Liew, A.C., Rahman, S.: Fuzzy Neural-Network and Fuzzy Expert-System for Load Forecasting. IEE Proceedings-Generation, Transmission and Distribution **143**(1), 106–114 (1996)
11. Kim, K.H., Park, J.K., Hwang, K.J., Kim, S.H.: Implementation of Hybrid Short Term Load Forecasting System Using Artificial Neural Networks and Fuzzy Expert Systems. IEEE Transactions on Power Systems **10**(3), 1534–1539 (1995)
12. Sarangi, P.K., Singh, N., Swain, D., Chauhan, R.K., Singh, R.: Short Term Load Forecasting Using Neuro Genetic Hybrid Approach: Results Analysis With Different Network Architectures. Journal of Theoretical and Applied Information Technology, 109–116 (2009)
13. Kandil, N., Sood, V., Saad, M.: Use of ANNs for short term load forecasting. In: Proceedings of the 1999 IEEE Canadian Conference on Electrical and Computer Engineering, pp. 1057–1061 (1999)
14. Zhang, S., Lian, J., Xu, H., Liu, J.: Grouping model application on artificial neural networks for short term load forecasting. In: Proc. of the 7th World Congress on Intelligent Control and Automation, pp. 6203–6206 (2008)
15. Bichpuriya, Y., Rao, M.S.S., Soman, S.A.: Combination approach for short term load forecasting. In: Proceedings of the 9th International Conference on Power and Energy, IPEC 2010, Singapore, pp. 818–823 (2010)
16. Barkitzis, A.G., Theocharis, J.B., Kiartzis, S.J., Satsios, K.J.: Short term load forecasting using fuzzy neural networks. IEEE Transactions on Power Systems **10**(3), 1518–1524 (1995)
17. Shayeghi, H., Shayanfar, H.A., Azimi, G.: STLF based on optimized neural network using PSO. World Academy of Science, Engineering and Technology **3**, 889–899 (2009)

18. Quaiyum, S., Khan, Y.I., Rahman, S., Barman, P.: Artificial neural network based short term load forecasting of power system. International Journal of Computer Applications (0975 – 8887) **30**(4), 1–7 (2011)

19. Subbaraj, P., Rajasekaran, V.: Evolutionary techniques based combined artificial neural networks for peak load forecasting. World academy of Science, Engineering and Technology **2**, 600–606 (2008)

20. Alshareef, A.: Next 24 hours load forecasting for the western area of Saudi Arabia using artificial neural network and particle swarm optimization. Journal of Engineering and Computer Sciences **3**(2), 97–117 (2010)

21. Jain, A., Jain, M.B., Srinivas, E.: A novel hybrid method for short term load forecasting using fuzzy logic and particle swarm optimisation. In: International Conference on Power System Technology, pp. 1–7 (2010)

22. Lee, K.Y., Cha, Y.T., Park, J.H.: Short term load forecasting using an artificial neural network. Trans. on Power Systems, **7**(1) (1992)

23. Mandal, J.K., Sinha, A.K., Parthasarathy, G.: Application of Recurrent Neural Network for Short Term Load Forecasting in Electric Power System. IEEE International Conference on Neural Networks **5**, 2694–2698 (1995)

24. Siddarameshwara, N., Yelamali, A., Byahatti, K.: Electricity short term load forecasting using elman recurrent neural network. In: Int. Conf. on Advances in Recent Technologies in Communication and Computing, pp. 351–354 (2010)

25. Riu, Y., AEl-Keib, A.A.: A review of ANN-based short term load forecasting models. In: Proc. of the 27th South Eastern Symposium on System Theory, pp. 78–82 (1995)

26. Liu, K., Subbarayan, S., Shoults, R.R., Manry, M.T., Kwan, C., Lewis, F.L., Naccarino, J.: Comparison of very short term load forecasting techniques. IEEE Trans. on Power Systems, **11**(2) (1996)

27. Folly, K.A., Venayagamoorthy, G.K.: Performance evaluation of a PBIL-based power system damping controller. In: Proc. Of the 2010 IEEE Congress on Evolutionary Computation (CEC), Barcelona, Spain (2010)

28. Kennedy, J., Eberhart, R.: Particle swarm optimization. In: Proc. of IEEE International Conference on Neural Networks IV, pp. 1942–1948 (1995)

29. MassioGrimaldi, E., Grimaccia, F., Mussetta, M., Zich, R.E.: PSO as an effective learning algorithm for neural network applications. In: Int. Conf. on Computational Electmmagnetics and Its Applications Proceedings, pp. 557–560 (2004)

30. Eberhart, R.C., Kennedy, J.: A new optimizer using particle swarm theory. In: Proceeding of the Sixth International Symposium on Micro Machine and Human Science, Nagoya, Japan, pp. 39–43 (1995)

31. Hu, X.: Particle Swarm Optimisation (2006). www.swarmintelligence.org/tutorials

32. Hernandez, L.: A study of the relationship between weather variables and electrical power demand inside a smart grid/smart world framework. Sensors, 11571–11591 (2012)

Using Big Data Technology to Contain Current and Future Occurrence of Ebola Viral Disease and Other Epidemic Diseases in West Africa

Foluso Ayeni, Sanjay Misra[(✉)], and Nicholas Omoregbe

Department of Computer and Information Sciences, Covenant University, Ota, Nigeria
{Foluso.Ayeni,Sanjay.Misra,Nicholas.
Omoregbe}@covenantuniversity.edu.ng

Abstract. West Africa is currently plagued with Ebola Viral Disease (EVD) and other minor epidemic diseases which has led to major economic meltdown and high mortality rate in countries like Guinea, Sierra Leone and Liberia as a result of immigration, emigration, foreign trade and investment, bilateral, poor health care issues amidst others. Harmonized EVD related data can help identify individuals who are at risk of contracting the terminal disease and at the same time controlling the outbreak which will in turn lower cost of health care across West Africa. This paper presents the significance, framework as well as an implementation plan and design for using Big Data Technologies (BDT) as an aid to prevent and control EVD in West Africa and the provision of how the principles of cloud computing could be applied to present and impending expectations of the West African Health sector.

Keywords: Ebola Viral Disease · Big data · Analytics · Electronic health records · Immigration and emigration · Data bank · Harmonized · West Africa

1   Introduction

Cloud computing is a model that enables on demand, convenient, ever present network access to a shared puddle of configurable computing resources that can be provisioned rapidly and released with insignificant management effort [1]. It can be squared back to the 1950's when enormous scale mainframes were made available to institutes and corporations [2]. According to [3] it is a platform in which services are carried out on behalf of client's resident on technologies that the clients do not own or manage. At the heart of cloud computing is the BDT which harmonizes pool of fragmented data.

BDT is amongst the eight technology-enabled business trends that has reshaped strategy across a wide range of industries [4]. It is a cloud computing techniques that provides storage as a service. It describes a new generation of technologies and architectures designed so organizations and institutions can economically extract value from very large volumes of a wide variety of data by enabling high velocity capture, discovery and analysis [5]. BDT requires a move in computer architecture so that

© Springer International Publishing Switzerland 2015
Y. Tan et al. (Eds.): ICSI-CCI 2015, Part III, LNCS 9142, pp. 107–114, 2015.
DOI: 10.1007/978-3-319-20469-7_13

clients can handle both data storage requirements and server processing needed to analyse and process large volumes of data economically. It is all about the ever increasing challenge that industries and organizations face as they deal with large and very fast growing sources of information and data that presents difficult range of analysis and use problem. It has been successfully proven in organizations and businesses such as financial institutions, logistics, telecommunications, oil and transportation. Despite the great characteristics of BDT, they haven't been utilized fairly in the health care industry, [6] most especially Africa which is commonly prone to terminal diseases such as the recent EVD upsetting West Africa.

EVD previously known as Ebola Haemorrhagic Fever is a severe, often fatal illness in humans [7]. It is a virus transmitted to people from wild animals and spreads in the human populace through human-to-human transmission. It was first discovered in 1976 in the Democratic Republic of Congo near the Ebola River [8]. The first set of outbreaks was in Central Africa but the most recent outbreaks in West Africa involve major countries such as Guinea, Sierra Leone, Liberia and Nigeria [9]. It is recorded that about fifty out of every 100 EVD patients die which is shockingly high. Although local and international agencies are not relenting in the fight against the EVD, technologies such as Big Data can be used as an aid to contain and control outbreaks of this deadly disease.

According to [9] effective and efficient outbreak control depends on application of interventions namely case management, surveillance and contact tracing, good laboratory service, safe burials and social mobilization.

This work is proposed to prove that EVD outbreak in West Africa can be contained and controlled using BDT. The study would adopt a combination of qualitative and quantitative research method.

2   Statement of Problem and Objective of Study

Technological advancements made in the world especially for the last ten to fifteen years have altered the entire scenario in every field including education, entertainment, health, workplace, environment and personal life [2]. There is also Zero Level of significance of BDT in EVD struck West African Countries. This is the major reason why BDT need to be implemented in order to aid EVD health care workers in their fight against the deadly disease. BDT is simply the collection of data sets so large or complex that it becomes difficult to process using the traditional data processing applications.

The main objective of this study is to explore the usefulness of BDT in fighting and containing EVD and also presents a technology based framework that allows stakeholders to harmonize all EVD related data towards containing the virus and the provision of how the principles of cloud computing could be applied to current and future expectations of the West African Health sector.

3   Review of Related Findings/Technologies

3.1   Existing Technologies

Wiser Together is leading the change in the field of BDT's in health care. It is a BDT based hospital resident on the web committed to motivate people to improve the quality of health care while reducing costs for themselves. Millions of people today are connected to this and also reaping the benefits of their products and services. In September 2014, Wiser Health is the only decision-making technology that compares treatment options using evidence-based clinical outcomes, individual consumer preferences, insurance coverage, cost and popularity of the treatment among physicians and patients in the local community.

3.2   Related Findings

The development of cloud computing has brought about information technology infrastructure developments for enterprises especially in the aspects of health care delivery systems of developed nations therefore causing medicals records to shift from manual processing to electronic. [2] tried to devise a means on how to overcome barriers of effective health care delivery in Nigeria using Socalized medicine while [10] also devised a means on how to maintain web service process models for effective mobile health care delivery in Nigeria. This connotes that internet based health care systems are existing in West Africa but not effective and efficient.

The major concerns of mobile health care service providers is security and privacy [11] which makes penetration of Big Data difficult for acceptance particularly in Africa. According to Odusote & Omoregbe, three models to curb security threats in health cloud were proposed though not implemented. [12] projected three schemes to enhance big data security because enterprises and health care providers need to ensure they have mechanisms in place which allows them to meet government compliancy regulations for data protection. First, Secure encryption technology in health information must be used to protect confidential data. Secondly, careful management of access to the cryptography keys which unlock the encrypted data must be put in place. Finally, Big Data Analytics can also secure Big Data by collecting all available digital evidence including raw packets, flow data and files, organizations can uncover advanced targeted attacks.

At the heart of all these transformations is the rising cost of Electronic Health Care Services which is a global challenge but [13] concluded that the future of these services is to seek for expansion and interoperability. BDT is a very important factor in shaping the future of health care and at the same time preventing the spread of EVD in West Africa.

BDTs in health care have not been fully implemented in developed countries owed to the challenges of access to necessary information. The power to access and analyse enormous data sets can improve ability to anticipate and treat illnesses [14]. According to Burg, aggregating massive amount of health data might sound easy but the challenge is maintaining patients integrity. EVD outbreak control in West Africa falls short of international disease control standards resulting from poor state of health care infrastructure, medical professionals and Information technology penetration. According to the [15], the driving push for big data is to create more value in health care. Advanced analytic approaches that use machine learning, predictive modelling, pattern detection, anomaly detection and other sophisticated techniques can successfully address the weaknesses of rule-based systems.

West Africa's major priority is to devise a means of making adequate use of Information Technology at our disposal to contain EVD that has savaged the entire region. Our interest is to develop a framework using an open source BDT tool to prove how BDT can be applied to the intervention and control schemes outlined by the WHO.

3.3   Economic Impact of EVD in West Africa

The Ebola epidemic continues to cripple the economies of Liberia, Guinea and Sierra Leone [16]. According to World Bank's report which was released in December 2014, Sierra Leone's Gross Domestic Product (GDP) growth was 11.3% before EVD outbreak in June 2014 and by December 2014 it had reduced to 4% which is about 60% decrease and if this persists, world bank projections shows that by October 2015, Sierra Leone's GDP growth will be 1%. Guinea's GDP growth was 4.5% before EVD outbreak in June 2014 and by December 2014 it had reduced to 0.5% and if this persists, world bank projections shows that by December 2015, Guinea's GDP growth will be -0.2%. Liberia's GDP growth was 5.9% before EVD outbreak in June 2014 and by December 2014 it had reduced to 2.2% and if this persists, world bank projections shows that by October 2015, Guinea's GDP growth will be 1.0%.

3.4   Motivation of the Work Based on the Existing Methodologies

The motivation for this work is to fill the gap in the existing technologies by focusing more on West Africa which is prone to terminal diseases and also using information technology tools available at our disposal to reduce mortality rate in West Africa.

4   Proposed Framework

There are over 50 existing open source tools for Big Data and the most interesting features of BDTs are that they have noSQL databases, presence of business intelligence tools , development tools and much more [17] .The very best known amongst these is the Hadoop MapReduce (HMR) which is spawning an entire industry of related services and products . It is a framework for managing big data by storing data on a large scale, organizing data so it can be accessible via a variety of different tools, it is also a set of tools that allows users to gain insights from the data. The HMR model is a BDT that has proven to be efficient over time in the business world such as Facebook, Twitter, ebay and google could also be applied to solve healthcare challenges most especially the EVD[18]. HMR model can be used to execute EVD case management, surveillance and contract tracing and social mobilization. HMR is a software framework for easily writing applications which process vast amounts of data in-parallel on large clusters (thousands of nodes) of commodity hardware in a reliable, fault-tolerant manner [19].

Fig 4(b) shows a typical collaborative virtual BDT based EVD reference framework/architecture for implementing an internet technology based integration solution between locations in West Africa.

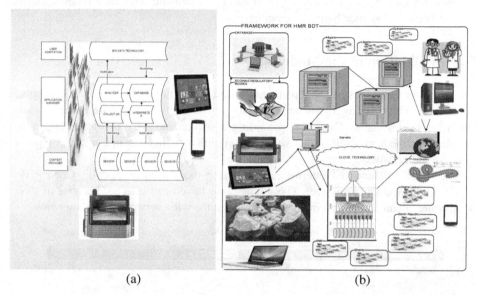

(a)   (b)

Fig. 1. (a) Context Diagram (b) Conceptual Framework

4.1   Context Diagram

Fig 1(a) shows a context diagram/architecture for implementing a BDT based Health care system. The context diagram is centered round the use of notification and monitoring systems which enables the provision of a number of services most especially mobile communication services and user adaptation services.

4.2   Conceptual Framework

The aggregated HMR BDT framework shows the communication among stakeholders and also the information technology infrastructures involved.

- The database and the ECOWAS regulatory bodies ensure proper accreditation and authentication of registered victims and also ensure efficiency, proper record keeping and security of the database including better knowledge of possible outcomes.
- The various remote locations indicated as discussed in the introduction and abstract shows that EVD based geographical areas can communicate with

each other real-time regardless of their geographical locations, for example, an EVD base station in Liberia, West Africa can connect to another EVD base station in Guinea, West Africa.

- International regulatory bodies such as the WHO can reach any station of his or her choice across the internet using any internet enabled device that is connected to the central server which will be domicile in the cloud as shown in Figure 3.
- Various Airports in West Africa can communicate with each other via Epidemic clearance barcode scanning.

5   Validation of the Model

The development of a system implementing new computer technologies that support big data in health care is a growing necessity just as continuity of care requires a cooperative environment. Therefore, providing a secure and easy to implement environment for medical applications will pose significant changes in current computer systems and networks.

The study will adopted internet programming tools such as HTML, CSS, PHP, JAVA Script and Dreamweaver. These languages have been selected because of their open source nature, platform independability and the wide acceptability they enjoy.

Fig. 2. Patients profile page which displays patient account information and also doctor's home page

Our designed system was implemented, evaluated and interpreted as shown in Figure 2. The sign up page provides a step-by-step account set-up process for patients, healthcare workers or regulatory bodies. The homepage is displayed upon successful account login (patient and physician). The patient homepage displays a list of suggested physicians that may be known to the logged in physician, enabling easy connection with other physicians. It also displays most recent unread messages from both patients and physicians. The patient homepage also displays a list of suggested physicians that might be close to the patients' location and shows most recent unread messages from physicians. Connection displays physicians and patients that are already connected. Also shows a list of suggested colleagues for physicians as well as suggested physicians for patients. Messages show interactions between physicians and patients. Provides displays physician and patient account information such as patient contact, medical information, and medical history. The comparative analysis in the Table 1 further proves that our proposed system is flexible; requires less time to use; provides reliable and robust performance; quick response and follow-up.

Table 1. A comparison of the proposed Big Data application with existing application

Functionalities	Wiser Together	Proposed Application (Medibook)
Cost (payment plans)	Expensive	Relatively Cheap
Covered Health Areas	6	20
Ease of Locating Physician	No	Yes
GPS technology required	Yes	No
Strictly designed for reporting emergencies in real-time	No	Yes
Social Networking Principles embedded	No	Yes
User Acceptance	Yes	Yes

6   Conclusion and Future Work

This study has strengthened and explored the need to better improve communication and information flow among stakeholders in the West African health care industry and the framework gives a pointer that BDT will definitely help contain current and future occurrence of EVD and other epidemic diseases in West Africa.

In any software/tool development, Maintainability will always be a process and performing this task usually comes with heavy cost [20]. Future research works tends to show a detailed maintainability model. The implementation of the system will also incorporate a more sophisticated security measure to prevent third party or cyber criminals from hacking into the system. An open source Apache HMR application for Big Data Databases which have no SQL will be integrated to the system. To address the security issue, encrypted authentications will be put in place to grant only role-based access to records. BDT in West Africa will also be evaluated by comparing it to the existing systems in the USA and other developed countries.

References

1. Mell, P., Grance, T.: The NIST definition of cloud computing. National Institute of Standards and Technology **53**(6), 1–7 (2009)
2. Ayeni, F., Misra, S.: Overcoming barriers of effective health care delivery and electronic health records in nigeria using socialized medicine. In: 2014 IEEE International Conference on Electronics Computer and Computation (ICECCO), pp. 1–4 (2014)
3. Pearson, S., Shen, Y., Mowbray, M.: A privacy manager for cloud computing. In: Jaatun, M.G., Zhao, G., Rong, C. (eds.) Cloud Computing. LNCS, vol. 5931, pp. 90–106. Springer, Heidelberg (2009)
4. Bughin, J., Chui, M., Manyika, J.: Clouds, Big Data and Smart Assets: Ten-tech Enabled Business Trends to Watch (2010). http://www.mckinseyquarterly.com (retrieved)
5. Villars, R., Olofson, C., Eastwood, M.: Big Data: what it is and Why you should Care. IDC (2011). http://www.idc.com (retrieved)
6. Groves, P., Kayyali, B., Knott, D., Van Kuiken, S.: The Big Data Revolution in Health Care: Accelerating Value and Innovation (2013). http://www.mckinseyquarterly.com (retrieved)
7. World Health Organization: EVD Implications of Introduction in the Americas (2014)
8. Minnesota Deapartment of Health. http://health.state.mn.us
9. World Health Organization Regional Office for Africa: Contract Tracing During an Outbreak of Ebola Disease (2014)
10. Olayinka, O., Misra, S., Olaniyi, M., Ahmed. A.: Towards an iterative maintainability web service model for effective mobile healthcare delivery. In: 2013 IEEE International Conference on Emerging & Sustainable Technologies for Power & ICT in a Developing Society (NIGERCON), pp. 177–181 (2013)
11. Odusote, B., Omoregbe, N.: Towards A Well-Secured Electronic Health Record in the Health Cloud. Journal of Computing **5**(1), 13–17 (2013)
12. Advancetech: Enhancing Big Data Security (2013). http://www.advantech.com/nc (retrieved)
13. Benson, M.B., Cole, A.: Hospital information systems in Nigeria: a review of literature. The Journal of Global Health Care Systems **1**(3), 3–26 (2011)
14. Burg, N.: How Big Data will help save Health Care (2014). http://www.forbes.com (retrieved)
15. Harvard Business Review: How Big Data Impacts Health Care (2014). http://www.hbr.org (retrieved)
16. The World Bank Group: Update on the Economic Impact of the 2014 Ebola Epidemic on Liberia, Sierra Leone and Guinea (2014)
17. Harvey, C: 50 open source tools for big data (2012). http://www.datamation.com/datacenter
18. Schouten, P.: Big Data in Health Care Solving Provider Revenue Leakage with Advanced Analytics. Health care financial management association (2013). http://www.hfma.org (retrieved)
19. Chen, H., Chiang, R., Storey, V.: Business Intelligence and Analytics: From Big Data to Big Impact. MIS Quarterly **36**(4), 1165–1188 (2012)
20. Omogbadegun, O., Adegboyega, A.: Framework for overcoming barriers of complementary and alternative medicine acceptance into conventional healthcare system. International Journal of Medicinal Plants and Alternative Medicine **1**(7), 118–136 (2013)

Real-Time Diagnosis and Forecasting Algorithms of the Tool Wear in the CNC Systems

Georgi M. Martinov, Anton S. Grigoryev, and Petr A. Nikishechkin[✉]

FSBEI HPE, MSTU "STANKIN", Moscow, Russia
pnikishechkin@gmail.com

Abstract. The article proposes concept of solution development for diagnosis and control of real-time cutting tool in for edge cutting machining. The functional model of a diagnosis subsystem based on data reading from sensors of various types established in a cutting zone is developed. Algorithms of subsystem accepted signals processing and averaging, allowing define condition of cutting tool with defined preciseness and it's condition forecast in future are offered. Architectural features of subsystem program realization are exposed and solutions for integration into CNC system are described. Testing results of the diagnosis subsystem and its main algorithms during manufacturing processes control on turning machine tools are presented.

Keywords: Diagnosis · Forecasting · Cutting tool · Sensor · Signal processing · Algorithm · CNC system

1   Introduction

One of the most important aspects of modern automated manufacturing is reliability and process control of machining. Operating experience of the technological systems that are based on automated machines shows that their reliability is often insufficient. Dimension wear or chipping of the cutting tool are the most common damage, which reduces the accuracy of the technological system.

Today, similar problems in EU countries and the USA are solved, mainly based on statistical methods, tracking the work tool and using tool resource for 80-90%. The main problem is that, even within a single batch, tool life of Russian production may vary widely by 15–40%. In this case, the use of a statistical approach the risk of defects in products or tool breakage is significantly higher. It can lead to big financial losses, especially during prolonged treatment of the critical parts [1].

In this regard, during operation of the technological systems solved the problem of increasing the reliability due to the technical diagnosis. Diagnosis provides timely termination of machine operation or making corrections in his work, at the expense of an operational definition of condition failure. Diagnosis of tool wear provides an operating time of each tool to the actual refusal. Besides, a problem of diagnosing is not only definition of technical condition in which there is an object at present, but also forecast of technical condition, in which the object will appear in the following

© Springer International Publishing Switzerland 2015
Y. Tan et al. (Eds.): ICSI-CCI 2015, Part III, LNCS 9142, pp. 115–126, 2015.
DOI: 10.1007/978-3-319-20469-7_14

interval of time. Implementation of solutions, allowing to determine the current state of the cutting tool and to predict its state in the future, will reduce the percentage of defects, to increase productivity and reliability of all technological system. This solution should provide a guaranteed ending process of transition, the ability to determine the critical situation and to make the necessary actions to stop the process of processing or removal tool and ensure maximum resource utilization of the cutting tool [1,2].

2   Analysis of Modern Diagnosis Systems of the Cutting Tool

Most of realized systems for diagnosis can recognize the current condition and cutting tool failure. For today, there are many diagnosis systems, which realize control of machining process products in the automated manufacturing. Table 1 summarizes commercially available diagnosis systems.

Table 1. Analysis systems for the diagnosis tools wear from international manufacturers

Characteristic	Diagnosis system			
	PROMETEC Promos (Germany)	Nordmann (Switzerland)	ARTIS Orantec (USA)	Brankamp CMS (Germany)
Real-time diagnosis	+	+	+	+
Diagnosis data	Force, acoustic emission, power, vibration	Force, acoustic emission, power, vibration	Force, power, vibration	Acoustic emission, longitudinal deformation
Real-time forecast of the cutting tool state	−	−	−	−
Capability of the integration in CNC system	SINUMERIK 810D/840D	SINUMERIK 840D, IndraMotion MTX, Fanuc	SINUMERIK 840D	− (autonomic module)
Capability of the using of different algorithms	−	−	−	−

Commercial systems mainly focus on the diagnostics and monitoring of tools, without remaining life forecasting, and can only be used with the numerically controlled systems for which they were developed. Also, there are no Russian commercial systems for real-time tool diagnosis. Our analysis permits the formulation of various requirements for a system capable of real-time diagnosis and forecast of the tool's wear in turning [3].

An actual task is developing a universal solution that operates in real time and allows the diagnosis and forecast of the residual resistance of the cutting tool, as well as having the ability to integrate in the control without changing the core of CNC system.

3   Constructing a Functional Model of the Solution for Tool Forecast and Diagnosis

Correct work of the diagnostic system requires the definition its basic modules, their functioning and interaction with each other. Fig. 1 shows a developed functional model solutions for the diagnosis and forecast of the cutting tool state, reflecting the architecture of its construction and the sequence of actions for it to work properly [2].

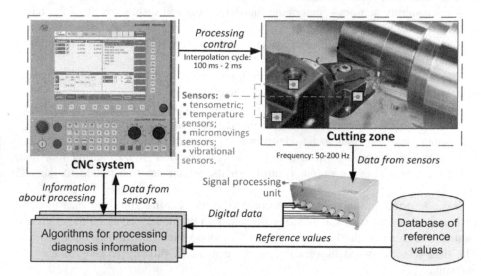

Fig. 1. Functional model of the solution for forecast and diagnosis of tool wear

The basis of the developed diagnostic solutions provided the use of a method based on the use of various types of sensors mounted in the cutting zone, and determining the current state of them, and the cutting tool in the forecast its remaining life (Fig. 2).

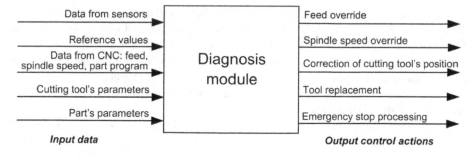

Fig. 2. Functional model of the solution for forecast and diagnosis of tool wear

The diagnostics and forecast of tool wear may be divided into four phases. The first involves data collection from the cutting zone, by means of sensors, which can characterize the current state of the cutting tool [4-6].

At present, it is proposed the use of four types of sensors:

– tensometric sensors – are used to determine the load on the cutting tool during machining in three axes;

– temperature sensors – are used for measuring temperatures in the cutting zone, as well as basic units of the machine;

– microsupply sensors – determine the deflection and displacement of the basic units of the machine during operation;

– vibration detector – are used to determine the level of vibration of the cutting tool and the basic units of the machine.

The second phase is digitization and preliminary analysis of the signals, by means of autonomous devices or circuits built into the computer.

Determination of the current condition of the cutting tool and forecast its of future condition is performed using specialized algorithms for information processing. The input data for diagnostic algorithms is a wide range of parameters and factors on which the algorithm determines the condition of the cutting tool and, if required, generates an output control signals, which are transmitted to the NC (Fig. 2).

Basic information are signals that come from the cutting zone and measured by various types of sensors. Calculations are based on previously obtained the reference values used to calculate the diagnostic coefficients, which in turn uses an algorithm. The data sets are stored in a database, where in block directly received diagnostic algorithms [7].

Control signals transmitted to the CNC system from diagnosis solutions could be:

– feed correction – is realized to maintain a constant load on the elastic elements of the machine in order to stabilize the elastic deformation and to reduce the risk of breakage of the cutting tool;

– spindle speed adjustment – to maintain the cutting speed in the optimal range for stabilizing the surface quality, stability of the cutting tool and to simplify procedures for the preparation of control programs;

– cutting tool's position correction – to improve the machining accuracy by compensating for the size of the cutting tool wear, thermal and elastic deformations of the machine manufacturing errors affecting the accuracy of movement of workers;

– tool replacement command– the command is executed when a critical tool wear to ensure the required quality of the product and to prevent accidents;

– emergency processing stop – the command is required to prevent damage to the mechanism of the machine and marriage details.

The proposed functional model allows diagnosis and monitoring of the cutting tool in accordance with various algorithms of data processing and transfer of control in the control system.

4   Development of Algorithms for Diagnostics and Forecast of the Cutting Tool State

Describing the operation of the cutting tool can be identified such a thing as its residual resistance, which is defined as the time it works until it reaches the limit value of wear. Work of the tool is possible and after time, however the probability of its refusal becomes high and unacceptable for the automated production. For determination of this parameter it is required to develop and realize the algorithm, allowing to estimate residual firmness of the tool at serial processing of a large number of the same details and, in case of impossibility of further processing with the set accuracy, to give out a signal to the operator and to stop processing.

In the presented work is offered the consideration of algorithm for determination of current state and forecast of residual firmness of the cutting tool by measurement of force operating on it at fair processing of a large consignment of the same products on machines of turning group.

When finishing the parts of details control algorithm should be applied only after detecting that the residual tool life is enough for the surface treatment once the details of the required accuracy and a given roughness.

Otherwise, the tool needs to be replaced before the end of treatment, leading to defects because of the trace left on the surface of the insertion of a new tool. Therefore, at fair processing, the algorithm of diagnosing, besides control, has to include actions of forecast of changes of a condition of the tool during processing of the following detail of the parts. Developed procedures necessary to control the cutting tool and forecast its residual resistance is presented in the Fig. 3.

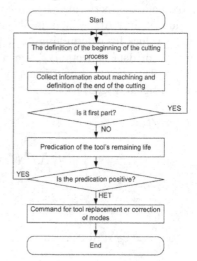

Fig. 3. Algorithm of the tool's monitoring and predict its remaining life

The processing of any part consists of the main periods: idling, digging, lead-out, and operating period – contact of cutting tool with part (Fig. 4). To solve the problem of determination of remaining life of cutting tool is necessary to analyze information from the sensors from the machine cutting zone during full contact of cutting tool with part (working period of machining). Processing data from the other zones is not required and can lead to significant errors in forecast of remaining life of cutting tool. To exclude unnecessary data from other processing zones are proposed algorithms to determine of the beginning and end of cutting another part.

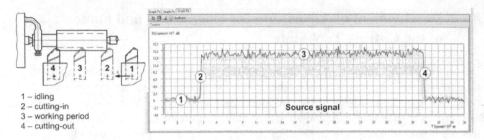

Fig. 4. Main periods of processing

Determination of the working period of the cutting tool requires finding zones change of a condition of the tool. To solve this problem, the received signal is approximated by the method of least square (OLS) and then made its differentiation. Approximation is performed by groups of digital signal values within the established period of time Δt_a, which are determined experimentally (Fig. 5, A). Group digital offset values for determining Δt_{a_1} produced by an amount equal to $1/f$ (f is a frequency of data collection). Then, the angular coefficient k is determined for each group. The set of all coefficients (k) is a differential signal from the originally received signal (Fig. 5, B).

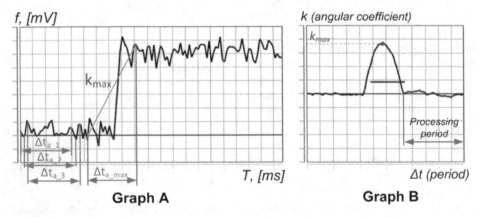

Fig. 5. Definition of the total working period of the signal by the method of least squares

Then, the diagnosis system receives two set points from the database: the minimum and maximum. If the value of the differentiated graph goes beyond of set points, the algorithm detects the initial and final positions of the entry and exit in the specified range of the signal level. During the received period of time is determined maximum and minimum signal level according to the cutting-in or cutting-out of the tool. The obtained values of the maximum and minimum suggests that the group of digital signal values for the period of time $\Delta t_{a\_max} - \Delta t_{a\_min}$ extreme values are in various states of the cutting tool [5,8].

The working period of the machining approximated by least-squares method and the general equation of a straight line is determined. Using of the slope of this straight line and the processing time of one part, remaining life of the cutting tool is determined.

When the level exceeds the limit value, the command a tool change is made. If the forecast for the next part is higher than a predetermined level adjustment mode, made a correction processing mode (Fig. 6).

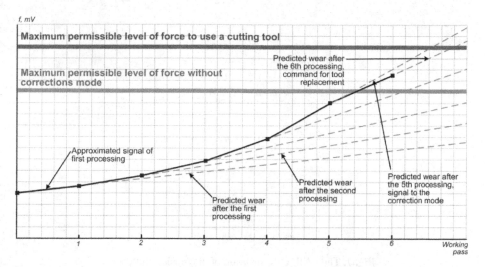

Fig. 6. Forecast of the state of the cutting tool using the approximate signal processing work periods

The developed algorithm of signal processing realizes the possibility of determining the current state of the cutting tool and forecast its resistance life in the future.

5   Software Implementation of Solutions for the Diagnosis of Tool Wear and Forecast it's Remaining Life

The variety of process equipment, types and modes of processing, types of cutting tools and materials of workpieces involves the development of a universal software solution with open architecture that allows the flexibility to modify the data processing algorithm, and integrate own algorithms. The proposed solution provides independence of external devices and data processing algorithms [2,4,9-11].

Software implementation of solutions for diagnosis and forecast of the resistance life of the cutting tool requires the determination of the basic modules and their interactions with each other. Develops solutions is offered to divide into four main modules: components manager, input module, the main diagnostic module, and output module. The UML class diagram of solution is presented in Fig. 7.

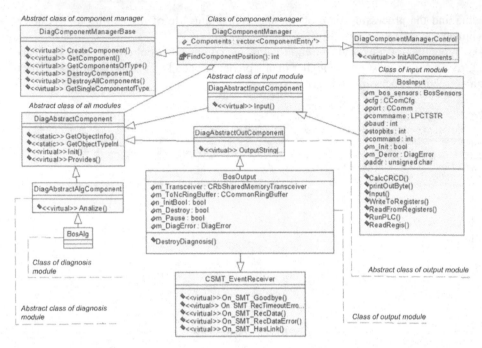

Fig. 7. The UML class diagram of diagnosis solution

Components manager is designed is intended for initialization of all modules solutions, and is configured by modifying the XML file. It allows to reconfigure the solution without changing the program code and recompiling the system.

Input module implements the interaction with external devices and produces initial processing of the received data. To facilitate communication system diagnostic modules produced cast data from the signal processing unit, to an internal format solutions.

Diagnostic module allows to process received from the input module information by diagnostic algorithms. Configure the diagnostic algorithm is also stored in the configuration XML-file, which allows to flexibly reconfigure the system without changing its programming code.

Output module implements the transfer of the core information and commands, received from the diagnostic module. In addition, it's able to receive data and instructions from the kernel the control system. This is required to be able to control the system through a terminal diagnosis of CNC system, as well as to provide data on the current operation of the system.

Diagnostic solutions can be implemented as integrated in the CNC system solution and as the external solution. Integrated solution involves the processing information of algorithms of diagnosing and forecast directly in the control system, by integrating into it all the software components. External solution provides independence from the CNC system [3].

The integration process involves the separation of the terminal part (machine time), and real-time component that implements the diagnostic algorithms and adaptive control of the machining process by interaction with the core of the system. Computing module of diagnostic solution operates on the same computer as the kernel of the CNC system and communicates with it using shared memory. The solution allows to transfer to the kernel of CNC system all the commands: processing modes correction, the tool replacement or emergency stop processing [7,11,12].

In the terminal part of the CNC system "AxiOMA Control" was implemented specialized diagnostic mode that allows to visualize the diagnostic process operator to make management of diagnostic solutions, as well as display the main parameters of the treatment process (Fig. 8).

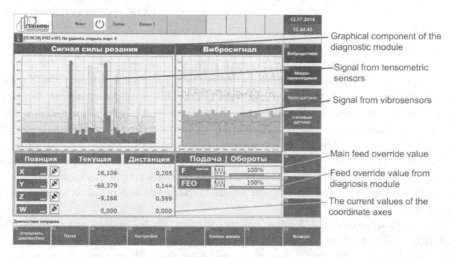

Fig. 8. The interface of terminal part of the diagnostic mode in the CNC system

The main terminal screen diagnostic solutions include visualization module signals coming from the cutting zone from different types of sensors, visualization component current command values of the coordinate axes, as well as a component to display the feed override / spindle speed override. The correction may be standard, and additional calculated in integrated in the core module according to the diagnostic algorithm [12-14].

6   Practical Tests of the Diagnosis Module Integrated into the CNC System

For justification of the practical importance of the developed solution, a series of tests, which included machining by pass-through cutter on the modernized lathe SA-700 (manufactured by Machine Tool "SASTA"). The machine is equipped with a complete control system: CNC system AxiOMA Control, the servodrives, the programmable-logic controller of electrics (PLC) and the developed in MSTU "STANKIN" data collection device (Fig. 9).

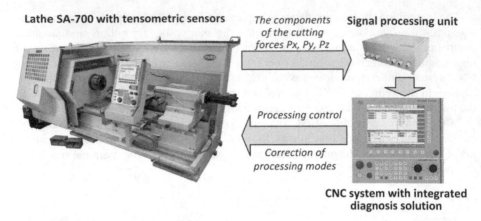

Fig. 9. Modernized lathe SA-700 with a diagnosis module integrated into the CNC system

Tensometric sensors were installed on the machine tool holder and that allow reading information about cutting forces in three axes. This information allows to carry out monitoring of the tool with high probability. For correct determination of the cutting force components is realized calibration of the sensors and the compilation of tables for conversion from the sensor readings to values of the forces [3,7,15].

Experiments consisted in serial processing of party of identical preparations until the message is displayed on the critical condition of tool wear. In this case, to demonstrate the efficiency of the system was made 12 passes to the full development of a resource tool (Fig. 10).

The graph shows that the loading which influence on the cutting tool is increased with time and indicating that the increasing tool wear. All the data were processed using the described diagnostic algorithm, and after the 12th pass was diagnosed the high level of tool wear.

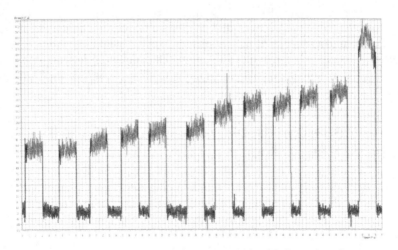

Fig. 10. Data from tensometric sensors during turning processing

To check the correctness of the tool wear definition after each pass measurement of tool wear was made by optical method and made the tables which confirmed association between wear size of a cutting tool and the received data from the sensors representing the making the cutting force.

Performed tests have clearly shown the accuracy of previously developed diagnostic algorithms and algorithms for forecast of the remaining tool life. The tests, which includes 12 passes, demonstrated that the operability of system and the predicted critical tool wear at the last pass [5,7,16].

7   Conclusions

The proposed solution for the real-time diagnosis and forecast of the remaining life of the cutting tool improves the dimensional precision of the machined blank and the final surface quality, with significant reduction in the rejection rate at quality control. The task of developing solutions for the real-time diagnosis and control of the cutting tool, and its integration to the CNC is solved. The chosen architecture is open and allows to expand the system to and integrate into it new algorithms for diagnosis and predicate of resistance tool life [7].

Integration of solution for cutting tool diagnosis in the CNC system "AxiOMA Control" extends its functionality and allows continuous monitoring the tool condition, predicting its resistance life, correction processing in real-time as well as the visualize diagnostic processes for operator.

Acknowledgements. This research was supported by the Ministry of Education and Science of the Russian Federation as a public program in the sphere of scientific activity and Russian Federal Program of Supporting Leading Scientific Schools (grant NSh-3890.2014.9).

References

1. Grigoriev, S.N., Martinov, G.M.: Scalable open cross-platform kernel of PCNC system for multi-axis machine tool. Procedia CIRP **1**, 238–243 (2012)
2. Martinov, G.M., Ljubimov, A.B., Grigoriev, A.S., Martinova, L.I.: Multifunction numerical control solution for hybrid mechanic and laser machine tool. Procedia CIRP **1**, 277–281 (2012)
3. Kozochkin, M.P., Sabirov, F.S.: Realtime diagnostics in metalworking, Vestn. MGTU Stankin **3**, 14–18 (2008)
4. Grigoriev, S.N., Martinov, G.M.: Decentralized CNC automation system for large machine tools. In: Proc. of COMA 13, International Conference on Competitive Manufacturing, Stellenbosch (South Africa), pp. 295–300 (2013). ISBN: 978-0-7972-1405-7
5. Nikishechkin, P.A., Grigoriev, A.S.: Practical aspects of the development of the module diagnosis and monitoring of cutting tools in the CNC. Vest. MGTU Stankin **4**, 65–70 (2013)
6. Martinov, G.M., Ljubimov, A.B., Martinova, L.I., Grigoriev, A.S.: Remote machine tool control and diagnostic based on web technologies. In: Proc. of COMA 13, International Conference on Competitive Manufacturing, Stellenbosch (South Africa), pp. 351–356 (2013). ISBN: 978-0-7972-1405-7

7. Grigoriev, S.N., Martinov, G.M.: Research and Development of a Cross-platform CNC Kernel for Multi-axis Machine Tool. Procedia CIRP **14**, 517–522 (2014)
8. Zoriktuev, V.T.S., Nikitin, Y.A., Sidorov, A.S.: Mechatronic machine tool systems. Russ Eng. Res. **1**, 69–73 (2008)
9. Martinov, G.M., Obuhov, A.I., Martinova, L.I., Grigoriev, A.S.: An Approach to Building Specialized CNC Systems for Non-traditional Processes. Procedia CIRP **14**, 511–516 (2014)
10. Timofeev, VYu., Zaitsev, A.A., Krutov, A.V.: Model of a diagnostic unit for a metalcutting tool based on thermosemf signals. Vestn. Voronezhsk. Gos. Tekhn. Univ. **5**(5), 42–45 (2009)
11. Martinova, L.I., Grigoryev, A.S., Sokolov, S.V.: Diagnostics and forecasting of cutting tool wear at CNC machines. Automation and Remote Control **73**(4), 742–749 (2012)
12. Kozochkin, M.P., Kochinev, N.A., Sabirov, F.S.: Diagnostics and monitoring of complex technological processes by means of vibroacoustic signals. Izmerit. Tekhn. **7**, 30–34 (2006)
13. Vereshchaka, A.S., Vereshchaka, A.A.: Functional coatings for cutting tools. Uprochn. Tekhnol. Pokryt. **6**, 28–37 (2010)
14. Martinov, G.M., Trofimov, E.S.: Modular configuration and structure of applied diagnostic applications in control systems. Prib. Sist., Upravl., Kontrol', Diagn. **7**, 44–50 (2008)
15. Martinov, G.M., Kozak, N.V., Nezhmetdinov, R.A., Pushkov, R.L.: Design principle for a distributed numerical control system with open modular architecture. Vestn. MGTU Stankin 4(12), 116–122 (2010)
16. Nezhmetdinov, R.A., Sokolov, S.V., Obukhov, A.I., Grigor'ev, A.S.: Extending the functional capabilities of NC systems for control over mechano-laser processing. Automation and Remote Control **75**(5), 945–952 (2014)

Classification

Wavelet Domain Digital Watermarking Algorithm Based on Threshold Classification

Zhiyun Chen[1], Ya Chen[1(✉)], Wenxin Hu[1], and Dongming Qian[2(✉)]

[1] Computing Center, East China Normal University, Shanghai 200062, China
{chenzhy,wxhu}@cc.ecnu.edu.cn, cy1201090128@126.com
[2] Shanghai Engineering Research Center of Digital Education Equipment,
East China Normal University, Shanghai 200062, China
dmqian@admin.ecnu.edu.cn

Abstract. This paper is to analyze the complexity of the images for the robustness of the watermark, and to propose a digital watermarking algorithm based on threshold classification in the wavelet domain combined with iterative threshold method. In this algorithm, the original image is divided into different parts of blocks. Then parts of the blocks are selected to be embedded with watermark respectively according to the image entropy. The watermark embedding strength for each block is set according to the image entropy. Optimal threshold of low-frequency sub-band from the two-stage decomposition is to be gained one by one with the iterative threshold method. With the optimal threshold derived, the DWT coefficient of the low-frequency sub-bands is to be classified. According to the result of the classification, different methods are used to overlay the watermark signal respectively. The experimental results have shown that, the algorithm has good imperceptibility and robustness to some common attacks.

Keywords: Threshold · Texture complexity · Discrete Wavelet Transform (DWT)

1 Introduction

Digital watermarking technology is one of the methods for protecting the copyright of digital products. It is also an important research direction of information hiding and multimedia information security technology [1]. The existing digital watermarking methods can be divided into the spatial domain and transform domain. The spatial domain methods calculate fast, but lack of robustness. Therefore, the transform domain watermarking technology gains more and more attention, such as Discrete Cosine Transform (DCT) [2], Discrete Fourier Transform (DFT) [3], Discrete Wavelet Transform (DWT) [4] and some other improved algorithms [5].Among all these methods in transform domain, Discrete Wavelet Transform gains more and more attention because of its high resolution characteristics.

In some existing algorithms, the watermark is embedded in accordance with the order of the DWT coefficients. These algorithms, for lacking of consideration of the characteristics of the DWT coefficients, may cause considerable change of the amplitude of the DWT coefficients of the embedded sub-band. In this paper, optimal threshold is to

© Springer International Publishing Switzerland 2015
Y. Tan et al. (Eds.): ICSI-CCI 2015, Part III, LNCS 9142, pp. 129–136, 2015.
DOI: 10.1007/978-3-319-20469-7_15

be gained by using the iterative threshold method, and the DWT coefficients are divided into two categories by using the optimal threshold. This algorithm has good robustness and imperceptibility.

2   Threshold Classification

2.1   Algorithm in DWT Domain

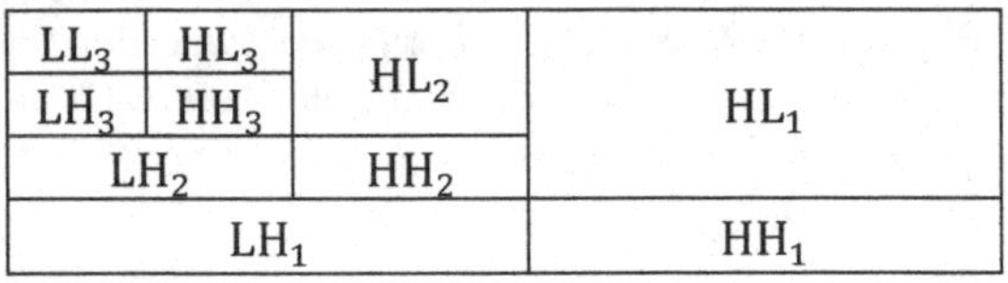

LL$_3$	HL$_3$	HL$_2$	HL$_1$
LH$_3$	HH$_3$		
LH$_2$		HH$_2$	
LH$_1$			HH$_1$

Fig. 1. Three-level wavelet decomposition of an image

Figure 1 is a schematic diagram which has finished a three-level wavelet decomposition of an image. One dimensional Discrete Wavelet Transform is used to transform rows and columns of an image in DWT algorithm. Each level of the wavelet decomposition is always to divide the low-frequency data from higher level into two categories: multi-high frequency sub-bands and one low-frequency sub-band. After the third-level wavelet decomposition, the high frequency sub-bands of the digital image are HH$_k$, LH$_k$ and HL$_k$(k=1, 2, 3) in sequence. The low-frequency sub-band remaining is the LL$_3$ band [6].

2.2   Classification of the DWT Coefficients

The essence of embedding watermark is using a watermark signal to modify the DWT coefficients of an original image. The watermark embedded in accordance with the order of the DWT coefficients may cause the considerable change of the amplitude of the DWT coefficients of the embedded sub-band. If the DWT coefficients of embedded sub-band can be divided into two categories by using an adaptive threshold, DWT coefficients in the same category will possess similar characteristics. According to the result of the classification, non-adaptive addition and subtraction to overlay the watermark signal is used respectively.

The iterative threshold method is an efficient method to get threshold. Iterative method can find the optimal threshold adaptively. The process is as follows. The average value of DWT coefficients is set as the initial threshold known as T_0. Then the DWT coefficients are divided into two parts by using T_0. One part called T_A which is less than T_0, and another is called T_B. The average values of the two parts are calculated respectively. The average value of T_A and T_B are chosen as the new threshold called T_1. Going on in this way until T_K is convergent. Then the T_K is chosen as the optimum threshold.

3   Image Complexity Analysis

Image complexity can be described from different perspectives. Gray histogram can reflect the frequency distribution of the images grayscale well, but it can't reflect the spatial distribution of the images grayscale. Gray-Level Co-occurrence Matrix (GLCM) reflects efficiently the spatial distribution of the images grayscale and texture with the statistical analysis method [7]. The image entropy mainly describes the texture complexity. Definition of image entropy is given as follow [8]:

$$S = -\sum_{i=0}^{255}\sum_{j=0}^{255} P(i,j) \log P(i,j) \tag{1}$$

Wherein, $P(i,j)$ is the value of GLCM P at the point(i,j).

Entropy is a measure of the amount of image information. Texture information is also the image information [9]. The more texture information the image contains, the greater the image entropy is. So that blocks which have more texture information are chosen to be embedded with the watermark.

4   The Watermark Embedding and Extraction Scheme

4.1   The Preprocessing of the Watermark Image

Before embedded into a carrier image, the watermark should be scrambled to be a messy image. The research object of this paper is two-dimensional images, so the two-dimensional Arnold transformation is adopted, and is defined as [10]:

$$\begin{pmatrix} x' \\ y' \end{pmatrix} = \begin{pmatrix} 1 & 1 \\ 1 & 2 \end{pmatrix} \begin{pmatrix} x \\ y \end{pmatrix} (mod\ N) \tag{2}$$

Wherein,$(x,\ y)$is the original image pixel, and$(x',\ y')$is the scrambled image pixels, N is the order of the watermark image.

4.2   Block Wavelet Transform

In this paper, the original image is divided into sixty-four blocks. Blocks dealt with Discrete Wavelet Transform can be divided into low-frequency and high-frequency sub-bands. Cox [11] et al. proposed that the watermark should be embedded into the low-frequency sub-band because the low-frequency sub-bands accumulate a lot of energy. They are not easy to be changed in signal processing, compared to the high frequency sub-bands. So the watermark embedded into the low-frequency sub-band will have stronger robustness. Therefore, the watermark set into the low-frequency sub-band after two-level decomposition is selected in this paper.

4.3   The Watermark Embedding Process

Step1: The size of the original image is 8M×8M. It is divided into sixty-four blocks in the size of M×M each. The entropy of each block is calculated for the purpose of selecting four blocks which have larger entropy from the upper left corner, the lower left corner, upper right corner and the lower right corner in sequence. Sixteen blocks are selected totally. Embedding strengths for the selected blocks are set suitably according to their entropies. The larger the entropies are, the larger the embedding strengths are set. The result is recorded in a matrix with the size of M×M, known as α.

Step2: The blocks selected are carried out for applying the two-level Discrete Wavelet Transform to get the low-frequency sub-bands for each. The optimal thresholds of each low-frequency sub-band are obtained by using the iterative threshold method. The DWT coefficient of each low-frequency sub-band is compared with their own threshold one by one. The comparison results were recorded in a matrix with the size of M×M, known as Mark. If the coefficient is larger than the threshold, the Mark is marked as 1, otherwise, Marked as 0.

Step3: The binary watermark image sized M×M is scrambled by Arnold transformation. The scrambling times can be kept as a key in the algorithm. Each low-frequency sub-band is carried out to form the low-frequency sub-band matrix with the size of M×M. According to the record of the comparison results in Mark matrix and the watermark embedding strength matrix α, the watermark can be embedded as the following formula.

$$X^r(i,j) = \begin{cases} X(i,j) - \alpha(i,j) * w(i,j), & Mark(i,j) = 1 \\ X(i,j) + \alpha(i,j) * w(i,j), & Mark(i,j) = 0 \end{cases} \tag{3}$$

Wherein, X is a DWT coefficient, X^r is the DWT coefficient of the watermarked image, the embedding strength matrix is α, the watermark is w.

Step4: Inverse wavelet transform is applied to those low-frequency sub-bands. And the watermarked image will be obtained by resuming their original position.

4.4   The Watermark Extraction Process

Step1: The watermarked image and the original image are divided equally into sixty-four blocks in the size of M×M each. The blocks selected are carried out to apply the Discrete Wavelet Transform respectively, getting the low-frequency sub-bands of the secondary decomposition one by one. Each low-frequency sub-band is saved to form a low-frequency sub-band matrix with the size of M×M.

Step2: According to the record of the comparison results in Mark and the watermark embedding strength matrix α, the watermark is extracted as following formula.

$$w^r(i,j = \begin{cases} (X(i,j) - X^r(i,j))/\alpha(i,j), & Mark(i,j) = 1 \\ (X^r(i,j) - X(i,j))/\alpha(i,j), & Mark(i,j) = 0 \end{cases} \tag{4}$$

The extracted watermark is w^r.

Step3: Arnold inverse transformation is applied to the extracted watermark to recover the watermark.

5   Experiments and Results Analysis

5.1   Imperceptibility and Robustness Assessment

In order to evaluate the quality of watermarked images objectively, two assessment indicators are generally used.

Peak Signal-to-noise Ratio (PSNR) is usually defined to measure the watermarked image. The larger PSNR is, the stronger imperceptibility is. PSNR is calculated as follow:

$$\begin{cases} \text{PSNR} = 10\log10\frac{255^2}{MSE} \\ \text{MSE} = \frac{1}{M*N}\sum_{i=1}^{M}\sum_{j=1}^{N}[\mathcal{F}_r(i,j) - \mathcal{F}(i,j)]^2 \end{cases} \tag{5}$$

Wherein, $\mathcal{F}_r$ is a watermarked image, $\mathcal{F}$ is an original image, M, N represents the image size. MSE is the Mean Square Error.

Normalized cross-correlation coefficient (NC) is usually used as the objective assessment of similarity between an extracted watermark and an original watermark [12]. The larger NC is, the more close to 1, the higher similarity is. NC can be calculated as follow:

$$NC = \frac{\sum_{i=1}^{M}\sum_{j=1}^{N}\mathcal{F}(i,j)\mathcal{F}_r(i,j)}{\sum_{i=1}^{M}\sum_{j=1}^{N}[\mathcal{F}(i,j)]^2} \tag{6}$$

Wherein, $\mathcal{F}_r$ is an extracted watermark, $\mathcal{F}$ is an original watermark.

5.2   Algorithm Simulation

Grayscale image Lena is adopted as the original image for the watermark embedded in this experiment. Its size is 256×256. The watermark with Chinese characters is adopted as a binary image, in size 32 × 32. The Chinese characters are "image processing". The experimental results are shown in Figure 2. The watermark applied 16 times Arnold transformation is displayed in Figure 2(d).

Visually, there is hardly any difference between the original watermark and extracted watermark without being attacked. The original image looks like the same as the watermarked image. The experimental result: PSNR=89.1481, NC=1.0000. When the PSNR is over 30, the watermarked image has got good imperceptibility.

5.3   Anti-Attack Experiment

To verify the robustness of the algorithm, attacks such as JPEG compression, salt & pepper noise, median filtering, rotation, stretching, and cropping are applied to the watermarked image.

(c) original
watermark

(d) scrambled
watermark

(e) extracted
watermark

(a) original image

(b) watermarked image

Fig. 2. The experimental results

Watermarked images through various attacks and extracted watermarks are displayed in Figure 3. Those attacks are JPEG compression with quality factor 80, salt & pepper noise with strength 0.01, 3×3 median filter, rotation with 2 degree, stretching ratio is1/2 and cropping ratio is 1/4. NC values are 0.9705, 0.9945, 0.9300, 0.8048, 0.8127 and 0.5755.

Table 1 and Table 2 display the values of PSNR and NC obtained by JPEG compression and salt & pepper noise of different degrees. Table 3 shows the value of PSNR and NC obtained by median filtering of different template.

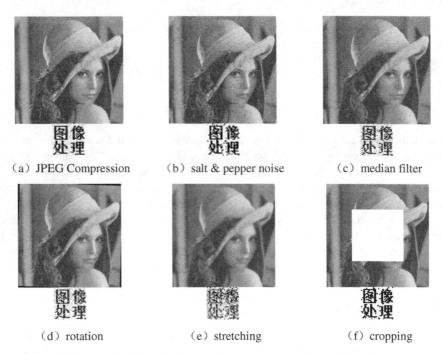

(a) JPEG Compression   (b) salt & pepper noise   (c) median filter

(d) rotation   (e) stretching   (f) cropping

Fig. 3. Watermarked image and extracted watermark against various attacks

Table 1. Records of robust experiment with JPEG attack

quality factor	80	60	40	20
PSNR	84.8865	81.5764	80.5472	78.8822
NC	0.9705	0.9615	0.9349	0.9284

Table 2. Records of robust experiment with salt & pepper noise

strength	0.01	0.02	0.03	0.05
PSNR	73.4638	70.3937	68.6322	66.3434
NC	0.9945	0.9611	0.9910	0.8721

Table 3. Records of robust experiment with median filter

template	PSNR	NC
[3,3]	80.4984	0.9300
[5.5]	77.5625	0.8669

It can be seen from Figure 3, the watermark can be well extracted from the watermarked image even being attacked variously. The extracted watermark can be identified with the human eyes. This indicates that the algorithm in this paper is robust to various attacks, especially to JPEG compression, salt & pepper noise, median filtering, rotation and stretching. Only being attacked by cropping, the value of NC obtained is smaller. But the extracted watermark has a good visual quality. This does not affect the plaintext meaning. Of course, that is still the work required to improve in the future.

It is shown that NC values are all above 0.86 in Table 1, Table 2 and Table 3. When NC is more than 0.7, it is generally considered that the extracted watermark is the embedded watermark [13]. So this is the convincible improvement that the algorithm in the paper has good robustness to the attacks of varying degrees, such as JEPG compression, salt & pepper noise, median filtering.

6   Conclusions

In this paper, texture complexity of images is analyzed and a digital watermarking algorithm based on threshold classification has been realized. The algorithm selects part of the blocks which have larger image entropies for embedding watermarks respectively. When the watermark is embedded, the algorithm divides the DWT coefficients into two categories by using the iterative threshold method, so as to embed watermark in different methods according to the characteristics of different categories of DWT coefficients, instead of in the order of DWT coefficients. In this paper, in order to improve the local embedding strength of watermark components as far as possible, the watermark embedding strength is controlled by using the image entropy. In this way, both watermark robustness and invisibility can be achieved under different attack schemes. The experimental results show that the extracted watermark has a good visual quality and the algorithm can resist different attack schemes, especially common attack from network, such as JEPG compression, salt & pepper noise and median filter. How to realize the blind watermark is suggested for further studies.

Acknowledgements. Thanks for the fund program of this paper: Shanghai science and technology commission project: 《Shanghai Engineering Research Center of Digital Education Equipment》 (44505300) in 2013 and Intel International cooperation program: 《eSchoolbag Std. and application Research》 in 2011.

References

1. Zhang, X.H., Wei, P.C.: Study of digital watermarking based on information theory. Computer Science **36**(3), 248–255 (2009)
2. Kii, H., Onishi, J., Ozawa, S.: The digital watermarking method by using both patchwork and DCT. In: IEEE International Conference on Multimedia Computing and Systems, pp. 895–899 (1999)
3. Kang, X.G., Huang, J.W., Shi, Y.Q., et al.: A DWT-DFT composite watermarking scheme robust to both affine transform and JPEG compression. IEEE Transactions on Circuits and Systems for Video Technology **13**(8), 776–786 (2003)
4. Ge, W.W., Cui, Z.M., Wu, J.: Robust gray level watermarking method based on wavelet. Journal of Information and Computational Science **5**(1), 391–396 (2008)
5. Bi, N., Sun, Q.Y., Huang, D.R., et al.: Robust image watermarking based on multiband wavelets and empirical mode decomposition. IEEE Transactions on Image Processing **16**(8), 1956–1966 (2007)
6. Ye, C.: Digital watermarking algorithm research based on discrete wavelet transform (in Chinese). Zhejiang University, Hangzhou (2012)
7. Hu, X.G., Wang, Y.: Digital watermarking algorithm based on image complexity (in Chinese). Journal of Chinese Computer Systems **33**(5), 1149–1152 (2012)
8. Gao, Z.Y., Yang, X.M., Gong, J.M., et al.: Research on image complexity description methods. Journal of Image and Graphics **15**(1), 129–135 (2010)
9. Xiao, Z.J., Tian, S.J., Chen, H.: Digital watermarking algorithm in wavelet domain based on image texture complexity (in Chinese). Computer Engineering **40**(6), 85–88 (2014)
10. Sun, X.D., Lu, L.: Research on application of Arnold transformation in digital image watermarking (in Chinese). Information Technology **10**(8), 129–132 (2006)
11. Cox, I.J., Kilian, J., Leighton, F.T., et al.: Secure spread spectrum watermarking for multimedia. IEEE Transactions on Image Processing **6**(12), 1673–1687 (1997)
12. Ye, C., Shen, Y.Q., Li, H., et al.: Digital watermarking algorithm based on human visual features (HVS) and discrete wavelet transform (DWT) (in Chinese). Journal of Zhejiang University: science edition **40**(2), 152–155 (2013)
13. Zhang, J.H.: Research on digital watermarking technology based on the wavelet transform domain (in Chinese). Tianjin University of Technology, Tianjin (2013)

A New Disagreement Measure for Characterization of Classification Problems

Yulia Ledeneva[1]([✉]), René Arnulfo García-Hernández[1], and Alexander Gelbukh[2]

[1] Autonomous University of the State of Mexico, Instituto Literario 100,
Toluca 50000, State of Mexico, Mexico
yledeneva@yahoo.com, renearnulfo@hotmail.com
[2] Natural Language and Text Processing Laboratory,
Center for Computing Research National Polytechnic Institute, DF 07738, Mexico
http://www.gelbukh.com/

Abstract. Robert P.W. Duin, Elzbieta Pekalska and David M.J. Tax proposed the characterization of classification problems by classifier disagreement. They showed that it is possible to use a standard set of supervised classification problems for constructing a rule that allows deciding about the similarity of new problems to the existing ones. The classifier disagreement could be used to group classification problems in a way which could help to select the appropriate tools for solving new problems. Duin et al proposed a dissimilarity measure between two problems taking into account only the full disagreement matrices. They used a measure of the disagreement based on the coincidence of the classifier output however the correctness was not considered. In this work, we propose a new measure of disagreement which takes into account the correctness of classification result. To calculate the disagreement each object is analyzed to verify if it was classified correctly or incorrectly by the classifiers. We use this new disagreement measure to calculate the dissimilarity between two problems. Some experiments were done and the results were compared against Duin's et al results.

1 Introduction

The characterization of classification problems by classifier disagreement was proposed in [1]. The authors of the paper used the problem characteristics to find the appropriate tools for solving it. They calculated the differences in classification results between pairs of individual classifiers and the results were defined as a measure of disagreement.

The research of the authors of the paper showed that it is possible to use a standard set of supervised classification problems for constructing a tool that allows deciding about the similarity of new problems to the existing ones. The performance of these tools gives a first indication about how to solve the problem, as they tell whether the chosen classifiers are appropriate.

Work done under partial support of Mexican Government (CONACyT).

Y. Tan et al. (Eds.): ICSI-CCI 2015, Part III, LNCS 9142, pp. 137–144, 2015.
DOI: 10.1007/978-3-319-20469-7_16

The disagreement between a set of classifiers could be used to group classification problems in a way which could help to select the appropriate tools for solving new problems. In turn the disagreement patterns point towards different types of classification problems and indicate the usefulness of a classifier with respect to a set of classification problems and classifiers.

We propose a new measure of disagreement which takes into account the correctness of classification result. The disagreement will be calculated analyzing each object to verify if it was classified correctly or incorrectly by the classifiers. We use this new disagreement measure to calculate the dissimilarity between two problems. We describe some experiments and the results are compared with previous results of Duin's et al.

The paper is organized as follows. In section 2, the set of classifiers, problems and the measure of disagreement proposed by Duin's et al are described. A new disagreement measure for characterization of classification problems is proposed in section 3. Some experiments are presented in section 4. Conclusions and future work are presented in section 5.

2   Related Work

The disagreement measures the difference, in classification results, between two standard classifiers. In [1] the way to calculate the disagreement between two classifiers C_i and C_k trained on a classification problem P_j is the disagreement $d(C_i, C_k)$ that is defined as follows:

$$d_j(C_i, C_k) = \frac{\sum_{x \in P} d(C_i, C_k, x)}{|P|} \tag{1}$$

where

$$d(C_i, C_k, x) = \begin{cases} 0, \text{ if the object } x \text{ was classified} \\ \quad \text{equally by } C_i \text{ and } C_k \,; \\ 1, \text{ if the object } x \text{ was classified} \\ \quad \text{differently.} \end{cases} \tag{2}$$

M classifiers constitute an $M \times M$ disagreement matrix for problem P_j, with elements

$$(m, n) = d_j(C_m, C_n) \tag{3}$$

They used the following set of 13 classifiers:

- NMC: the nearest mean classifier;
- Fisher: Fisher's linear discriminant;

- UNormalBC: the Bayes classifier assuming uncorrelated normal densities;
- NormalBC: the Bayes classifier based on 10-bin histograms per feature;
- *NaiveBC: the naive Bayes classifier based on 10-bin histograms per feature;*
- ParzenC: the Parzen classifier, using a leave-one-out optimization of the smoothing parameter;
- 1-NN: the one nearest neighbor rule;
- k-NN: the k-nearest neighbor rule. The value of k is optimised for the leave-one-out classification error;
- LogC: the logistic classifier;
- SVC-1: the support vector classifier using a linear kernel, and with regularization parameter c=1;
- SVC-2: the support vector classifier using a quadratic kernel, and with regularization parameter c=1;
- LM-NeurC: a neural net with one hidden layer with 5 neurons, trained by the Levenberg-Marquart rule;
- CART: a CART decision tree, maximizing the purity and using early pruning.

The set of classifiers was selected arbitrary by the authors [1]. Other classifiers can be added or eliminated from the list. It is recommended to analyze the set of problems which will be classified and select the appropriate set of classifiers. They indicated another way to assure the choice of the classifiers is revalidate them after distinguishing the difference between the problems.

Duin et al used a set of 18 2-class problems. For each of the problems they estimated the disagreement matrix. And then to compare the full disagreement matrices, they used a dissimilarity measure. To obtained the matrix they compared the problems P_r and P_s using the following formula

$$D(P_r, P_s) = \sum_{m,n} \left| d_r(m,n) - d_s(m,n) \right|$$ (4)

The measure of the disagreement in [1] is based on the coincidence of the classifier output however the correctness was not considered.

3   Measure of Disagreement

In this paper we propose an alternative way to measure the difference between two classifiers C_i and C_k trained on a classification problem P_j, the disagreement $\partial(C_i, C_k)$. This disagreement assigns to an object different values. These values take into account if the object were classified correctly or incorrectly, different or equal by two classifiers. So the values assigned to the object can have the following values:

$$\partial(C_i, C_k, x) = \begin{cases} \text{1, if the object } x \text{ was classified equally} & (5) \\ \quad \text{and correctly;} \\ \text{2, if the object } x \text{ was classified equally} \\ \quad \text{and incorrectly;} \\ \text{3, if the object } x \text{ was classified} \\ \quad \text{differently and incorrectly by one} \\ \quad \text{classifier, and correctly by another;} \\ \text{4, if the object } x \text{ was classified} \\ \quad \text{differently and incorrectly by both} \\ \quad \text{classifiers.} \end{cases}$$

Then the measure of disagreement can be represented as follow:

$$\partial j(C_i, C_k) = \frac{\sum_{x \in P} \partial(C_i, C_k, x)}{|P|} \qquad (6)$$

$\partial_j(C_i, C_k)$ returns the values for the object x according to the classifier (C_i, C_k). In this case the dissimilarity measure is calculated as

$$D(P_r, P_s) = \sum_{m,n} |\partial_r(m, n) - \partial_s(m, n)| \qquad (7)$$

4   Experimental Results

In this section we present some experiments with the Duin's et al disagreement measure and the one proposed in this paper.

4.1   The Set of Classifiers

We used the next set of 11 classifiers:
- NMC: the nearest mean classifier;
- LDC: normal densities linear classifier;
- QDC: normal densities based quadratic classifier;
- Fisher: Fisher's linear discriminant;
- NaiveBC: the naive Bayes classifier based on 10-bin histograms per feature;
- ParzenC: the Parzen classifier, using a leave-one-out optimization of the smoothing parameter;
- k-NN: the k-nearest neighbor rule. The value of k is optimised for the leave-one-out classification error;
- LogC: the logistic classifier;
- SVC: the support vector classifier;

–   BpxnC: train neural network classifier by back-propagation;
–   TreeC: construct binary decision tree classifier.

4.2   Problems for Evaluation

A set of seven 2-class problems were used in this work. These datasets were taken from the UCI repository [2].

Table 1. Description of the problems

Name of Dataset	#features	$K_1 + K_2$
Biomed	5	127 + 67
Sonar	60	97+ 111
Diabetes	8	500 + 268
Auto-mpg	6	229 + 169
Ionosphere	34	225 + 126
Liver	6	145 + 200
Breast	9	444 + 239

For each of the problems P_j ($j = 1$,..., N; N is the size of the set of problems) the disagreement matrix D_j was estimated.

For the shorthand notation we will use the following symbols for the problems:
P1 – Biomed,
P2 – Sonar,
P3 – Diabetes,
P4 – Auto-mpg,
P5 – Ionosphere,
P6 – Liver,
P7 – Breast.

4.3   Experimental Results

In a first series of experiments we estimated the disagreement matrices for each problem. And then we calculated the problem dissimilarity matrix using the new disagreement measure and the measure of disagreement which do not take into account the correctness of classification.

In tables 2 and 3 we show the problem dissimilarity matrix, in others words each value represents the dissimilarity measure between two problems. In the table 2 the problem dissimilarity matrix calculated without taking into account the correctness of classification result and in the table 3 using the proposed measure of disagreement, are shown.

According to the results presented in the tables 2 and 3 we can see that the dissimilarity changes when we take into account the correctness of classification result. For example, the dissimilarity between P1 and P7 is 0.12 that is P1 y P7 were the less dissimilar objects. But when we use the correctness of classification result, the dissimilarity changes to 0.69 that is P1 and P7 are two of the most dissimilar objects. This difference is very significant. Another way to illustrate the dissimilarity is given in figure 1, 2.

Table 2. Problem dissimilarity matrix without taking into account the correctness of classification

	P1	P2	P3	P4	P5	P6	P7
P1	0	0.47	0.44	0.12	0.38	0.91	0.12
P2	0.47	0	0.34	0.5	0.44	0.72	0.56
P3	0.44	0.34	0	0.38	0.34	0.44	0.53
P4	0.12	0.5	0.38	0	0.34	0.84	0.19
P5	0.38	0.44	0.34	0.34	0	0.62	0.47
P6	0.91	0.72	0.44	0.84	0.62	0	1
P7	0.12	0.56	0.53	0.19	0.47	1	0

Table 3. Problem dissimilarity matrix taking into account the error of classification

	P1	P2	P3	P4	P5	P6	P7
P1	0	0.51	0.28	0.54	0.18	0.35	0.69
P2	0.51	0	0.28	0.3	0.62	0.74	0.36
P3	0.28	0.28	0	0.3	0.41	0.57	0.44
P4	0.54	0.3	0.3	0	0.69	0.86	0.15
P5	0.18	0.62	0.41	0.69	0	0.22	0.84
P6	0.35	0.74	0.57	0.86	0.22	0	1
P7	0.69	0.36	0.44	0.15	0.84	1	0

Figure 1 shows dissimilarity between the problems calculated without taking into account the error of classification and in figure 2 using a new measure of disagreement. In the scale of colors we can observe what dissimilarity has each color and in the graphic it is possible note the regions of dissimilarity.

To group the problems by the dissimilarity we used the method of β_0 connected sets. This method allows grouping the problems which are less dissimilar. This method compare two objects x_1 and x_2, and if they are similar (that means less dissimilar), the method will be locate them in the same group. In tables 4 y 5 we show the problems grouped using this method for a new measure of disagreement and the disagreement which does not take into account the correctness of classification result.

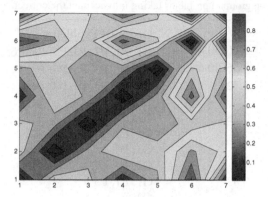

Fig. 1. Dissimilarity between the problems without taking into account the correctness of classification result

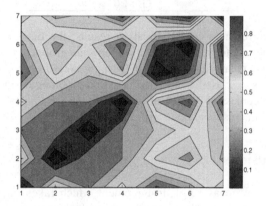

Fig. 2. Dissimilarity between the problems taking into account the correctness of classification result

Table 4. The grouped problems without taking into account the correctness of classification

β_0	Grouped Problems
0.16□0.53	G1: P1, P4, P7, P5, P3, P2, P6
0.56	G1: P1, P3, P4, P5, P7, P2, P6
0.62	G1: P1, P4, P5, P7, P3, P2
	G2: P6
0.66	G1: P1, P4, P7, P5, P3, P2
	G2: P6
0.81□0.88	G1: P1, P4, P7
	G2: P2
	G3: P3
	G4: P5
	G5: P6

Table 5. The grouped problems taking into account the correctness of classification

β_0	Grouped Problems
0.16□0.49	G1: P1, P2, P3, P4, P5, P6, P7
0.56□0.65	G1: P1, P3, P5, P6, P2, P4, P7
0.72	G1: P1, P3, P5, P2, P6
	G2: P4, P7
0.78	G1: P1, P5, P6
	G2: P2
	G3: P3
	G4: P4, P7
0.82	G1: P1, P5
	G2: P2
	G3: P3
	G4: P4, P7
	G5: P6

Using the correctness of classification results, the dissimilarity of the problems changes, and these determine how the problems are grouped.

The experimental results show that the new measure proposed in this work is a new way to measure classifier disagreement and problem dissimilarity which take into account the correctness of classification results.

All the experiments were done with PRTools4 [3].

5   Conclusions

We defined a new measure of disagreement which takes into account the error of classification. To calculate the disagreement each object is analyzed to verify if it was classified correctly or incorrectly by the classifiers. We use this new disagreement measure to calculate the dissimilarity between two problems. Some experiments were done and the results were compared against Duin's et al results.

However, this new disagreement measure could be used to search for a suitable tool for solving a problem and group classification problems in a consistent way.

References

1. Duin, R.P.W., Pekalska, E., Tax, David M.J.: The Characterization of Classification Problems by Classifier Disagreement. In: Proceedings of the ICPR 2004, in CD (2004)
2. Blake, C.L., Merz, C.J., UCI Repository of machine learning databases. Univ. of California, Irvine, CA (1998). http://www.ics.uci.edu/~mlearn/MLRe-pository.html
3. Duin, R.P.W., Juszczak, P., Paclik, P., Pekalska, E., Ridder, D., Tax, D.M.J., PRTools4, a Matlab toolbox for Pattern Recognition. Delft Univ. of Techn. (2004). http://prtools.org

Tracking Analysis

User Intention Mining: A Non-intrusive Approach to Track User Activities for Web Application

Piyush Yadav[✉], Kejul Kalyani, and Ravi Mahamuni

Tata Research Development and Design Centre (TRDDC),
Tata Consultancy Services Limited, Pune, India
{piyush.yadav1,kejul.kalyani,ravi.mahamuni}@tcs.com

Abstract. Monitoring user interaction with web applications is of vital importance as it helps in finding the user's cognitive behavior towards applications and also helps in analyzing various web metrics. Understanding activity data and finding user insights is next challenge to the web world. User intention can be understood if the user activity logs are associated appropriately with application context. In current scenario, web applications are using crawlers, scrappers, bugs, bots etc. to track user activities and extract out relevant information. Thus, by modifying the source code, it is easy to track and maintain the logs of user activities. But, many organization use third party web applications, where changing the source code is out of reach, making it trickier to maintain the logs of user activities. In this paper, we present an innovative way of logging, processing and extracting out meaningful information by tracking user's activities on web application.

Keywords: Web Content Mining · User activity tracking · User monitoring · DOM Parsing · Visualization · Browser · Semantic Analysis · User intention

1 Introduction

In recent years, there has been an enormous advent in the web technologies leading to its wide spread adoption. Monitoring user activities is becoming more challenging with increase in web usage. In an enterprise, where multiple people access key applications, it becomes a big challenge to investigate and find out their malicious activities. Monitoring can also help in generating user specific profiles, considering their interest and search history.

Effective monitoring of user activities gives insights about many interesting information such as what exactly user did to perform various actions such as Signing Up, Searching a product, responding to surveys etc., i.e. deducing the user intention. User Intention can further be explained as the purpose behind the user's basic actions. Many models have been proposed to monitor user activity. Earlier only navigational data was captured, and as the web complexity grew, data was captured page wise and semantically.

In this paper, we propose a way to capture the semantics of user activities. Our solution is based on the notion that, semantics of the user actions are on the webpage

© Springer International Publishing Switzerland 2015
Y. Tan et al. (Eds.): ICSI-CCI 2015, Part III, LNCS 9142, pp. 147–154, 2015.
DOI: 10.1007/978-3-319-20469-7_17

itself. The semantics of the user activities are extracted out by parsing the DOM page which is then categorized logically. It then explains how these data can be used to find out the user intent.

1.1   Related Work

In recent years, there has been lot of study on user activity tracking. User interaction with the system has been studied for wide variety of investigative purposes such as usability metrics, website performance, debugging etc.

Various methods have been proposed to track user actions on website and these tracked data is used for calculating various metrics. One such method tracks user actions by using USA proxy which modifies the code by inserting javascript. It collects data about the mouse movements, keyboard inputs etc. [1][10][11][12] and then uses it to deduce navigational behavior and time based metrics. Later a different concept, accessibility-in-use was coined to differentiate real world data from laboratory studies [2]. Easily deployable tool was developed by modifying USA Proxy which captures longitudinal data unobtrusively. WebQuilt [6], a web logging and visualization tool, was introduced to show various user activity results such as the web pages people viewed, or the most common paths taken through the website for a given task or the optimal path for that task as designated by the designer. Letizia[7], another user agent tool that tracks user behavior and attempts to anticipate items of interest by doing concurrent, autonomous exploration of links from the user's current position. Later another method was proposed to study user behavior from a longitudinal perspective [3].

To understand the user behavior, user task models were defined and compared with actual user activity on the website. As explained in [5], user interactions were captured and analyzed, and the results were compared with the task model of the website supporting remote usability evaluation of websites. But, in all these studies, less amount of work was performed to understand the semantics of the captured user interaction.

Later, as logging user activities achieved maturity, advancement in associating the user activity with semantics was needed. A study on understanding of semantics of user interaction was performed, which associated the semantics using microformats [4]. Microformats are the simple codes embedded inside the HTML tags which tracks the user interaction and extracts out semantic information using the same. Major drawbacks of using it are that not every element on a website can be given microformats. Also, very less meta-data information can be extracted out from same. User intention mining was also done to predict user actions based on the past recorded data [8]. In contrast to all these studies, our approach not only tracks the user activities but also associates semantics which is used to find user intention.

2   Method: Determining User Intention

Our solution is based on the fact that, user interacts with webpage and user intentions can be deduced from the webpage itself. HTML is the basic building block of webpage and its tags can be broadly categorized into three components viz. interactive, decorative and structural as shown in Fig. 1. Interactive refers to those tags which are capable of taking inputs from the user or to invoke some action or change the state of the web-page. Examples of interactive tags are input, submit etc. Decorative refer to those tags which try to add meaning to the input tags such as label or span. These tags are used to justify the existence of the input tags. Structural tags are the ones which are used for laying out the HTML pages. They define the structure of the HTML pages.

To track user activities, we have extended browser so that whenever any page is loaded various listeners are attached to the elements present in that page. Event listeners such as tab out, mouse click and double click are attached only to interactive elements.

Appropriate semantics of the user interacted elements are extracted out in two steps i.e. semantics extraction and aggregation.

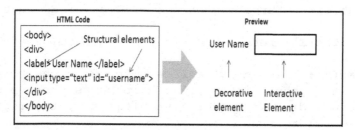

Fig. 1. HTML Components: Interactive, Decorative and Structural

2.1   Semantic Extraction

Semantics or meta-data of user activities can be determined from the webpage decorator elements itself. We aim to find the correct decorator tag by parsing the structural tags for the interactive component. Layout design on various webpages was studied. We found that the placing of the interactive component and the decorator occurs in few defined formats. Many visual forms were studied to find different ways a decorator element can be placed around an interactive element. Fig. 2 lists various visual forms where a decorator element is placed around interactive elements. If there is more than one way of decorator element being placed around an interactive element, then relevant decorator element is selected depending on its most occurring position.

To find the correct decorator element for the interactive element, DOM tree is parsed. Parsing the DOM tree to obtain the decorator element had lot of complexities,

○ Agree ○ Disagree	Search	Date ☐--☐--☐
Login	Visit : www.xyz.com	Select → Option 1 / Option 2
☐ Remember my credentials	Details: ○ Yes ○ No	🔍

Fig. 2. Few samples of decorative and interactive elements and their relative placement

as structural elements used to place the decorator and interactive element is implementation specific. A tree parsing algorithm was followed, wherein the starting point is interactive element.

Let E: Any DOM element
Let I: Interactive element
Let D: Decorative element
isDecorator(E): returns true, if element E is decorative otherwise return false
getPrevSibling(E): returns previous sibling of element E
getNextSibling(E): returns next sibling of element E
getParent(E): returns the parent element of E in the DOM tree
getLastChild(E): returns the last child element of E in the DOM tree
isRegular(E): returns false if element E is checkbox, radio etc. i.e. those interactive components whose probability of finding the decorative element to the right is high else return true
getDecorativeAttributeValue(E): returns the decorative attribute value of the element E. Decorative attribute value are the elements such as alt, HTML inner text etc.
getType(E): returns the type of the element E. HTML element types can be radio, checkbox, submit etc.
getNextChild(E): returns the child element one by one starting from first child element, every time when this function is called

```
getPreviousDecoratorElement(I)
BEGIN
Let E1 = I
Let E2 = getPrevSibling(E1)
A :
IF E2 IS NOT NULL
THEN
            IF isDecorator(E2)
            THEN
                      return E2
            ELSE
                      E6 = getDecoratorInSequence(I)
                      return E6
            END IF
ELSE
            E3 = getParent(E1)
            E1 = E3
            E2 = getPrevSibling(E3)
            GOTO LABEL A
END IF
END
```

```
getDecoratorInSequence(I)
BEGIN
        Let E1 = getParent(I)
        Let T1 = getType(I)
        Let E2 = getNextChild(E1)
        IF isDecorator(E2)
        THEN
                REPEAT
                            E3 = getNextChild(E1)
                            T2 = getType(E3)
                            IF T2 IS NOT EQUAL TO T1
                            THEN
                                      return "No decorator element found"
                            END IF
                UNTIL E3 IS NOT EQUAL TO I
                return E2
        END IF
END
```

Fig. 3. Algorithm to find previous decorator element for element I

If element I is regular, then its decorative element is present before its location in the DOM tree and vice versa. Fig. 3 shows algorithm to get the previous decorator element of I. To find decorative element if element I is not regular replace *getPreviousSibling* by *getNextSibling* in algorithm *getPreviousDecoratorElement* and *getNextChild* by *getLastChild at line 4* in *getDecoratorInSequence* .

After predicting the decorator element, next step Aggregation is performed to segregate various actions and group similar elements and find the final intent.

2.2   Aggregation

To gain insight into user activities, aggregation of the activity logs is performed. Thus, it is important to establish the relation among the data. Relation among the decorative elements can bind or categorize these elements in a group. To achieve this, we have extracted the parent decorator element of the decorator element. After extracting the parent elements, a tree is created to get the common parent decorator. Thus, the parent at the top links the data. Logged data can create more than one tree if user worked in more than one task on the same webpage. Fig. 4 shows the normal "Sign Up" page and the result tree which is created to establish relation among various decorator elements. As shown in Fig. 4, the decorator element for interactive elements textbox1, text-box2, textbox3, textbox4, combobox1, male, female are email id, password, re-enter password, secret question, city, gender respectively. Data is extracted out in the semantic extraction step. Then the association between this data is identified again by parsing the DOM tree and then association tree is created.

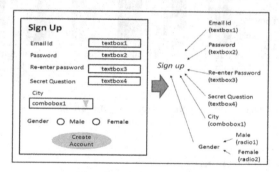

Fig. 4. Aggregated Data

3   Experimental Results

We conducted few experiments to find out the accuracy of the above mentioned algorithm and results are presented below.

3.1   Semantic Extraction Algorithm Results

Semantic Extraction algorithm was executed on various interactive elements to find accuracy in each category. Interactive elements were categorized as input, link, image, span, select and button. Algorithm was executed on more than 300 elements in each category spanning around more than 50 websites from various domains like banking, finance, travel, technology etc. having rich content. Fig. 5 showcases the accuracy in finding out the decorator element for various HTML tags. Consider input, link, span (also consists of header, font, strong, td, tr etc. tags), select and button tags accuracy was above 85 percent. Decorator element for an image achieved 76 percent as most of the times, the decorator attribute for an image is on the image itself and to find the text from the image is not the part of the algorithm.

Finding the correct decorator element has many challenges, as placing the structural element depends completely on the user's creativity. After analyzing the results we found that if the algorithm be further modified with the incremental learning mechanism, and UI guidelines are followed by websites, more accurate results can be achieved.

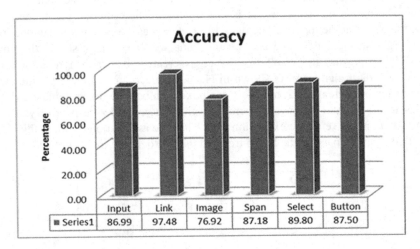

Fig. 5. Experimental Data showing percentage of accuracy achieved for various HTML interactive elements

3.2   Inferring User Actions

Fig. 6 shows an instance of user activity on a corporate "Facebook" account which is handled by multiple users. User activities were tracked for a day. Activity Score is no. of times user visits certain links or items. Similarly horizontal axis represents the labels of the interactive elements. It can be easily inferred that Password /Change Password link is used many times by User2.   Now, for an enterprise account this can be a malicious activity where that user is not allowed to change the password. It can also be inferred that   User1 is interested in GIS related groups.

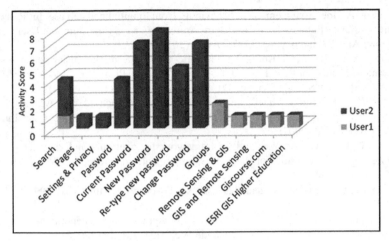

Fig. 6. Tracked user activities

4   Limitations

Since the repository of HTML elements is too big, so it is difficult to cover all the elements.We have focused on the most occurring HTML elements. The algorithm can become more robust by adding different elements. The experimental result in Fig. 5 shows that on an average, more than 87% percent of the time, we have found the correct decorator label for the element. Our algorithm is based on standard web practices which are being followed. If web pages are coded and designed poorly then it may happen that the results come up with wrong labels as every data is retrieved from DOM tree itself. We have tried to preserve the privacy by not tracking *labels* specified by user.

5   Conclusion and Future Work

This paper discusses various methods developed to track user interaction and its various applications. It explains how user interaction logging can be enhanced further to find the user intention. It then explains the method of semantic extraction of the user interacted components and aggregation to find the user intention. The method is demonstrated through various experiments and accuracy and useful charts are shown. Our future work is dedicated in finding insights which involves the process flow spanning multiple web pages.

References

1. Atterer, R., Wnuk, M., Schmidt, A.: Knowing the user's every move: user activity tracking for website usability evaluation and implicit interaction. In: Proceedings of the 15th International Conference on World Wide Web, pp. 203–212. ACM (2006)
2. Apaolaza, A., Harper, S., Jay, C.: Understanding users in the wild. In: Proceedings of the 10th International Cross-Disciplinary Conference on Web Accessibility, p. 13. ACM (2013)

3. Apaolaza, A.: Identifying emergent behaviours from longitudinal web use. In: Proceedings of the Adjunct Publication of the 26th Annual ACM Symposium on User Interface Software And Technology, pp. 53–56. ACM (2013)

4. Plumbaum, T., Stelter, T., Korth, A.: Semantic web usage mining: using semantics to understand user intentions. In: Houben, G.-J., McCalla, G., Pianesi, F., Zancanaro, M. (eds.) UMAP 2009. LNCS, vol. 5535, pp. 391–396. Springer, Heidelberg (2009)

5. Paganelli, L., Paternò, F.: Intelligent analysis of user interactions with web applications. In: Proceedings of the 7th International Conference On Intelligent User Interfaces, pp. 111–118. ACM (2002)

6. Hong, J.I., Landay, J.A.: WebQuilt: a framework for capturing and visualizing the web experience. In: Proceedings of the 10th International Conference on World Wide Web, pp. 717–724. ACM (2001)

7. Lieberman, H.: Letizia: an agent that assists web browsing. In: IJCAI 1995, vol. 1, pp. 924–929 (1995)

8. Chen, Z., Lin, F., Liu, H., Liu, Y., Ma, W.-Y., Wenyin, L.: User intention modeling in web applications using data mining. World Wide Web 5(3), 181–191 (2002)

9. Bigham, J.P., Prince, C.M., Ladner, R.E.: Web Anywhere: a screen reader on-the-go. In: Proceedings of the 2008 International Cross-Disciplinary Conference on Web Accessibility (W4A), pp. 73–82. ACM (2008)

10. Atterer, R., Schmidt, A.: Tracking the interaction of users with AJAX applications for usability testing. In: Proceedings of the SIGCHI Conference on Human Factors In Computing Systems, pp. 1347–1350. ACM (2007)

11. Guo, Q., Agichtein, E.: Exploring mouse movements for inferring query intent. In: Proceedings of the 31st Annual International ACM SIGIR Conference on Research and Development in Information Retrieval, pp. 707–708. ACM (2008)

12. Kammenhuber, N., Luxenburger, J., Feldmann, A., Weikum, G.: Web search clickstreams. In: Proceedings of the 6th ACM SIGCOMM Conference On Internet Measurement, pp. 245–250. ACM (2006)

A Multiple Objects Tracking Method
Based on a Combination of Camshift
and Object Trajectory Tracking

Guo-wu Yuan[1(✉)], Ji-xian Zhang[1], Yun-hua Han[2], Hao Zhou[1], and Dan Xu[1]

[1] College of Information Science and Engineering,
Yunnan University, Kunming 650091, China
yuanguowu@sina.com
[2] Yunnan Shui On Construction Materials Investment Holding Co. LTD,
Kunming 650041, China

Abstract. Multiple objects tracking in dynamic background is one of the key techniques in computer vision. An improved method of multiple objects tracking based on a combination of Camshift and object trajectory tracking is presented in this paper. The algorithm uses Harris corner matching to estimate background movement parameters, adopts two-frame difference to detect moving objects, combines object trajectory tracking with Camshift track moving objects. Our improved algorithm can achieve satisfactory effect not only in tracking multiple objects, but also in tracking continuously the objects which are static, re-enter the current scene or recover motion. The experiments show that the improved algorithm can achieve better result in the accuracy and robustness of detecting and tracking moving objects for dynamic background.

Keywords: Multiple objects tracking · Object trajectory tracking · Camshift · Harris corner

1   Introduction

Tracking multiple objects in dynamic background is a challenging research in computer vision. It includes several key steps: background correction between adjacent frames, moving object detection, establishing the corresponding relationship between moving objects and acquiring the motion parameters of moving objects.

In recent years, more and more tracking algorithms have been proposed, for example, Kalman filtering [1,2], Camshift[3,4], particle filter [5,6]. These algorithms can achieve good effect in simple scenes. However, particle filter leads to incorrect estimation of posteriori probability because of particle degeneracy phenomenon; Kalman filtering is not suitable for real-time tracking of moving objects owing to high computational complexity; Camshift is usually used in a single object tracking and is not suitable for multiple objects tracking.

In the reference [7], the robustness of tracking in static background has been greatly improved, but the color characteristics of moving objects are lost and the algorithm

© Springer International Publishing Switzerland 2015
Y. Tan et al. (Eds.): ICSI-CCI 2015, Part III, LNCS 9142, pp. 155–163, 2015.
DOI: 10.1007/978-3-319-20469-7_18

is not suitable for dynamic background. In the reference [8], Camshift and Kalman filter is combined to track moving object, but it does not achieve good tracking effect in a dynamic background. The method in the reference [9] can track single moving object in dynamic background, but cannot track multiple objects in dynamic background. The reference [10] can achieve good effect in complex scenes, but fails to track when a moving object becomes static or re-enters current scene after going away. The algorithm in the reference [11] can track objects when a moving object re-enters current scene after going away, but it can not obtain satisfied effects in the complex scenes.

The above algorithms have some limitations in practical applications. In order to improve further tracking effect in dynamic background and multiple objects scene, this paper proposes an improved method of multiple objects tracking based on a combination of Camshift and object trajectory tracking. The algorithm detects foreground objects in dynamic background using frame difference after background correction, and then tracks multiple moving objects using a combination of object trajectory tracking and Camshift. It overcomes a limitation of the traditional Camshift with tracking only a single object. At the same time, it can continuously track multiple objects when the objects are static, re-enter the current scene or recover moving.

2   Moving Objects Detection for Dynamic Background

Moving objects detection is the foundation of tracking. It is more difficult to detect moving objects from dynamic background than from fixed background. Optical flow can relatively detect moving objects in dynamic background, but it has high algorithm complexity. The traditional algorithms based on frame difference and background subtraction cannot get accurate moving objects' region in a dynamic background. Frame difference is very sample, and it can achieve good effect as long as the background is corrected. Harris corner matching is used to estimate background movement parameters, and this method is described as follows.

2.1   To Estimate Background Movement Parameters

Assuming that the background's motion includes only translation and rotation, the movement can be estimated using the following affine coordinate transformation model with 6 parameters [12].

$$\begin{cases} x^{'} = a_1 x + b_1 y + c_1 \\ y^{'} = a_2 x + b_2 y + c_2 \end{cases} \tag{1}$$

In the formula 1, (x', y') is the corresponding pixel's coordinate in the current frame when (x, y) is a pixel coordinate in the previous frame. The 6 parameters $a1$, $a2$, $b1$, $b2$, $c1$, $c2$ can be calculated by substituting three pairs of corresponding pixels' coordinate into the formula 1 [13].

Now, the key issue is how to select the three pairs of corresponding pixels. Because Harris corner is a kind of representative feature point, it was selected as the matching points. The three pairs of Harris corners are selected according to the method in the reference [13].

2.2 To Correct the Current Frame's Background According to the Background Movement Parameters

The movement distance of background can be estimated by the formula 2 according to the background movement parameters from the above.

$$\begin{cases} \mu = x' - x \\ v = y' - y \end{cases} \tag{2}$$

According to the movement distance of background (μ, v), each pixel in the current frame can be returned to the previous frame's position. If there are no corresponding pixels in the corrected frame, the pixels' gray-scale value are filled by zero.

| (a) the k-th frame | (b) the $k+1$-th frame | (c) the $k+1$-th frame corrected |

Fig. 1. Correct background according to the background movement parameters

2.3 To Detect Moving Objects Using Frame Difference

After correcting background, moving objects can been detected using frame difference. Figure 2a is the result without background correction, and then Figure 2b is the result after background correction. It is shown that the moving objects detection with background correction is more accurate than one without background correction.

| (a) before background correction | (b) after background correction |

Fig. 2. Result of moving objects detection

3   An Improved Method of Multiple Object Tracking Based on a Combination of Camshift and Object Trajectory Tracking

Multiple objects tracking algorithm can track the interested objects until the objects disappear from the scenes. In practical application, the motion states of multiple objects are diversified in video scenes. For example, an object, which stopped after entering a scene, is required to maintain tracking focus in order to be tracked continuously; an object, which recovered to move after stopping, is required to be identified and be tracked continuously; an object, which has exited from the video scene and re-entered the current scene, is required to be identified, be matched with the previous motion parameters, and be tracked continuously. At the same time, it is a challenging problem to track multiple objects effectively.

In order to meet above tracking requirements, an improved method of multiple objects tracking based on a combination of Camshift and object trajectory tracking is proposed in this paper. Object trajectory tracking algorithm can achieve satisfactory effect in tracking multiple objects, but the algorithm cannot continuously track the objects which are static, re-enter the current scene or recover motion. Camshift can track single object accurately, but can be good at tracking multiple objects.

Object trajectory tracking algorithm is described in the chapter 3.1. Because Camshift is a common algorithm in moving objects tracking, it will not go into details and can be referred to the reference [3, 4].

3.1   Algorithm Principle of Object Trajectory Tracking

After moving objects were extracted from video frames, they are matched and related with the existing paths using moving trajectory tracking [14]. If there is only one object in the scene and this object is not new one, then this object is directly related with the only existing trajectory. However, if there are multiple objects in the current scene, the correspondence between the moving objects and the moving trajectories is established through distance matrix and correlation matrix in order to track continuously moving objects. The basic steps of moving trajectory tracking algorithm are as follows.

(1) Calculate the distance matrix D_E^k

Assuming that the formula 3 indicates the centroid coordinates of n moving objects at the time k:

$$M(k) = \{M_1(k), M_2(k), \ldots, M_n(k)\} \tag{3}$$

where n is the total number of the moving objects detected at the time k.

The existing m trajectories are expressed as the formula 4:

$$T(k) = \{T_1(k), T_2(k), \ldots, T_m(k)\} \tag{4}$$

Then, the distance matrix between the n moving objects and the m trajectories is obtained by the formula 5.

$$D_E^k(i,j) = \sqrt{(T_{ix}(k) - M_{jx}(k))^2 + (T_{iy}(k) - M_{jy}(k))^2}$$

(5)

where $(T_{ix}(k), T_{iy}(k))$ is the coordinate of the trajectory T_i at the time k, $(M_{jx}(k), M_{jy}(k))$ is the centroid coordinates of the object M_j at the time k, $i=1,\ldots$ m and $j=1,\ldots$ n.

(2) Calculate the correlation matrix C_E^k

a. Initialize all the elements in C_E^k by 0;

b. Location the smallest element in each row and each column, and the smallest element position is respectively shown by the vector $\alpha = \{\alpha_1, \ldots \alpha_m\}$ and $\beta = \{\beta_1, \ldots \beta_n\}$. The two vectors must be accorded with the formula 6:

$$\begin{cases} D_E^k(i, \alpha_i) = \min_{1 \le j \le n}(D_E^k(i,j)) \\ D_E^k(\beta_j, j) = \min_{1 \le i \le m}(D_E^k(i,j)) \end{cases}.$$

(6)

c. The corresponding position elements of the correlation matrix C_E^k are added by 1 according to the formula 7;

$$\begin{cases} C_E^k(i, \alpha_i) = C_E^k(i, \alpha_i) + 1, i = 1, \ldots, m \\ C_E^k(\beta_j, j) = C_E^k(\beta_j, j) + 1, j = 1, \ldots, n \end{cases}$$

(7)

(3) Establish the relationship between the moving objects and the trajectories

The elements $C_E^k(i,j)$ have three possible values: 0, 1 or 2. If $C_E^k(i,j) = 0$, it is indicated that the i-th trajectory and the j-th moving object are not matched with each other; If $C_E^k(i,j) = 1$, it is indicated that only one of them is matched with the other; If $C_E^k(i,j) = 2$, it is indicated that the trajectory and the moving object are matched with each other. Therefore, the whole matrix C_E^k may have three possible cases:

a. If the i-th row has several nonzero elements, it is indicated that the i-th trajectory matches with several moving objects. Two types are divided again as follows:

- if the i-th row has a value of 2, for the other $C_E^k(i,j) = 1$, then the j-th moving object is a new object entered into the scene;
- Otherwise, if the i-th row has only several value of 1 and assuming $C_E^k(i,j) = 1$, it is divided into two cases:

    i) If the j-th column has an element with a value of 2, the i-th trajectory is a trajectory away from the scene;

    ii) If the j-th column has not an element with a value of 2, the j-th moving object is a new object entered into the scene.

b. If the j-th column has several nonzero elements, it is indicated that the j-th moving object matches with several trajectory. Two types are divided again as follows:

- if the j-th column has a value of 2, for the other $C_E^k(i,j) = 1$, then the i-th trajectory is a trajectory away from the scene;

- Otherwise, if the the *j-th* column has only several value of 1 and assuming $C_E^k(i,j)=1$, the *j-th* moving object is a new object entered into the scene.

c. If $C_E^k(i,j)=2$, it is indicated that the *i-th* trajectory matches strictly with the *j-th* moving object. Two types are divided again as follows:

- if $C_E^k(i,j)=1$ and $D_E^k(i,j) \leq T_0$ (T_0 is the maximum displacement of a moving object between adjacent frames), the *i-th* trajectory is related with *j-th* moving object;
- Otherwise, if $C_E^k(i,j)=1$ and $D_E^k(i,j) > T_0$, it is considered that the *i-th* trajectory has a false relation with the *j-th* moving object. It is possible that the *i-th* trajectory, which is related to the object away from the previous frame, is associated with the *j-th* moving object which enters into the current frame.

3.2　Algorithm Combined Camshift with Object Trajectory Tracking

The combination of Camshift and moving trajectory tracking can achieve better tracking effect. When the objects are moving in the scene, they are tracked using the trajectory tracking method; and when the objects are motionless, have moved to the outside of current scene, or re-enter the scene, trajectory tracking algorithm will lose the focus of moving objects, but the moving objects can be tracked continuatively by searching in a limited range using Camshift algorithm and the tracking information before the focus is lost. If the moving objects is still unable to be tracked using Camshift in a limited range, they are considered to move out of the scene. In addition, when some objects recover motion and are tracked using Camshift from static state, the Camshift will be stopped, and the objects are tracked again using trajectory tracking method after retrieving the object index.

In order to facilitate analyzing, the storage structure of moving objects *Area* {*rect, object, trace, old*} is defined, where *Area $_i(k)$* represents the region of the *i-th* object in the *k-th* frame, *rect* represents the rectangle of moving object, *object* represents the index of moving object, *trace* represents the trajectory of moving object, and *old* represents the aging ratio. Then the steps of the combination algorithm of trajectory tracking and Camshift are described as follows:

(1) The *m* moving objects are detected using the movement parameters estimation of background;

(2) For the *m* moving objects detected, the moving regions are matched with the trajectory of moving object in the previous frame, and then the regions matched successfully are marked with an index. According to each moving object's centroid position in the current frame, the trajectory of moving object corresponding to the index is updated. At the same time, if there are the objects which enter or exit from the current frame, they are stored.

(3) For the new object entered the scene, the object region is matched and compared with all regions tracked using Camshift in order to judge whether it is a previous object re-entered the scene; if it is a previous object, Camshift tracking is aborted, it is numbered using the previous index retrieved, the moving object's record

is updated, and this algorithm goes to the step 5; if it is not a previous object but is a whole new one, this algorithm goes to the step 4.

(4) For a whole new object, the corresponding parameters are added to the structure record of moving objects according the next formula.

$$\begin{cases} Area_n(0)\text{-}>object = n \\ Area_n(0)\text{-}>rect = \mathrm{Re}\,ct_n(0) \end{cases}.$$ (8)

where, n represents the total number of the moving object in the current frame. If there is an object that has not been processed yet, this algorithm goes to the step 3 to continue processing. If the objects have processed, this algorithm goes to the step 5.

(5) For the moving objects that are still not matched in the previous frame, in order to avoid the individual moving objects disappear temporarily, the objects are added to the list of the moving objects area, and the aging value $Area_i(n)\text{-}>old$ is increased by 1. If the aging value does not exceed a beforehand threshold, the object is considered to loss temporarily tracking focus in the scene, and this algorithm goes to the step 6. If the aging value exceeds a beforehand threshold, the object is considered to leave from the scene and is cleared from the objects' record.

(6) For the object that loss temporarily tracking focus in the current scene, its motion region parameters will be transferred to Camshift; then, the moving object's color probability distribution is calculated, the object's feature template is obtained, and the matched object is searched in the nearby area; if the matched object is obtained, the matched object is regarded as static state in the current frame; at this time, the information of motion area will be transmitted to Camshift, and the static object is locked and tracked using Camshift in subsequent frames until it regains moving. If the matched object is not found, the object is regarded to leave from the scene, and the motion parameters are temporarily stored in order to prevent re-enter the scene in the subsequent frames.

(7) If there is an object that is not matched and is not processed, this algorithm goes to the step 5.

(8) All trajectories of moving object are updated.

4   Experiments and Analysis

In the programming environment with VC++ and OpenCV, our improved algorithm is verified using our videos shot outdoors, in which there are multiple objects and complex moving background with a video resolution of 640 x 480. Experimental results show that multiple objects in the scene, including moving objects, static objects, renewing moving objects, the objects re-entered the scene and the objects left from the scene, are tracked robustly and accurately.

Figure 3 shows the tracking effect using Camshift. We find that Camshift tracks only a single object and is inaccurate in the video.

Figure 4 shows the tracking effect using trajectory tracking. We find that a static object is lost to track in the scene, at the same time, and an object re-entered the scene is defined as a new object to continue tracking, which is not reasonable.

Figure 5 shows the tracking effect using our improved method. We can see that our method has more satisfactory tracking effect compared with alone Camshift and trajectory tracking algorithms, can track multiple objects accurately and can correctly track a static object and an object re-entered the scene.

Fig. 3. The tracking effect using Camshift

Fig. 4. The tracking effect using trajectory tracking [14]

Fig. 5. The tracking effect using our improved method

From the above experimental results, we can see, Camshift and trajectory tracking algorithm based on combination of paper made reference [14] more accurate, more ideal tracking effect, which not only realizes the Camshift and accurate tracking of multiple targets, but also can effectively to a stationary target, in the scene again into the target, and then moving target and exit the scene target effectively tracking, meet the requirement of the practical application of the tracking.

5   Conclusion

In the paper, through analyzing the advantages and disadvantages of Camshift and object trajectory tracking, we discover that the two methods just can well overcome each other's shortcomings. Therefore, this paper combines effectively the two methods so as to realize multiple objects tracking in complex and dynamic scenes.

The experiments show that the improved algorithm can achieve better result in detecting and tracking moving objects for dynamic background.

Acknowledgments. This work is supported by the Natural Science Foundation of China (61163024, 61262067), the Application and Foundation Project of Yunnan Province (2011FB019, 2013FB010) and the Young Teachers Training Program of Yunnan University.

References

1. Ristic, B., Arulampalam, S., Gordon, N.: Beyond the Kalman Filter: Particle Filters for Tracking Applications. Artech House, Boston (2004)
2. Moussakhani, B., Flam, J.T., Ramstad, T.A., Balasingham, I.: On change detection in a Kalman filter based tracking problem. Signal Processing **105**, 268–276 (2014)
3. Dorin, C., Visvanathan, R., Peter, M.: Kernel-based object tracking. IEEE Transactions on Pattern Analysis and Machine Intelligence **25**(5), 564–577 (2003)
4. Hong-zhi, Z., Jin-huan, Z., Hui, Y., Shi-lin, H.: Object Tracking Algorithm Based on CamShift. Computer Engineering and Design **27**(11), 108–110 (2006)
5. Nummiaro, K., Koller-Meier, E., Van Gool, L.: Object tracking with an adaptive color-based particle filter. In: Van Gool, L. (ed.) DAGM 2002. LNCS, vol. 2449, pp. 353–360. Springer, Heidelberg (2002)
6. Rui, T., Zhang, Q., Zhou, Y., Xing, J.C.: Object tracking using particle filter in the wavelet subspace. Neurocomputing **119**(7), 125–130 (2013)
7. Nouar, O.D., Ali, G., Raphael, C.: Improved object tracking with Camshift algorithm. In: Proceedings of the IEEE International Conference on Acoustics, Speech, and Signal Processing, pp. 657-660. IEEE, Piscataway (2006)
8. Kai, S., Shi-rong, L.: Combined algorithm with modified Camshift and Kalman Filter for multi-object tracking [J]. Information and Control **38**(1), 11–16 (2009)
9. Yuan, G.-w., Gao, Y., Dan, X.: A Moving Objects Tracking Method Based on a Combination of Local Binary Pattern Texture and Hue. Procedia Engineering, Elsevier **15**, 3964–3968 (2011)
10. Xin-jun, S., Jin-bo, C., Fa-wei., X.: Multiple Targets Tracking in Traffic Image Sequence. Journal of Changshu Institute Technology (Nature Sciences) **23**(4), 82–86 (2009)
11. Xue, L., Fa-liang, C., Hua-jie, W.: An Object Tracking Method Based on Improved Camshift Algorithm. Microcomputer Information **23**(21), 304–306 (2007)
12. Zhang, L.Y., Zhang, G.L.: Background motion model parameters estimation based on extended Kanade-Lucas tracker. Computer Applications **25**(8), 1946–1947 (2005)
13. Yan, X.L., Liang, B., Zeng, G.H.: Object tracking method based on block motion estimation. Journal of Image and Graphics **13**(10), 1869–1872 (2008)
14. Xin-jun, S., Jin-bo, C., Fa-wei, X.: Multiple Targets Tracking in Traffic Image Sequence. Journal of Changshu Institute Technology (Nature Sciences) **23**(4), 82–86 (2009)

An Access Point Trajectory Tracking Method by the Weight Update

Lin Chen[✉], Guilin Cai, and Yun Lan

School of Computer, National University of Defense Technology, Changsha, China
agnes_nudt@qq.com, agnes_nudt@hotmail.com

Abstract. In recent years, wireless access technology is quite popular for being convenient, fast and flexible. However, due to the openness of wireless network, this technology is also faced with a number of security challenges, one of which is how to deal with the unauthorized access point effectively. As we all know, the unauthorized access point leads to not only increasing interference between signals induced by the fierce competition of wireless channel resources, but also data leakage resulting in ''wireless phishing''. In response to these security threats, much importance has been put on the research of unauthorized access point location and trajectory tracking. This paper firstly proposes an optimization model of wireless signal propagation. Then an access point location and tracking method called APL_T is put forward, which supports three-dimensional location based on the weight update improving the location accuracy effectively and raises the trajectory tracking of the access point in the light of the three-dimensional location. Finally, the experimental results show that APL_T has high accuracy and can meet practical requirements.

Keywords: Wireless network · Wireless location · Trajectory tracking · Weight update · RSSI

1 Introduction

Wireless network technology gains rapid growth and begins to be applied to enterprises, government departments, private households and other fields. It brings a lot of convenience to working, studying and everyday life. However, the wireless LAN uses air as the transmission medium, introducing greater difference in access control, authentication and other security technologies compared with the traditional LANs whose transmission medium are cables. The wireless LAN is easy to cause data leakage resulting in "wireless phishing" and other information security issues. Among these issues, unauthorized access points (APs) problem are particularly prominent, which affect national security, network security, privacy and economic security. Therefore, unauthorized APs problem has become one of the most important security issues that affect the security of wireless LAN.

The APs (Access Point) are major wireless LAN access equipments. The unauthorized APs are illegal wireless access points without authorization [1].

© Springer International Publishing Switzerland 2015
Y. Tan et al. (Eds.): ICSI-CCI 2015, Part III, LNCS 9142, pp. 164–171, 2015.
DOI: 10.1007/978-3-319-20469-7_19

The unauthorized APs are divided into two categories. The first one is built by the staff of companies. The APs cannot be monitored by the network security system, through which the malicious users can invade the internal network, and present a huge security risk to the internal networks [2]. The other one is the malicious users that get users' information by a variety of means of attack to paralyze and destroy the wireless LAN. In this case unauthorized APs are used to create a fake wireless LAN. When users access this fake wireless LAN, the malicious users can intercept the information of users through wireless transmission networks, and engage in criminal activities such as theft of state secrets, commercial secrets and personal privacy, et al. The unauthorized APs are hazardous in the following aspects.

- The malicious ones use the software tools such as mdk2 to fake AP signals, which occupy the regular channels and interfere with the normal radio communication [3].
- The unauthorized APs can intercept the wireless network communication data, invade personal computers, and divulge personal information such as personal privacy and personal electronic account information.
- The unauthorized APs can be built on the internal networks inside the companies and bring the serious threats to the network security.
- The unauthorized APs will consume the limited resources of the wireless network.
- The unauthorized AP will bring the problem of the co-channel interference.
- In summary, it is important to accurately locate and track the unauthorized APs, by which we can manage, control, or remove the unauthorized APs conveniently.

The rest of the paper is organized as follows. We briefly review the related works in Section 2 and present an optimization model of wireless signal propagation in Section 3. In Section 4, we describe an access point location and tracking method named APL_T. We show that the scheme can achieve fairness, and we present the evaluation results in Section 5. Finally, we draw conclusions and point out potential future works in Section 7.

2   Related Research

The wireless location methods have been studied intensively for a long time.

Reference [4] presents the location method based on the time of arrival (TOA) which utilizes the propagation of electromagnetic waves to measure the distance between the transmitter and the receiver. Assum that the propagation time of electromagnetic waves from the beacon nodes to the mobile nodes is t. Since the speed of light is equal to the speed of wave, the speed of wave is c. So the distance between the beacon node to the node to be located is $t \times c$. Because the speed of light is very fast, TOA is very sensitive to deviations.

Reference[5] introduces a location method based on the signal arrival of angle (AOA). It uses an antenna array to estimate the angle of arrived signals, and locates the unknown node by the angles. An angle of the source can be obtained by an AOA measurement. If effective measurements of two or more different places are acquired, the location of the signal source can be determined by the intersection of two angles.

Usually, multiple AOA values of different locations can be utilized to improve the accuracy of measurements. Since the indoor environment is complex, the walls, the plants and the other obstructions will affect the propagation of radio signals. Because of reflection, refraction, multipath propagation and multipath interference, AOA-based location method is not suitable for indoor wireless location.

Reference[6] introduces a location method based on time difference of arrival (TDOA). It reduces the requirements for the clock synchronization, and it locates moving objects by measuring the time difference of arrival between two base stations. The moving object is located on the hyperbolic whose focus are the base stations. To locate the moving object, we need at least three base stations, and create two hyperbolic equations. Then the equation can be solved to obtain the position coordinates of the moving object. However, due to the fact that ultrasonic propagation distance is limited, the network must be densely built. Compared with TOA-based location method, this method is more convenient to use, and it can achieve better locating accuracy. The TDOA-based method can also be affected by multipath propagation and NLOS propagation.

Reference[7] introduces a location method based on the Received Signal Strength Indication (RSSI). The basic idea of this location method is as follows: if the transmission power of the transmitting node is already known, the received power at the receiving node can be measured, and the propagation loss of the signal is therefore calculated. We turn the propagation loss into the distance by the theoretical model of the signal propagation, then the triangular location method or the maximum likelihood method are applied to locate the receiving node. This method does not require high performance hardware, and its algorithm is relatively simple. In the experimental environment, this method achieves fairly good performance. However, due to the changes in environmental factors, it often needs to be improved in practice. The main reason for the deviation is reflection, multipath propagation, non-line dissemination et al.

Reference[8] introduces a range-free location method. This method does not measure the absolute distance and the angle between nodes, instead it calculates the location of nodes by using the network connectivity information. algorithms of the range-free method including the centroid algorithm, the dv-hop algorithm, the amorphous location algorithm, the convex programming location algorithm et al.

Nowadays the wireless location technologies are confronted with many difficulties and challenges. How to eliminate these negative factors and improve the location accuracy will determine whether the wireless location can be widely used and developed.

3   Access Point Location and Trajectory Tracking Method Based on the Weight Update

3.1   The Optimization Model of Signal Propagation

The propagation distance of the wireless signal is related to the transmission power, the receiver sensitivity and the operating frequency. Assuming d is the transmission distance, its unit by km, f is the card working frequency, its unit by MHz, Lfs is the transmission loss, its unit by dBm, and n and c are constants. We can induce the following formula.

$$Lfs = c + 10n \lg d + 10n \lg f \qquad (1)$$

Assume that the transmission power of the wireless device is F, its units by dBm. $P(d)$ is the received signal strength when the signal transmission distance is d, its units by dBm. Because the signal transmission loss equals the transmission power minus the received signal strength, we can induce the formula $Lfs = F - P(d)$.

From the above equation , we can induce the following formula (2).

$$F - P(d) = c + 10n \lg d + 10n \lg f \qquad (2)$$

When the signal transmission distance is d_0, we can get the following formulas.

$$F - P(d_0) = c + 10n \lg d_0 + 10n \lg f \qquad (3)$$

$$P(d) = P(d_0) - 10n \lg(d / d_0) \qquad (4)$$

When d equals $1m$, $P(d_0)$ is the received signal strength when the signal transmission distance is $1m$. Assume that A is $P(d_0)$. Assume that n is the propagation attenuation constant. Equation (4) can be expressed as follow.

$$P(d) = A - 10n \lg d \qquad (5)$$

Equation (5) is used to calculate the signal propagation distance in the free space. But in practice, we have to consider such factors as obstacles. $\sum WAF$ means the impact of all the obstacles, we can get $P(d) = A - 10n \lg d - \sum WAF$.

3.2 The Access Point Location

To describe the proposed location method, we give the following definitions.

Definition 1. Anchor Node: It is also called beacon node, which can be self-located some way.

Definition 2. Unknown Node: Its location cannot be known in advance. We need to use the location information of the anchor nodes, and some algorithms to calculate its location.

Using any four distances between the anchor nodes and point P, we can calculate the theoretical coordinates P1, P2, P3, P4 of point P. Finally, we calculate the average of these four theories coordinates as the coordinate of P.

Having four anchor nodes to locate an unknown node, the most common mathematical method is the least squares method. However, the least squares method leads to large deviations in practical. Therefore we proposed a least squares method based on weight update. Assume that (x, y, z) is the coordinates of the unknown node. The coordinates and the RSSI of four anchor nodes are: $(a_1, b_1, c_1), RSSI_1$;

$(a_2, b_2, c_2), RSSI_2;$ $(a_3, b_3, c_3), RSSI_3;$ and $(a_4, b_4, c_4), RSSI_4$. We can calculate four distance information respectively d_1, d_2, d_3, d_4 from the equation $RSSI = A + 10n \lg d$.

$$(x-a_1)^2 + (x-b_1)^2 + (x-c_1)^2 = d_1 \quad (x-a_2)^2 + (x-b_2)^2 + (x-c_2)^2 = d_2$$
$$(x-a_3)^2 + (x-b_3)^2 + (x-c_3)^2 = d_3 \quad (x-a_4)^2 + (x-b_4)^2 + (x-c_4)^2 = d_4 \qquad (6)$$
$$r^2 = d_i^2 - (a_i^2 + b_i^2 + c_i^2) \quad i = 1,2,3,4 \quad R^2 = x^2 + y^2 + z^2$$

Equation (6) is converted to

$$\begin{bmatrix} -2a_1 & -2b_1 & -2c_1 & 1 \\ -2a_2 & -2b_2 & -2c_2 & 1 \\ -2a_3 & -2b_3 & -2c_3 & 1 \\ -2a_4 & -2b_4 & -2c_4 & 1 \end{bmatrix} \begin{bmatrix} x \\ y \\ z \\ r^2 \end{bmatrix} = \begin{bmatrix} r_1^2 \\ r_2^2 \\ r_3^2 \\ r_4^2 \end{bmatrix}$$

We obtain the result by the least squares method $\hat{\theta} = (Q^T Q)^{-1} Q^T b$.

3.3   The Access Point Trajectory Tracking

We further establish the tracking model based on location. Assume that the target i does uniform motion in a certain period of time, then we can create the tracking model as shown by equation (7).

$$x_i(k+1) = f(k)x_i(k) + rw(k) \qquad (7)$$

Assume that r is the propagation radius of the particle. At time k, assume that w(k) is the white Gaussian noise of particles. Its average value is 0. We further get:

$$x(k) = \begin{bmatrix} x_1(k) & v_{x1}(k) & y_1(k) & y_{y1}(k) \\ x_2(k) & v_{x2}(k) & y_2(k) & y_{y2}(k) \\ \vdots & \vdots & \vdots & \vdots \\ x_n(k) & v_{xn}(k) & y_n(k) & y_{yn}(k) \end{bmatrix} \qquad (8)$$

In equation (8), $x(k)$ is the state vector. $(x_t(k), y_t(k))$ is the location of the unknown node at time k. $(v_x(k), v_y(k))$ is the speed of the unknown node at time k. t is the interval.

$$f = \begin{bmatrix} 1 & t & 0 & 0 & 1 & t & 0 & 0\cdots \\ 0 & 1 & 0 & 0 & 0 & 1 & 0 & 0\cdots \\ 0 & 0 & 1 & t & 0 & 0 & 1 & t\cdots \\ 0 & 0 & 0 & 1 & 0 & 0 & 0 & 1\cdots \end{bmatrix} \qquad (9)$$

Observation equation based on the theory of particle filter can be represented by equation (10). $(\hat{x}(k+1)/\hat{y}(k+1))$ is the location of the unknown node at time $k+1$.

$$z(k+1) = \tan^{-1}(\hat{y}(k+1)/\hat{x}(k+1)) \tag{10}$$

We will adopt the predictive filtering method and the data association method. When there is more than one unknown node aggregating in the same area, it would result in relevance between unknown nodes. As a result we need to consider the interaction between unknown nodes, and use cross-motion tracking algorithm for this case. By this way we can solve the association problem between the measured values and objectives probabilistic data. First, we determine the matrix by calculating the relationship between the location of the unknown node and tracking range.

$\Omega w_j^i (i=0,1...m; j=0,1...m_2)$ is represented by the equation (11).

$$\begin{bmatrix} 1 & w_1^1 & w_1^2 & \cdots & w_1^n \\ 1 & w_2^1 & w_2^2 & \cdots & w_2^n \\ \vdots & \vdots & \vdots & \vdots & \vdots \\ 1 & w_{m_2}^1 & w_{m_2}^2 & \cdots & w_{m_2}^n \end{bmatrix} \tag{11}$$

w_j^i is a binary variable. w_j^1 represents the unknown node j within the tracking range of anchor node. w_j^0 represents the unknown node j which is not within the tracking range of anchor node. Assume that n is the quantity of detected unknown node and that m_k is the quantity of detected nodes after data fusion.

The matrix above reflects the relationship between measured values and unknown nodes. Then we can get feasible joint events and corresponding index, which depends on two basic assumptions. Firstly, assume that the only source of the signal strength is either unknown node or a clutter. Each unknown node has at most one measured value whose source is the unknown node itself. By searching the signal strength of the unknown nodes, we can find the possible event index. So the associated probability between the signal strength and unknown nodes can be expressed as equation (12).

$$\beta_k^{j,i} = P(\theta_k^{j,i}/Z(k)) = \sum_{j=1}^{n_i} P(\theta_{k,i}/Z(k))w_j^i(\theta_{k,i}) \tag{12}$$

We present data integration with data association method based on particle filter. The randomness produced by the noise and the multipath makes the value of the node tracking range very random. Therefore we apply the particle filter optimization to each measurement value within the tracking range of anchor nodes. Then we integrate the data using probabilistic data association, update particles, and determine the location and status of the unknown node at the next moment.

4   Performance Analysis

We use the Mat lab software to simulate this location and trajectory tracking method. We build five anchor nodes, and their communication radius is 20m. The monitoring distance is half the communication radius, and the sampling period is 5s. The initial location of the target is (0,0), and the amount of the effective particles N are 2 and 4. Ignoring the fixed obstacle, the distance can be calculated by equation (6). First we need to calculate the equation parameters A and n.

$A = P(d_0)$ represents the received signal strength when the distance between the unknown node and anchor nodes is 1 meter. Anchor nodes scan channels once per minute, and collect the transmission signal strength of anchor node. We count the results 50 times, and obtain the distribution about $A = P(d_0)$ to calculate its average value, which is -40dBm. As A is already known, we set the distance between anchor nodes and the unknown node 2m and 3m respectively. The unknown node scan the channels once per minute and collect the transmission signal strength of anchor nodes. We count the results 120 times. When d equals 2m, we calculate that P(d2) equals -49dBm. According to equation (6), we can obtain n_2 equals 3.08. When d equals 3m, we calculate the result that P(d3) equals -55dBm. According to equation (6), we can obtain n_3 equals 3.12. At last, the average value of n is 3.1. In summary, the signal propagation model can be presented as $P(d) = 40 - 31\lg d$.

Using several experiments datasets, we compared the location data with and without weight updating. The experimental data is shown in Figure 2.

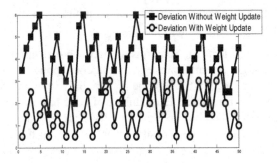

Fig. 1. The positioning accuracy comparison

5   Conclusions and Future Work

The paper first proposes a wireless signal propagation optimization model. Then it proposes an access point location and tracking method APL_T (Access Point Location and Trajectory Tracking Method based on the Weight Update). APL_T supports three-dimensional locating based on the weight update, which effectively improves the accuracy of location. APL_T further obtains the trajectory tracking of the access point. The experiment result demonstrates the effectiveness of the method.

APL_T is primarily used for the location and trajectory tracking of a single target. However, there is usually more than one target that appears in a wireless network environment at the same time. Therefore in the future we will study how to solve the problem of multi-targets location and tracking as well as the problem of multi-targets trajectory cross-cutting.

Acknowledgments. This work is supported by the Natural Science Foundation of China (No.61379148).

References

1. Rappaport, T.S.: Wireless Communications Principles and Practices Second Edition [M], pp. 42–56. Publishing House of Electronics Industry, Beijing (2011)
2. Priyantha, N., Chakraborthy, A., Balakrishnan, H.: The Cricket location-support system[J]. Mobile Computing and Networking 1(30), 140–147 (2010)
3. Savvides, A., Park, H., Srivastava, M.B.: The bits and flops of the N-hop multilateration primitive for node localization problems [J]. Mobile Networks and Applications 8, 443–451 (2003)
4. Harter, A.: A distributed location system for the active office[J]. IEEE Network 8(1), 62–70 (2007)
5. Cao, Y.: Target Localization Based on AOAs[J]. Journal of Electronic Science and Technology of China 5(2), 172–174 (2007)
6. Sayed, A.H., Tarighat, A., Khajehnouri, N.: Network-based wire-less location[J]. IEEE Signal Processing Magazine 22(4), 24–40 (2005)
7. Paul, A.S., Wan, E.A.: RSSI-Based Indoor Localization and Tracking Using Sigma-Point Kalman Smoothers[J]. IEEE Journal of Selected Topics in Signal Proeessing 3(5), 860–873 (2009)
8. Savvides, A., Park, H., Srivastava, M.B.: The bits and flops of the N-hop multilateration primitive for node localization problems[J]. Wireless Sensor Networks and Application, 112–121 (2011)
9. Xin, J., Sano, A.: Computationally efficient subspace-based method for direction-of-arrival estimation without eigen decomposition
10. Li, J., Li, J., Guo, L., Wang, P.: Power-efficient node localization algorithm in wireless sensor networks. In: Shen, H.T., Li, J., Li, M., Ni, J., Wang, W. (eds.) APWeb Workshops 2006. LNCS, vol. 3842, pp. 420–430. Springer, Heidelberg (2006)

Urban Principal Traffic Flow Analysis
Based on Taxi Trajectories Mining

Bing Zhu[✉] and Xin Xu

Science and Technology on Information Systems Engineering Lab, Nanjing, China
bingzhazha@126.com

Abstract. The understanding of urban traffic pattern can benefit the urban operation a lot, including the traffic forecasting, traffic jam resolution, emergency response and future infrastructure planning. In modern cities, thousands of taxicabs equipped with GPS can be considered as a large number of ubiquitous mobile probes traversing and sensing in the urban area, whose trajectories will bring great insight into the urban traffic management. Thus, in this paper we investigate the urban traffic pattern based on the taxi trajectories, especially the principal Origin-Destination traffic flow (OD flow) extraction. Focusing on the picking-up and dropping-off events, the issue is solved by a spatiotemporal density-based clustering method. The OD flow analysis is formulated as a 4-D node clustering problem and the relative distance function between two OD flows is defined, including a clustering preference factor which is adjustable according to the observation scale favor. Finally, we conduct the method on the taxi trajectory dataset generated by 28,000 taxicabs in Beijing from May 1st to May 30th, 2009 to evaluate its performance and interpret some underlying insights of the time-resolved results.

Keywords: Urban computing · Principal traffic flow analysis · Density-based clustering

1 Introduction

Nowadays, taxi is one of the most important and widely used public transportation tool. In modern cities, taxicabs are always equipped with GPS, which can be considered as ubiquitous mobile probes traversing and sensing in the urban areas. There are more than 10,000 taxicabs in New York, Tokyo and London and the number is even larger in big cities of developing countries. For instance, there are as many as 66,600 licensed taxicabs in Beijing in 2012, which generates over 1.2 million ridden trips per day with a $30\% - 40\%$ vacant ratio. In fact, the taxi-net has a very good coverage of the city in both space and time. Therefore the trajectories of taxis can provide great insights into the urban traffic pattern, which will serve the Location Based Service(LBS) development, transport management, as well as urban planning.

In this paper, we investigate the most important urban traffic pattern, the principal traffic flow distribution, based on the taxi trajectories. We extract the

© Springer International Publishing Switzerland 2015
Y. Tan et al. (Eds.): ICSI-CCI 2015, Part III, LNCS 9142, pp. 172–181, 2015.
DOI: 10.1007/978-3-319-20469-7_20

taxi's Origin-Destination(OD) flows from the vast amount of trajectories data and then explore the principal OD traffic flows utilizing a high dimensional spatiotemporal clustering method. Among the enormous trajectory data, we particularly pay attention to the picking-up and dropping-off points of taxis, which correspond to certain locations of the origins and destinations of traffic flows. Considering the principal traffic flows are these large traffic OD flow clusters whose members share similar geographical origins and destinations, the principal OD flow extraction are formulated as a spatiotemporal clustering problem and some density-based clustering techniques are utilized to solve it. The contribution of this paper lies in two aspects:

- *OD traffic flow clustering:* We formulate the principal traffic flow extraction issue as a 4-dimensional spatiotemporal clustering problem. The coordinates of both the origin and destination constitute the 4-D coordinate of the OD flow. The *distance* between OD flows with a clustering preference factor is defined to utilize the density-based clustering algorithm, which can be tuned to favor long or short trips.
- *Real evaluation:* We evaluate our method on a large-scale GPS trajectory dataset generated by 28,000 taxis in Beijing from May 1st to May 30th, 2009. Combining the running results with Beijing's geographic-social background information, we can find that we successfully identify the principal traffic flows of Beijing, justifying the effectiveness of our methods.

The remainder of the paper is organized as follows. In Sect. 2, we review some related work. We introduce our method to mine the principal OD flow clusters in Sect. 4 and its fundamental algorithm Spatial-temporal DBSCAN in Sect. 3. In Sect. 5, we evaluate our method on a real large-scale dataset and discuss the results. In Sect. 6, we draw a conclusion to this paper.

2   Related Work

In this section, we briefly review some related work. There is a growing interest in utilizing taxi trip data to analyze and understand traffic patterns and urban environment. We refer to [13], [12], [10], [9], [8], [17], [14], [18], [16] for some recent advances in this domain. The previous related works always solve the problem by counting the traffic volume of flows between hotzones, which degenerated the principal traffic flows to the traffic flows between principal nodes [15,18]. Also, some classic spatiotemporal traffic flow analysis methods in P2P network, e.g. Random Matrix Theory(RMT) and Principal Component Analysis(PCA), are utilized in [11], [6], [7]. In addition, the OD pair cluster extraction problem can also be addressed using the machinery of well-separated pair decomposition in the computational geometry community (See Chapter 3 of [5]). In our setting, we propose to formulate it as solving a clustering problem due to the noise presented in the input data.

3   Spatial-Temporal DBSCAN

DBSCAN proposed by Ester in 1996 [3] is a density-based clustering algorithm which is designed to discover arbitrary-shaped high-density clusters on a spatial plane and can distinguish noise points [2]. Thus, the distance function $dist(p, q)$ between two points p and q is obviously a key definition in the method. In fact, there are two parameters in DBSCAN, *Eps* and *MinPts*. *Eps* is a spatial distance threshold to delimit the *neighborhood* of a point while *MinPts* is a density measurement to define the *core point*.

Definition 1 (Neighborhood). *The* Neighborhood *of a point* p, *denoted by* $N_{Eps}(p)$, *is defined as* $N_{Eps}(p) = \{q \in D | dist(p, q) \leq Eps\}$, *where* D *is the whole dataset.*

Definition 2 (Core point). *A point* p *is defined as a* core point, *if* $|N(q)| \geq$ *MinPts, otherwise it is called a* border point.

Considering that principal traffic flows are observed between areas on which taxi picking-up or dropping-off events occur intensively in space and frequently in time, the clusters should be dense in both spatial and temporal dimensions. Therefore, we modify the DBSCAN to be a spatial-temporal density-based clustering algorithm by utilizing two distance metrics Eps_s and Eps_t to measure the *similarity* between two points, where Eps_s is used to measure the geographical closeness $dist_s()$ while Eps_t is used to measure the temporal closeness $dist_t()$. Thus, q is p's neighbor if $dist_s(p, q) < Eps_s$ and $dist_t(p, q) < Eps_t$, where $dist_t(p, q) = |t_p - t_q|$.

As Algorithm. 1 shows, the algorithm starts with an arbitrary point p in the database D, and aggregates all its neighbors within distance Eps_s and Eps_t. If the the number of neighbors is greater than $MinPts$, then p is a core point and a new *cluster* is created based on p and its neighbors. Then, iteratively check all the neighbors, and add all the core points' neighbors into the cluster. Repeat the process until all the points in D have been processed. The point does not belong to any cluster is called *noise*.

DBSCAN is an unsupervised learning process without any requirement of prior knowledge. The cluster of arbitrary shape can be found among noisy points and no predetermination of the number clusters is needed. This is reason why we choose it to solve the traffic flow analysis problem. The runtime complexity of the DBSCAN is $O(nlogn)$, where n is the number of points to be clustered and it has been proven to be able to handling very large-scale dataset [1,4].

4   Principal Origin-Destination Traffic Flow Clustering

In this section, we look into the passengers' trips from the taxi GPS data to extract the principal origin-destination traffic flow graph. We formulate the principal traffic flow extraction issue as a 4-D node clustering problem. The coordinates of both the origin and destination constitute the 4-D coordinate of the

---

Algorithm 1. Spatio-temporal DBSCAN

---

input : $D = \{p_1, \ p_2, \ ..., \ p_n\}$
output: $C = \{C_1, \ C_2, \ ..., \ C_m\}$

Cluster_Label $= 0$;
Noise $= \emptyset$;
for $i = 1$ **to** n **do**
    if p_i *is neither in a cluster nor in* Noise **then**
        $N =$FindNeighbors(p_i, Eps_s, Eps_t);
        if $|N| \leq MinPts$ **then** AddTo$(p_i,$ Noise$)$;
        ;
        else Cluster_Label $=$Cluster_Label $+1$;
        New a cluster $C_{\text{Cluster\_Label}}$;
        AddTo$(p_i, C_{\text{Cluster\_Label}})$;
        forall the $q \in N$ **do**
            AddTo$(q, C_{\text{Cluster\_Label}})$;
            if $q \notin$ Noise **then** Push(q);
            ;

        //To check whether the stack is empty;
        while *not* IsEmpty$()$ **do**
            $Cur =$Pop$()$;
            $M =$FindNeighbors(Cur, Eps_s, Eps_t); **if** $|M| \geq MinPts$ **then**
            forall the $o \in M$ **do**
                AddTo$(o, C_{\text{Cluster\_Label}})$;
                if $o \notin$ Noise **then** Push(o);
                ;

        ;

    ;

OD flow, and the distance between two OD pairs with a clustering preference factor is defined to utilize the density-based clustering algorithm.

The OD flow O_iD_j is projected to the point $(x_O^i, y_O^i, x_D^j, y_D^j)$ in the 4-D space, and the spatial distance between OD flow O_1D_1 and O_2D_2 is defined as:

$$dist_s(O_1D_1, O_2D_2) = \frac{(||O_1 - O_2||_2^2 + ||D_1 - D_2||_2^2)^{\frac{1}{2}}}{(||O_1 - D_1||_2 + ||O_2 - D_2||_2)^\alpha} \tag{1}$$

where, $\alpha(\geq 0)$ is a factor indicating the clustering preference to longer trips and the 2-norm $|| \cdot ||_2$ is defined as

$$||O_1 - O_2||_2 = \sqrt{(x_O^1 - x_O^2)^2 + (y_O^1 - y_O^2)^2}, \tag{2}$$

where, $O_1(x_O^1, y_O^1), O_2(x_O^2, y_O^2)$.

When $\alpha > 0$, $dist_s(\cdot)$ is a relative distance, which is normalized by the exponential length of the trips. Because under the clustering framework, if the absolute distance between two OD pairs are comparable, then the OD pairs with longer travel length are regarded more similar to each other. Thus α is an adjustable parameter, whose value can be adjusted to find different groups of clusters with different travel length scale. As α goes larger, the absolute distance between two long trips will be more unconspicuous and the long trips will be more preferred to be clustered.

When $\alpha = 0$, the distance metric degenerates to the absolute Euclidean distances between two OD pairs in 4-D space. In this case, the length of trip is not taken into account.

Based on the above definition of distance, DBSCAN can be used to solve the 4-D node clustering problem. In order to prevent the chain phenomenon in clustering, we restrict the area of the clustered origins and destinations and modify the DBSCAN into an iterative algorithm. As the flowchart in Fig. 1 shows, we check the distribution of both the origins and destinations in each OD flow cluster C_i, if the area $S(C_{iO}) > \theta$ or $S(C_{iD}) > \theta$, then do DBSCAN w.r.t. cluster C_i iteratively, halving the value of parameter Eps_s, until the output clusters satisfy the area restraint.

5   Evaluation

In this section, we carry out our methods on a real large-scale GPS trajectory dataset and evaluate their performances.

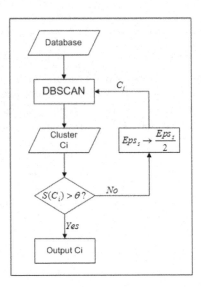

Fig. 1. The iterative DBSCAN flowchart

5.1   Settings

The trajectory dataset was generated by 28,000 taxicabs in Beijing from May 1st to May 30th, 2009. The cabs transmitted their service data about every one minute and generated over 20 billion records per day. In the data pre-processing, taxis whose valid records were less than 200 in a whole day have been omitted. Table. 1 shows a piece of data sample, including the *ID* of the cab, the sampling time stamp *UTC*, the geographical location latitude *Lat* and longitude *Long*, and the occupation state *Heavy*.

Table 1. A piece of data sample

ID	UTC	Lat	Long	Heavy
1	1241107203	4025640	11673435	1

Table 2. OD flow clustering results w.r.t. α

α	Number of clusters	Average length of trips(km)
0	26	7.8
0.5	24	16.1
1	20	21.5

5.2   Result Analysis

First of all, we did a set of experiments to verify the impact of α on the clustering results. The OD flow clustering algorithm was conducted on the data generated from 8:00 to 9:00 a.m., May 1st, 2009, with the distance function defined as Equation. 1. The results are shown in Table. 2 and Fig. 2. We can see that, the clustering preference factor α does have great control over the clustering results. When $\alpha = 0$, that is under the absolute distance definition, clusters grouped by the algorithm are all constituted of short trips within $10km$(except the one to the airport). When $\alpha = 0.5$, long trips from $9km$ to $21km$ emerge over the short trips. When α goes larger to 1, longer trips are clustered successfully. Thus, benefiting from α, we can easily control the algorithm with different clustering preference to investigate the traffic pattern in different observation scale.

Furthermore, since the traffic pattern is time-dependent, we did some more experiments based on datasets generated in different time period with the same algorithm parameter settings($\alpha = 0.5$). The following are some interesting results and their semantic interpretations, which can rationally verify the effectiveness of the OD flow clustering algorithm.

- *Different time periods through a day*

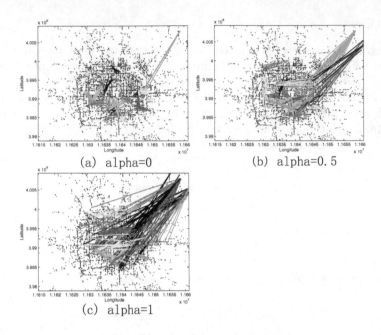

(a) alpha=0

(b) alpha=0.5

(c) alpha=1

Fig. 2. Clustering results w.r.t. α

The results of the principal OD flow clustering conducted in different time periods on May 1st are shown in Table. 3 and Fig. 3, where the point depicted in dot is the origin while the one depicted in star is the destination. Comparing the results on May 1st, we can find some interesting hints. First, the urban traffic are more active during the daytime than early morning and late night, because more picking events occur and more OD flow clusters are grouped. Second, higher ratio of OD flows are included into the clusters in the active hours, so we can infer that people's traffic behaviors are more similar and converge to the principal traffic flows, especially the ones involve business districts and transportation centers, which brings great challenges for the traffic management. Third, apart from the level of activity corresponding to the traffic volume, in different time periods through a day, the specific traffic pattern varies. For example, in the early morning (Fig. 3(a)), the main flows are those connecting the residential areas and transportation centers. At late night, a large flow cluster from the East 3rd Ring Road CBD to the suburban residential area (the red cluster in the right side of Fig. 3(c)) emerges, which never appears before 11:00 p.m.. Because the route is exactly along the subway Line 1, people always prefer to take the subway than the taxi until the subway goes out of service.

– *Workdays and holidays*

Different patterns also can be found between workdays and holidays. In contrast to the workday afternoon on May 25th, 2009 (Fig. 3(d)), the traffic is much heavier in the holiday afternoons such as Fig. 3(b) shows on the Labor Day.

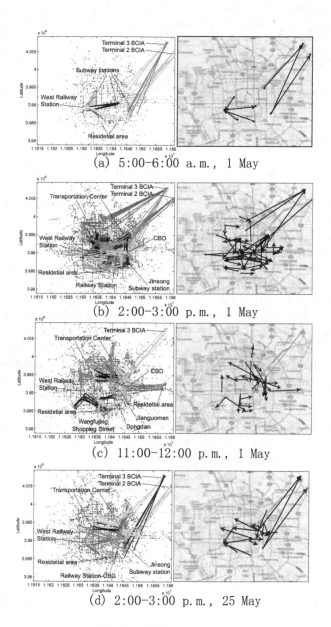

Fig. 3. OD flow clusters

Table 3. OD flow clustering results

Time period	Number of trips	Number of clusters	Ratio of trips included in clusters
05:00-06:00 a.m.,1 May	1093	7	4.10%
08:00-09:00 a.m.,1 May	3627	24	5.97%
02:00-03:00 p.m.,1 May	4865	67	10.14%
11:00-12:00 p.m.,1 May	3974	30	6.12%
02:00-03:00 p.m.,25 May	1772	22	6.02%

6   Conclusion

In this paper, we investigate the urban principal traffic flow based on the taxi trajectories. The principal traffic flow extraction is formulated as a 4-D node clustering problem and the relative distance between two OD flows is defined, including a clustering preference factor which is adjustable according to the observation scale favor. Experiments based on the Beijing taxi trajectory dataset have been done to evaluate the performances of the method we proposed. The trajectory dataset was generated by 28,000 taxicabs in Beijing from May 1st to May 30th, 2009. The results show that, first the principal traffic flow clusters are successfully grouped by the 4-D node clustering algorithm, with the locations and shapes of origin and destination areas matching perfectly to certain functional zones in Beijing. Second, the clustering preference factor does have a great control over the clustering results. Third, combining the time-dependent clustering results with Beijing's geographic-social background information, some insights underlying the results are interpreted, which can justify the effectiveness of the method.

Acknowledgments. This research work was supported by National Natural Science Foundation of China under Grant No. 61402426 and partially supported by Collaborative Innovation Center of Novel Software Technology and Industrialization.

References

1. Aoying, Z., Shuigeng, Z.: Approaches for scaling dbscan algorithm to large spatial database. Journal of Computer Science and Technology **15**(6), 509–526 (2000)
2. Birant, D., Kut, A.: St-dbscan: An algorithm for clustering spatial-temporal data. Data and Knowledge Engineering **60**(1), 208–221 (2007)
3. Ester, M., Kriegel, H., Sander, J.: A density-based algorithm for discovering clusters in large spatial databases with noises. In: Proceedings of Second International Conference on Knowledge Discovery and Data Mining, pp. 226–231 (1996)
4. Ester, M., Kriegel, H., Sander, J.: Clustering for mining in large spatial database. KI-Journal(Artificial Intelligence) Sepcial Issue on Data Mining **12**(1), 18–24 (1998)
5. Har-Peled, S.: Geometric approximation algorithms. No. 173, American Mathematical Soc. (2011)

6. Lakhina, A., Papagiannaki, K., Crovella, M., Diot, C.: Structural analysis of network traffic flows. In: SIGMETRICS 2004/Performance 2004 Proceedings of the Joint International Conference on Measurement and Modeling of Computer Systems, pp. 61–72 (2004)
7. Reades, J., Calabrese, F., Ratti, C.: Eigenplaces: Analysing cities using the space-time structure of the mobile phone network. Environment and Planning B: Planning and Design **36**, 824–836 (2009)
8. Rosenfeld, A.: Connectivity in digital pictures. Journal of the ACM (JACM) **17**(1), 146–160 (1970)
9. Wei, L., Zheng, Y., Peng, W.: Constructing popular routes from uncertain trajectories. In: Proceedings of the 18th ACM SIGKDD International Conference on Knowledge Discovery and Data Mining, pp. 195–203 (2012)
10. Wei, L., Zheng, Y., Zhang, L.: T-finder: A recommender system for finding passengers and vacant taxis. IEEE Transctions on Knowledge and Data Engineering PP(99), 1 (2012)
11. Yuan, J., Mills, K.: A cross-correlation based method for spatial-temporal traffic analysis. Performance Evaluation **61**(2–3), 163–180 (2005)
12. Yuan, J., Zheng, Y., Xie, X.: Discovering regions of different functions in a city using human mobility and pois. In: Proceedings of the 18th ACM SIGKDD International Conference on Knowledge Discovery and Data Mining, pp. 186–194 (2012)
13. Yuan, J., Zheng, Y., Zhang, L.: Where to find my next passenger? In: Proceedings of the 13th International Conference on Ubiquitous Computing, pp. 109–118 (2011)
14. Yue, Y., Zhuang, Y., Li, Q.: Mining time-dependent attractive areas and movement patterns from taxi trajectory data. In: 17th International Conference on Geoinformatics, pp. 1–6 (2009)
15. Zhang, D., Li, N., Zhou, Z.: ibat: Detecting anomalous taxi trajectories from gps traces. In: Proceedings of the 13th International Conference on Ubiquitous Computing, pp. 99–108 (2011)
16. Zhang, D., Guo, B., Yu, Z.: The emergence of social and community intelligence. Computer **44**(7), 21–28 (2011)
17. Zhang, W., Zhu, B., Zhang, L.: Exploring urban dynamics and pervasive sensing: correlation analysis of traffic density and air quality. In: Sixth International Conference on Innovative Mobile and Internet Services in Ubiquitous Computing (IMIS), pp. 9–16 (2012)
18. Zheng, Y., Liu, Y., Yuan, J.: Urban computing with taxicabs. In: Proceedings of the 13th International Conference on Ubiquitous Computing, pp. 89–98 (2011)

Simulation

Building Ontologies for Agent-Based Simulation

Sergey Gorshkov[✉]

TriniData, Mashinnaya 40-21, 620089 Ekaterinburg, Russia
serge@trinidata.ru

Abstract. Using ontologies for simulation models construction has some advantages that cannot be underestimated. Building the ontology, a modeler has to choose conceptualization method, which significantly affects the structure and usability of resulting models. A tendency of using standard ontologies without critically estimating their applicability for particular tasks may even lead to the loss of the model's efficiency and reliability. In this work, we are considering a simple criterion which may be used to pragmatically assess applicability of particular modeling techniques for building ontologies for simulation models. We will specially focus on the temporal aspect, states and events representation methods in the model. A fragment of ontology for the city social infrastructure optimization modeling will be considered.

Keywords: Ontologies modeling · Ontology assessment · Collaborative modeling

1 Introduction

Semantic Web technologies became attractive for the implementers of agent-based simulation methods since they have emerged. Semantic Web–oriented software is providing a convenient way for digital representation of conceptual models. The further development of this idea leads to the representation of the simulation model itself in the ontological form, and building a model execution framework around the semantic data storage [11]. This approach has many advantages, among which are:

— Simplification of the information gathering from various sources. A semantic model can be used for merge heterogeneous data into a single representation.
— Ease of the model management during its development and exploration. Adding new object types, attributes and interaction rules into semantic model is simple. It is much more flexible that the relational data oriented model.
— The ontologies are very convenient for expressing agent behavior and interaction rules in simulation models – especially the logical behavioral patterns.
— The reasoning software can automate rules execution during the simulation.

It is natural that the use of rather complex and flexible modeling methods and information representations is producing some uneasy problems. To ensure that the use of semantic technologies will benefit modeling process and result, it is necessary to adequately construct the target model ontology, and implement it. The term

© Springer International Publishing Switzerland 2015
Y. Tan et al. (Eds.): ICSI-CCI 2015, Part III, LNCS 9142, pp. 185–193, 2015.
DOI: 10.1007/978-3-319-20469-7_21

"adequately" means that the chosen ontology building principles, which implicitly or explicitly defines modeling methods, should bring the tangible benefit for the expected results. In other words, the only criterion for "correctness" of chosen conceptualization and modeling approach is its impact on the pragmatic goals of the modeling, in the aspects of:

— Effectiveness (positive influence on the goals achievement),
— Reliability (ontology stability in case of model's application conditions changes),
— Safety (absence of negative impact risks of the ontology structure on the modeling process and result).

In ontology-driven simulation modeling the ontology quality strongly affects the overall model quality, and ontology building method is strictly bound with the modeling method. The goal of our work is to resolve some particular questions of ontology construction for multi-agent simulation application. We are aiming to:

— Formulate criterion of ontology building methods correctness, according to the simulation modeling tasks;
— Analyze the approaches of simulation model's temporal aspect, object states and events representation;
— Describe some aspects of software implementation of the ontology for infrastructure investment optimization simulation model.

2   Related Work

As mentioned above, the use of semantic technologies in simulation modeling was attracting researcher's attention for a long time. An example of the application framework structure, based on the ontological data model, is given in [12]. The more or less deep ontology use for multi-agent modeling is described also in [9], [6] and other works. Some efforts were made to create standard ontologies for use in simulation modeling [6]. Ontologies may be used for bridging various models: for example, one model may provide source data for another, exposing its interface as a service having ontological description [1].

OWL, the core standard for semantic knowledge representation, offers the requisite level of expressiveness to capture the rich relationships, and can be easily extended [12]. OWL also offers the great flexibility in ontology modeling, which brings many benefits and some problems. One of the main problem is the correct choice or definition of the model's top level. This level is usually consisting of the highly abstract concepts, but it seriously affects the overall model structure. For example, the higher level of the model is often started with distinction between physical and nonphysical object classes [12], which immediately rises some important questions. We will review some of them later.

The detailed description of the use of ontologies in agent-based modeling is given in [2]. This book also discusses conceptualization methods, and ontology quality evaluation. The ontologies and models are observer-dependent, which may somehow contradict the task of ontologies use as the "common language" for merging all the information on the modeled system. The authors are considering verification of the model, but not of the high-level conceptual ontology on which it is based. In its turn, ontology validation methods are usually focused on the formal measurement of their quality aspects, such as structural coherence, intended purpose matching, and methodological support completeness [4]. Such methods are undoubtedly useful, but they are not always answering the principal questions of how adequate are the chosen methods for target reality part conceptualization.

There is a trend to use high-level or widespread ontologies as a basis for custom ontology creation, or at least to align with them. The general purpose ontologies, such as Dublin Core, John Sowa's ontology, REA [5], or domain-specific ones, such as ISO 15926, CIM etc, may serve such role [3]. Their use has two general advantages – enabling several knowledge engineers to interact in the standardized modeling method context, and the support of the common high-level conceptual semantic core. By the other hand, such ontologies often bring implicitly or explicitly defined conceptualization and knowledge representation method, which affects the whole model, but may have no value for resolving the model's particular task. The weak pragmatic of standard ontologies use is outlined by the researchers, who are offering to use narrow domain-specific ones instead [3]. However, as we will show below, domain-specific ontologies may bring the same problems as the high-level ones, when used without strong justification. They are usually intended for resolution of the specific scope of tasks, and might not show any benefits for other goals, even within the same subject matter area.

3   Ontology Building Method Correctness Criterion

Finding out a reasonable abstraction level is a key question of the modeling [8]. Abstraction means simplification of model elements description, relative to the corresponding real world objects. A greater simplification grade allows to manipulate model's elements easily, but leads to the lack of meaningfulness in its result. The possible ways of simplification are numberless. Anyway, this process is consisting of:

a) selection of the objects and phenomena which cannot be neglected in the model,
b) choice of methods of their meaningful aspects representation in the model.

Opportunities and drawbacks of particular ontology construction and model conceptualization methods are often discussed. This may lead to an impression that some methods are more "right" than others. It is more correct to say that there are methods that allow to resolve some task classes, and methods that are not, or are excessively complex for this.

Considering modeling costs, it is important not to fall for seeking maximal depth of phenomenon coverage, and precision of obeying chosen method guidelines, in pursuit of model "correctness". The rational level of model's specification and complexity can be found relative to the expected economic effect of its application, and model building/running cost. It is the better criterion of modeling assumptions permissibility.

Surely, the only fundamental criterion of the model correctness is the conformity of its results with the behavior of the modeled system and its elements [2]. Let us define this criterion in a more formal and useful way. The reality (or, better say, our image of the reality) within the modeled scope has the similarity relations with the model. These relations are expressing the chosen modeling method rules. The modeling method has to be chosen accordingly to the acceptable abstraction level, as described above.

Simulation model are usually representing some processes, which have the initial and final states. These states can be reflected from the image of the real world to the model using the mentioned similarity relations. If the final state, calculated by the model, and the final state, reflected to the model from the real world, are differing no more than by acceptable level of deviation – then the model is correct.

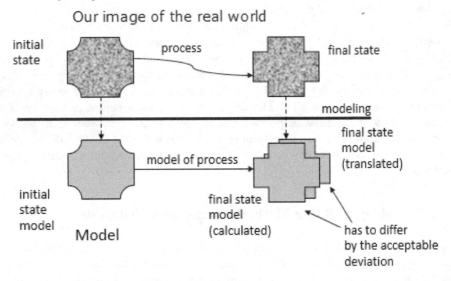

Fig. 1. Model correctness criterion

This illustrates that the model may be as simple as possible. This is acceptable while it gives an appropriate pragmatic result within the task scope, including the interested extremal cases. Surely, we can only use this method on the historical data.

This principle can be extended to the ontology correctness assessment, as in the semantic-driven modeling the ontology defines the structure of the model. An ontology may be considered as the formal expression of the mentioned similarity rules.

An ontology is <u>effective</u>, if it allows building models matching the described criterion, with the minimal possible expenses.

An ontology is <u>reliable</u>, if it is keeping efficiency within the extensible modeling task scope. This includes possible change of the modeled real world fragment (and its structure reflected in the model), of the initial state, duration of the simulation and pragmatic task evolution.

An ontology is <u>safe</u>, if the chosen conceptualization and modeling methods are not affecting the modeling results negatively. They should not produce additional risks in case of modeling task scope changes, and new data sources incorporation.

The proposed criterion is rather intuitive, and we are sure that a lot of researchers are assessing their models from this point of view, formally or not. The thing we want to emphasize is that there is always a decision to make: shall the model be aligned with some standard conceptualization theories in the quest of "correctness", or it just has to be proven by practice, and convenient to use. We will offer some arguments for the second point of view in the remaining part of our work.

4   States and Events in the Model

Let us apply the proposed criterion to the particular task: temporal aspect formalization methods assessment and choice. Multi-agent simulation models always have discrete time, as it increments on some small value on each modeling step. This stimulates to define "state" of the modeled object as a complex of its properties and relations at some modeling step, and "state" of the model – as a complex of states of all its elements. If the properties are not changing during several steps, a state may have non-zero duration. Then it is convenient to define the "event" as a fact of transition from one state to another.

One of the methods of states and events formalization is their description as separate objects. This approach is used, for example, by ISO 15926 standard [7]. The temporal evolution of an object is described as a sequence of "low level" sub-objects – its "temporal parts", related with the source object. Each event in this method is also represented as an object, related with the initiated or finished state, and, probably, the actor who has triggered the state change. Surely, temporal parts distinction is a prerogative of the modeler, as the object allocation itself. The events distinction is even more subjective, and completely depends on the point of view. For example, ISO 15926 standard distinct physical and functional events at the root level of modeling ontology. The first are describing changes of the physical properties of the objects (which are having spatial localization), the last are describing results of influence of one objects to another. In most models, such distinction may have no pragmatic sense. Moreover, the event perceived as single in the real world may be forcedly split to several events following these modeling rules.

In the business process modeling systems, the events and states are considered from another point of view (example: [3]). The state is interpreted as the change of some object's attribute (often having numeric or at least literal representation), and, thus, does not require formalization as a separate object. An event is interpreted primarily as a volition, manipulation of the subjects which are performing process, or as an environment change which has to be reacted. This event, surely, has to be represented as an object.

In the ontology offered by Guizzardi et al in [6] as a universal framework for simulation models formalization, events are also treated as objects. This methodology also offers further events classification. The events are considered primarily as the interactions of several subjects (participants), and formalized accordingly. The events of objects emergence, destruction and evolution are classified separately. The states of the objects before and after the events are also different.

In physical systems modeling, a model may have continuous temporal aspect instead of discrete. For example, some object property's value may be defined using mathematical equation. Then it is possible to find out its value at any moment of time, disregarding modeling steps. In such models an event may have non-zero duration, especially if common terms were used in model formalization. For example, describing the orbiter landing we may define a subjective, observer-dependent event "atmosphere entry". The only pragmatic of such events is in their tracking while observing the model. The modeled objects interaction, represented in analytical form, may be described without using the notion of "event".

So, generally, the events distinction have sense only in case if the modeling result implies observation of such events, and is represented in the same context in which these events are defined. In other words – if events perception and interpretation are valuable for the model's user. The user is always asking questions to the model, explicitly or not. If the question is formulated as "when the orbiter will enter the atmosphere?", such event distinction is necessary. If the question is "at which moment the orbiter will land?", or "how long the landing will take?", then this event is excessive. We may conclude that there is no reliable way of distinction between "objective", or "physical" events, and "subjective" ones.

As another example, let us consider the program which monitors some industrial equipment functioning. It registers an event: the sensor have reported that the pressure in some unit has exceeded 1 MPa. At a first glance, this may be considered a physical event. But from the application's functionality point of view we are not interested in particular pressure value, and in the physical process at all. We just have to know that the defined critical pressure level was exceeded. In the terms of application usage context, this may be named as "Violation of the normal operations mode of the unit that carries the risk of destroying its structure". This formulation is most likely given in some regulatory documents. So, we are shifting from the physical context to the external subjective one. In this context, an application has to react accordingly with the regulations: notify the duty staff, turn on emergency pressure relief, etc. In this process, we will never be interested in the initial pressure value reported by the sensor, which have triggered process execution.

In addition, it is obvious that the "pressure has exceeded 1 MPa" is just an abstraction. The sensor readings represented using any measurement scale are just a very simplified echo of the processes that are going on in the unit. Therefore, there is no way to "exactly" describe the "real physical" aspect of the process, and it may not be defined as a goal of ontology or model building. The goal shall be formulated as building of the ontology and model which will allow pragmatic task resolution – in our case, to prevent unit explosion due to excessive pressure. Distinction of the objects, events and states shall be considered only from this point of view.

ISO 15926 and other standard or widespread, high-level or domain-specific modeling methods are providing ways for the several subjects to agree on the common context, in which they are manipulating objects, and on the common meanings of the ontology terms. However, this may restrict model's efficiency, reliability and safety. It is worth having external high-level ontology on top of the model only when it may be used for representing modeling results to the user, when it is required for collaboration on the model, or data exchange with other models. It is worth using domain-specific ontology, if it is related with the task that is resolved – not just the subject matter area itself. If these preconditions are not met, the better way is to build an ontology based on the subjective context of the model use, despite its possible inconsistences with the concepts defined by standards. If this may lead to erroneous modeling results – then the subjective context has to be revised, as it is the source of errors.

5   Evaluation

Let us consider a very simplified fragment of the ontology that we have built in the project aimed to plan an optimal program of municipal kindergarten construction and reconstruction. Classes, properties and relations of the static objects and agents in this ontology are shown below:

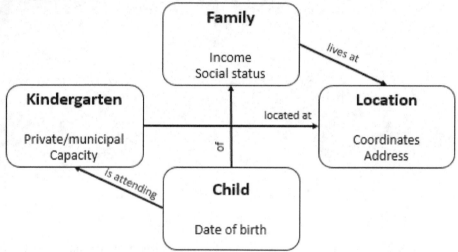

Fig. 2. Fragment of the ontology for kindergarten construction optimization task

Our primary task is to predict births in different parts of the city for several years. In our simulation, the children (agents) are born accordingly to the model of repro-ductive behavior of various groups of population. This model is based on a number of properties, reflecting socio-demographic characteristics of inhabitants. When a child reaches an appropriate age, the need arises to find a kindergarten for him. The fact and the moment when this need emerges is defined according to the family's social status, which has rather complex definition and distribution.

In this fragment of model, the families' emergence, children birth and school-age achievement, needs emergence and fulfillment may be considered as events. Children birth and school-age achievement seems to be "physical" events, having exact temporal localization, while family and needs-related events are much more diffuse. It is more useful to consider having family status, or having some need, as the states of the subjects, without exactly defined time boundaries. The degree of having these statuses may be represented using some weight coefficients, which leads to the fuzzy logic.

The convenience of semantic models use for simulation modeling is that we do not have a need to keep the model's evolution description, i.e. include the temporal dimension inside the model. It is enough to refresh agents' state during the simulation, and make its snapshots as separate static models for further analysis. This leads to the absence of necessity to use such a cumbersome modeling concepts as temporal parts.

In our model, the "decision" is the most obvious pretender to be represented as an "event". We have modeled a set of factors leading to the decision to have a child, for various social groups. As the agent's behavior simulation is necessary for our model, it is justified to define the events meaningful in the subjective context of the person's thinking. This model has the software implementation, based on the RDF triple store. Its user interface is shown below:

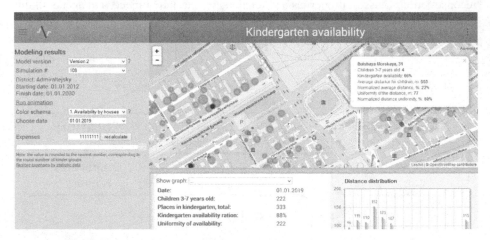

Fig. 3. User interface of the kindergarten construction optimization model

The yearly economic forecast is one of the inputs of the model, and can be adjusted. It is affecting the agents' behavior, which are used to obtain the detailed demographic forecast. In its turn, this forecast is used to calculate load of the kindergartens and some key metrics. A user can adjust operational expenses level on the social infrastructure, which affects existing kindergartens capacity, and see feedback from the model. The level of capital expenses affects implementation of the program of new kindergartens development and reconstruction. Expenses level may be set by the user, or, in contrary, calculated in the model according to the required key metrics values. The construction program itself can also be adjusted, or calculated optimally according to the expected children population distribution.

The ontology used in this model is completely based on the subject matter area concepts, including those defined in regulatory documents and demographic research reports. This simplifies model structure, makes it easier to define the agents' behavior, to interact with the user allowing him to control the model, and view the results.

6 Conclusions

We have formulated a simple criterion of sufficiency of ontology and model abstraction level, explained in terms of efficiency, reliability and safety. This criterion gives a useful point of view for evaluation of practicability of the standard ontologies usage for building semantic models. It provides a starting point for proving the right choose of conceptual basis for an ontology, instead of doing this intuitively. We have considered temporal aspect of the system's dynamic representation in the ontology, made conclusions on the events distinction criteria, and events conceptualization, particularly for the models intended for use in multi-agent simulation modeling environments. The logical structure of the sample ontology, and the mode of its use in the city's social infrastructure investment optimization model was described, proving the idea that a rational way of events definition is completely related with the subjective context of the modeled phenomena. The general conclusion of our work is that the model, being more an image of the human consciousness that of the "real world", has better be aligned with the subjective context of its use – when this gives better practical results – than with the abstract conceptual theories or standards.

References

1. Bell, D., de Cesare, S., Lycett, M.: Semantic web service architecture for simulation model reuse. In: 11th IEEE Symposium on Distributed Simulation and Real-Time Applications (2007)
2. van Dam, K., Nikolic, I., Lukszo, Z. (eds.): Agent-Based Modelling of Socio-Technical Systems. Springer (2013)
3. Gailly, F., Poels, G.: Conceptual modeling using domain ontologies: improving the domain - specific quality of conceptual schemas. In: Proceedings of the 10th Workshop on Domain-Specific Modeling (2010)
4. Gangemi, A., Catenacci, C., Ciaramita, M., Lehmann, J.: A theoretical framework for ontology evaluation and validation. In: SWAP, vol. 166 (2005)
5. Geerts, G., McCarthy, W.E.: An Accounting Object Infrastructure for Knowledge Based Enterprise Models. IEEE Intelligent Systems and their Applications 14(4), 89–94 (1999)
6. Guizzardi, G., Wagner, G.: Towards an ontological foundation of agent-based simulation. In: Proceedings of the 2011 Winter Simulation Conference (2011)
7. ISO 15926-1:2004. Industrial automation systems and integration – Integration of lifecycle data for process plants including oil and gas production facilities. http://www.iso.org/iso/home/store/catalogue_tc/catalogue_detail.htm?csnumber=29556
8. Miller, J., He, C., Couto, J.: Impact of the semantic web on modeling and simulation. In: Fishwick, P.A. (ed.) Handbook of Dynamic System Modeling. Chapman & Hall (2007)
9. Simeone, D.: An ontology-based system to support agent-based simulation of building use. Journal of Information Technology in Construction (2012)
10. Sowa, J.: Knowledge Representation: Logical, Philosophical, and Computational Foundations. Brooks/Cole, Pacific Grove (2000)
11. Zhou, Q., Bakshi, A., Prasanna, V., et al.: Towards an integrated modeling and simulation framework for freight transportation in metropolitan areas. In: IEEE Intl. Conf. on Information Reuse and Integration (2008)
12. Zhou, Q., Prasanna, V., Chang, H., Wang, Y.: A semantic framework for integrated modeling and simulation of transportation systems. Relation 10(1.116), 3458 (2010)

Simulation with Input Uncertainties Using Stochastic Petri Nets

Sérgio Galdino$^{(\boxtimes)}$ and Francisco Monte$^{(\boxtimes)}$

Polytechnic School of Pernambuco University, Recife, PE, Brazil
sergio.galdino@ieee.org, fmss@ecomp.poli.br

Abstract. Simulation with input uncertainties using stochastic Petri nets (SPN) takes into account the effects of uncertainties in exponential transition rates when calculating performance indices. The proposed simulation method, using interval arithmetic, allows the computing with simultaneous influence of uncertainties and variabilities of parameters. A case study is presented for processing uncertainty from stochastic Petri nets models. Interval SPN simulation do not compute the reachability graph, but follows the proposed Monte Carlo interval style simulation.

Keywords: Simulation and modelling · Discrete event systems · Monte Carlo · Interval arithmetic · Petri nets

1 Introduction

Model-based performance and dependability evaluation constitute an important aspect of the design of systems such as Data Center , Local Area Networks, and Manufacturing Systems. In any case, the evaluations results should be interpreted with care. When uncertainties or variabilities are associated with systems parameters, single points characterization of parameters is inadequate. Uncertainties may be associated to parameters that are not known in advance especially in early stages of system design. Nevertheless, designers may have a good idea about the range of values associated to these parameters considering previous experiences with similar systems [1]. System parameters variabilities may represent different mean demands at a given system component during various time periods. Therefore, when parameters are not precisely known, but known to lie within a certain interval, algorithms may be implemented using interval arithmetic.

Interval SPN simulation models takes into account the effects of uncertainties in exponential transition rates when calculating performance indices. The proposed model and respective simulation method, using interval arithmetic, allows the computing with simultaneous influence of uncertainties and variabilities of parameters. Since the interval SPN simulation do not compute the reachability graph the technique is useful for large systems. In such cases (large systems), discrete event simulation is the preferred tool for performance evaluation. In this paper, we show how simulation with input uncertainties using stochastic Petri nets models can be used for conducting discrete event simulations.

© Springer International Publishing Switzerland 2015
Y. Tan et al. (Eds.): ICSI-CCI 2015, Part III, LNCS 9142, pp. 194–202, 2015.
DOI: 10.1007/978-3-319-20469-7_22

This paper focuses on simulating interval SPN models. The next section briefly introduces interval arithmetic. Section 3 gives some background on interval simulation. Section 4 describes the Interval SPN in the context of discrete event interval simulation. Section 5 presents a case study and the respective analysis results. Finally, Sect. 6 concludes and suggests future studies.

2   Preliminary Considerations

Prior to present the Interval SPN simulation, this section introduces some basic concepts needed to understand how interval arithmetic may be used for evaluating system's metrics. Hence, we initially introduce some concepts on interval arithmetic.

2.1   Interval Arithmetic

Interval analysis is a relatively young branch of mathematics, originating with the work of R. E. Moore [2]. It is based on an extension of the real number system to a system of closed intervals on the real line. If $\underline{x}$ and $\overline{x}$ are real numbers with $\underline{x} \leq \overline{x}$, then the set of real numbers x satisfying $\underline{x} \leq x \leq \overline{x}$ is the closed interval $[\underline{x}, \overline{x}]$. In this paper, the term interval will mean closed (rather than open or half-open) interval. Each real number x can be identified with a degenerate interval $[x, x]$.

Throughout this paper, all matrices are denoted by bold capital letters ($\mathbf{A}$), vectors by bold lowercase letters ($\mathbf{a}$), and scalar variables by ordinary lowercase letters (a). Interval variables are enclosed in square brackets ($[\mathbf{A}]$, $[\mathbf{a}]$, $[a]$). Underscores and overscores denote lower and upper bounds, respectively. Angle brackets $\langle\ ,\ \rangle$ are used for intervals defined by a midpoint and a radius. A real interval $[x]$ is a nonempty set of real numbers

$$[x] = [\underline{x}, \overline{x}] = \{\tilde{x} \in \mathbb{R} : \underline{x} \leq \tilde{x} \leq \overline{x}\} \tag{1}$$

where $\underline{x}$ and $\overline{x}$ are called the *infimum (inf)* and *supremum (sup)*, respectively, and $\tilde{x}$ is a point value belonging to an interval variable $[x]$.

The midpoint of $[x]$ is defined as the real number,

$$mid[x] = \check{x} = \frac{1}{2}\left(\underline{x} + \overline{x}\right) \tag{2}$$

and the width and radius of $[x]$ are defined as,

$$\begin{aligned} width([x]) &= (\overline{x} - \underline{x})\ and \\ rad([x]) &= \Delta x = \tfrac{1}{2}w([x]). \end{aligned} \tag{3}$$

The radius may be used to define an interval $[x] \in I(\mathbb{R})$. Then $\langle \check{x},\ \Delta x\rangle$ denotes an interval with midpoint $\check{x}$ and radius Δx. The $[x, x] \equiv x$ is called point interval or thin interval. A point or thin interval has zero radius and a thick interval has a radius greater than zero.

The elementary operations on intervals $[a] = [\underline{a}, \overline{a}]$ and $[b] = [\underline{b}, \overline{b}]$ are calculated explicitly as

1. $[\underline{a}, \overline{a}] + [\underline{b}, \overline{b}] = [\underline{a} + \underline{b}, \overline{a} + \overline{b}]$
2. $[\underline{a}, \overline{a}] - [\underline{b}, \overline{b}] = [\underline{a} - \overline{b}, \overline{a} - \underline{b}]$
3. $[\underline{a}, \overline{a}] \cdot [\underline{b}, \overline{b}] = [min(\underline{a} \cdot \underline{b}, \underline{a} \cdot \overline{b}, \overline{a} \cdot \underline{b}, \overline{a} \cdot \overline{b}), max(\underline{a} \cdot \underline{b}, \underline{a} \cdot \overline{b}, \overline{a} \cdot \underline{b}, \overline{a} \cdot \overline{b})]$
4. $[\underline{b}, \overline{b}]^{-1} = [1/\overline{b}, 1/\underline{b}]$ if $0 \notin [\underline{b}, \overline{b}]$
5. $[\underline{a}, \overline{a}] / [\underline{b}, \overline{b}] = [\underline{a}, \overline{a}] \cdot [\underline{b}, \overline{b}]^{-1}$

These operations are called interval arithmetic. Besides these operations, interval functions ($exp[x], log[x], sin[x], cos[x], ...$) return both upper and lower bounds. The interval arithmetic operations are defined for exact calculation [2]. Machine computations are affected by rounding errors. Therefore, the formulas were modified in order to consider the called directed rounding [3].

An important result is the inclusion property theorem . Rall aptly calls this the fundamental theorem of interval analysis [4,5].

Theorem 1 (Fundamental Theorem).
If the function $f([x]_1, \cdots [x]_n)$ is an expression with a finite number of intervals $[x]_1, \cdots [x]_n \in I(\mathbb{R})$ and interval operations $(+, -, \times, \div)$, and if $[w]_1 \subseteq [x]_1, \cdots, [w]_n \subseteq [x]_n$ then $f([w]_1, [w]_2, [w]_3, \cdots, [w]_n) \subseteq f([x]_1, [x]_2, [x]_3, \cdots, [x]_n)$.

3 Interval Simulation

The interval simulation is based on intervals instead of the traditional floating-point numbers. However, when the limitations mentioned in Sect. 1 are encountered and input uncertainties cannot be ignored, one needs to find approaches to extend the current fitting method based on precise parameters [6].

3.1 Random Interval Variate Generation

The parameters of statistical distributions in interval simulation are intervals instead of real numbers [9]. So, generate a uniform random number, u, in $[0, 1)$, then calculate $[x]$ by:

$$[x] = -log(u)/[\lambda], \tag{4}$$

where $[\lambda]$ is the interval rate parameter of the exponential distribution. Now, $[x]$ is a interval random number with an interval exponential distribution.

4 Interval SPN

Petri Nets (PNs) and their extensions are a family of formalisms of graphical representation for description of systems whose dynamics are characterized by concurrency, synchronization, mutual exclusion, and conflict, which are typical features of distributed environments. PNs incorporate a notion of local state and a rule for state change (transition firing) that allow them to capture both the static and the dynamic characteristics of a real system being explicit about time considerations. Stochastic Petri Nets (SPNs) have emerged as a important performance modelling tool.

Interval SPN is an extension of GSPN (*Generalized Stochastic Petri Nets*) model in order to introduce the interval analysis with practical applications in performance and dependability evaluation [8].

A formal definition of Interval SPN (*ISPN*) is provided below. This definition keeps to the SPN definition presented in [10], but considers real intervals assigned to transition delays and weights instead of adopting real single values.

Let $ISPN = (P, T, I, O, \Pi, G, M_0, Atts)$ be an interval stochastic Petri net, where

- $P = \{p_1, p_2, \cdots p_n\}$ is the set of places,
- $T = \{t_1, t_2, \cdots t_m\}$ is the set of transitions,
- $I \in (\mathbb{N}^n \to \mathbb{N})^{n \times m}$ is a matrix of marking-dependent multiplicities of input arcs, where the i_{jk} entry of I gives the possibly marking-dependent arc multiplicity of input arcs from place p_j to transition t_k $[A \subset (P \times T) \cup (T \times P) - set\ of\ arcs]$,
- $O \in (\mathbb{N}^n \to \mathbb{N})^{n \times m}$ is a matrix of marking-dependent multiplicities of output arcs, where o_{jk} entry of O gives the possibly marking-dependent arc multiplicity of output arcs from transition t_j to place p_k,
- $H \in (\mathbb{N}^n \to \mathbb{N})^{n \times m}$ is a matrix of marking-dependent multiplicities of inhibitor arcs, where h_{jk} entry of H gives the possibly marking-dependent arc multiplicity of inhibitor arcs from place p_j to transition t_k,
- $G \in (N^n \to \{true, false\})^m$ is a vector that assigns a guard condition related to place markings to each transition,
- $M_0 \in \mathbb{N}^n$ is a vector that assigns the initial marking of each place (initial state),
- $Atts = (Dist, W, Markdep, Police, Concurrency)^m$ comprises the set of attributes for transitions, where
  - $[Dist] \in \mathbb{N}^m \to [\mathcal{F}]$ is a possible marking-dependent firing interval distribution function (the domain of $[\mathcal{F}]$ *is* $[0, \infty)$),
  - $[W] \in \mathbb{N}^m \to I(\mathbb{R}^+)$ is a possible marking-dependent weight,
  - $Markdep \in \{constant, enabdep\}$ where the transition interval firing timing distribution could be marking independent (*constant*) or enabling dependent (*enabdep* - the distribution depends on the actual enabling condition),
  - $Police \in \{prd, prs\}$ is the preemption policy (*prd- preemptive repeat different* means that when a preempted transition becomes enabled again the previous elapsed firing time is lost; *prs- preemptive resume*, in which the firing time related to a preempted transition is resumed when the transition becomes enabled again),

- *Concurrency* ∈ {*ss, is*} is the degree of concurrency of transitions, where *ss* represents single server semantics and *is* depicts infinite-server semantics.

ISPN models deal with system uncertainties by considering intervals for representing time assigned to transition models. The proposed model and the respective methods, adapted to take interval arithmetic into account, allow the influence of simultaneous parameters and variabilities on the computation of metrics to be considered, thereby providing rigorously bounded metric ranges.

ISPN is a model that might be considered for both stochastic simulation and numerical analysis. In an early work in particular, ISPN is considered to be a high-level formalism for ICTMC (Interval Continuous-Time Markov Chain) [9]. Simulation algorithms, as an alternative approach, follows the proposed Monte Carlo interval style simulation with some refinements. Since the ISPN tool [8] is based on the generation of the entire state space, the technique becomes intractable for large systems. In such cases, discrete event simulation is the preferred tool for performance evaluation. Now we show how ISPNs can be used as a simulation model.

4.1   Simulative Analysis of ISPN

Discrete-event simulation accompanies the whole design and evaluation process of the ISPN model. The simulative approach is simple from a logical point of view, and is easily implementable in a computer program, so that ISPN can also be considered as a possible general simulative language. The automated simulation for measure estimations is realized.

In general, an ISPN simulation is performed by executing the following steps:
Algorithm I
Input data: It contains matrix description of a ISPN model. These matrices are used in recursive simulation algorithms.

1. $clock = [0, \ 0]$, a degenerated interval;
2. Find enabled transitions;
3. • If immediate transitions are enabled: according to priorities and weights choose one (random) immediate transition to fire;
   • If timed transitions $[\theta]_k$ are enabled: according to the (generate a interval random sample $[T]_k = -log(u)/[\lambda_k]$, $u \in [0,1)$) firing delays choose one timed transition to fire (find minimum $\inf[T]_k$);
4. Fire the selected transition, i.e. compute the new marking;
5. $clock = clock + min([T]_k)$;
6. Update the statistics of the measures (if necessary);
7. Repeat cyclically the steps $2 - 6$ until the terminating condition is fulfilled.

This discrete-event simulation is a stochastic experiment where samples drawn are interval random variates. Measures can be obtained by estimating the mean value of the sampled data. Precision of this estimate are usually based on confidence intervals derived from the sample variance.

5   Results(Case Study)

In this section ISPN (Algorithm I) simulation is used for computation of interval metrics . These originate from the steady state analysis of ISPN models. The MATLAB toolbox INTLAB framework is used by the ISPN simulation. The comparative study is focused on the relative quality of resulting interval probabilities. Two case studies are presented. In the first we see that ICTMC simulation and ISPN simulation gives similar results of both mean values and interval width of computed indexes. The second case show the usefulness of ISPN simulation for complex systems.

5.1   Case Study: ISPN Analysis of a Parallel System Model

The usefulness of simulation for the ISPN analysis of complex systems can be demonstrated with the use of the simple parallel processing example of Fig. 1, by assuming that a large number of customers are considered in the system [12].

The ISPN comprises nine places and nine transitions. The characteristics of transitions for this model are summarized in Table 1. The initial markings are ($p_1 = N$, $N = 2$, 8, 20, 50).

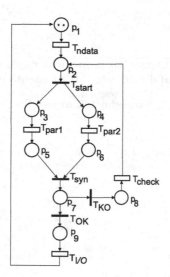

Fig. 1. The ISPN description of a parallel system

Table 2 summarizes the results of an experiment in which estimates for the throughput and for the processing power (i.e., the mean number of tokens in place p_1) of the system are computed with verified results in the case of small populations.

The results presented in Table 3 show the propagation effects of the interval parameters with the growth of the states space. We see from Table 2 and Table 3

Table 1. Specifications of the interval parameters transitions of ISPN of Fig. 1

Transition	Interval Rate	Semantics
t_{ndata}	$0.2 \pm 1e-4$	Infinite-Server
t_{par1}	$2.0 \pm 1e-3$	Single-Server
t_{par2}	$1.0 \pm 1e-3$	Single-Server
t_{check}	$0.1 \pm 1e-4$	Single-Server
$t_{I/O}$	$5.0 \pm 1e-3$	Single-Server

Transition	Weight	Priority
t_{start}	1	1
t_{syn}	1	1
t_{KO}	0.01	1
t_{OK}	0.99	1

Table 2. Performance and state space cardinality for various ISPN/GSPN models for Parallel system of Fig. 1. The T is the number of tangible markings. N is the number of tokens.

N	T	Processing Power (verified result)	System Throughput (verified result)
2	20	$< 1.50555067214211, 0.00000000000001 >$	$< 0.30111013442842, 0.00000000000001 >$
8	825	$< 4.49775685335993, 0.00000000000023 >$	$< 0.89955137067199, 0.00000000000004 >$
20	19481	4.9500 [1]	0.9900 [1]
50	609076	5.060861 ∓ 0.035105 [2]	0.994953 ∓ 0.006787 [2]

1 - **computed with numerical method [12]**.
2 - **mean and 95 % confidence interval Student's t distribution of 10 measures
(TimeNeT Satandard/Default Simulation).**

Table 3. Performance from algorithm 2 simulation of various ISPN models for the Parallel system of Fig. 1. N is the number of tokens (customers).

N	Processing Power[1]	System Throughput[1]	Simulation Time (s)
2	$< 1.49939470982660 \mp 0.01263389801653,$ $0.00169611864064 \mp 0.00001315425440 >$	$< 0.30019665075241 \mp 0.00668571969967,$ $0.00030649512514 \mp 0.00006704634124 >$	31.10 ∓ 0.85
	$< 1.50660524485266 \mp 0.00240050004822,$ $0.00170524761188 \mp 0.00002469958150 >$	$< 0.302053470069916 \mp 0.003205167157467,$ $0.000308757790009 \mp 0.000033205268453 >$	325.43 ∓ 8.34
8	$< 4.45927719509260 \mp 0.13706766176585,$ $0.00556655445414 \mp 0.0001593739519 0 >$	$< 0.896430073576824 \mp 0.017774762608343,$ $0.001065821843289 \mp 0.000021332639384 >$	27.97 ∓ 3.70
	$< 4.50318839248343 \mp 0.02952967685400,$ $0.00562211674757 \mp 0.00003819920759 >$	$< 0.90221016280905 7 \mp 0.004483941134655,$ $0.001071185959894 \mp 0.000069582214517 >$	269.86 ∓ 13.38
20	$< 4.88292749572942 \mp 0.10779415561462,$ $0.00613467987262 \mp 0.00013707231290 >$	$< 0.966866108486971 \mp 0.029578863125779,$ $0.001168722525812 \mp 0.000039356611371 >$	29.27 ∓ 0.97
	$< 4.94798740425193 \mp 0.04492733125670,$ $0.00619710132538 \mp 0.00005252627289 >$	$< 0.986325383146433 \mp 0.011260106249735,$ $0.001188061868799 \mp 0.000014419230155 >$	278.63 ∓ 9.84
50	$< 5.08267896677375 \mp 0.15092684769951,$ $0.00637136469359 \mp 0.00021820979865 >$	$< 0.981042258576016 \mp 0.011949354786812,$ $0.001183091739681 \mp 0.000014814795795 >$	29.78 ∓ 1.76
	$< 4.97141093878957 \mp 0.03930101615189,$ $0.00622327391604 \mp 0.00004143052367 >$	$< 0.991866541100299 \mp 0.008769904830827,$ $0.001195910561842 \mp 0.000012260280340 >$	293.53 ∓ 17.97
	$< 4.95341558712124 \mp 0.01328266145083,$ $0.00620249250325 \mp 0.00001520132824 >$	$< 0.989029148510779 \mp 0.002769710251423,$ $0.001191382160659 \mp 0.000033270477812 >$	10460.52 ∓ 8757.82 [3]

1 - **10000/ 100000/ (1000000) iterations.**
2 - **mean and 95 % confidence interval Student's t distribution of 10 measures.**
3 - **Windows hibernation occurred in some simulations .**

that mean/mid interval value are in agreement. The growth of interval width is observed with the increase of N (number of customers). There exist no significant differences in interval widths with 10000/100000/(1000000) iterations, that can be interpreted as steady state was reached in all simulations. Finally, is also observed uncertainty decreasing to the increasing number of interactions.

6   Conclusions

In this paper we apply an interval based simulation mechanism, which simulates based on intervals instead of floating-point numbers. A random interval variate generating method was used to generate random intervals from a pair of cdf's corresponding to the lower and upper bounds of the exponential input parameter.

In this paper we show how ISPNs can be used as a simulation model to large systems. ISPN simulation (algorithm II) do not compute the reachability graph, but follows the Monte Carlo interval style simulation.

The future work includes the systematic study of the ISPN simulation. We also need to study other distribution functions with interval parameters, e.g. Weibull, Erlang and Normal functions for systems simulation. Inverval simlation should be explored to sensitivity studies in variabilities of models parameters .

References

1. Majumdar, S., Lüthi, J., Haring, G., Ramadoss, R.: Characterization and analysis of software and computer systems with uncertainties and variabilities. In: Dumke, R., Rautenstrauch, C., Schmietendorf, A., Scholz, A. (eds.) Performance Engineering. LNCS, vol. 2047, pp. 202–221. Springer, Heidelberg (2001)
2. Moore, R.E.: Interval Analysis. Prentice Hall, Englewood Clifs (1966)
3. Kulish, U.W., Miranker, W.L.: The Arithmetic of the Digital Computers: A New Approach. SIAM Review **28**, 1 (1986)
4. Rall, L.B.: Computacional Solution of Nonlinear Operator Equtions. Wiley, New York (1969)
5. Hansen, E.R., Walster, G.W.: Global Optimization Using Internal Analysis, 2nd edn. Marcel Dekker, Inc., New York (2004). Revised and Expanded
6. Batarseh, O.G., Wang, Y.: Reliable simulation with input uncertainties using an interval-based approach. In: Mason, S.J., Hill, R.R., Münch, L., Rose, O., Jefferson, T., Fowler, J.W. (eds.) Winter Simulation Conference, WSC, pp. 344–352 (2008)
7. Balakrishnan, N., Rao, C.R.: Handbook of Statistics 20: Advances in Reliability. Handbook of Statistics Series, vol. 20. Elsevier North Holland (2001)
8. Galdino, S.M.L., Maciel, P.R.M.: ISPN: modeling stochastic with input uncertainties using an interval- based approach. In: Michalowski, T. (ed.) Applications of MATLAB in Science and Engineering, vol. 20, pp. 409–432. InTech (2011). ISBN: 978-953-307-708-6. http://www.intechopen.com/books/applications-of-matlab-in-science-and-engineering/ispn-modeling-stochastic-with-input-uncertainties-using-an-interval-based-approach
9. Galdino, S.M.L.: Interval continuous-time markov chains simulation. In: 2013 International Conference on Fuzzy Theory and Its Applications (iFUZZY2013). National Taiwan University of Science and Technology (NTUST), Taipei (2013). http://isdlab.ie.ntnu.edu.tw/ifuzzy2013/Parallel_Oral_Sessions_III.html#IJFS2

10. German, R.: Performance Analysis of Communicating Systems - Modeling with Non-Markovian Stochastic Petri Nets. Wiley (2000)
11. Desrochers, A.A., Al-Jaar, R.Y.: Applications of Petri nets in manufacturing systems: modeling, control, and performance analysis. IEEE Press, Piscataway (1994)
12. Marsan, M.A., Balbo, G., Conte, G., Donatelli, S., Franceschinis, G.: Modelling with Generalized Stochastic Petri Nets. Wiley Series in Parallel Computing. John Wiley and Sons (1995)

Image and Texture Analysis

Single Image Dehazing
Based on Improved Dark Channel Prior

Taimei Zhang[(⊠)] and Youguang Chen[(⊠)]

Computer Center, East China Normal University, Shanghai 200062, China
zhangtm552@163.com, ygchen@cc.ecnu.edu.cn

Abstract. The sky region of restored images often appears serious noise and color distortion using classical dark channel prior algorithm. To address this issue, we propose an improved dark channel prior algorithm which recognizes the sky regions in hazy image by gradient threshold combined with the absolute value of the difference of atmospheric light and dark channel. And then we estimate the transmission in sky and non-sky regions separately. At last, we enhance the brightness and contrast of results. Experimental results show that our restored images are more natural and smooth in sky regions.

Keywords: Dark channel prior · Image dehazing · Sky regions

1 Introduction

Images of outdoor screen are usually degraded in condition of fog and haze weather, how to remove fog effectively is important to improve the robustness and reliability of visible systems which is badly affected by the bad weather.

Recently, many single image dehazing algorithms have been proposed [4] [5]. Tan's [2] and Fattal's [3] algorithms are typical image restoration methods. Tan finds that haze-free image must have higher contrast compared with hazy image and airlight tends to be smooth, then he uses the way of maximizing local image contrast to achieve haze removal. The results are compelling but may not be physically valid. Fattal assumes that the transmission and the surface shading are locally uncorrelated, and then he estimates the reflectivity of the scene and the medium transmission. However, his method can't handle images got in the heavy fog condition.

In 2009, He [1] proposed the dark channel prior, which is based on the observation that most local patches in outdoor haze-free images contain some pixels whose intensity is very low in at least one color channel. Using this prior with the haze imaging model, we can recover a high-quality haze-free image. However, in sky regions, the assumption is broken and the restored images of sky regions tend to appear some serious noise and color distortion.

To address this issue, Sun [6] has proposed an improved dehazing algorithm by identifying sky regions and redefined the transmission. However, he judged the sky region only by the absolute value of the difference of atmospheric light and dark channel, which will lead some area with high intensity being judged as sky wrongly.

© Springer International Publishing Switzerland 2015
Y. Tan et al. (Eds.): ICSI-CCI 2015, Part III, LNCS 9142, pp. 205–212, 2015.
DOI: 10.1007/978-3-319-20469-7_23

In addition, Sun's new transmission of sky region sometimes doesn't work well. In this paper, we add gradient information as another criteria standard to judge sky regions, at the same time, we use a constant transmission in sky regions. At last, we enhance the brightness and contrast of the restored image.

The remaining of this paper is organized as follows. In section 2, we introduce the image model, and describe haze removal using dark channel prior. In section 3, our improved method is described in detail. Experimental results and analysis are provided in section 4. At last, we have a conclusion in section 5.

2   Dark Channel Prior

2.1   Atmospheric Scattering Model

In computer vision and computer graphics [11], the atmospheric scattering model widely used to describe the formation of a hazy image is as (1):

$$I(x) = J(x)t(x) + A(1 - t(x)) \tag{1}$$

Where I is the observed intensity, J is scene radiance, A is global atmospheric light, and t is the medium transmission describing the portion of the light that is not scattered and reaches the camera. The goal of haze removal is to recover J, A, and t from I.

2.2   Haze Removal Using Dark Channel Prior

In most of non-sky regions, at least one color channel has some pixels whose intensity are very low and close to zero [1]. In other word, the minimum intensity in such a region is close to zero. For a haze-free image J, expect for the sky regions, the intensity of its dark channel J^{dark} is close to zero, the formula is shown as (2).

$$J^{dark}(x) = \min_{y \in \Omega(x)}\left(\min_{c \in \{R,G,B\}} J^c(y)\right) \to 0 \tag{2}$$

Where J^c is a color channel of J and $\Omega(x)$ is a local patch centered at x.

We assume the atmospheric light A is given and the transmission in a local patch $\Omega(x)$ is constant. $\tilde{t}(x)$ is estimated transmission and we divide (1) by A and put the minimum operators on both sides

$$\min_{y \in \Omega(x)}\left(\min_c \frac{I^c(y)}{A^c}\right) = \tilde{t}(x)\min_{y \in \Omega(x)}\left(\min_c \frac{J^c(y)}{A^c}\right) + 1 - \tilde{t}(x) \tag{3}$$

As A^c is always positive, so we get

$$\min_{y \in \Omega(x)}(\min_c(J^c(y)/A^c)) \to 0 \tag{4}$$

Putting (4) into (3), we can estimate the transmission $\tilde{t}$ simply by

$$\tilde{t}(x) = 1 - \omega \min_{y \in \Omega(x)}(\min_c(I^c(y)/A^c)) \tag{5}$$

The constant parameter ω ($0 < \omega < 1$) in (5) is to keep a very small amount of haze for distant objects. In this paper, we fix ω to 0.95 for all the results.

Finally, the dehazing image can get from

$$J(x) = \frac{I(x)-A}{\max(\tilde{t}(x),t_0)} + A \tag{6}$$

The dehazing image is prone to noise when the transmission is close to zero, so we restrict the transmission with the lower bound t_0. A typical value of t_0 is 0.1.

3   Improved Dark Channel Prior

3.1   Reasons of Color Distortions in Sky Regions

Accurate transmission function[8] should be as (7).

$$\tilde{t}(x) = \frac{1-\min_{y\in\Omega(x)}(\min_c(I^c(y)/A^c))}{1-\min_{y\in\Omega(x)}(\min_c(J^c(y)/A^c))} \tag{7}$$

Since the dark channel prior is not a good prior for sky regions, so in the regions, $J^{dark}(x)$ may not be close to zero. That means the denominator part of (7) may be smaller than 1, consequently, the final transmission $\tilde{t}(x)$ we get is smaller than which actually is. Because the intensity of three color channels in bright area are similar, so $I(x) - A$ of three channels are in small distance. However, after divided by $\tilde{t}(x)$ in (6), the small distance would be amplified, which leads to color distortion.

3.2   Improved Dark Channel Prior Algorithm to Avoid Color Distortion

Because it is bright and smooth in sky regions, so we combine parameter k and gradient threshold r to recognize the sky region more accurately, in which k is the absolute value of difference between atmospheric light A and dark channel J^{dark} .

Detailed steps are as follows: First, we transform the hazy image into a gray image, then we get the gradient map $G(x)$ of gray image with sobel calculate operator [7]. Second, for each pixel in gradient image, if $G(x)$ is smaller than r and $|A - I^{dark}|$ is smaller than k, then we think it is the sky region and set $D(x)$ to 255, otherwise, $D(x)$ is 0. After that, we take a median filter operation to remove noise. At last, we get

$$D(x) = \begin{cases} 0 & \text{others} \\ 255 & |A - I^{dark}| < k \text{ and } G(x) < r \end{cases} \tag{8}$$

Finally, we estimate the transmission map adaptively as (9).

$$\tilde{t}(x) = \begin{cases} 1 - \omega\min_{y\in\Omega(x)}(\min_c(I^c(y)/A^c)) & D(x) \neq 255 \\ t_{sky} & D(x) = 255 \end{cases} \tag{9}$$

Where t_{sky} is a constant value.

3.3   Adjust Brightness and Contrast of Recovered Image

Since the scene radiance is usually not as bright as the atmospheric light, the restored image looks dim [9]. To solve this issue, we transform the haze-free image into HSV color space, and then enhance the brightness and contrast in V channel. At last, we transform the enhanced HSV image to RGB color space to get the final restored image.

On one hand, we adjust the brightness with a nonlinear method, which is to increase or decrease a constant value on original brightness. On the other hand, to adjust the contrast of images, we expand or narrow the difference between bright and dark pixels on the basis that the average brightness is unchanged [10].

The comprehensive adjustment formula is shown as (10).

$$out(x) = (1 + \alpha) * (V(x) - (\mu + \beta)) + (\mu + \beta) + \beta \tag{10}$$

Where $V(x)$ is the V channel of a haze-free image, $out(x)$ is enhanced result, μ is the average brightness. While β is the brightness increment ranging from -255 to 255, and α is the contrast increment, which ranges from -1 to 1.

Fig.1 shows the comparison of restored images before and after enhancement.

(a)   (b)   (c)

Fig. 1. (a) A hazy image. (b) Our result before enhancement. (c) Our result after enhancement

4   Experimental Results and Analysis

In our experiment, we implement the improved method by Microsoft Visual Studio 2012 on a PC with 2.8 GHz Intel Core 2 Duo Processor.

We use a patch of 15×15 in (2) to get dark channel image. Our atmospheric light A is estimated in the same way as He's [1]. After getting the transmission map, we use guided filter [12] [13] to refine it. At last, we enhance the restored image, for fair comparison, b and c in (11) are the same for restored images with different methods.

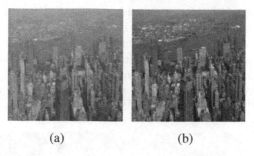

Fig. 2. (a) A input no sky image. (b) The result of no sky image

Experimental results show that more natural and smooth recovered images are got when t_{sky} ranges from 0.2 to 0.6, the parameter k ranges from 60 to 80, and gradient threshold r is fixed to 5. However, k and r are set as 0 when the input hazy image has no sky regions, then our method for images with no sky regions is actually the same as He's, the result is shown as Fig.2.

Fig. 3. (a) Input hazy image. (b) Sun's binary image with recognized sky region. (c) Our binary image (d) He's result. (e) Sun's result. (f) Our result.

Fig.3 shows the comparison of results. Fig.3d shows He's recovered image, which appears color distortion in sky regions. While the sky region of Fig.3e and Fig.3f are more natural than Fig.3d. Because in [6], the sky region is judged only by the value of $|A - I^{dark}|$, which leads some most hazy objects whose texture still can be seen to be judged as sky wrongly. Fig. 3b shows the binary image with recognized sky region using sun's method, while Fig.3c shows our binary image, in these images the sky region are displayed in white color. Apparently, our method can distinguish most hazy branches from sky accurately.

The areas judged as sky wrongly tend to have a larger transmission according to the modified transmission in [5], consequently, the restored result of the area is still hazy. As Fig.4a shows, the details of branches in red rectangle are not obvious, while in Fig.4b, the restored results in red rectangle are better using our method.

Fig.5 shows more haze-free images. Fig. 5b is the result of He's method, in which

(a)   (b)

Fig. 4. (a) Sun's detail image of result. (b) Our detail image of result

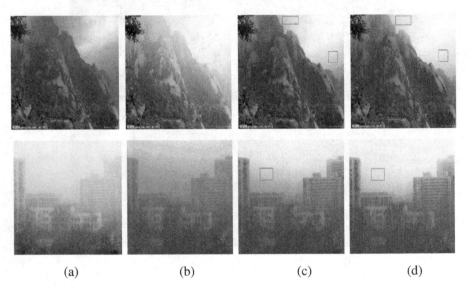

(a)   (b)   (c)   (d)

Fig. 5. (a) Input hazy image. (b) He's result. (c) Sun's result. (d) Our result

the sky region is distorted obviously. While, Sun's method and ours can avoid the distortion of sky region well as shown in Fig.5c and Fig.5d.

However, compared to our result in Fig.5d, Sun's result still has some areas distorted in sky regions. This is because we take a constant value as transmission for sky regions, while Sun's transmission in sky regions is $k / |A - I^{dark}|$ multiplied by original transmission. Some regions in sky where the intensity in dark channel is low, have the value of $|A - I^{dark}|$ close to k, then the final estimated transmission is still small, so the recovered result of these regions is still distorted.

Additionally, as shown in Fig.6c, instead of distortion, the detail of Sun's result in sky is not so obvious compared with our result in Fig.6d.This is because in Fig.6a, the value of dark channel in sky regions is close to the estimated atmospheric light A, which result in a larger transmission, then the recovered result in sky region will lose some details such as clouds. As a consequence, our method has a better result. The average gradient, variance and average intensity of sky parts of results are shown in Table1, it is consistent with our analysis.

(a)   (b)   (c)   (d)

Fig. 6. (a) Input hazy image. (b) He's result. (c) Sun's result. (d) Our result

Table 1. Average gradient, Variance and Average intensity of Results

		averGra	variance	averIn			averGra	variance	averIn
Fig.5 (up)	a	1.86	1390.52	244.00	Fig.3	a	13.09	266.21	170.00
	b	5.63	1465.83	231.00		d	20.39	1055.33	90.00
	c	2.09	1587.03	243.00		e	21.21	1631.47	124.00
	d	2.06	1578.20	242.00		f	21.10	1535.09	129.00
Fig.5 (down)	a	3.34	274.21	226.00	Fig.6	a	2.11	53.34	224.00
	b	9.24	1675.97	160.00		b	8.53	639.78	168.00
	c	6.83	2039.98	209.00		c	5.07	382.64	203.00
	d	6.12	1601.57	221.00		d	7.34	664.39	177.00

5   Conclusions

In this paper, we analyze the reasons of color distortion in sky regions using classic dark channel prior algorithm, and then we solve it effectively with an improved algorithm, which recognizes sky regions in a haze image and estimates the transmission separately in sky and non-sky regions. Furthermore, we enhance the brightness and contrast of restored images. Experimental results show that our method is effective.

However, there are still some disadvantages of our method. First, to get more natural recovered images, we use some parameters which need human intervention. In a consequence, the restoration process is not applicable for real-time systems. Second, some heavily hazy regions may be judged as sky regions wrongly. Therefore, more work should be done to improve the method further.

References

1. He, K.M., Sun, J., Tang, X.O.: Single image haze removal using dark channel prior. In: IEEE Conference on Computer Vision and Pattern Recognition, pp. 1956–1963 (2009)
2. Fattal, R.: Single image dehazing. ACM Transactions on Graphics **27**, 721–729 (2008)
3. Tan, R.: Visibility in bad weather from a single image. In: IEEE Conf. Computer Vision and Pattern Recognition (June 2008)
4. Yu, J., Xu, D.B., Liao, Q.M.: Image defogging: a survey. Journal of Image and Graphics **16**, 1561–1576 (2011). (in Chinese)
5. Guo, F., Cai, Z.X., Xie, B., Tang, J.: Review and prospect of image dehazing techniques. Journal of Computer Applications **30**, 2417–2421 (2010). (in Chinese)
6. Sun, X.M., Sun, J.X., Zhao, L.R., Zhao, Y.G.: Improved algorithm for single image haze removing using dark channel prior. Journal of Image and Graphics **19**, 381–385 (2014). (in Chinese)
7. Wang, G.Y., Ren, G.H., Jiang, L.H., Quan, T.F.: Single image dehazing algorithm based on sky region segmentation. Information Technology Journal **12**, 1168–1175 (2013)
8. Jiang, J.G., Hou, T.F., Qi, M.B.: Improved algorithm on single image haze removal using dark channel prior. Journal of Circuits and Systems **16**, 7–12 (2011). (in Chinese)
9. Zhang, B.B., Dai, S.K., Sun, W.Y.: Fast image haze-removal algorithm based on the prior dark-channel. Journal of Image and Graphics **18**, 184–188 (2013). (in Chinese)
10. Streidt, W.D.: Digital Image Filter Processing with FilterMeister. http://www.filter meister.com/docs/da.pdf
11. Fang, R.: Research on enhancement methods for weather degraded images. East China Normal University, Shanghai (2014). (in Chinese)
12. He, K.M., Sun, J., Tang, X.O.: Guided image filtering. IEEE Transactions on Pattern Analysis and Machine Intelligence, 1397–1409 (2013)
13. Tang, J.B., Jiang, B., Wang, T.: Research on single image dehazing using guided filtering. Science Technology and Engineering **13**, 3021–3025 (2013). (in Chinese)

Rotation Invariant Texture Analysis
Based on Co-occurrence Matrix and Tsallis Distribution

Mateus Habermann[1], Felipe Berla Campos[1,2], and Elcio Hideiti Shiguemori[1(✉)]

[1] Institute for Advanced Studies, São José dos Campos, Brazil
elcio@ieav.cta.br
[2] Federal University of São Paulo, São José dos Campos, Brazil

Abstract. This article addressed some extensions of a texture classifier invariant to rotations. Originally, that classifier is an improvement of the seminal Haralick's paper in a sense that the former is rotation invariant due to a circular kernel, which encompasses two concentric circles with different radii and then the co-occurrence matrix is formed. It is not considered only pixels falling exactly on the circle, but also others in its vicinity according to a Gaussian scattering. Firstly, 6 attributes are computed from each of the 18 texture patterns, after that texture patterns are rotated and a correct classification, considering Euclidian distance, is sought. The present paper assesses the performance of the aforementioned approach with some alterations: Tsallis rather than Gaussian distribution; addition of noise to rotated images before classification; and Principal Components Analysis during the extraction of features.

Keywords: Co-occurrence matrix · Image texture · Tsallis distribution

1   Introduction

Texture is an innate property of objects [1], and it can be seen as the distinctive physical composition or structure of objects, especially with respect to the size, shape and arrangement of its parts. Despite of a lack of consensus on a definite characterization in the literature, it is possible to state the texture is perceived by humans and contains an array of information cues that help humans infer facts about the environment that surrounds them [2].

The main texture recognition approaches can be categorized into four groups in terms of their different theoretical backgrounds. The first approach defines texture features that are derived from Fourier power spectrum of the image via frequency domain filtering. The second approach is based on statistics that measure local properties that are thought to be related to texture, for instance, mean or standard deviation. The third approach is based on modeling the image using assumptions, such as the image being processed possessed fractal properties, or it can be modeled using random field model. The last approach is the use of the joint gray-level probability density, as proposed by Haralick, that describes some easily computable texture features extracted from a co-occurrence matrix. [3][1]. However, that app-roach is not invariant to rotations of the image.

© Springer International Publishing Switzerland 2015
Y. Tan et al. (Eds.): ICSI-CCI 2015, Part III, LNCS 9142, pp. 213–221, 2015.
DOI: 10.1007/978-3-319-20469-7_24

Some alternatives that are meant to be invariant to rotations have been proposed, including an algorithm that is, supposedly, a robust multiresolution approach to texture classification, based on a local binary descriptor [4]. In [5] the rotation invariance is achieved by means of image descriptors that interpret variations as shifts in the feature vector, that are modeled as covariate shift in the data. In [6] it is suggested a modification in the co-occurrence matrix proposed by Haralick, using geometrical forms over patch of pixels instead of considering gray-levels of pairs. It is used two concentric circles in order to obtain a co-occurrence matrix that is invariant to rotations. That same method is ameliorated by considering a Gaussian distribution around the pixels that fall over the circle [2].

The present paper performs exploratory further analysis of the latter approach by considering a Tsallis distribution around the pixels. Furthermore, it is done feature extraction by means of Principal Components Analysis from 14 textural features proposed by Haralick, rather than only 6, as done in [6] and [2]. Finally we added noise to rotated images during the classification process.

In the next section, it is described the methods used in this paper, including a brief explanation about the method that is the motivation of this article; the texture features proposed by Haralick; the set of texture images used here and their classification; and the Tsallis distribution. In Section 3 it is shown some results with different parameters, and in Section 4 there are some discussion about the results and suggestions for further developments of this paper.

2   Materials and Methods

The contributions of this paper are related to the works of [6] and [2], which, in turn, are an extension of the Haralick´s article [3].

2.1   Haralick´s Grey Level Co-occurrence Matrix

The classical grey level co-occurrence matrix texture descriptors are easily computable statistical features that rely on grey level spatial relationships. It consists on generating a matrix that captures the grey level spatial dependencies. After that, it is possible to calculate some characteristics of that matrix.

Mathematically, an image I is a function that associates a grey level to each pixel, i.d., $I: D \rightarrow G$, where D is the domain of the image and $G = \{1, 2, \ldots, L_G\}$ is the set of grey levels. The grey level co-occurrence matrix $P_{(\Delta x, \Delta y)} \in \mathbb{N}^{L_G \times L_G}$ parameterized by $(\Delta x, \Delta y)$ is given by

$$P_{(\Delta x, \Delta y)}(i, j) = \sum_{p=1}^{n} \sum_{q=1}^{m} 1, \text{ if } I(p, q) = i \text{ and } I(p + \Delta x, q + \Delta y) = j. \quad (1)$$

where n is the number of columns and m, number of rows of the image patch. The offset $(\Delta x, \Delta y)$ means an angle parameter, which may take, normally, the values 0, 45, 90 and 135 degrees, according to Fig. 1.

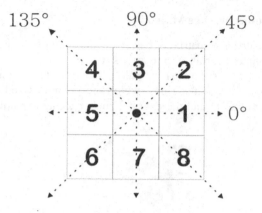

Fig. 1. Angles for calculation of the co-occurrence matrix [6]

Furthermore, Haralick proposed 14 texture descriptors; however, after deep examination, only 6 of them are statistically important [7]. They are:

$$f_{hom} = \sum_{i=1}^{L_G} \sum_{j=1}^{L_G} \frac{p(i,j)}{1+(i-j)^2}. \tag{2}$$

$$f_{con} = \sum_{n=0}^{L_G-1} n^2 \left\{ \sum_{i=1}^{L_G} \sum_{j=1}^{L_G} p(i,j) \right\}_{|i-j|=n}. \tag{3}$$

$$f_{ent} = -\sum_{i=1}^{L_G} \sum_{j=1}^{L_G} p(i,j) \ln p(i,j). \tag{4}$$

$$f_{corr} = \sum_{i=1}^{L_G} \sum_{j=1}^{L_G} \frac{(ij)p(i,j) - \grave{\i}_x \grave{\i}_y}{\acute{o}_x \acute{o}_y}. \tag{5}$$

$$f_{ener} = \sum_{i=1}^{L_G} \sum_{j=1}^{L_G} p^2(i,j). \tag{6}$$

$$f_{var} = \sum_{i=1}^{L_G} \sum_{j=1}^{L_G} (i - \grave{\i})^2 p(i,j). \tag{7}$$

where $p(i,j)$ is the normalized co-occurrence matrix; $p_x(i,j) = \sum_{j=1}^{L_G} p(i,j)$ and $p_y(i,j) = \sum_{i=1}^{L_G} p(i,j)$ are the marginal distributions; $\grave{\i}_x$ and $\grave{\i}_y$ are the mean of the marginal distributions and $\acute{o}_x$ and $\acute{o}_y$ are their standard deviations; and $\grave{\i} = \sum_{i=1}^{L_G} \sum_{j=1}^{L_G} i\, p(i,j)$.

2.2　Circular Co-occurrence Matrix

The circular co-occurrence matrix is based on measurements over circular shapes, which feature an innate invariance to rotations [6].

The circular co-occurrence matrix is calculated from the mean grey scale occurrences of two concentric circular rings, which are parameterized by their radii. The internal ring has a radius r_i and the external ring, r_o. One example of this is seen in Fig. 2.

Fig. 2.　An illustration of concentric rings [6]

Let $F_r(p, q)$ be the mean grey level on the circle ring centered at (p, q) with radius r. The circular co-occurrence matrix $P_{(r_i r_o)}$, with $r_o > r_i$, is thus

$$P_{(r_i r_o)}(i, j) = \sum_{p=1}^{n} \sum_{q=1}^{m} 1, \text{ if } F_{r_i}(p, q) = i \text{ and } F_{r_o}(p, q) = j. \tag{8}$$

In order to improve this approach, pixels in the vicinity of the circle can also be considered, according to a Gaussian distribution, and the kernel $K_{r,6}$ is thus calculated:

$$K_{r,6}(i, j) = e^{-\frac{\left(r - \sqrt{i^2 + j^2}\right)^2}{26^2}}. \tag{9}$$

In this way, pixels that are nearer the radius have greater importance.

Thus, the image patch I can be transformed according to (10).

$$I'(u, v) = \sum_{i,j} I(u - i, v - j) . K(i, j). \tag{10}$$

And, finally, the co-occurrence matrix originating from two images of kernels $K_{r_i,6}$ and $K_{r_o,6}$ is calculated according to (8).

2.3　Images under Analysis and Classification Method

2.3.1　Images Used in this Paper

The images of the Fig. 3 are referred to according to the nomenclature adopted in its database provided by University of Southern California [8]. In total, there are 18 texture patterns.

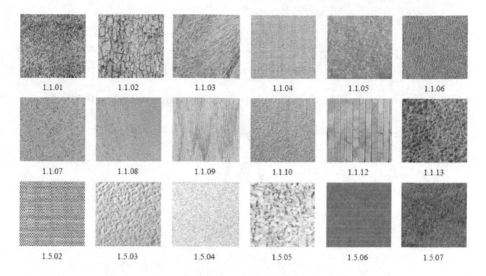

1.1.01	1.1.02	1.1.03	1.1.04	1.1.05	1.1.06
1.1.07	1.1.08	1.1.09	1.1.10	1.1.12	1.1.13
1.5.02	1.5.03	1.5.04	1.5.05	1.5.06	1.5.07

Fig. 3. Texture patterns evaluated [6]

2.3.2   Classification Method

All the images of Fig. 3, one by one, are submitted to the texture features extraction method described earlier, what yields an 18x6 matrix M, whose lines correspond to the texture features of the images.

Then, each image is rotated in 6 different angles è, according to (11).

$$\grave{e}_i = \frac{i}{7}360, \qquad \text{for } i = 1, 2, ..., 6. \tag{11}$$

From each rotated image $r_k \in R$ the texture features are extracted, forming the vector $v_k \in \mathbb{R}^6$, with $k = 1, 2, ..., 108$. Those rotated images can be subdivided into 18 groups, and each group g_l corresponds to a non-rotated image. r_k is said to be correctly classified if $d(v_r, row_l)$ is minimum *and* $ceil\left(\frac{r}{6}\right) = l$, where $d(.)$ is the Euclidian distance; $ceil(.)$ operator rounds the input to the next integer; and row_l is the lth row of matrix M.

2.4   Tsallis Distribution

Tsallis distribution is a generalization of Gaussian distribution, and the latter is recovered when $q \to 1$ [9].

When $q < 1$ we have

$$p_q(x) = \frac{1}{\acute{o}}\sqrt{\left[\frac{1-q}{\eth(3-q)}\right]}\frac{\Gamma\left(\frac{5-3q}{2(1-q)}\right)}{\Gamma\left(\frac{2-q}{1-q}\right)}\left[1 - \frac{(1-q)\,x^2}{(3-q)\,\acute{o}^2}\right]^{\frac{1}{1-q}}. \tag{12}$$

If $|x| < \acute{o}[(3-q)/(1-q)]^{1/2}$.

When $q > 1$

$$p_q(x) = \frac{1}{\acute{o}} \sqrt{\left[\frac{q-1}{\acute{o}(3-q)}\right]} \frac{\Gamma\left(\frac{1}{q-1}\right)}{\Gamma\left(\frac{3-q}{2q-2}\right)} \frac{1}{\left(1 + \frac{(q-1)x^2}{(3-q)\acute{o}^2}\right)^{\frac{1}{q-1}}} \qquad (13)$$

In Fig. 4 it is emphasized the difference between Gaussian and Tsallis distribution.

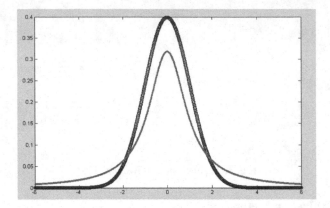

Fig. 4. Gaussian (blue), Tsallis (red)

In Fig. 4, the Gaussian distribution has $\acute{o} = 1$; whereas Tsallis distribution has $\acute{o} = 1$ and $q = 2$.

3   Results

Considering that the experiments are run using Tsallis distribution, there are, in total, 5 parameters to be adjusted. They are: r_i, r_o, $\acute{o}$, q and radiometric resolution (in bits) of the image. However, r_i and r_o will be fixed in 2 and 5, respectively, because those values yielded good results in [6] and [2].

3.1   Tsallis Distribution with $q = 2$ and $q = 0$

Firstly, it will be done a comparison between $q = 2$ and $q = 0$. Table 1 shows the results.

Table 1. Comparison between $q = 0$ and $q = 2$

$\acute{o}$	bits	accuracy (%) $q = 2$	accuracy (%) $q = 0$
1	4	87.04	89.81
2	4	81.48	90.74
3	4	71.3	96.3

Table 1 shows that, when $q = 0$, the overall accuracy of the classification is better, especially with ó = 3. Therefore, new classifications are performed keeping those parameters.

3.2   Tsallis Distribution with $q = 0$ and other Bit Values

Table 2 exhibits some results with $q = 0$ and different bits.

Table 2. Results for $q = 0$ and other bit values

ó	bits	accuracy (%) $q = 0$
3	5	92.48
3	6	90.74
3	7	89.81
3	8	90.74

According to Table 2, there is a tendency of decreasing accuracy with the increase of the radiometric resolution of the image.

3.3   Tsallis Distribution with $q = 0$ and Principal Components Analysis

In [6] and [2] only 6 textural features were calculated from the co-occurrence matrix, namely those considered by [7] as the most statistically significant. In this section, it is performed a feature extraction by means of Principal Components Analysis (PCA) [10] taking into account all the 14 texture attributes suggested by Haralick. In order to maintain the same reasoning as before, 6 principal components are retained, which, in turn, embrace 99,31% of the total variance of data.
In Table 3 the results are shown.

Table 3. Results for $q = 0$ and Principal Components Analysis

ó	bits	accuracy (%) $q = 0$ and PCA
3	4	97.22
3	5	96.3
3	6	92.59
3	7	91.67
3	8	90.74

It is evident, from Table 3, that the use of PCA improved the accuracy of the classifier.

3.4   Tsallis Distribution with $q = 0$, PCA and Addition of Noise

In Table 3 we see the best result attained thus far, that is, 97.22% of accuracy, with $q = 0$, $ó = 3$ and Principal Components Analysis to extract the attributes.

The intention now is to add some *salt and peppers* noise to rotated images and check the performance of the classifier. Table 4 shows the results.

Table 4. $q = 0$, PCA and salt and peppers

$ó$	bits	accuracy (%) 1% of noise	accuracy (%) 5% of noise
3	4	92.59	70.37

It is perceived that, with the increase of noise, the accuracy decreases, according to Table 4.

4   Discussion

4.1   Conclusions

The purpose of this letter was to perform further analysis of the works [6] and [2]. Tsallis distribution was considered instead of Gaussian, and fair results (see Table 1) were obtained mainly with $q = 0$ and $ó = 3$. Different curve shapes give different weights to pixels falling on the circle and their neighbors, and this is the cause of the variation in results.

Keeping $q = 0$ and $ó = 3$ new tests were made changing the radiometric resolution, *i.d.*, the number of bits. It could be noticed that the increase in the number of bits did not enhance the classification accuracy, something that can be rather counterintuitive, as exhibited by Table 2.

New tests were performed with $q = 0$ and $ó = 3$, but this time proceeding to feature extraction by means of Principal Components Analysis using all the 14 texture features proposed by Haralick. According to Table 3, the results were sensibly better than the ones with same configuration, but without PCA. This means that, at least when using Tsallis distribution, the results achieved in [7] are not valid any longer.

With the parameters of the last test — PCA, $q = 0$ and $ó = 3$ — it was added noise only to images during the classification step. In Table 4 it was apparent that the more noise, the less accurate becomes the classifier.

4.2   Further Works

For further realistic analysis, one could employ aerial images in order to recognize objects by their texture patterns. This could be useful, for instance, in aerial navigation of unmanned aerial vehicles.

References

1. Brandt, T., Mather, P.M.: Classification Methods for Remotely Sensed Data. CRC Press (2009)
2. Acunã, M.A.B.: Rotation-Invariant Texture Classification Based on Grey Level Co-occurrence Matrices. Universidade de São Paulo (2013)
3. Haralick, R.M., Shanmugam, M., Dinstein, I.: Texture feature for image classification. IEEE Transactions on Systems, Man and Cybernetics **3**, 610–621 (1973)
4. Liu, L., Long, Y., Fieguth, P.W., Lao, S., Zhao, G.: BRINT: Binary Rotation Invariant and Noise Tolerant Texture Classification. IEEE Transactions on Image Processing **23**(7), 3071–3084 (2014)
5. Hassan, A., Riaz, F., Shaukat, A.: Scale and Rotation Invariant Texture Classification Using Covariate Shift Methodology. IEEE Signal Processing Letters **21**(3), 321–324 (2014)
6. Ito, R.H., Kim, H.Y., Salcedo, W.J.: Classificação de Texturas Invariante a Rotação Usando Matriz de Co-ocorrência. In: 8th International Information and Telecommunication Technologies Symposium (2009)
7. Baraldi, A., Parmiggiani, F.: An Investigation of the Textural Characteristics Associated with Gray Level Co-occurrence Matrix Statistical Parameters. IEEE Transactions on Geoscience and Remote Sensing **33**(2), 293–304 (1995)
8. http://sipi.usc.edu/database/database.php?volume=textures
9. Tsallis, C.: Possible generalization of boltzmann-gibbs statistics. Journal Statistical Physics **52**, 479 (1988)
10. Bishop, C.M.: Pattern Recognition and Machine Learning. Springer (2006)

Dimension Reduction

A Fast Isomap Algorithm
Based on Fibonacci Heap

Taiguo Qu[✉] and Zixing Cai

School of Information Science and Engineering,
Central South University, Changsha 410083, Hunan, China
qutaiguo88@sohu.com

Abstract. For the slow operational speed problem of Isomap algorithm in which the Floyd-Warshall algorithm is applied to finding shortest paths, an improved Isomap algorithm is proposed based on the sparseness of the adjacency graph. In the improved algorithm, the runtime for shortest paths is reduced by using Dijkstra's algorithm based on Fibonacci heap, and thus the Isomap operation is speeded up. The experimental results on several data sets show that the improved version of Isomap is faster than the original one.

Keywords: Isomap · Fibonacci heap · Shortest path · Dijkstra's algorithm

1   Introduction

Dimensionality reduction has been widely applied to many fields of information processing, such as machine learning, data mining, information retrieval, and pattern recognition.

As one of the nonlinear dimensionality reduction techniques, Isomap [1] has drawn great interests in recent years. It combines the major algorithmic features of principal component analysis (PCA) [2] and multidimensional scaling (MDS) [3]—computational efficiency, global optimality, and asymptotic convergence guarantees—with the flexibility to learn a broad class of nonlinear manifolds. Due to its excellent performance, it has aroused interests of researchers from many fields.

In order to preserve the global geometric properties of manifold, Isomap extends MDS by using the geodesic distance metric instead of the Euclidean distance metric. For neighboring points, input-space distance provides a good approximation to geodesic distance. For faraway points, geodesic distance can be approximated by adding up a sequence of "short hops" between neighboring points. These approximations are computed efficiently by finding shortest paths in an adjacency graph with edges connecting neighboring data points.

By calling the Floyd-Warshall algorithm [4] to compute the shortest paths, the classical Isomap(C-Isomap) [1] is very slow. For sparse graphs such as the adjacency graph in Isomap, Dijkstra's algorithm based on Fibonacci heap [4] is faster than the Floyd-Warshall algorithm. So an improved version of Isomap with Dijkstra's

© Springer International Publishing Switzerland 2015
Y. Tan et al. (Eds.): ICSI-CCI 2015, Part III, LNCS 9142, pp. 225–231, 2015.
DOI: 10.1007/978-3-319-20469-7_25

algorithm based on Fibonacci heap is proposed in this paper. The new algorithm is called Fib-Isomap, and the Dijkstra's algorithm based on Fibonacci heap is abbreviated to Fib-Dij.

2   The Floyd-Warshall Algorithm and Fib-Dij

As mentioned in section 1, in Isomap algorithm, shortest paths between all pairs of vertices need to be found in the adjacency graph, where each vertex represents a data point. The adjacency graph can be represented with a connected, undirected graph $G = (V, E)$, where V is the set of vertices, E is the set of possible edges between pairs of vertices. Let $n = |V|$ be the number of vertices, $m = |E|$ be the number of the edges. The vertices are numbered $1, \cdots, n$. For each $(i, j) \in E$, a weight w_{ij} is given specifying the distance between i and j. Let $p = (i_1, \cdots, i_k)$ be a path from i_1 to i_k, the weight $w(p)$ is defined as the sum of the weights of its constituent edges:

$$w(p) = \sum_{u=2}^{k} w(i_{u-1}, i_u) \tag{1}$$

Among all the paths from i to j, any path with the minimum weight is defined as the shortest path, and the weight is denoted by $\delta_{i,j}$.

2.1   The Floyd-Warshall Algorithm

For any pair of vertices $i, j \in V$, consider all paths from i to j whose intermediate vertices are all drawn from $\{1, ..., k\}$. Let p be a minimum-weight path from among them, $d_{ij}^{(k)}$ be the weight of p. $d_{ij}^{(k)}$ can be recursively computed by

$$d_{ij}^{(k)} = \begin{cases} w_{ij}, & k = 0 \\ \min(d_{ij}^{(k-1)}, d_{ik}^{(k-1)} + d_{kj}^{(k-1)}), & k \geq 1 \end{cases} \tag{2}$$

Because for any path, all intermediate vertices are in the set $\{1, ..., n\}$, $d_{ij}^{(n)}$ is the distance of the shortest path from i to j, i.e., $\delta_{i,j} = d_{ij}^{(n)}$. The runtime is $O(n^3)$.

2.2   Fib-Dij Algorithm

Fibonacci Heap.
A Fibonacci heap is a collection of rooted trees that are min-heap ordered. That is, for each tree, the key of a node is greater than or equal to the key of its parent. Each node x contains a pointer $x.p$ to its parent and a pointer $x.child$ to any one of its children. The children of x are linked together in a circular, doubly linked list, which is called

the child list of x. A given Fibonacci heap H is accessed by a pointer $H.min$ to the root of a tree containing the minimum key. Fig. 1 shows an example of Fibonacci heap.

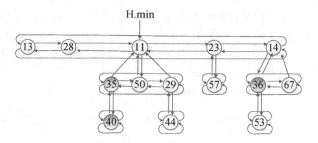

Fig. 1. A Fibonacci heap

Dijkstra's Algorithm.
Dijkstra's algorithm can solve both single-source shortest-paths problems and all-pairs shortest-paths problems.

In the single-source shortest-paths problem, Dijkstra's algorithm maintains a set S of vertices whose final shortest-path weights from the source s have already been determined. The running time depends on how the min-priority queue $Q = V - S$ is implemented. For sparse graph, the Fibonacci heap implementation is the fastest and its runtime is $O(V \lg V + E)$. It's easy to see the runtime of Dijkstra's algorithm based on Fibonacci heap (Fib-Dij) for all-pairs shortest-paths problem is $O(V^2 \lg V + VE)$.

Another two popular algorithms for shortest-paths problem are the Bellman-Ford algorithm and Johnson's algorithm [4]. For a graph like the adjacency graph in Isomap, where no edge has negative weight, Johnson's algorithm is the same as Dijkstra's algorithm. The runtime of the four algorithms are summarized in table 1:

Table 1. Runtime of the four shortest-paths algorithms

Algorithm	Runtime
Floyd-Warshall	$O(n^3)$
Fib-Dij	$O(n^2 \lg n + mn)$
Bellman-Ford	$O(mn^2)$
Johnson	$O(n^2 \lg n + mn)$

3   C-Isomap and Fib-Isomap

C-Isomap consists of the following steps:

Step 1: Construct adjacency graph $G = (V, E)$

Let d_{ij} be the Euclidean distance between i and j. Define the adjacency graph G over all data points by connecting i and j if they are closer than ε (ε-nn) or if i is one of the k nearest neighbors of j (k-nn). Let $w_{ii} = 0$; $w_{ij} = d_{ij}$ if i, j are linked by an edge; $w_{ij} = \infty$ otherwise. Obviously the weight matrix $W = (w_{ij})$ is symmetric.

Step 2: Compute all-pairs shortest-paths

For any pair of vertices $i, j \in V$, C-Isomap calculates the shortest path distance δ_{ij} by calling the Floyd-Warshall algorithm.

Step 3: Construct d-dimensional embedding

Let $A = [-\delta_{i,j}^2 / 2]$, $H = I - \frac{1}{n} 1_n 1_n^T$, $B = HAH$. Let λ_p be the p-th eigenvalue (in decreasing order) of B, and v_p^i be the i-th component of the p-th eigenvector. Then set the p-th component of the d-dimensional coordinate vector y_i equal to $\sqrt{\lambda_p} v_p^i$.

We take k-nn as an example to illustrate that adjacency graph is a sparse graph. To begin with, we define $(i, j) \in E$ as one of j's k-nn edge if i is one of the k nearest neighbors of j. Following this definition, each vertex has k k-nn edges and any edge is a k-nn edge of at least one of its two endpoints, therefore, $m = O(n)$. In other words, the adjacency graph is sparse.

Now, let's compare the runtime of the four aforementioned algorithms on the sparse graph in Isomap. For the Bellman-Ford algorithm, the time is

$$O(mn^2) = O(n^3) \tag{3}$$

For Fib-Dij algorithm, the time is

$$O(n^2 \lg n + mn) = O(n^2 \lg n + n^2) = O(n^2 \lg n) \tag{4}$$

As mentioned before, the runtime of the Floyd-Warshall algorithm is also $O(n^3)$, and Johnson's algorithm is the same as Dijkstra's algorithm. So, Fib-Dij is the fastest to compute the shortest-paths.

In Fib-Isomap, Fib-Dij is used to replace the Floyd-Warshall algorithm in the second step of C-Isomap, thus speeding up Isomap.

4   Experiments

4.1   The Comparison of Speed

In this experiment, five data sets are involved to compare the speed of Fib-Isomap and C-Isomap. The data sets are "Swiss roll" [5], S-curve, "2"s from the MNIST database [6], Frey face[7], and the face images[8]. They are all projected onto two-dimensional subspace by C-Isomap and Fib-Isomap, i.e., d=2. The other parameters for the data sets are shown in table 2. For each data set, four runtimes are generated. Let Fib-Isomap,C-Isomap be the overall runtime of Fib-Isomap and C-Isomap respectively, let Fib-Dij and Floyd be the runtime of finding the shortest-paths. They are shown in Fig. 2, where Fib-Dij, Fib-Isomap, Floyd and C-Isomap are arranged from left to right for each data set.

Table 2. Parameters of the five data sets

Data set	S-curve	Swiss roll	MNIST("2")	The face images	Frey face
n	1500	1500	1000	698	1965
Original dimensionality	3	3	784	4096	560
Neighborhood size k	10	10	6	6	5

As Fig.2 shows, for each data set, (1) finding the shortest-paths costs most of the runtime of C-Isomap; (2) Fib-Dij algorithm greatly reduces the time of finding shortest-paths, and hence improves the overall speed of Isomap.

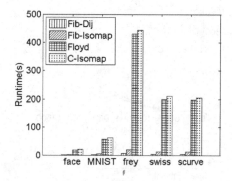

Fig. 2. The runtimes of C-Isomap and Fib-Isomap on the five data sets

4.2   The Quantitative Analysis

In this experiment, C-Isomap and Fib-Isomap are   applied to the S-curve, with n varying from 100 to 2800. The runtime of C-Isomap, Fib-Isomap, Floyd and Fib-Dij are shown in fig.3(a), and the runtime for finding shortest-path by the Floyd-Warshall algorithm and the Fib-Dij algorithm are plotted in Fig.3(b) and 3(c) respectively. The figures show that when n is big enough, the runtime of the Floyd-Warshall algorithm falls between $t_1 = 6.8 \times 10^{-8} \times n^3$ and $t_2 = 5.4 \times 10^{-8} \times n^3$; and that of the Fib-Dij algorithm is less than $t_3 = 2.8 \times 10^{-7} \times n^2 \lg n$. The experimental results validate the above theoretical analysis, that is, the runtime of the Floyd-Warshall algorithm for the shortest-paths in Isomap is $O(n^3)$, and the runtime of the Fib-Dij is $O(n^2 \lg n)$.

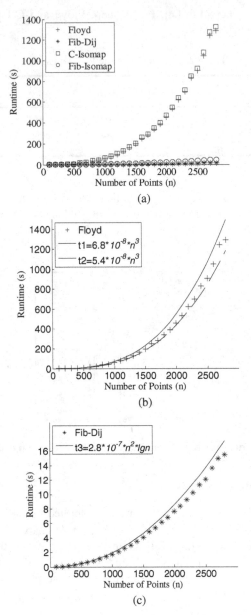

Fig. 3. (a) Runtime of Floyd, Fib-Dij, C-Isomap and Fib-Isomap vs. n (b)Runtime of the Floyd-Warshall algorithm vs. n (c)Runtime of the Fib-Dij algorithm vs. n

5   Conclusions

Isomap is one of the most important methods for dimensionality reduction. Due to the Floyd-Warshall algorithm for finding the shortest path, C-Isomap is very slow. In order to overcome this bottleneck, considering the sparseness of the adjacency graph, this paper uses Dijkstra's algorithm based on Fibonacci heap to calculate the shortest paths. Both theoretical analysis and experimental results show that Fib-Isomap greatly decreases the runtime for the shortest path, thus speeds up the overall Isomap operation. This facilitates applications of Isomap to more fields, especially to large data sets.

References

1. Tenenbaum, J.B., de Silva, V., Langford, J.: A global geometric framework for nonlinear dimensionality reduction. J. Science **290**, 2319–2323 (2000)
2. Jolliffe, I.T.: Principal component analysis, 2nd edn. Springer, New York (2002)
3. Cox, T., Cox, M.: Multidimensional scaling, 2nd edn. Chapman & Hall/CRC, London (2001)
4. Cormen, T.H., Leiserson, C.E., Rivest, R.L., et al.: Introduction to algorithms, 3rd edn. MIT Press, Cambridge (2009)
5. Tenenbaum, J.: Mapping a manifold of perceptual observations. In: Advances in Neural Information Processing Systems, pp. 682–688. MIT Press, Cambridge (1998)
6. The MNIST database of handwritten digits. http://yann.lecun.com/exdb/mnist
7. Data for MATLAB hackers. http://www.cs.toronto.edu/~roweis/data.html
8. Data sets for nonlinear dimensionality reduction. http://isomap.stanford.edu/datasets.html.

A Novel Wavelet Transform – Empirical Mode Decomposition Based Sample Entropy and SVD Approach for Acoustic Signal Fault Diagnosis

Jiejunyi Liang and Zhixin Yang[✉]

Department of Electromechanical Engineering,
Faculty of Science and Technology, University of Macau,
Macau, SAR China
{mb25403,zxyang}@umac.mo

Abstract. An advanced and accurate intelligent fault diagnosis system plays an important role in reducing the maintenance cost of modern industry. However, a robust and efficient approach which can serve the purpose of detecting incipient faults still remains unachievable due to weak signals' small amplitudes, and also low signal-to-noise ratios (SNR). One way to overcome the problem is to adopt acoustic signal because of its inherent characteristic in terms of high sensitive to early stage faults. Nonetheless, it also suffers from low SNR and results in high computational cost. Aiming to solve the aforesaid problems, a novel wavelet transform - empirical mode decomposition (WT-EMD) based Sample Entropy (SampEn) and singular value decomposition (SVD) approach is proposed. By exerting wavelet analysis on the intrinsic mode functions (IMFs), the end effects, which decreases the accuracy of EMD, is significantly alleviated and the SNR is greatly improved. Furthermore, SampEn and SVD, which function as health indicators, not only help to reduce the computational cost and enhance the SNR but also indicate both irregular and periodic faults adequately.

Keywords: Incipient fault diagnosis · Acoustic signal · Wavelet transform-Empirical mode decomposition · Sample entropy · Singular value decomposition

1 Introduction

Reducing the downtime and the maintenance cost of functioning machine is an efficient way to maximize productivity and has gained considerable attention. Nowadays, fault diagnostics is mainly based on the fundamental of characterizing a fault signal which is acquired by various sensors. The source of the required signals varies from vibration, acoustic signal to temperature and so on[1]. Although vibration analysis still accounts for a large proportion in the field of fault diagnosis, there is a tendency to employ acoustic analysis considering the merits of acoustic signal. There are three advantages which make acoustic signal desirable. First, acoustic signal is non-directional so that one acoustic sensor can fulfill the acquisition adequately. Second, acoustic signal is relatively independent of structural resonances. Thirdly, acoustic

© Springer International Publishing Switzerland 2015
Y. Tan et al. (Eds.): ICSI-CCI 2015, Part III, LNCS 9142, pp. 232–241, 2015.
DOI: 10.1007/978-3-319-20469-7_26

signal has a relative high sensitiveness[2] which provides the ability of identifying incipient faults. However, acoustic signal has several disadvantages such as low SNR and high computational cost. It is essential to find a way both can reduce the computational cost and enhance the SNR.

Several signal processing techniques such as Fourier transform (FFT) and wavelet transform (WT) have been employed and investigated[3, 4]. But FFT fails to identify the whole and local features of the signal in time and frequency domain, and WT also suffers the defects of energy leakage and a biased fundamental because of the selection of the base function. EMD, which is developed as a time-frequency signal processing technique, solves the aforesaid problems by being capable of processing nonlinear and non-stationary signals[5]. It is based on the local characteristic time scales of a signal and could generate a set of intrinsic mode functions which are theoretically orthogonal by decomposing the original signal. The IMFs, which have real physical significance and are not determined by pre-determined kernels but the signal itself, represent different components of the signal. Therefore, EMD is a self-adaptive time frequency analysis method. The milestone of this approach is the paper of Huang et al.[6]. Unfortunately, the classical EMD is degraded because of end effects associated with EMD processes[7]. Large swings occur in the interpolation fitting process, which eventually propagate inward and corrupt the whole data span, especially in the low-frequency components. In order to overcome the shortcomings, many approaches have been investigated [8, 9]. These approaches extended the signal sides using different algorithms, while as these predictions are mostly empirical, they introduce new distortions that could not throughout eliminate the end effect. In this paper, a wavelet transform assisted EMD is proposed to alleviate the problem of end effects and to enhance the ability to detect weak signals buried in strong background noise.

Feature extraction is another crucial procedure of fault diagnosis. Different fault types require different indicators as the characteristics of each fault are not the same. Traditionally, statistic feature such as peakiness, amplitude level, deviation from the mean and so on[10]. These features only provide the physical characteristics of time series data, and have limited performance dealing with acoustic signal. Therefore, a novel hybrid feature extraction module via combination of SampEn and SVD is proposed to function as the proper condition indicators. Richman and Moorman proposed SampEn[11] to express the complexity of time series by a non-negative data which has great significance in representing irregular faults. The more complex the time series was, the greater corresponding approximate entropy would be. Singular value decomposition of matrix has been widely applied in the fields of noise reduction, regularization, signal detection, signal estimation etc.[12]. When a local fault starts to emerge in rotating machinery, periodic impulses usually occur in the acquired signals. Considering the fact that singular value is the nature characteristic of matrix and owns favorable stability, it is preferable to apply it to periodic fault diagnosis. According to the fundamental of both methods, the hybrid approach would generate indicator matrix both can identify irregular faults and periodic faults with high accuracy. After been processed by SampEn and SVD, the obtained feature vectors could be further applied to classifiers to generate a desirable outcome.

2   Proposed Wavelet Transform EMD

EMD is a direct, posteriori, and adaptive method for signal decomposition. It is derived from signal to be analyzed itself and accomplished through a sifting process which employs an iterative algorithm. However, end effects occur because of the lack of enough extrema at both ends of the signal. As a result, the fitting error propagates inward during each iteration process and corrupts each IMF with complex frequency components[7]. In order to reduce the fitting errors on each IMF and in the meantime reduce the effect of noise, wavelet denoising is performed during the sifting process. Fig.1 shows the flowchart of the proposed WT-EMD, and the different steps between EMD and WT-EMD are summarized as follows.

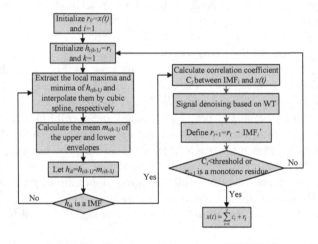

Fig. 1. The flowchart of the proposed WT-EMD

(1) Apply Eq.(1) on each IMF_i, where $W_{j,k}$ is the kth wavelet coefficient at jth level, $j,k \in Z$. And $\psi_{j,k}(t)$ is the commonly supported orthogonal wavelet functions:

$$W_{j,k} = \int_{-\infty}^{+\infty} \mathrm{IMF}_i(t)\psi_{j,k}(t)dt \tag{1}$$

(2) Calculate the denoising threshold T and generate the noise variance at each level.

(3) Compute new wavelet coefficients $W_{j,k}$ by threshold T.

$$d_t(W_{j,k}) = \mathrm{sgn}(W_{j,k})(|W_{j,k}| - T)_+ \tag{2}$$

(4) Through the inverse wavelet transform, reconstruct IMF_i' according to the denoised detail coefficients and approximate coefficients.

In order to verify the effectiveness of the proposed method, a simulated signal which consists of a sinusoidal component and an impulse component is investigated. The simulated signal and its decomposition using original EMD and WT-EMD, and their time-frequency analysis are shown in Fig.2.

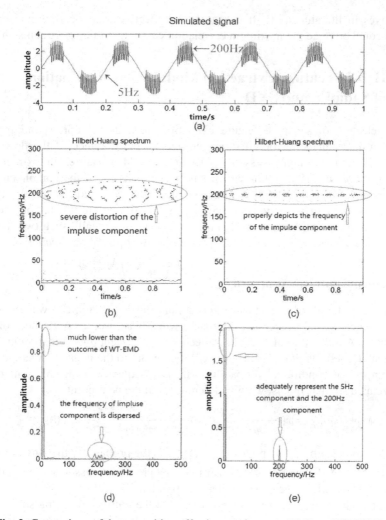

Fig. 2. Comparison of decomposition effectiveness between EMD and WT-EMD

The frequencies of the sinusoidal component and the impulse component are 5Hz and 200Hz, respectively. It can be seen in Fig.2(b) that the HH-spectrum of the impulse component is severely distorted due to the end effects of EMD. By adopting WT-EMD, the end effects is significantly alleviated, which is shown in Fig.2(c). It is the same situation with the frequency spectrum that in Fig.2(d) both the sinusoidal component and the impulse component are dispersed, on the contrary, in Fig.2(e) there is little leakage for both components.

The selection of basis mother wavelet filter and the recovery of signal are also crucial in this procedure. The wavelet parameters indicate in what extent the extracted IMFs match the real modes. It can be decided through a proper selection of commonly supported orthogonal wavelet functions such as Daubechies wavelet filter, which depends on signal under investigation. As to the threshold, there are several eligible alternatives in literatures[13]. It proved to be effective that the rigorous threshold method could produce the maximum correlation coefficient in most circumstances.

3   Hybrid Feature Extraction Module via Combination of SampEn and SVD

Proper selection of sensitive feature extraction methods for corresponding failure types is a determinant factor affecting the general performance of the downstream classifier in the fault diagnosis system. Once those hidden features are extracted, it is needed to observe their evolutionary performance in terms of the completeness and effectiveness of feature extraction from raw data. The extracted features are targeted be sensitive to one or more defects, and be less sensitive to the fluctuation caused by running conditions. The results of SampEn and SVD are actually such two indicators that complement each other which are illustrated in *section 3.1* and *section 3.2*.

3.1   Sample Entropy Approach

In [11], a new family of statistics called Sample Entropy (SampEn) was introduced, It's a nonlinear signal processing approach, which was used as a measure of signal complexity. A low value of SampEn reflects more frequent and more similar epochs in a signal. Generally, SampEn is approximately equal to the negative average natural logarithm of the conditional probability that two sequences that are similar for m points remain similar, that is, within a tolerance r, at the next point.

$$\text{SampEn}(m,r,N) = \ln[\frac{1}{N-m+1}\sum_{i=1}^{N-m+1} C_r^m(i) - \frac{1}{N-(m+1)+1}\sum_{i=1}^{N-(m+1)+1} C_r^{m+1}(i)] \tag{3}$$

where N is the length of data points, and $C_r^m(i)$ is the probability that any two epochs match each other. It computes the logarithm of a probability associated with the time series as a whole when one epoch find a match of length $m+1$.

It is possible to achieve a robust estimation of the entropy by using short data sequences, and process high resistance to short strong transient interference. By choosing a proper noise filter parameter r, it is possible to eliminate the effect of noise.

3.2   Singular Value Decomposition Approach

SVD is a technique for analysis of multivariate data, and can detect and extract small signals from noisy data, which is perfectly suitable in solving our problem.

If A is a $N \times M$ real matrix, then there exists

$$A = USV^T \tag{4}$$

where $U = [u_1,...,u_N] \in R^{N \times N}$, $U^T U = I$; $V = [v_1,...,v_M] \in R^{M \times M}$, $V^T V = I$; $S \in R^{N \times M}$, $S = [\text{diag}\{\delta_1,...,\delta_p\}:0]$; $p = \min(N,M)$; $\delta_1 \geq,...,\geq \delta_p \geq 0$. The values of δ_i are the singular values of matrix A and the vectors u_i and v_i are respectively the i-th left and right singular vectors

IMFs are sorted by frequencies from high to low and are orthogonal, which properly serve as initial feature vector matrices. Since the IMFs represent the natural oscillatory mode embedded in the signal, it can reveal the nature characteristic of the potential fault. Before applying SVD to the matrices formed by IMFs, N IMFs should be obtained by WT-EMD, which is denoted by $n_1, n_2, ..., n_N$, and $n = \max(n_1, n_2, ..., n_N)$. As different samples may have different amount of IMF, in other words, $n < n_{max}$, it can be padded with zero in order to achieve a matrix of n_{max} components. After conducting SVD with the pre-processed initial matrix, the singular value $\delta_{A,j}$ can be got.

$$\delta_{A,j} = [\delta_{A,j}^1, \delta_{A,j}^2, ..., \delta_{A,j}^n] \tag{5}$$

where $\delta_{A,j}^1 \geq \delta_{A,j}^2 \geq ... \geq \delta_{A,j}^n$, $j = 1, 2, ..., m$ denotes the m different faults, respectively. Finally, the singular value $\delta_{A,j}$ we got, which could describe the characteristic of the fault signal, could be used as fault feature vectors.

3.3   Novelty of Combination

According to the fundamental of SampEn(m, r, N), a lower value indicates high degree of regularity. This characteristic is especially suitable for the signal with low regularity or, in other words, unpredictable, such as the acoustic signal of wear of outer race of a bearing. The acoustic signal of wear of outer race of the bearing has no obvious regularity because the contact area between the rolling elements and the outer race is changing all the time, making it hard to extract a proper vector. SampEn, as an indicator, solve the problem properly via directly measure the regularity. But SampEn has its shortcomings. In the circumstance where the fault presents as high regularity signal or a periodic signal such as broken tooth of a gearbox, SampEn will produce a relatively lower classification accuracy than for irregular faults. It is because that Entropy is under a low boundary at zero. For those periodic faults that are presented as high regularity signals, their SampEns are uniformly approaching to zero, which make it difficult to identify the difference among them. Another reason is that the periodic faults, such as broken tooth and chipped tooth, have similar periods which make the difference of SampEn even smaller.

On the other hand, SVD could perfectly solve the problems encountered by SampEn. SVD extracts the singular value matrix of the input acoustic signal and is well used in principal component analysis (PCA). Although the aforesaid periodic signal of broken tooth presents in high regularity and results in a low level of SampEn, it does generates a regular impulse. It could perfectly represent the fault information and can be detected by SVD easily, through figuring out the singular value matrix of the impulse signal. As to the irregular faults, there is no such representative main fault component, which makes it unlikely for SVD to extract a value to summarize the fault information. Thus, the hybrid of SampEn and SVD serves the purpose of detecting various faults properly. The flow chart of the proposed approach is shown in Fig.3.

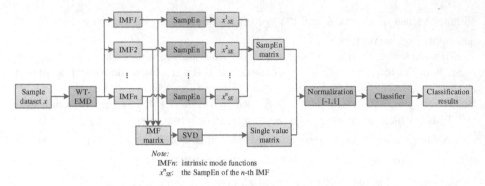

Fig. 3. The flow chart of the proposed fault diagnosis system

4   Experimental Setup and Data Analysis

To verify the effectiveness of the proposed framework in real working condition, representative samples are acquired and processed. The following sub-sections depict the details of the experiment and the data analysis.

The test rig, which is shown in Fig.4, includes a computer for data acquisition, an electric load simulator a prime mover, a gearbox, a flywheel and an asynchronous generator. The test rig can simulate many common periodic faults and irregular faults in a gearbox such as broken tooth, chipped tooth, wear of outer race of the bearing and so on. A total of 4 cases are simulated in order to generate the required dataset.

Fig. 4. The fault simulator

The samples of acoustic signals are acquired by an acoustic sensor located about 20cm above the floor and 10cm away from the input shaft. To construct and test the diagnostic framework, the test rig was kept running at a frequency of 20Hz for several hours, for each simulated fault, 100 samples were taken out randomly. Each time 1 second of acoustic data was recorded with a sampling rate of 25.6 kHz. There were 400 fault samples (i.e. (broken tooth, chipped tooth, wear of outer race of bearing, wear of the rolling elements) × 100 samples). Fig.5 shows segments of the acquired raw signals mentioned before.

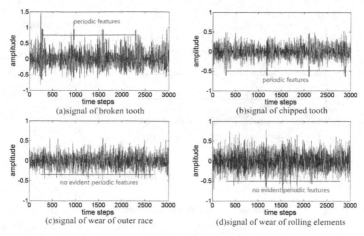

Fig. 5. The acquired signal segments of four different faults

It can be seen from Fig.5(a) and Fig.5(b) that the signals of periodic faults contain regular impulse and have evident periodic features. In Fig.5(c) and Fig.5(d), no such periodic features is available. These characteristics entail an approach both can depict periodic signal and irregular signal. The proposed module of SampEn and SVD could match such requirements.

In this paper, WT-EMD is applied to the acquired raw signals to generate a dataset of IMF matrixes, and then SampEn and SVD are conducted to extract irregular features and periodic features, respectively. Table 1 shows the data flow of this experiment and demonstrates the computational efficiency of the proposed approach.

Table 1. Data flow of the experiment for one sample

	Raw data	WT-EMD	SampEn	SVD	Total features
Size of the data matrix	$1*25600$	$n*25600$	$n*1$	$n*1$	$n*2$

Note: n denotes the number of IMFs and is usually not greater than twenty.

From Fig.6 we can see that the condition of wear of the rolling elements has the highest value of SampEn which complies with the fact that the acoustic signal of wear of rolling elements has the lowest regularity as the contact area between rolling elements and the outer race is always changing and the rolling elements rotate itself while rolling with the cage. Following the wear of rolling elements is the wear of outer race of the bearing because the rolling element which contacts the wear on the outer race is shifting all the time. The difference of SampEn between wear of rolling elements and wear of outer race is obvious as the extent of the irregularity is quite different. As to the periodic faults of broken tooth and chipped tooth, the SampEn of these two faults are quite similar. There are two reasons for that. First, although the impulses of broken tooth and chipped tooth are different, they have similar regularity as they have a similar period. The second one is entropy is always greater than zero, if two signals both have high regularity, the SampEn of them are both approaching to zero, which makes the difference between them approaching to zero.

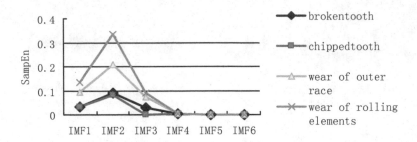

Fig. 6. The SampEn of each IMF for different fault types

Singular value matrix represents the periodic features in signals such as broken-tooth and chippedtooth, and cannot adequately rendering reliable indications for irre-gular signals, for example, wear of outer race and wear of rolling elements. It can be seen that irregular patterns can be drawn out with a low classification accuracy.

In order to shade light upon the effectiveness of the proposed approach, support vec-tor machine (SVM) is employed as a classifier to evaluate the extracted features of the 400 samples. The final outcome accuracy is around 94%. It can be seen that the pro-posed approach can significantly enhance the classification accuracy for both periodic faults and irregular faults.

5   Conclusion

Fault diagnosis has gathered soaring attention as they can greatly enhance the produc-tiveness in modern industry. In this paper, acoustic signals are introduced other than vibration signals for the advantage that acoustic signals are more sensitive to incipient faults because its transmission does not depend on the solid structures which will make low energy vibration undetectable. In order to enhance the classification accu-racy, it is imperative to choose the most proper condition indicators to detect different types of faults. This paper innovatively combines two complementary approaches of condition indicators on the fundamental of signal decomposition module which is WT-EMD. The proposed fault diagnosis system proved to be a promising way to both detect periodic faults and irregular faults accurately.

The research originalities of this paper are summarized as follows:

The hybrid SampEn and SVD methods have been proposed to further process the outcome of WT-EMD, which has been verified to be effective for solving the prob-lems in terms of high computational cost, low S/N ratio, and the end effects problems suffered by EMD.

(1) To solve the problems of high noise ratio of acoustic signals and the end effects problem of EMD, the WT-EMD has been proposed and proven to be effective.

(2) It is the first in literature that compares the effectiveness of SampEn and SVD for irregular faults and periodic faults, respectively.

(3) The proposed WT-EMD-based SampEn and SVD system provides a promising way that enhances both the effectiveness of signal decomposition process and the representativeness of corresponding condition indicators.

Acknowledgments. The authors would like to thank the funding support by the University of Macau, under grant numbers: MYRG079(Y1-L2)-FST13-YZX, and MYRG2015-00077-FST.

References

1. Tan, C.K., Irving, P., Mba, D.: A Comparative Experimental Study on The Diagnostic and Prognostic Capabilities of Acoustics Emission, Vibration and Spectrometric Oil Analysis for Spur Gears. Mechanical Systems and Signal Processing **21**, 208–233 (2007)
2. Tandon, N., Mata, S.: Detection of Defects in Gears by Acoustic Emission Measurements. Journal of Acoustic Emission **17**, 23–27 (1999)
3. El Hachemi Benbouzid, M.: A Review of Induction Motors Signature Analysis as a Medium for Faults Detection. IEEE Transactions on Industrial Electronics **47**, 984–993 (2000)
4. Zou, J., Chen, J.: A Comparative Study on Time-Frequency Feature of Cracked Rotor by Wigner-Ville Distribution and Wavelet Transform. Journal of Sound and Vibration **276**, 1–11 (2004)
5. Tsakalozos, N., Drakakis, K., Rickard, S.: A Formal Study of The Nonlinearity and Consistency of The Empirical Mode Decomposition. Signal Processing **92**, 1961–1969 (2012)
6. Huang, N.E., Shen, Z., Long, S.R., Wu, M.C., Shih, H.H., Zheng, Q., et al.: The empirical mode decomposition and the hilbert spectrum for nonlinear and non-stationary time series analysis: proceedings of the royal society of london. In: Series A: Mathematical, Physical and Engineering Sciences, vol. 454, pp. 903–995 (1998)
7. Yan, J., Lu, L.: Improved Hilbert-Huang Transform Nased Weak Signal Detection Methodology and Its Application on Incipient Fault Diagnosis and ECG Signal Analysis. Signal Processing **98**, 74–87 (2014)
8. Xun, J., Yan, S.: A Revised Hilbert-Huang Transformation Based on The Neural Networks and Its Application in Vibration Signal Analysis of A Deployable Structure. Mechanical Systems and Signal Processing **22**, 1705–1723 (2008)
9. Cheng, J., Yu, D., Yang, Y.: Application of Support Vector Regression Machines to The Processing of End Effects of Hilbert-Huang Transform. Mechanical Systems and Signal Processing **21**, 1197–1211 (2007)
10. Yang, Z., Wong, P.K., Vong, C.M., Zhong, J., Liang, J.: Simultaneous-Fault Diagnosis of Gas Turbine Generator Systems Using A Pairwise-Coupled Probabilistic Classifier. Mathematical Problems in Engineering 2013 (2013)
11. Richman, J.S., Moorman, J.R.: Physiological Time-Series Analysis Using Approximate Entropy and Sample Entropy. American Journal of Physiology-Heart and Circulatory Physiology **278**, H2039–H2049 (2000)
12. De Lathauwer, L., De Moor, B., Vandewalle, J.: A Multilinear Singular Value Decomposition. SIAM Journal on Matrix Analysis and Applications **21**, 1253–1278 (2000)
13. Joy, J., Peter, S., John, N.: Denoising Ssing Soft Thresholding. International Journal of Advanced Research in Electrical, Electronics and Instrumentation Engineering **2**, 1027–1032 (2013)

Unsatisfiable Formulae of Gödel Logic
with Truth Constants and $=$, $\prec$, Δ
Are Recursively Enumerable

Dušan Guller$^{(\boxtimes)}$

Department of Applied Informatics, Comenius University,
Mlynská dolina, 842 48 Bratislava, Slovakia
guller@fmph.uniba.sk

Abstract. This paper brings a solution to the open problem of recursive enumerability of unsatisfiable formulae in the first-order Gödel logic. The answer is affirmative even for a useful expansion by intermediate truth constants and the equality, $=$, strict order, $\prec$, projection Δ operators. The affirmative result for unsatisfiable prenex formulae of G_∞^Δ has been stated in [1]. In [7], we have generalised the well-known hyperresolution principle to the first-order Gödel logic for the general case. We now propose a modification of the hyperresolution calculus suitable for automated deduction with explicit partial truth.

Keywords: Gödel logic · Resolution · Many-valued logics · Automated deduction

1 Introduction

Current research in many-valued logics is mainly concerned with left-continuous t-norm based logics including the fundamental fuzzy logics: Gödel, Łukasiewicz, and Product ones. Most explorations of t-norm based logics are focused on tautologies and deduction calculi with the only distinguished truth degree 1, [8]. However, in many real-world applications, one may be interested in representation and inference with explicit partial truth; besides the truth constants 0, 1, intermediate truth constants are involved in. In the literature, two main approaches to expansions with truth constants, are described. Historically, first one has been introduced in [10], where the propositional Łukasiewicz logic is augmented by truth constants $\bar{r}$, $r \in [0,1]$, Pavelka's logic (PL). A formula of the form $\bar{r} \rightarrow \phi$ evaluated to 1 expresses that the truth value of ϕ is greater than or equal to r. In [9], further development of evaluated formulae, and in [8], Rational Pavelka's logic (RPL) - a simplification of PL, are described. Another approach relies on traditional algebraic semantics. Various completeness results for expansions of t-norm based logics with countably many truth constants are investigated, among others, in [2–6,11].

D. Guller—This work is partially supported by VEGA Grant 1/0592/14.

Y. Tan et al. (Eds.): ICSI-CCI 2015, Part III, LNCS 9142, pp. 242–250, 2015.
DOI: 10.1007/978-3-319-20469-7_27

Concerning the fundamental first-order fuzzy logics, the set of logically valid formulae is Π_2-complete for Łukasiewicz logic, Π_2-hard for Product logic, and Σ_1-complete for Gödel logic, as with classical first-order logic. Among these fuzzy logics, only Gödel logic is recursively axiomatisable. Hence, it was necessary to provide a proof method suitable for automated deduction, as one has done for classical logic. In contrast to classical logic, we cannot make shifts of quantifiers arbitrarily and translate a formula to an equivalent (satisfiable) prenex form. In [7], we have generalised the well-known hyperresolution principle to the first-order Gödel logic for the general case. Our approach is based on translation of a formula of Gödel logic to an equivalent satisfiable finite order clausal theory, consisting of order clauses. An order clause is a finite set of order literals of the form $\varepsilon_1 \diamond \varepsilon_2$ where ε_i is either an atom or a quantified atom; and $\diamond$ is a connective either $\eqcirc$ or $\prec$. $\eqcirc$ and $\prec$ are interpreted by the equality and standard strict linear order on $[0,1]$, respectively. On the basis of the hyperresolution principle, a calculus operating over order clausal theories, has been devised. The calculus is proved to be refutation sound and complete for the countable case with respect to the standard $\boldsymbol{G}$-algebra $\boldsymbol{G} = ([0,1], \leq, \vee, \wedge, \Rightarrow, \bar{}, \eqcirc, \prec, 0, 1)$ augmented by binary operators $\eqcirc$ and $\prec$ for $\eqcirc$ and $\prec$, respectively. As another step, one may incorporate a countable set of intermediate truth constants $\bar{c}$, $c \in (0,1)$, to get a modification of our hyperresolution calculus suitable for automated deduction with explicit partial truth. We shall investigate the so-called canonical standard completeness, where the semantics of the first-order Gödel logic is given by the standard $\boldsymbol{G}$-algebra $\boldsymbol{G}$ and truth constants are interpreted by themselves. We say that a set $\{\boldsymbol{0}, \boldsymbol{1}\} \subseteq X$ of truth constants is admissible with respect to suprema and infima if, for all $\emptyset \neq Y_1, Y_2 \subseteq X$ and $\bigvee Y_1 = \bigwedge Y_2$, $\bigvee Y_1 \in Y_1$, $\bigwedge Y_2 \in Y_2$. Then the hyperresolution calculus is refutation sound and complete for a countable order clausal theory if the set of all truth constants occurring in the theory is admissible with respect to suprema and infima. This condition obviously covers the case of finite order clausal theories. As an interesting consequence, we get an affirmative solution to the open problem of recursive enumerability of unsatisfiable formulae in the first-order Gödel logic with intermediate truth constants and the operators $\eqcirc$, $\prec$, Δ over $[0,1]$:

$$a \eqcirc b = \begin{cases} 1 \ \textit{if } a = b, \\ 0 \ \textit{else,} \end{cases} \qquad a \prec b = \begin{cases} 1 \ \textit{if } a < b, \\ 0 \ \textit{else,} \end{cases} \qquad \Delta a = \begin{cases} 1 \ \textit{if } a = 1, \\ 0 \ \textit{else,} \end{cases}$$

which strengthens a similar result for prenex formulae of G_∞^Δ, stated in Conclusion of [1].

2   First-Order Gödel Logic

Throughout the paper, we shall use the common notions and notation of first-order logic.[1] By $\mathcal{L}$ we denote a first-order language. We assume truth

---

[1] Cf. http://ii.fmph.uniba.sk/~guller/cci15.pdf, Section 2.

constants - nullary predicate symbols $0, 1 \in Pred_{\mathcal{L}}$, $ar_{\mathcal{L}}(0) = ar_{\mathcal{L}}(1) = 0$; 0 denotes the false and 1 the true in $\mathcal{L}$. Let $C_{\mathcal{L}} \subseteq (0,1)$ be countable. In addition, we assume a countable set of nullary predicate symbols $\overline{C}_{\mathcal{L}} = \{\bar{c} \mid \bar{c} \in Pred_{\mathcal{L}}, ar_{\mathcal{L}}(\bar{c}) = 0, c \in C_{\mathcal{L}}\} \subseteq Pred_{\mathcal{L}}$. 0, 1, $\bar{c} \in \overline{C}_{\mathcal{L}}$ are called truth constants. We denote $Tcons_{\mathcal{L}} = \{0, 1\} \cup \overline{C}_{\mathcal{L}} \subseteq Pred_{\mathcal{L}}$. Let $X \subseteq Tcons_{\mathcal{L}}$. We denote $\overline{X} = \{0 \mid 0 \in X\} \cup \{1 \mid 1 \in X\} \cup \{c \mid \bar{c} \in X \cap \overline{C}_{\mathcal{L}}\} \subseteq [0,1]$. In addition, we introduce a new unary connective Δ, delta, and binary connectives $=$, equality, $\prec$, strict order. By $OrdForm_{\mathcal{L}}$ we designate the set of all so-called order formulae of $\mathcal{L}$ built up from $Atom_{\mathcal{L}}$ and $Var_{\mathcal{L}}$ using the connectives: $\neg$, negation, Δ, $\wedge$, conjunction, $\vee$, disjunction, $\rightarrow$, implication, $\leftrightarrow$, equivalence, $=$, $\prec$, and the quantifiers: $\forall$, the universal one, $\exists$, the existential one. In the paper, we shall assume that $\mathcal{L}$ is a countable first-order language; hence, all the above mentioned sets of symbols and expressions are countable.

Gödel logic is interpreted by the standard G-algebra augmented by the operators $=$, $\prec$, Δ for the connectives $=$, $\prec$, Δ, respectively.

$$G = ([0,1], \leq, \vee, \wedge, \Rightarrow, \overline{}, =, \prec, \Delta, 0, 1)$$

where $\vee \mid \wedge$ denotes the supremum $\mid$ infimum operator on $[0,1]$;

$$a \Rightarrow b = \begin{cases} 1 & \text{if } a \leq b, \\ b & \text{else,} \end{cases} \qquad \overline{a} = \begin{cases} 1 & \text{if } a = 0, \\ 0 & \text{else.} \end{cases}$$

Let $\mathcal{I}$ be an interpretation for $\mathcal{L}$. Notice that $0^{\mathcal{I}} = 0$, $1^{\mathcal{I}} = 1$, for all $\bar{c} \in \overline{C}_{\mathcal{L}}$, $\bar{c}^{\mathcal{I}} = c$. Let $\phi \in OrdForm_{\mathcal{L}}$ and $T \subseteq OrdForm_{\mathcal{L}}$. We denote $tcons(\phi) = \{0, 1\} \cup (preds(\phi) \cap \overline{C}_{\mathcal{L}}) \subseteq Tcons_{\mathcal{L}}$ and $tcons(T) = \{0, 1\} \cup (preds(T) \cap \overline{C}_{\mathcal{L}}) \subseteq Tcons_{\mathcal{L}}$.

3   Translation to Clausal Form

We firstly define a notion of quantified atom.[2] Let $a \in OrdForm_{\mathcal{L}}$. a is a quantified atom of $\mathcal{L}$ iff $a = Qx\, p(t_0, \ldots, t_\tau)$ where $p(t_0, \ldots, t_\tau) \in Atom_{\mathcal{L}}$, $x \in vars(p(t_0, \ldots, t_\tau))$, either $t_i = x$ or $x \notin vars(t_i)$. $QAtom_{\mathcal{L}} \subseteq OrdForm_{\mathcal{L}}$ denotes the set of all quantified atoms of $\mathcal{L}$. Let $Qx\, p(t_0, \ldots, t_\tau) \in QAtom_{\mathcal{L}}$ and $p(t'_0, \ldots, t'_\tau) \in Atom_{\mathcal{L}}$. Let $I = \{i \mid i \leq \tau, x \notin vars(t_i)\}$ and $r_1, \ldots, r_k$, $r_i \leq \tau$, $k \leq \tau$, be the increasing sequence of I. We denote $freetermseq(Qx\, p(t_0, \ldots, t_\tau)) = t_{r_1}, \ldots, t_{r_k}$ and $freetermseq(p(t'_0, \ldots, t'_\tau)) = t'_0, \ldots, t'_\tau$.

We further introduce order clauses in Gödel logic. Let $l \in OrdForm_{\mathcal{L}}$. l is an order literal of $\mathcal{L}$ iff $l = \varepsilon_1 \diamond \varepsilon_2$, $\varepsilon_i \in Atom_{\mathcal{L}} \cup QAtom_{\mathcal{L}}$, $\diamond \in \{=, \prec\}$. The set of all order literals of $\mathcal{L}$ is designated as $OrdLit_{\mathcal{L}} \subseteq OrdForm_{\mathcal{L}}$. An order clause of $\mathcal{L}$ is a finite set of order literals of $\mathcal{L}$. An order clause $\{l_1, \ldots, l_n\}$ is written in the form $l_1 \vee \cdots \vee l_n$. We designate the set of all order clauses of $\mathcal{L}$ as $OrdCl_{\mathcal{L}}$.

Let $C \in OrdCl_{\mathcal{L}}$, $S \subseteq OrdCl_{\mathcal{L}}$, $\mathcal{I}$ be an interpretation for $\mathcal{L}$, $e \in \mathcal{S}_{\mathcal{I}}$. C is true in $\mathcal{I}$ with respect to e, written as $\mathcal{I} \models_e C$, iff there exists $l^* \in C$ such that $\mathcal{I} \models_e l^*$. We denote $tcons(S) = \{0, 1\} \cup (preds(S) \cap \overline{C}_{\mathcal{L}}) \subseteq Tcons_{\mathcal{L}}$. Let $\tilde{f}_0 \notin Func_{\mathcal{L}}$; $\tilde{f}_0$

---

[2] Cf. http://ii.fmph.uniba.sk/~guller/cci15.pdf, Section 3.

is a new function symbol. Let $\mathbb{I} = \mathbb{N} \times \mathbb{N}$; $\mathbb{I}$ is an infinite countable set of indices. Let $\tilde{\mathbb{P}} = \{\tilde{p}_i \,|\, i \in \mathbb{I}\}$ such that $\tilde{\mathbb{P}} \cap Pred_{\mathcal{L}} = \emptyset$; $\tilde{\mathbb{P}}$ is an infinite countable set of new predicate symbols.

We assume the reader to be familiar with the standard notions and notation of substitutions.[3] Let $Qx\,a \in QAtom_{\mathcal{L}}$ and ϑ be a substitution of $\mathcal{L}$. ϑ is applicable to $Qx\,a$ iff $dom(\vartheta) \supseteq freevars(Qx\,a)$ and $x \notin range(\vartheta|_{freevars(Qx\,a)})$. We define the application of ϑ to $Qx\,a$ as $(Qx\,a)\vartheta = Qx\,a(\vartheta|_{freevars(Qx\,a)} \cup x/x) \in QAtom_{\mathcal{L}}$. We denote $Inst_{\mathcal{L}}(S) = \{C \,|\, C \text{ is an instance of } S \text{ of } \mathcal{L}\} \subseteq OrdCl_{\mathcal{L}}$ and $Vrnt_{\mathcal{L}}(S) = \{C \,|\, C \text{ is a variant of } S \text{ of } \mathcal{L}\} \subseteq OrdCl_{\mathcal{L}}$.

3.1 A Formal Treatment

Translation of an order formula or theory to clausal form is based on the following lemma:

Lemma 1. *Let $n_\phi, n_0 \in \mathbb{N}$, $\phi \in OrdForm_{\mathcal{L}}$, $T \subseteq OrdForm_{\mathcal{L}}$.*

(I) *There exist either $J_\phi = \emptyset$ or $J_\phi = \{(n_\phi, j) \,|\, j \leq n_{J_\phi}\}$, $J_\phi \subseteq \{(n_\phi, j) \,|\, j \in \mathbb{N}\}$, and $S_\phi \subseteq_{\mathcal{F}} OrdCl_{\mathcal{L} \cup \{\tilde{p}_j \,|\, j \in J_\phi\}}$ such that*

   (a) *there exists an interpretation $\mathfrak{A}$ for $\mathcal{L}$ and $\mathfrak{A} \models \phi$ if and only if there exists an interpretation $\mathfrak{A}'$ for $\mathcal{L} \cup \{\tilde{p}_j \,|\, j \in J_\phi\}$ and $\mathfrak{A}' \models S_\phi$, satisfying $\mathfrak{A} = \mathfrak{A}'|_{\mathcal{L}}$;*

   (b) *$|S_\phi| \in O(|\phi|^2)$; the number of all elementary operations of the translation of ϕ to S_ϕ, is in $O(|\phi|^2)$; the time and space complexity of the translation of ϕ to S_ϕ, is in $O(|\phi|^2 \cdot (\log(1 + n_\phi) + \log|\phi|))$;*

   (c) *$tcons(S_\phi) \subseteq tcons(\phi)$.*

(II) *There exist $J_T \subseteq \{(i, j) \,|\, i \geq n_0\}$ and $S_T \subseteq OrdCl_{\mathcal{L} \cup \{\tilde{p}_j \,|\, j \in J_T\}}$ such that*

   (a) *there exists an interpretation $\mathfrak{A}$ for $\mathcal{L}$ and $\mathfrak{A} \models T$ if and only if there exists an interpretation $\mathfrak{A}'$ for $\mathcal{L} \cup \{\tilde{p}_j \,|\, j \in J_T\}$ and $\mathfrak{A}' \models S_T$, satisfying $\mathfrak{A} = \mathfrak{A}'|_{\mathcal{L}}$;*

   (b) *if $T \subseteq_{\mathcal{F}} OrdForm_{\mathcal{L}}$, then $J_T \subseteq_{\mathcal{F}} \{(i, j) \,|\, i \geq n_0\}$, $\|J_T\| \leq 2 \cdot |T|$, $S_T \subseteq_{\mathcal{F}} OrdCl_{\mathcal{L} \cup \{\tilde{p}_j \,|\, j \in J_T\}}$, $|S_T| \in O(|T|^2)$; the number of all elementary operations of the translation of T to S_T, is in $O(|T|^2)$; the time and space complexity of the translation of T to S_T, is in $O(|T|^2 \cdot \log(1 + n_0 + |T|))$;*

   (c) *$tcons(S_T) \subseteq tcons(T)$.*

Proof. By interpolation. Cf. http://ii.fmph.uniba.sk/~guller/cci15.pdf, Subsection 3.3, Lemma 1, for a proof. □

Corollary 1. *Let $n_0 \in \mathbb{N}$, $\phi \in OrdForm_{\mathcal{L}}$, $T \subseteq OrdForm_{\mathcal{L}}$. There exist $J_T^\phi \subseteq \{(i, j) \,|\, i \geq n_0\}$ and $S_T^\phi \subseteq OrdCl_{\mathcal{L} \cup \{\tilde{p}_j \,|\, j \in J_T^\phi\}}$ such that*

(i) *$T \models \phi$ if and only if S_T^ϕ is unsatisfiable;*

---

[3] Cf. http://ii.fmph.uniba.sk/~guller/cci15.pdf, Appendix, Subsection 5.1.

(ii) *if* $T \subseteq_{\mathcal{F}} OrdForm_{\mathcal{L}}$, *then* $J_T^{\phi} \subseteq_{\mathcal{F}} \{(i,j) \mid i \geq n_0\}$, $\|J_T^{\phi}\| \in O(|T| + |\phi|)$, $S_T^{\phi} \subseteq_{\mathcal{F}} OrdCl_{\mathcal{L} \cup \{\tilde{p}_j \mid j \in J_T^{\phi}\}}$, $|S_T^{\phi}| \in O(|T|^2 + |\phi|^2)$; *the number of all elementary operations of the translation of T and ϕ to S_T^{ϕ}, is in $O(|T|^2 + |\phi|^2)$; the time and space complexity of the translation of T and ϕ to S_T^{ϕ}, is in* $O(|T|^2 \cdot \log(1 + n_0 + |T|) + |\phi|^2 \cdot (\log(1 + n_0) + \log|\phi|))$;

(iii) $tcons(S_T^{\phi}) \subseteq tcons(\phi) \cup tcons(T)$.

Proof. Straightforward. Cf. http://ii.fmph.uniba.sk/~guller/cci15.pdf, Subsection 3.3, Corollary 1, for a proof.   □

4   Hyperresolution Over Order Clauses

In this section, we propose an order hyperresolution calculus and prove its refutational soundness and completeness.

4.1   Order Hyperresolution Rules

At first, we introduce some basic notions and notation concerning chains of order literals. A chain Ξ of $\mathcal{L}$ is a sequence $\Xi = \varepsilon_0 \diamond_0 \upsilon_0, \ldots, \varepsilon_n \diamond_n \upsilon_n$, $\varepsilon_i \diamond_i \upsilon_i \in OrdLit_{\mathcal{L}}$, such that for all $i < n$, $\upsilon_i = \varepsilon_{i+1}$. Let $\Xi = \varepsilon_0 \diamond_0 \upsilon_0, \ldots, \varepsilon_n \diamond_n \upsilon_n$ be a chain of $\mathcal{L}$. Ξ is an equality chain of $\mathcal{L}$ iff, for all $i \leq n$, $\diamond_i = \mathbf{=}$. Ξ is an increasing chain of $\mathcal{L}$ iff there exists $i^* \leq n$ such that $\diamond_{i^*} = \prec$. Ξ is a contradiction of $\mathcal{L}$ iff Ξ is an increasing chain of $\mathcal{L}$ of the form $\varepsilon_0 \Xi 0$ or $1 \Xi \upsilon_n$ or $\varepsilon_0 \Xi \varepsilon_0$. Let $S \subseteq OrdCl_{\mathcal{L}}$ be unit and $\Xi = \varepsilon_0 \diamond_0 \upsilon_0, \ldots, \varepsilon_n \diamond_n \upsilon_n$ be a chain | an equality chain | an increasing chain | a contradiction of $\mathcal{L}$. Ξ is a chain | an equality chain | an increasing chain | a contradiction of S iff, for all $i \leq n$, $\varepsilon_i \diamond_i \upsilon_i \in S$.

Let $\tilde{W} = \{\tilde{w}_i \mid i \in \mathbb{I}\}$ such that $\tilde{W} \cap (Func_{\mathcal{L}} \cup \{\tilde{f}_0\}) = \emptyset$; $\tilde{W}$ is an infinite countable set of new function symbols. Let $P \subseteq \tilde{\mathbb{P}}$ and $S \subseteq OrdCl_{\mathcal{L} \cup P}$. We denote $ordtcons(S) = \{0 \prec \bar{c} \mid \bar{c} \in tcons(S) \cap \overline{C}_{\mathcal{L}}\} \cup \{\bar{c} \prec 1 \mid \bar{c} \in tcons(S) \cap \overline{C}_{\mathcal{L}}\} \cup \{\bar{c}_1 \prec \bar{c}_2 \mid \bar{c}_1, \bar{c}_2 \in tcons(S) \cap \overline{C}_{\mathcal{L}}, c_1 < c_2\} \subseteq OrdCl_{\mathcal{L}}$.

An order hyperresolution calculus is defined in Table 1. The first rule is a central order hyperresolution one with obvious intuition.

A notion of deduction by order hyperresolution is defined in the standard manner.[4] By $clo^{\mathcal{H}}(S)$ we denote the set of all order clauses of $\mathcal{L} \cup \tilde{W} \cup P$ deduced from S by order hyperresolution.

4.2   Refutational Soundness and Completeness

We are in position to state the refutational soundness and completeness of the order hyperresolution calculus. Let $\{0, 1\} \subseteq X \subseteq Tcons_{\mathcal{L}}$. X is admissible with respect to suprema and infima iff, for all $\emptyset \neq Y_1, Y_2 \subseteq \overline{X}$ and $\bigvee Y_1 = \bigwedge Y_2$, $\bigvee Y_1 \in Y_1, \bigwedge Y_2 \in Y_2$.

---

[4] Cf. http://ii.fmph.uniba.sk/~guller/cci15.pdf, Subsection 4.1.

Table 1. Order hyperresolution calculus

(Order hyperresolution rule) (1)

$$\frac{\bigvee_{j=0}^{k_0} \varepsilon_j^0 \diamond_j^0 v_j^0 \vee \bigvee_{j=1}^{m_0} l_j^0, \ldots, \bigvee_{j=0}^{k_n} \varepsilon_j^n \diamond_j^n v_j^n \vee \bigvee_{j=1}^{m_n} l_j^n \in S_{\kappa-1}^{Vr}}{\left(\bigvee_{i=0}^{n} \bigvee_{j=1}^{m_i} l_j^i\right)\theta \in S_\kappa};$$

for all $i < i' \le n$,

$freevars(\bigvee_{j=0}^{k_i} \varepsilon_j^i \diamond_j^i v_j^i \vee \bigvee_{j=1}^{m_i} l_j^i) \cap freevars(\bigvee_{j=0}^{k_{i'}} \varepsilon_j^{i'} \diamond_j^{i'} v_j^{i'} \vee \bigvee_{j=1}^{m_{i'}} l_j^{i'}) = \emptyset,$

$\theta \in mgu_{\mathcal{L}_{\kappa-1}}\big(\bigvee_{j=0}^{k_0} \varepsilon_j^0 \diamond_j^0 v_j^0, l_1^0, \ldots, l_{m_0}^0, \ldots, \bigvee_{j=0}^{k_n} \varepsilon_j^n \diamond_j^n v_j^n, l_1^n, \ldots, l_{m_n}^n,$
$\{v_0^0, \varepsilon_0^0\}, \ldots, \{v_0^{n-1}, \varepsilon_0^n\}, \{a, b\}\big),$

$dom(\theta) = freevars(\{\varepsilon_j^i \diamond_j^i v_j^i \mid j \le k_i, i \le n\}, \{l_j^i \mid 1 \le j \le m_i, i \le n\}),$

$a = \varepsilon_0^0, b = 1$ *or* $a = v_0^n, b = 0$ *or* $a = \varepsilon_0^0, b = v_0^n,$

there exists $i^* \le n$ *such that* $\diamond_0^{i^*} = \prec.$

(Order trichotomy rule) (2)

$$\frac{a, b \in atoms(S_{\kappa-1}^{Vr}), \{a, b\} \not\subseteq Tcons_{\mathcal{L}}}{a \prec b \vee a \approx b \vee b \prec a \in S_\kappa};$$

$vars(a) \cap vars(b) = \emptyset.$

(Order ∀-quantification rule) (3)

$$\frac{\forall x\, a \in qatoms^\forall(S_{\kappa-1})}{\forall x\, a \prec a \vee \forall x\, a \approx a \in S_\kappa}.$$

(Order ∃-quantification rule) (4)

$$\frac{\exists x\, a \in qatoms^\exists(S_{\kappa-1})}{a \prec \exists x\, a \vee a \approx \exists x\, a \in S_\kappa}.$$

(Order ∀-witnessing rule) (5)

$$\frac{\forall x\, a \in qatoms^\forall(S_{\kappa-1}^{Vr}), b \in atoms(S_{\kappa-1}^{Vr}) \cup qatoms(S_{\kappa-1}^{Vr})}{a\gamma \prec b \vee b \approx \forall x\, a \vee b \prec \forall x\, a \in S_\kappa};$$

$freevars(\forall x\, a) \cap freevars(b) = \emptyset,$

$\tilde{w} \in \tilde{W} - Func_{\mathcal{L}_{\kappa-1}}, ar(\tilde{w}) = |freetermseq(\forall x\, a), freetermseq(b)|,$

$\gamma = x/\tilde{w}(freetermseq(\forall x\, a), freetermseq(b)) \cup id|_{vars(a)-\{x\}} \in Subst_{\mathcal{L}_\kappa},$

$dom(\gamma) = \{x\} \cup (vars(a) - \{x\}) = vars(a).$

(Order ∃-witnessing rule) (6)

$$\frac{\exists x\, a \in qatoms^\exists(S_{\kappa-1}^{Vr}), b \in atoms(S_{\kappa-1}^{Vr}) \cup qatoms(S_{\kappa-1}^{Vr})}{b \prec a\gamma \vee \exists x\, a \approx b \vee \exists x\, a \prec b \in S_\kappa};$$

$freevars(\exists x\, a) \cap freevars(b) = \emptyset,$

$\tilde{w} \in \tilde{W} - Func_{\mathcal{L}_{\kappa-1}}, ar(\tilde{w}) = |freetermseq(\exists x\, a), freetermseq(b)|,$

$\gamma = x/\tilde{w}(freetermseq(\exists x\, a), freetermseq(b)) \cup id|_{vars(a)-\{x\}} \in Subst_{\mathcal{L}_\kappa},$

$dom(\gamma) = \{x\} \cup (vars(a) - \{x\}) = vars(a).$

Theorem 1 (Refutational Soundness and Completeness). *Let* $P \subseteq \tilde{\mathbb{P}}$, $S \subseteq OrdCl_{\mathcal{L} \cup P}$, $tcons(S)$ *be admissible with respect to suprema and infima.* $\square \in clo^{\mathcal{H}}(S)$ *if and only if* S *is unsatisfiable.*

Proof. By forcing. Cf. http://ii.fmph.uniba.sk/~guller/cci15.pdf, Subsection 4.2, Theorem 4, for a proof. $\square$

Table 2. An example: $\phi = \forall x\,(q_1(x) \to \overline{0.3}) \to (\exists x\,q_1(x) \to \overline{0.5})$

$\phi = \forall x\,(q_1(x) \to \overline{0.3}) \to (\exists x\,q_1(x) \to \overline{0.5})$

$\{\tilde{p}_0(x) \prec 1, \tilde{p}_0(x) \leftrightarrow \underbrace{(\underbrace{\forall x\,(q_1(x) \to \overline{0.3})}_{\tilde{p}_1(x)} \to \underbrace{(\exists x\,q_1(x) \to \overline{0.5})}_{\tilde{p}_2(x)})}\}$

$\{\tilde{p}_0(x) \prec 1, \tilde{p}_1(x) \prec \tilde{p}_2(x) \vee \tilde{p}_1(x) \approx \tilde{p}_2(x) \vee \tilde{p}_0(x) \approx \tilde{p}_2(x), \tilde{p}_2(x) \prec \tilde{p}_1(x) \vee \tilde{p}_0(x) \approx 1,$
$\tilde{p}_1(x) \leftrightarrow \forall x\,\underbrace{(q_1(x) \to \overline{0.3})}_{\tilde{p}_3(x)}, \tilde{p}_2(x) \leftrightarrow \underbrace{(\exists x\,q_1(x)}_{\tilde{p}_4(x)} \to \underbrace{\overline{0.5}}_{\tilde{p}_5(x)})\}$

$\{\tilde{p}_0(x) \prec 1, \tilde{p}_1(x) \prec \tilde{p}_2(x) \vee \tilde{p}_1(x) \approx \tilde{p}_2(x) \vee \tilde{p}_0(x) \approx \tilde{p}_2(x), \tilde{p}_2(x) \prec \tilde{p}_1(x) \vee \tilde{p}_0(x) \approx 1,$
$\tilde{p}_1(x) \approx \forall x\,\tilde{p}_3(x), \tilde{p}_3(x) \leftrightarrow \underbrace{(q_1(x) \to}_{\tilde{p}_6(x)} \underbrace{\overline{0.3}}_{\tilde{p}_7(x)}),$
$\tilde{p}_4(x) \prec \tilde{p}_5(x) \vee \tilde{p}_4(x) \approx \tilde{p}_5(x) \vee \tilde{p}_2(x) \approx \tilde{p}_5(x), \tilde{p}_5(x) \prec \tilde{p}_4(x) \vee \tilde{p}_2(x) \approx 1,$
$\tilde{p}_4(x) \leftrightarrow \underbrace{\exists x\,q_1(x)}_{\tilde{p}_8(x)}, \tilde{p}_5(x) \approx \overline{0.5}\}$

$\{\tilde{p}_0(x) \prec 1, \tilde{p}_1(x) \prec \tilde{p}_2(x) \vee \tilde{p}_1(x) \approx \tilde{p}_2(x) \vee \tilde{p}_0(x) \approx \tilde{p}_2(x), \tilde{p}_2(x) \prec \tilde{p}_1(x) \vee \tilde{p}_0(x) \approx 1,$
$\tilde{p}_1(x) \approx \forall x\,\tilde{p}_3(x), \tilde{p}_6(x) \prec \tilde{p}_7(x) \vee \tilde{p}_6(x) \approx \tilde{p}_7(x) \vee \tilde{p}_3(x) \approx \tilde{p}_7(x), \tilde{p}_7(x) \prec \tilde{p}_6(x) \vee \tilde{p}_3(x) \approx 1,$
$\tilde{p}_6(x) \approx q_1(x), \tilde{p}_7(x) \approx \overline{0.3},$
$\tilde{p}_4(x) \prec \tilde{p}_5(x) \vee \tilde{p}_4(x) \approx \tilde{p}_5(x) \vee \tilde{p}_2(x) \approx \tilde{p}_5(x), \tilde{p}_5(x) \prec \tilde{p}_4(x) \vee \tilde{p}_2(x) \approx 1,$
$\tilde{p}_4(x) \approx \exists x\,\tilde{p}_8(x), \tilde{p}_8(x) \approx q_1(x), \tilde{p}_5(x) \approx \overline{0.5}\}$

$S^\phi = \Bigg\{\; \boxed{\tilde{p}_0(x) \prec 1}$ [1]

$\tilde{p}_1(x) \prec \tilde{p}_2(x) \vee \tilde{p}_1(x) \approx \tilde{p}_2(x) \vee$
$\boxed{\tilde{p}_0(x) \approx \tilde{p}_2(x)}$ [2]

$\tilde{p}_2(x) \prec \tilde{p}_1(x) \vee \boxed{\tilde{p}_0(x) \approx 1}$ [3]

$\boxed{\tilde{p}_1(x) \approx \forall x\,\tilde{p}_3(x)}$ [4]

$\tilde{p}_6(x) \prec \tilde{p}_7(x) \vee \tilde{p}_6(x) \approx \tilde{p}_7(x) \vee$
$\boxed{\tilde{p}_3(x) \approx \tilde{p}_7(x)}$ [5]

$\tilde{p}_7(x) \prec \tilde{p}_6(x) \vee \tilde{p}_3(x) \approx 1$ [6]

$\boxed{\tilde{p}_6(x) \approx q_1(x)}$ [7]

$\boxed{\tilde{p}_7(x) \approx \overline{0.3}}$ [8]

$\boxed{\tilde{p}_4(x) \prec \tilde{p}_5(x) \vee \tilde{p}_4(x) \approx \tilde{p}_5(x)} \vee$
$\tilde{p}_2(x) \approx \tilde{p}_5(x)$ [9]

$\tilde{p}_5(x) \prec \tilde{p}_4(x) \vee \boxed{\tilde{p}_2(x) \approx 1}$ [10]

$\boxed{\tilde{p}_4(x) \approx \exists x\,\tilde{p}_8(x)}$ [11]

$\boxed{\tilde{p}_8(x) \approx q_1(x)}$ [12]

$\boxed{\tilde{p}_5(x) \approx \overline{0.5}}\Bigg\}$ [13]

Rule (1) : [1][3] :

$\boxed{\tilde{p}_2(x) \prec \tilde{p}_1(x)}$ [14]

Rule (1) : [10][14] :

$\boxed{\tilde{p}_5(x) \prec \tilde{p}_4(x)}$ [15]

repeatedly **Rule (1)** : [9][15] :

$\boxed{\tilde{p}_2(x) \approx \tilde{p}_5(x)}$ [16]

Rule (3) : $\forall x\,\tilde{p}_3(x)$:

$\boxed{\forall x\,\tilde{p}_3(x) \prec \tilde{p}_3(x) \vee \forall x\,\tilde{p}_3(x) \approx \tilde{p}_3(x)}$ [17]

repeatedly **Rule (1)** : [4][5][8][13][14][16][17] :

$\boxed{\tilde{p}_6(x) \prec \tilde{p}_7(x) \vee \tilde{p}_6(x) \approx \tilde{p}_7(x)}$ [18]

Rule (6) : $\exists x\,\tilde{p}_8(x), \overline{0.5}$:

$\overline{0.5} \prec \tilde{p}_8(\tilde{w}_{(0,0)}) \vee$
$\boxed{\exists x\,\tilde{p}_8(x) \prec \overline{0.5} \vee \exists x\,\tilde{p}_8(x) \approx \overline{0.5}}$ [19]

repeatedly **Rule (1)** : [11][13][15][19] :

$\boxed{\overline{0.5} \prec \tilde{p}_8(\tilde{w}_{(0,0)})}$ [20]

repeatedly **Rule (1)** : [7][8][12][18]; $\tilde{w}_{(0,0)}$: [20] :

$\square$ [21]

Consider $S = \{0 \prec a\} \cup \{a \prec \frac{1}{n} \mid n \in \mathbb{N}\} \subseteq OrdCl_\mathcal{L}$, $a \in Pred_\mathcal{L} - Tcons_\mathcal{L}$, $ar_\mathcal{L}(a) = 0$. $tcons(S)$ is not admissible with respect to suprema and infima; for $\{0\}$ and $\{\frac{1}{n} \mid n \in \mathbb{N}\}$, $\bigvee\{0\} = \bigwedge\{\frac{1}{n} \mid n \in \mathbb{N}\} = 0$, $0 \in \{0\}$, $0 \notin \{\frac{1}{n} \mid n \in \mathbb{N}\}$. S is unsatisfiable; both the cases $\|a\|^{\mathfrak{A}} = 0$ and $\|a\|^{\mathfrak{A}} > 0$ lead to $\mathfrak{A} \not\models S$ for every interpretation $\mathfrak{A}$ for $\mathcal{L}$. However, $\square \notin clo^{\mathcal{H}}(S)$; which contains the order clauses from S, from $ordtcons(S)$, and some superclauses of them produced by Rules (2), (1). So, the condition on $tcons(S)$ being admissible with respect to suprema and infima, is necessary.

Corollary 2. *Let $n_0 \in \mathbb{N}$, $\phi \in OrdForm_{\mathcal{L}}$, $T \subseteq OrdForm_{\mathcal{L}}$, $tcons(T)$ be admissible with respect to suprema and infima. There exist $J_T^\phi \subseteq \{(i,j) \mid i \geq n_0\}$ and $S_T^\phi \subseteq OrdCl_{\mathcal{L} \cup \{\tilde{p}_j \mid j \in J_T^\phi\}}$ such that $tcons(S_T^\phi)$ is admissible with respect to suprema and infima; $T \models \phi$ if and only if $\square \in clo^{\mathcal{H}}(S_T^\phi)$.*

Proof. An immediate consequence of Corollary 1 and Theorem 1.   □

In Table 2, we show that $\phi = \forall x \, (q_1(x) \rightarrow \overline{0.3}) \rightarrow (\exists x \, q_1(x) \rightarrow \overline{0.5}) \in OrdForm_{\mathcal{L}}$ is logically valid using the translation to order clausal form and the order hyperresolution calculus.

5   Conclusions

In the paper, we have proposed a modification of the hyperresolution calculus from [7] which is suitable for automated deduction with explicit partial truth. The first-order Gödel logic is expanded by a countable set of intermediate truth constants $\bar{c}$, $c \in (0,1)$. We have modified translation of a formula to an equivalent satisfiable finite order clausal theory, consisting of order clauses. An order clause is a finite set of order literals of the form $\varepsilon_1 \diamond \varepsilon_2$ where $\diamond$ is a connective either $\equiv$ or $\prec$. $\equiv$ and $\prec$ are interpreted by the equality and standard strict linear order on $[0,1]$, respectively. We have investigated the so-called canonical standard completeness, where the semantics of the first-order Gödel logic is given by the standard G-algebra and truth constants are interpreted by themselves. The modified hyperresolution calculus is refutation sound and complete for a countable order clausal theory if the set of all truth constants occurring in the theory is admissible with respect to suprema and infima. This condition covers the case of finite order clausal theories. As an interesting consequence, we get an affirmative solution to the open problem of recursive enumerability of unsatisfiable formulae in the first-order Gödel logic with intermediate truth constants and the equality, $\equiv$, strict order, $\prec$, projection Δ operators.

Corollary 3. *The set of unsatisfiable formulae of $\mathcal{L}$ is recursively enumerable.*

Proof. Let $\phi \in OrdForm_{\mathcal{L}}$. Then ϕ contains a finite number of truth constants and $tcons(\{\phi\})$ is admissible with respect to suprema and infima. The statement ϕ is unsatisfiable, is equivalent to $\{\phi\} \models 0$. Hence, the problem that ϕ is unsatisfiable can be reduced to the deduction problem $\{\phi\} \models 0$ after a constant number of steps. Let $n_0 \in \mathbb{N}$. By Corollary 2 for n_0, 0, $\{\phi\}$, there exist $J_{\{\phi\}}^0 \subseteq \{(i,j) \mid i \geq n_0\}$, $S_{\{\phi\}}^0 \subseteq OrdCl_{\mathcal{L} \cup \{\tilde{p}_j \mid j \in J_{\{\phi\}}^0\}}$ and $tcons(S_{\{\phi\}}^0)$ is admissible with respect to suprema and infima, $\{\phi\} \models 0$ if and only if $\square \in clo^{\mathcal{H}}(S_{\{\phi\}}^0)$; if $\{\phi\} \models 0$, then $\square \in clo^{\mathcal{H}}(S_{\{\phi\}}^0)$ and we can decide it after a finite number of steps. This straightforwardly implies that the set of unsatisfiable formulae of $\mathcal{L}$ is recursively enumerable. The corollary is proved.   □

References

1. Baaz, M., Ciabattoni, A., Fermüller, C.G.: Theorem proving for prenex Gödel logic with delta: checking validity and unsatisfiability. Logical Methods in Computer Science 8(1) (2012)
2. Esteva, F., Gispert, J., Godo, L., Noguera, C.: Adding truth-constants to logics of continuous t-norms: axiomatization and completeness results. Fuzzy Sets and Systems 158(6), 597–618 (2007)
3. Esteva, F., Godo, L., Noguera, C.: On completeness results for the expansions with truth-constants of some predicate fuzzy logics. In: Stepnicka, M., Novák, V., Bodenhofer, U. (eds.) New Dimensions in Fuzzy Logic and Related Technologies. Proceedings of the 5th EUSFLAT Conference, Ostrava, Czech Republic, September 11-14, vol. 2: Regular Sessions, pp. 21–26. Universitas Ostraviensis (2007)
4. Esteva, F., Godo, L., Noguera, C.: First-order t-norm based fuzzy logics with truth-constants: distinguished semantics and completeness properties. Ann. Pure Appl. Logic 161(2), 185–202 (2009)
5. Esteva, F., Godo, L., Noguera, C.: Expanding the propositional logic of a t-norm with truth-constants: completeness results for rational semantics. Soft Comput. 14(3), 273–284 (2010)
6. Esteva, F., Godo, L., Noguera, C.: On expansions of WNM t-norm based logics with truth-constants. Fuzzy Sets and Systems 161(3), 347–368 (2010)
7. Guller, D.: An order hyperresolution calculus for Gödel logic - General first-order case. In: Rosa, A.C., Correia, A.D., Madani, K., Filipe, J., Kacprzyk, J. (eds.) IJCCI 2012, Proceedings of the 4th International Joint Conference on Computational Intelligence, Barcelona, Spain, October, 5–7, pp. 329–342. SciTePress (2012)
8. Hájek, P.: Metamathematics of Fuzzy Logic. Trends in Logic. Springer (2001). http://books.google.sk/books?id=Eo-e8Pi-HmwC
9. Novák, V., Perfilieva, I., Močkoř, J.: Mathematical Principles of Fuzzy Logic. The Springer International Series in Engineering and Computer Science. Springer, US (1999). http://books.google.sk/books?id=pJeu6Ue65S4C
10. Pavelka, J.: On fuzzy logic I, II, III. Semantical completeness of some many-valued propositional calculi. Mathematical Logic Quarterly 25(2529), 45–52, 119–134, 447–464 (1979)
11. Savický, P., Cignoli, R., Esteva, F., Godo, L., Noguera, C.: On Product logic with truth-constants. J. Log. Comput. 16(2), 205–225 (2006)

System Optimization

An Intelligent Media Delivery Prototype System with Low Response Time

Jinzhong Hou[1]([✉]), Tiejian Luo[1], Zhu Wang[2], and Xiaoqi Li[1]

[1] University of Chinese Academy of Sciences, Beijing, China
{houjinzhong13,tjluo,lixiaoqi13}@mails.ucas.ac.cn
[2] Data Communication Technology Research Institute, Beijing, China
wangzhu09@mails.ucas.ac.cn

Abstract. Streaming media has been increasing in the Internet as a popular form of content. However, the streaming media delivery between server and client browser still has problems to be solved, such as the poor processing efficiency in dealing with large concurrent access and the high usage of bandwidth. In particular, large scale video site usually has a large number of distributed server nodes. It is of great importance for the system to respond to users' request rapidly by choosing the proper video source for users and handling resource cache problem properly. To solve the above problems in streaming media delivery, this paper proposes a content delivery solution. The system consists of several nodes that differ in role function, and streaming media content is stored in these nodes. Users will get an optimized response, and the system selects the nearest node that has the requested video according to logical distance intellectually. The selected node will provide video stream for users. In addition, the system is equipped with a high performance content indexing and searching mechanism. The index is able to retrieve users' requested resource rapidly and therefore guarantees a good performance in selecting nodes.

Keywords: Streaming media · Content delivery · Intelligent routing · High-performance index

1 Introduction

With the development of Internet and the wide spread of Web 2.0, the content on the Internet has become more and more rich in variety. Streaming media gains popularity from users as it has vivid manifestation. As a result, it uses up more and more Internet bandwidth. At the same time, because of the larger size compared with other internet content, streaming media often needs more internet resource. Those reasons lead to the high cost of streaming media delivery system. A high performance content delivery mechanism is required to achieve availability and good economical efficiency.

Direct Access Network (DAN) [1] responses to user's request with nearby servers, and meets customer needs at a relatively lower cost. At the same time

© Springer International Publishing Switzerland 2015
Y. Tan et al. (Eds.): ICSI-CCI 2015, Part III, LNCS 9142, pp. 253–264, 2015.
DOI: 10.1007/978-3-319-20469-7_28

it has good performance to handle sudden traffic, and uses resource reasonably. Streaming media delivery system based on DAN intelligently selects a nearby server to respond to user with the proper video streaming which user required. This leads to shorter time delay and less bandwidth occupation, satisfying user requests rapidly at a lower cost. This system aims at solving the video streaming delivery problem, trying to find a solution to big concurrency and high bandwidth problems in practical content delivery. To achieve this goal, we designed a node selecting algorithm, a high performance index and search method, and some other components in the system.

This work gives a solution to streaming media delivery. It may be a good reference to large scale video site and meaningful for the increasing streaming media's delivery performance. In addition, the content indexing and searching method can also be applied to other types of data that need to do fast searching task.

2   Related Work

Streaming media delivery has attracted much research interest nowadays. There have been several models for that system. Traditional CS (Client/Server) model comes at first, then P2P (Peer to Peer) network and CDN (Content delivery Network) follows.

In CS architecture, the streaming media resource is transmitted to user by a server directly. User and server get connected by socket, the video is transmitted by UDP protocol and the control information is transmitted by TCP protocol [2]. The advantage of this approach is that the architecture is quite concise and resource management, user management is relatively easy to implement; the disadvantage is that it will pay much bandwidth cost, user may get response at a longer delay time. In addition this structure is not able to handle with large number of requests at sudden.

In P2P network, a user is both service user and service provider, it plays both two roles. The users get resources from other nodes and share their own resources to other users at the same time [3]. The system nodes are often organized as a tree or no structure. The advantage of P2P is that it has high flexibility in establishing network and the speed is fast [4]. But it often uses too much common bandwidth, causing others' download speed slow down. The network is also not reliable comparing to system managed absolutely by a formal organization. Another disadvantage is that users often need to download the whole video before watching it. In addition, the P2P application usually occupies uses' machine resource, affecting user's benefits. As a result, some users will forbid the P2P application to protect their profit. To deal with this problem, the service provider often has to design motivational strategy [5].

CDN adds a new layer to current network. Through this layer, it publishes the web content to the network edge that nearest to users [6]. Then the content can be reached by user from nearby servers, thus relieving network congestion problem, and getting a fast response speed when users browsing the site. The advantage of

CDN is that it can reduce public bandwidth when transmitting the same size of content, thus providing users with in time, fluent playing experience. But it must consider many factors comprehensively to select proper location to put servers, and the resources uploaded by users is not well handled [7].

In a new media delivery model DAN, there are three kinds of nodes. Different kinds of nodes have differences in system roles and functionalities. They work together to achieve system goal. The DAN adds a new streaming media delivery layer in the network without changing the underlying architecture of current network. The system responds to a particular user according to the distance of user and server nodes. User will get response from the nearest available server node. By this method, a user will get a shorter response time, thus promoting VOD experience, maximizing profit of end user, streaming media service provider and internet provider. DAN adopts Bloom Filter based index and search mechanism. Bloom filter is suitable for distributed systems and can get quite good efficiency in both time and space [8]. Bloom Filter uses hash. It can find whether or not an element belongs to a particular collection. Because of the usage of hash, the searching process is very efficient [9]. But in extreme situation it may make a wrong judgment. Bloom Filter is applied in searching resources in many domains and achieves a good performance [10].

3   DAN Prototype System

To implement the streaming media delivery system based on DAN, the server nodes should be set up. According to DAN model, three kinds of nodes are needed. They correspond the three layers of DAN architecture. The first layer node works as Service Mediator, we refer to it as MainCMPE (Main Central Media Process Engine); The second layer node is called SlaveCMPE (Slave Central Media Process Engine); And the last layer node is EMPE (Edge Media Process Engine). The architecture of our prototype system is shown in Fig 1.

Every layer of DAN model should have at least one server node. Considering nearest selection mechanism, if we only have one SlaveCMPE, it's not enough to prove that the system select a nearer node to a particular user. And because of the limitation of our current experiment environment, four servers are set up in our prototype system: one MainCMPE node, two SlaveCMPE nodes and an EMPE node.

The edge server EMPE is nearest to users. It interacts with user directly. At the same time, it can provide media stream to users. Another important job EMPE does is that it caches some video resource. If the video user requested by user is already cached in the EMPE, it will directly provide user with the cached video, speeding up the response time. To achieve this mechanism, the EMPE should have the ability to index and retrieve video resources. The searching process should be fast enough to ensure good performance.If the user requested video is not yet cached in EMPE, there will be a mechanism to redirect user's request to another node that has the video and meanwhile nearer to user. This node then provide media stream to user. Another functionality EMPE has is

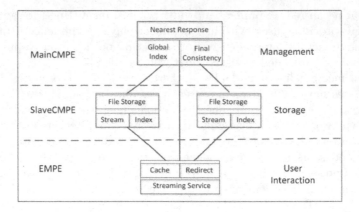

Fig. 1. Architecture of streaming media delivery prototype system

that when a user uploads a video file, the EMPE should assign the uploaded file to the proper storage server.

MainCMPE maintains global video list and Bloom Filter data. It has index of the whole system and all nodes' information. The recorded information uses time stamp to keep consistency. Time stamp represents the last time the data modified. By comparing two time stamps, we can judge whether video list or Bloom Filter data is already updated in other server. If the data is outdated, we can update it and by this way we maintain system consistency.

SlaveCMPE is the storage node of the system. All the video files are distributed by particular strategy to different SlaveCMPE. SlaveCMPE participate in uploaded file store and resources cache. If needed, the SlaveCMPE will provide media stream directly for user, as a method to improve VOD experience.

4   Video Storage and Index

4.1   Video Storage

Users can upload video files to the system. The video files are firstly uploaded to EMPE, and then EMPE modifies its index and send it to a particular SlaveCMPE to store. MainCMPE should also change its index and video list to maintain the consistency among different system nodes.

In DAN network, the MainCMPE is responsible for maintaining the video list. Moreover, MainCMPE also inserts an element to Bloom Filter index and update the timestamp, the timestamp is the basic for other nodes to check whether there is update in system. The video list and Bloom Filter index need to be transferred from different system nodes, this leads to communication between nodes. To solve this question, we use serialization [11] to transfer data to other nodes. By using serialization, the data in memory becomes file, than we transfer this file to other nodes.The algorithm of this method is summarized in Algorithm *StoreIndex*.

STOREINDEX(*filename*)

```
1    ▷ select k nodes from the N SlaveCMPEs.
2    for i ← 1 to k
3        do
4            node[i] ← Random(N)
5    for j ← 1 to k
6        do
7            ▷ SendFile send file filename to SlaveCMPE with node id node[j]
8            SendFile(filename,node[j])
9            ▷ insert filename to BF index
10           BF_node[j] ← filename
11   update BF in MainCMPE
```

In this media streaming delivery prototype system, the video files are stored in SlaveCMPE nodes. They are stored in several nodes according to a redundant strategy. As a result, there may be some common video files in different nodes. Because of the limitations in our machine number, the redundant strategy is not implemented in prototype system. Redundancy in video file storage is meant to push resources to places near users. Simultaneously, it is also a backup mechanism. The file system we use in SlaveCMPE is simply the operating system's file system.

4.2   Video File Index

In order to establish index for the video resources, we use both database and Bloom Filter. Information of each video is inserted in the database table as a basic index. More importantly, the information is inserted to Bloom Filter data structure as a fast index.

Bloom Filter is a data structure that can be used to data index and element query. It can be transferred from one node to other node without too much internet bandwidth. At the same time, it uses hash functions in insert and query process, thus gains good time efficiency.

The disadvantage of Bloom Filter is that it may give false positive answer [12]. In other words, it may return true (i.e. the element is in the set) when an element is not in the set. But the possibility can be controlled by parameters well designed. We assume the Bloom Filter has an m bits array with k hash functions and n elements have been inserted. The optimal hash function number is

$$k = \frac{m}{n} \ln 2 \tag{1}$$

Given the false positive possibility p, the optimal m is

$$m = -\frac{n \ln p}{(\ln 2)^2} \tag{2}$$

In addition, given m, n, k, we can get the false positive possibility is not large than

$$p = \left(1 - e^{-k(n+0.5)/(m-1)}\right)^k \tag{3}$$

Because Bloom Filter doesn't explicitly store the video name, it is secure. But on the other hand, it may lack information about the video stored at this server. Database is a make up for this disadvantage. We should note that Bloom Filter has a very little opportunity to give wrong judgment for whether a particular video exists in the system. At this time, getting particular video may be failed. If this situation occurs, we will get time exceeds error, then redirect user to another nodes that has the requested video. The Bloom Filter index data will spread to other nodes, and then other nodes can use this index to search resources. Eventual consistency is maintained during the index spreading process.

5   Nearest Response

For a particular user's video request, the system first checks whether local EMPE has cached this video. If EMPE fails (i.e. EMPE does not have that video), user's request will redirected to other server node that contains the video. The best node that can serve the user may be the node that is nearest to the user among the servers which contains the requested video. Selecting that server to provide user streaming media service can lead to shorter response time because of the shorter distance [13]. The system uses this mechanism to improve streaming media service experience. We refer to this approach as nearest response. We have designed a nearest response mechanism in prototype system. Because the time and experimental environment limitations, we haven't locate the servers at different area, instead we assume that each node has the virtual coordinates (x,y). The coordinates represent the server's location. By computing euclidean distance between two nodes, we get their distance. For example, given server A and B, the distance is:

$$dist(A, B) = \sqrt{(x_1 - x_2)^2 + (y_1 - y_2)^2} \tag{4}$$

For a particular node, we can get the distance from it to all the other nodes. By comparing distance, it's possible to select a nearest node. Assume that total nodes number is n, and there are m random nodes stored the requested video files. The Expectation node number to check is

$$E_1 = mP(m) + (m+1)P(m+1) + ... + nP(n) \tag{5}$$

$$P(x) = C_x^{m-1}/C_n^m, \quad m \le n \tag{6}$$

By equation (5) and (6),we have

$$E_1 \ge \frac{n}{2} \tag{7}$$

We use an optimization when the system starts. We order the server node from the nearest to the furthest. Using this strategy, when we search an eligible node, we search it also from the nearest to the furthest. The first node that contains requested video is the nearest, this node will be the choice. When check one node, the probability that it is the node we want is

$$P = \frac{m}{n}, \quad m \leq n \tag{8}$$

Its obvious that the expectation nodes amount we checked until find the best is

$$E_2 = \frac{n}{m} \leq E_1, \quad 2 \leq m \leq n \tag{9}$$

In most cases, E_1 is far larger than E_2. This means that the optimized method probably checks less nodes when find the best node. The algorithm is shown in function "SelectServer".

SELECTSERVER(*CoordinateSet*,*BFSet*,*filename*)

```
 1    ▷ each node only computes the distance between itself and other nodes
 2    for each node nd in CoordinateSet
 3        do
 4            if nd ≠ local
 5            then
 6                    DistanceSet ← (nd, (local.x − nd.x)² + (local.y − nd.y)²)
 7    Sort(DistanceSet)
 8    for each pair (nd, distance) in DistanceSet
 9        do
10            bf ← BFSet.get(nd)
11            ▷ if bf.query returns true, it means that the file exists
12            if bf.query(filename) = true
13            then
14                    return nd
```

In most conditions, the system need not search all the servers to find the optimal server node, thus ensures high efficiency.

6   Video Caching

When the requested video is not cached in EMPE that is nearest to user, video caching process will take place. The prototype system first find the optimal node to serve the user, then invokes video caching. The requested video file will be transferred to EMPE as a fast cache. When user requests the same video second time, it is served locally. We expect that this method can cache the popular video in EMPE and get shorter delay time. This approach is based on the fact that the video people watched before will be more likely to get watched again.

Assume that hit rate is h, search time in one node is t_{nd}, network transfer time is t_{short} when file cached in EMPE, otherwise t_{long}. EMPE will takes t_{nd} time to search whether it has the requested file. The average time spend is

$$t_{avg} = h(t_{nd} + t_{short}) + (1 - h)(t_{nd}E_2 + t_{long}) \tag{10}$$

In Equation (10), there is

$$t_{nd} + t_{short} < t_{nd}E_2 + t_{long} \qquad (11)$$

So from Equation (10) and (11), we can get the conclusion that when the hit rate h is higher, the average time will be shorter.

After video caching, EMPE serves user with video stream directly when user request it again.

7   Experiments

7.1   Video Search Time

We deploy our prototype system on four nodes. All of the nodes are virtual machines with one CPU core and 1GB memory.

We do two experiments on video search time. In the first one, we request only the videos that already cached by local EMPE, then measure the time used. The second experiment uses another way to test search time, we request only the video not have been cached by EMPE. In this condition the system will use our nearest response mechanism to find the proper node and provide media stream. We measure the search time 100 times in each experiment.

In the first experiment, the average search time for cached video is 0.123567 milliseconds, the standard error of the data is 0.117982. In the second experiment. The average search time is 0.236811, which is longer than the first experiment. The standard error is 0.060323. We can see from the search time that

Table 1. Video search time. avg is average search time, σ is standard error

	avg (ms)	σ
cached video	0.123567	0.117982
remote video	0.236811	0.060323

when a requested video is not cached in EMPE, the search time is longer. That is because when EMPE misses, the system will search the video on other servers, and this takes up more time. As shown in the result, although both standard errors are small, the standard error in first experiment is less than in the second one. In the second experiment, we may find the requested video after searching one or two nodes. That leads to the difference in time cost. Both standard errors are small because Bloom Filter computes hash functions with fixed number. The compute time is then quite stable.

In the searching process, hash function finishes in constant time. Assume that we use k hash functions in one Bloom Filter, then each query is of constant time complexity. Firstly, the EMPE checks whether the request video exists in its file system, the time is constant. If it does not exist in the EMPE, it looks for the video in other nodes. The total time complexity is liner to the nodes number in the system. Assume we have n nodes in system, the time complexity is O(n). In our experiments, the second one will search more nodes, so its search time doubles. That result is caused by the liner time complexity indeed.

7.2 EMPE Hit Rate

We focus on the performance of different cache strategies, algorithm is more important. So the experiment is a simulation. We use different numbers of total videos stored in the system to test the EMPE hit rate. The EMPE can cache limited size of video resources, so changing the total video number may change the percentage of file cached in EMPE. In this experiment, we use 10000, 50000 and 100000 total video files and do experiment separately. According to Zipf's law [14], the frequency that each video is requested is different. There are frequently requested videos and seldomly visited videos. We generate simulation data of users' requests based on Zipf's law and measure the hit rate of the EMPE cache. Each measuring under particular setting is repeated 10 times.

In practice, a common video file is not very large. We assume it to be 100MB, a relative large size. So the video storage capacity of EMPE is measurable. We assume that 3000 video files can be cached in EMPE. Three replacement algorithms are used in our experiment, they are First In First Out(FIFO), Second Chance(SC) and Least Recently used(LRU).

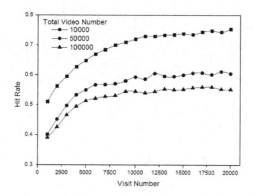

Fig. 2. Hit rate when visit number changes, using FIFO

FIFO is the most basic replace algorithm, it swaps the oldest item in cache. As visit amount increases, the hit rate increases fast at the beginning, slows down after about 10000 visit. Total video number also affects the hit rate.

The three replacement algorithms have similar feature. There are two rules we can get from the figure:

1. When the total number of videos is permanent, the more user visit number, the more hit rate.
2. When user visit number is permanent, the more video files, the less hit rate.

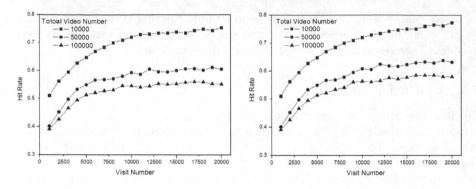

Fig. 3. Hit rate using SC   **Fig. 4.** Hit rate using LRU

The three algorithms use different swap strategies, we hope to select a best one. In order to get an intuitive sense of the different performance, we compare them in Fig5. This figure uses data with total video number of 50000.

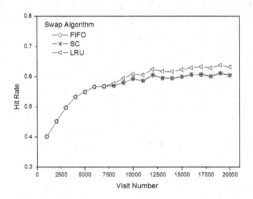

Fig. 5. Hit rate when use different replacement algorithms

LRU algorithm is concerned as most effective in page replacement. It also achieves the best result in our experiment. In addition, FIFO and SC are very close to each other. It is worth noting that SC is improvement of FIFO, it gives the frequently used item a second chance to stay at cache. In fact, the SC algorithm result is slightly better than FIFO in our result data. But it is hard to distinguish from the figure. The hit rate of different total video numbers from Fig2 shows the relative relations of the three different conditions.

8   Conclusion and Future Work

We implemented a streaming media delivery system based on DAN. The system comprises of four nodes, one MainCMPE node, two SlaveCMPE nodes, and one

EMPE node. Each node has its own role and functionality. They work together to provide streaming media to users.

The system has a distributed video storage and management mechanism. Videos are stored in different nodes, and may have multiple copies if we have enough nodes. To manage the video resources, we designed and implemented Bloom Filter based index. Because of the usage of Bloom Filter, the searching process becomes faster. To improve the user experience, the system response to user from the nearest eligible node which is referred to as Nearest Response in our paper. Considering the different frequency a particular video be requested, we use video file caching to maintain the frequently accessed video file available at EMPE. We measured the system by its search time and EMPE hit rate. The result shows that our work is effective and efficient.

Moreover, it is found that some topics in the prototype system are still worth further study. Because of the limitations of our experiment environment, we only have four nodes. It is important to add more nodes to the system and test its performance. Currently, video is uploaded to EMPE, and then delivered to SlaveCMPE. If we can upload the file directly to SlaveCMPE, it will be more efficient. Although we use sorting to improve our searching speed, it still searchs nodes one by one from the nearest to the furthest. We will continue our work to find a heuristic search method. We hope the method can decrease the nodes we have to search.

References

1. Zhu, W., Tiejian, L.: Video content routing in a direct access network. In: Proceedings 2011 3rd Symposium on Web Society (SWS 2011), pp. 147–152. IEEE Press, New York (2011)
2. Jian, E., Qian, J.: Research of Video Data Transmission Based on the C/S system. Information Technology **28**(1), 35–41 (2004)
3. Ma, R.T., Lee, S., Lui, J., Yau, D.K.: Incentive and service differentiation in P2P networks: a game theoretic approach. IEEE/ACM Transactions on Networking (TON) **14**(5), 978–991 (2006)
4. Wang, Y., Nakao, A., Vasilakos, A.V., Ma, J.: On the effectiveness of service differentiation based resource-provision incentive mechanisms in dynamic and autonomous P2P networks. Computer Networks **55**(17), 3811–3831 (2011)
5. Alex, S., Jason, N., Clifford, S.: FairTorrent: Bringing fairness to peer-to-peer systems. In: Proceedings of the 2009 ACM Conference on Emerging Networking Experiments and Technologies, pp. 133–144 (2009)
6. Pathan, A.M.K., Buyya, R.: A taxonomy and survey of content delivery networks. University of Melbourne,Technical Report, Grid Computing and Distributed Systems Laboratory(2007)
7. Content Delivery Network. http://en.wikipedia.org/wiki/Content_delivery_network
8. Tarkoma, S., Rothenberg, C.E., Lagerspetz, E.: Theory and practice of bloom filters for distributed systems. IEEE Communications Surveys and Tutorials **14**(1), 131–155 (2012)

9. Yuen, W.H., Schulzrinne, H.: Improving search efficiency using Bloom filters in partially connected ad hoc networks: A node-centric analysis. Computer Communications **30**(16), 3000–3011 (2007)
10. Broder, A., Mitzenmacher, M.: Network applications of bloom filters: A survey. Internet Mathematics **1**(4), 485–509 (2004)
11. Alboaie, S., Buraga, S.C., Alboaie, L.: An XML-based serialization of information exchanged by software agents. Informatica (Slovenia) **28**(1), 13–22 (2004)
12. Bloom, B.H.: Space/time trade-offs in hash coding with allowable errors. Communications of the ACM **13**(7), 422–426 (1970)
13. Zhuang, Z., Guo, C.: Optimizing CDN infrastructure for live streaming with constrained server chaining. In: Proceedings of 2011 IEEE 9th International Symposium on Parallel and Distributed Processing with Applications (ISPA), pp. 183–188. IEEE Press, New York (2011)
14. Wolf, D.N., Serpanosatand, W.H.: Caching Web objects using Zipf's law. Multimedia Storage and Archiving Systems **3527**, 320 (1998)

Incomplete Distributed Information Systems Optimization Based on Queries

Agnieszka Dardzinska[✉] and Anna Romaniuk

Department of Mechanics and Computer Science, Bialystok University of Technology,
ul. Wiejska 45a, 15-351 Bialystok, Poland
a.dardzinska@pb.edu.pl, a.romaniuk@doktoranci.pb.edu.pl

Abstract. In this paper we assume there is a group of connected distributed information systems (DIS). They work under the same ontology. Each information system has its own knowledgebase. Values of attributes in information system form atomic expressions of a language used for communication with others. Collaboration among systems is initiated when one of them (called a client) is asked to resolve a query containing nonlocal attributes for. In such case, the client has to ask for help other information systems to have that query answered. As the result of its request, knowledge is extracted locally in each information system and sent back to the client. The outcome of this step is a knowledgebase created at the client site, which can be used to answer given query. In this paper we present a method of identifying which information system is semantically the closest to client.

Keywords: Query answering system · Knowledge extraction · Query · Client · Chase algorithm

1 Introduction

In this paper we assume that there is a group of collaborating information systems, which can be incomplete, coupled with a Query Answering System (QAS) and a knowledgebase which is initially empty. By incompleteness we mean a property which allows us use attributes and their values with corresponding weights [4]. Additionally, we assume that the sum of these weights for one particular attribute value for one object has to be equal 1. The definition of an information system of type λ given in this paper was initially proposed in [10]. The type λ was introduced with a purpose to check the weights assigned to values of attributes by Chase algorithm [4],[9]. If a weight is less than λ, then the corresponding attribute value is ruled out and weights assigned to the remaining attribute values are equally adjusted so its sum is equal one again.

Semantic inconsistencies are due to different interpretations of attributes and their values among sites (for instance one site can interpret the concept healthy or good differently than other sites). Different interpretations are also implied by the fact that each site may differently handle incompleteness.

Null value replacement by a value suggested either by statistical or some rule-based methods is quite common before a query is answered by QAS. Ontologies

© Springer International Publishing Switzerland 2015
Y. Tan et al. (Eds.): ICSI-CCI 2015, Part III, LNCS 9142, pp. 265–274, 2015.
DOI: 10.1007/978-3-319-20469-7_29

[1], [2], [6], [7], [8], [13], [14], [15], [16] are widely used as a part of semantical bridge between agents built independently so they can collaborate and understand each other. In [10], the notion of the optimal rough semantics and a method of its construction was proposed. The rough semantics can be used to model and nicely handle semantic inconsistencies among sites due to different interpretations of incomplete values.

As the result of collaborations among information systems, a knowledgebase of any system is updated and it contains rules extracted from information systems representing other systems. Although the names of attributes can be the same among information systems, their granularity levels or their interpretation may differ. As the result of these differences, the knowledgebase has to satisfy certain properties in order to be used by Chase [4],[5]. Also, the semantic differences between information systems may influence the precision and recall of a query answering system. We will show that it is wise to use the knowledge obtained from systems which are semantically close to the client while solving a query. This way, the precision and recall is getting improved [3],[4].

2   Incomplete Information System

In reality, different data are often collected and stored in information systems residing at many different locations, built independently, instead of collecting and storing them at a single location. In this case we talk about distributed information systems. It is very possible that some attributes are missing in one of these systems and occur in others.

Definition 1:
We say that $S = (X, A, V)$ is an incomplete information system of type λ, if S is an incomplete information system introduced by Pawlak in [9] and the following four conditions hold:

1. X is a set of objects, A is a set of attributes, and $V = \cup \{V_a : a \in A\}$ is a set of values of attribute a
2. $(\forall x \in X) (\forall a \in A)[a_S(x) \in V_a$ or $a_S(x) = \{(a_i, p_i) : a_i \in V_a \wedge p_i \in [0,1] \wedge 1 \leq i \leq m\}]$
3. $(\forall x \in X) (\forall a \in A)[(a_S(x) = \{(a_i, p_i) : 1 \leq i \leq m\}) \Rightarrow \sum_{i=1}^{m} p_i = 1]$
4. $(\forall x \in X) (\forall a \in A)[(a_S(x) = \{(a_i, p_i) : 1 \leq i \leq m\}) \Rightarrow (\forall i)(p_i \geq \lambda)]$

Now, let us assume that (S_1, K) and (S_2, K) are incomplete information systems, both of type λ, the knowledgebase K (empty at the beginning). The same set of objects X is stored in both systems and the same set of attributes A is used to describe them. The meaning and granularity of values of attributes from A in both systems S_1 and S_2 is also the same.

In this paper we concentrate on granularity – based semantic inconsistencies. Assume now that (S_i, K) for any $i \in I$, I is an information system and all information systems form distributed information system (DIS). Additionally, we assume that, if $a \in A_i \cap A_j$, then only the granularity levels of attribute a in S_i and S_j may differ but conceptually its meaning, both in S_i and S_j is the same. Assume now that $D = \cup L(D_i)$ is a set of rules which can be used by Chase algorithm [3],[4], associated with any of the sites of DIS, and $L(D_i)$ contains rules extracted from S_i. Now, let us say that system S is a client. Chase algorithm, to be applicable to S, has to be based on rules from D which satisfy the following conditions:

— attribute value used in the decision part of a rule from D has the granularity level either equal to or finer than the granularity level of the corresponding attribute in S.
— the granularity level of any attribute used in the classification part of a rule from D is either equal or softer than the granularity level of the corresponding attribute in S.
— attribute used in the decision part of a rule from D either does not belong to A or is incomplete in S.

Assuming that a match between the attribute value in IS and the attribute value used in a description of a rule is found, the following two cases can be taken into consideration:

— attribute involved in matching is the decision attribute in rule. If two attribute values, involved in that match, have different granularity, then the decision value d has to be replaced by a softer value which granularity will match the granularity of the corresponding attribute in S.
— attribute a involved in matching is the classification attribute in rule. If two attribute values, involved in that match, have different granularity, then the value of attribute a has to be replaced by a finer value which granularity will match the granularity of a in S.

In paper we also assume that:

Definition 2:
Information system S_1 can be transformed into S_2 by containment mapping ψ if two conditions hold:
— $(\forall x \in X)(\forall a \in A)[card(a_{S_1}(x)) \geq card(a_{S_2}(x))]$
— $(\forall x \in X)(\forall a \in A)[card(a_{S_1}(x)) = card(a_{S_2}(x)) \Rightarrow [\sum_{i \neq j}|p_{2i} - p_{2j}| > [\sum_{i \neq j}|p_{1i} - p_{1j}|]]$

So, if containment mapping ψ converts an information system S_1 to S_2, then S_2 is more complete than S_1. Saying another words, for a minimum one pair $(a, x) \in A \times X$, either ψ has to decrease the number of attribute values in $a_{S_1}(x)$ or the average difference between confidences assigned to attribute values in $a_{S_2}(x)$ has to be increased by ψ.

To give an example of a containment mapping ψ, let us take two information systems S_1, S_2 both of the type λ, represented as Table 1 and Table 2.

Table 1. Information System S_1

X	a	b	c	d	e
x1	$\{(a_1,\frac{1}{3}),(a_2,\frac{2}{3})\}$	$\{(b_1,\frac{2}{3}),(b_2,\frac{1}{3})\}$	c_1	d_1	$\{(e_1,\frac{1}{2}),(e_2,\frac{1}{2})\}$
x2	$\{(a_2,\frac{1}{4}),(a_3,\frac{3}{4})\}$	$\{(b_1,\frac{1}{3}),(b_2,\frac{2}{3})\}$		d_2	e_1
x3		b_2	$\{(c_1,\frac{1}{2}),(c_3,\frac{1}{2})\}$	d_2	e_3
x4	a_3		c_2	d_1	$\{(e_1,\frac{2}{3}),(e_2,\frac{1}{3})\}$
x5	$\{(a_1,\frac{2}{3}),(a_2,\frac{1}{3})\}$	b_1	c_2		e_1
x6	a_2	b_2	c_3	d_2	$\{(e_2,\frac{1}{3}),(e_3,\frac{2}{3})\}$
x7	a_2	$\{(b_1,\frac{1}{4}),(b_2,\frac{3}{4})\}$	$\{(c_1,\frac{1}{3}),(c_2,\frac{2}{3})\}$	d_2	e_2
x8		b_2	c_1	d_1	e_3

It can be easily checked that the values of attributes assigned to $e(x1), b(x2), c(x2), a(x3), e(x4), a(x5), c(x7)$, and $a(x8)$ in S_1 are different than the corresponding values in S_2.

In each of these eight cases, an attribute value assigned to an object in S_2 is less general than the value assigned to the same object in S_1. It means that $\psi(S_1) = S_2$.

Table 2. Information System S_2

X	a	b	c	d	e
x1	$\{(a_1,\frac{1}{3}),(a_2,\frac{2}{3})\}$	$\{(b_1,\frac{2}{3}),(b_2,\frac{1}{3})\}$	c_1	d_1	$\{(e_1,\frac{1}{3}),(e_2,\frac{2}{3})\}$
x2	$\{(a_2,\frac{1}{4}),(a_3,\frac{3}{4})\}$	b_1	$\{(c_1,\frac{1}{3}),(c_2,\frac{2}{3})\}$	d_2	e_1
x3	a_1	b_2	$\{(c_1,\frac{1}{2}),(c_3,\frac{1}{2})\}$	d_2	e_3
x4	a_3		c_2	d_1	e_2
x5	$\{(a_1,\frac{3}{4}),(a_2,\frac{1}{4})\}$	b_1	c_2		e_1
x6	a_2	b_2	c_3	d_2	$\{(e_2,\frac{1}{3}),(e_3,\frac{2}{3})\}$
x7	a_2	$\{(b_1,\frac{1}{4}),(b_2,\frac{3}{4})\}$	c_1	d_2	e_2
x8	$\{(a_1,\frac{2}{3}),(a_2,\frac{1}{3})\}$	b_2	c_1	d_1	e_3

3   Query Processing Based on Collaboration Between Information Systems

Assume now that user submits a query to one of the information system (a client), which cannot be answered because some of the attributes used in a query are unknown or hidden in the information system representing the client site.

Assume, we have a set of collaborating information systems (DIS) working under the same ontology, and user submits a query $q(B)$ to an information system (S, K) from DIS, where $S = (X, A, V)$, $K = \emptyset$, B are the attributes used in $q(B)$, and $A \cap B \neq \emptyset$. All attributes in $B \setminus [A \cap B]$ are called foreign for (S, K). Since (S, K) can collaborate with other information systems in DIS, definitions of hidden or missed attributes for (S, K) can be extracted from other information systems in DIS. In [8], it was shown that (S, K) can answer the query $q(B)$ assuming that definitions of all values of attributes from $B \setminus [A \cap B]$ can be extracted at the remote sites for S and used to answer $q(B)$.

Hidden attributes for S, can be seen as attributes with only null values assigned to all objects in S. Assume now that we have three collaborating information systems with knowledgebase connected with them: (S, K), (S_1, K_1), (S_2, K_2), where $S = (X, A, V)$, $S_1 = (X_1, A_1, V_1)$, $S_2 = (X_2, A_2, V_2)$, and $K = K_1 = K_2 = \emptyset$. If the consensus between (S, K) and (S_1, K_1) on the knowledge extracted from $S(A \cap A_1)$ and $S_1(A \cap A_1)$ is closer than the consensus between (S, K) and (S_2, K_2) on the knowledge extracted from $S(A \cap A_2)$ and $S_2(A \cap A_2)$, then (S_1, K_1) is chosen by (S, K) as the closer information system, more helpful in solving user queries. Rules defining hidden attribute values for S are then extracted at S_1 and stored in K.

Assuming that systems S_1, S_2 store the same sets of objects and use the same attributes to describe them, system S_1 is more complete than system S_2, if $\psi(S_1) = S_2$. The question remains, if the values predicted by the imputation process are really correct, and if not, how far they are (assuming that some distance measure can be set up) from the correct values? Classical approach is to start with a complete information system and remove randomly some percent of its values and next run the imputation algorithm on the resulting system. The next step is to compare the descriptions of objects in the system which is the outcome of the imputation algorithm with descriptions of the same objects in the original system. But, before we can continue any further this discussion, we have to decide on the interpretation of functors "or" (+) and "and" (*). We will propose the semantics of terms presented in [10] since their semantics preserves distributive property, which means:

$t_1 * (t_2 + t_3) = (t_1 * t_2) + (t_1 * t_3)$, for any queries t_1, t_2, t_3.

So, let us assume that $S = (X, A, V)$ is an information system of type λ and t is a term constructed in a standard way from values of attributes in V seen as constants and from two functors + and *. By $N_s(t)$, we mean the standard interpretation of a term t in S [12] defined as:

- $N_s(v) = \{(x, p) : (v, p) \in a(x)\}$, for any $v \in V_a$
- $N_s(t_1 + t_2) = N_s(t_1) \oplus N_s(t_2)$
- $N_s(t_1 * t_2) = N_s(t_1) \otimes N_s(t_2)$

where:

- $N_s(t_1) \oplus N_s(t_2) = \{(x_i, p_i)\}_{i \in I \setminus J}\} \cup \{(x_j, q_j)\}_{j \in J \setminus I} \cup \{(x_i, \max(p_i, q_i))\}_{j \in I \cap J}$
- $N_s(t_1) \otimes N_s(t_2) = \{(x_i, p_i \cdot q_i)\}_{i \in I \cap J}$

Assume that (S, K) is an information system, where $S = (X, A, V)$ and K contains definitions of attribute values in B. Clearly $A \cap B = \emptyset$. The null value imputation algorithm Chase converts information system $S(A \cup B)$ of type λ to a new more complete information system Chase$(S(A \cup B))$ of the same type. Initially null values are assigned to all attributes in B for all objects in $S(A \cup B)$.

The proposed algorithm is new in comparison to known strategies for chasing null values in relational tables because of the assumption about partial incompleteness of data (sets of weighted attribute values can be assigned to an object as its value). Algorithm ERID [3],[4] is used by Chase algorithm to extract rules from this type of data.

4   Searching The Closest Information System

Assume again that information system (S, K) represents the client site. As we already pointed out, the knowledge base K, contains rules extracted from information systems representing other agents. Our goal is to find optimal i-information system (S_i, K) for client (S, K), where by optimal we mean an agent of maximal precision and recall. The distance between two agents is calculated using the formula:

$$d(S_i, S_j) = \frac{\sum_r d_r(S_i \to S_j) + \sum_r(S_j \to S_i)}{\sum_r \sup r\, S_i \cdot conf\, rS_i + \sum_r \sup r\, S_j \cdot conf\, rS_j}$$

where

$$d_r(S_i \to S_j) = |\frac{\sup rS_j \cdot conf\, rS_j}{\max(\sup r\, S_i, 1)} - \frac{\sup rS_i \cdot conf\, rS_i}{\max(\sup r\, S_j, 1)}|$$

From all the distributed systems we have, we choose the one with the minimal value of $d(S_i \to S_j)$, which corresponds to the closest information system to the client. The definition of hidden attribute d discovered at the closest system to the client and stored in KB of the client will guarantee that Query Answering System connected with the client has maximal precision and recall in group of agents having attribute d.

Example
Let us assume we have three information systems S, S_1, S_2, as represented as Table 3, Table 4 and Table 5.

Table 3. Information System S

X	a	b	c	d
x1	1	2	L	
x2	1	1	H	
x3	2	1	H	
x4	0	2	H	
x5	2	2	L	
x6	0	3	L	

Table 4. Information System S_1

X	a	b	c	d	f
x7	0	1	H	2	−
x8	0	2	L	2	+
x9	1	3	L	2	+
x10	1	1	H	2	−
x11	0	2	L	1	+
x12	1	3	L	1	+
x13	2	1	H	2	−

Table 5. Information System S_2

X	a	b	c	d
x14	1	1	H	1
x15	1	1	H	2
x16	0	2	L	1
x17	0	3	L	2
x18	2	2	L	1
x19	2	3	H	2

Information system S received a query $q(B) = a_2 * b_1 * d_2$ and has no information about hidden attribute d, which appears in other systems such as S_1 and S_2. Our goal is to choose one of the system either S_1 or S_2, from which we will be able to predict values of attribute d in system S and next to answer query $q(B)$.

Because attributes a, b, c are common for all the systems, first we extract rules describing them in terms of other attributes. For each rule we calculate support and confidence in a standard way [4].

For system S_1 we have:

$$b_1 \rightarrow a_0, \sup = 1, conf = \frac{1}{3}$$

$$b_1 \rightarrow a_1, \sup = 1, conf = \frac{1}{3}$$

$$b_1 \rightarrow a_2, \sup = 1, conf = \frac{1}{3}$$

$$b_2 \rightarrow a_0, \sup = 2, conf = 1$$

$$b_3 \rightarrow a_1, \sup = 2, conf = 1$$

$$c_H \rightarrow a_0, \sup = 1, conf = \frac{1}{3}$$

$$c_H \rightarrow a_1, \sup = 1, conf = \frac{1}{3}$$

$$c_H \rightarrow a_2, \sup = 1, conf = \frac{1}{3}$$

$$a_0 \rightarrow b_1, \sup = 1, conf = \frac{1}{3}$$

$$a_0 * c_H \rightarrow b_1, \sup = 1, conf = 1$$

$$a_0 * c_L \rightarrow b_2, \sup = 2, conf = 1$$

For system S_2 we have:

$$b_1 \rightarrow a_0, \text{sup} = 2, conf = 1$$
$$b_1 \rightarrow a_1, \text{sup} = 2, conf = 1$$
$$b_2 \rightarrow a_0, \text{sup} = 1, conf = \frac{1}{2}$$
$$b_2 \rightarrow a_2, \text{sup} = 1, conf = \frac{1}{2}$$
$$b_3 \rightarrow a_0, \text{sup} = 1, conf = \frac{1}{2}$$
$$b_3 \rightarrow a_2, \text{sup} = 1, conf = \frac{1}{2}$$
$$c_H \rightarrow a_1, \text{sup} = 2, conf = \frac{2}{3}$$
$$c_H \rightarrow a_2, \text{sup} = 1, conf = \frac{1}{3}$$
$$c_L \rightarrow a_0, \text{sup} = 2, conf = \frac{2}{3}$$
$$c_L \rightarrow a_2, \text{sup} = 1, conf = \frac{1}{3}$$
$$a_0 \rightarrow b_2, \text{sup} = 1, conf = \frac{1}{2}$$
$$a_2 * b_3 \rightarrow c_H, \text{sup} = 1, conf = 1$$
$$a_2 * d_1 \rightarrow c_L, \text{sup} = 1, conf = 1$$

We do the same for system S.

Next the distance between S and S_1 is calculated:

$$d(S \rightarrow S_1) = \frac{33.36 + 20.31}{30 + 34.66} = 0.83$$

and between S and S_1:

$$d(S \rightarrow S_2) = \frac{33.36 + 17.17}{30 + 29.66} = 0.85.$$

Because the distance between systems S and S_1 is smaller than between systems S and S_2 (the factor is smaller), we choose S_1 as the better, closer information system for contact with S. From chosen information system S_1, rules describing attribute d in terms of a, b and c are extracted. If we take into consideration incomplete information systems in DIS we can use algorithm ERID [3] for extracting rules from incomplete information system and next load them into knowledgebase K. These rules allow us to uncover some hidden attribute values in information system S. Therefore the submitted query $q(B)$ can be answered in next step.

5   Conclusions

In this paper we proposed the method of finding the closest information system to the client. Our goal was to find the best information system, which help to find objects satisfying submitted query. Once we find the closer information system to the given one, we are able to build knowledgebase consisting of rules extracted in distributed systems. These rules can be applied, so changes of attributes and their values in a query $q(B)$ can be made. Therefore we have chance to replace unknown attributes by mixture of attributes which are present in both of the systems, and the query can be answered.

References

1. Benjamins, V.R., Fensel, D., Prez, A.G.: Knowledge management through ontologies. In: Proceedings of the 2nd International Conference on Practical Aspects of Knowledge Management, Basel, Switzerland (1998)
2. Chandrasekaran, B., Josephson, J.R., Benjamins, V.R.: The ontology of tasks and methods. In: Proceedings of the 11th Workshop on Knowledge Acquisition, Modelling and Management, Ban, Alberta, Canada (1998)
3. Dardzinska, A., Ras, Z.W.: Extracting Rules from Incomplete Decision Systems: System ERID. In: Foundations and Novel Approaches in Data Mining. Studies in Computational Intelligence, vol. 9, pp. 143–153 (2006)
4. Dardzinska, A., Ras, Z.W.: Chase2, Rule based chase algorithm for information systems of type lambda. In: Tsumoto, S., Yamaguchi, T., Numao, M., Motoda, H. (eds.) AM 2003. LNCS (LNAI), vol.~3430, pp. 255–267. Springer, Heidelberg (2005)
5. Dardzinska, A., Ras, Z.W: On Rules Discovery from Incomplete Information Systems. In: Proceedings of ICDM 2003 Workshop on Foundations and New Directions of Data Mining, Melbourne, Florida, pp. 31–35. IEEE Computer Society (2003)
6. Fensel, D.: Ontologies: A silver bullet for knowledge management and electronic commerce. Springer (1998)
7. Guarino, N.: Formal Ontology in Information Systems. IOS Press, Amsterdam (1998)
8. Guarino, N., Giaretta, P.: Ontologies and knowledge bases, towards a terminological clarification. In: Towards Very Large Knowledge Bases: Knowledge Building and Knowledge Sharing. IOS Press (1995)
9. Pawlak, Z.: Information systems - theoretical foundations. Information Systems Journal **6**, 205–218 (1981)
10. Ras, Z.W., Dardzinska, A.: Cooperative Multi-hierarchical Query Answering Systems. In: Encyclopedia of Complexity and Systems Science, pp. 1532–1537 (2009)
11. Ras, Z.W.: Dardzinska, A: Solving Failing Queries through Cooperation and Collaboration. World Wide Web Journal **9**(2), 173–186 (2006)
12. Ras, Z.W., Joshi, S.: Query approximate answering system for an incomplete DKBS. Fundamenta Informaticae Journal **30**(3/4), 313–324 (1997)
13. Sowa, J.F.: Ontology, metadata, and semiotics. In: Ganter, B., Mineau, G.W. (eds.) ICCS 2000. LNCS (LNAI), vol. 1867, pp. 55–81. Springer, Berlin (2000)
14. Sowa, J.F.: Knowledge Representation: Logical, Philosophical, and Computational Foundations. Brooks/Cole Publishing Co., Pacific Grove, CA (2000)

15. Sowa, J.F.: Ontological categories. In: Albertazzi, L. (ed.) Shapes of Forms: From Gestalt Psychology and Phenomenology to Ontology and Mathematics, pp. 307–340. Kluwer Academic Publishers, Dordrecht (1999)
16. Van Heijst, G., Schreiber, A., Wielinga, B.: Using explicit ontologies in KBS development. International Journal of Human and Computer Studies **46**(2/3), 183–292 (1997)

Strategies for Improving the Profitability of a Korean Unit Train Operator: A System Dynamics Approach

Jae Un Jung[1(✉)] and Hyun Soo Kim[2]

[1] BK21Plus Groups, Dong-A University, Busan, Korea
imhere@dau.ac.kr
[2] Department of MIS, Dong-A University, Busan, Korea
hskim@dau.ac.kr

Abstract. A unit train (UT) has been developed primarily in countries that have wide or long territories, to move freight quickly over long distances. In South Korea, UTs have contributed to the facilitation of the overland export/import logistics for the last decade. However, UT operators in South Korea, which is a small country surrounded by North Korea and bodies of water, suffer from low profitability when competing with trucking companies because of diverse reasons that they cannot control. On this account, this research aims to develop business strategies for improving the profitability of a Korean UT operator. We analyzed both the revenues and expenses of a representative operator in Korea, and found simple but meaningful financial circular causality, using the system dynamics methodology. Thus, we presented and scientifically reinterpreted two strategies that might be acceptable alternatives: the internalization of shuttle carriers and the securing of more freight.

Keywords: Profitability · Rail freight · Strategy · System dynamics · Unit train

1 Introduction

A UT is composed of railcars that carry the same commodity from the same origin to the same destination, without stopping en route [1]. Since it was introduced by American railroad companies in the 1950s, the UT has been preferred in countries where territories are wide, and where quick transportation is required over long distances with beneficial shipping rates [2]. Recently, the environmental and social issues caused by road transportation, such as carbon emission and traffic congestion, redound to the benefits of the UT [3].

In South Korea, the UT was adopted for the first time, by Korail in 2004, under the name of 'block train', and 'Block Train' was trademarked in 2007 by KORAILLOGIS, a subsidiary company [4, 5]. During the past decade, UTs have played an outstanding role in the facilitation of the overland logistics, especially for exporting/importing in the area between Busan (the largest port in Korea) and Uiwang (the largest inland container depot (ICD) in Korea). Thus, in the entirety of overland freight, the utility of the UT deserves to be praised; unfortunately, this is not true in each case. UT operators licensed from Korail, which has a monopoly on rail operations, suffer from low profitability

© Springer International Publishing Switzerland 2015
Y. Tan et al. (Eds.): ICSI-CCI 2015, Part III, LNCS 9142, pp. 275–283, 2015.
DOI: 10.1007/978-3-319-20469-7_30

because the short transport distance of 400–500km at most, between an origin and destination, pushes them into competition with trucking companies [6, 7].

In 2014, Korail resolutely shut down 47 of its 127 freight stations that were performing in the red, and terminated its temporary freight service (non-UT service), which had low punctuality [8]. Through this enthusiastic reorganization of the rail operation policy, Korail experienced a gain for the first time since it changed its name from Korean National Railroad in 2005 [9]. However, Korail-centered rail operations, as well as the sudden collapse in oil prices in 2014, resulted in the modal shift that consignors, who wanted to carry their commodities by the irregular or temporary, yet cheaper than UT, freight service, and who did not need to utilize railroads, changed their transport mode to the truck. Even consignors that could not secure a freight rail schedule in the closed sections, did not want to change their transport mode to the truck more expensive but they should do that [10].

In such a situation, recent changes seem unsympathetic to UT operators in Korea. As such, this research aims to explore a strategy to improve the profitability of a UT operator, specifically in Korea. Therefore, we analyze both the external and internal business environments of a Korean UT operator, using the system dynamics methodology, thereby discovering the loss causation and the consequent improvement strategies.

2   Business Environment Analysis

2.1   Rail Freight Traffic

According to statistics [11] from the International Transport Forum at the Organization for Economic Co-operation and Development (OECD), worldwide rail freight traffic decreased with the global financial crisis in 2009; however, most countries or groups experienced an increase soon enough.

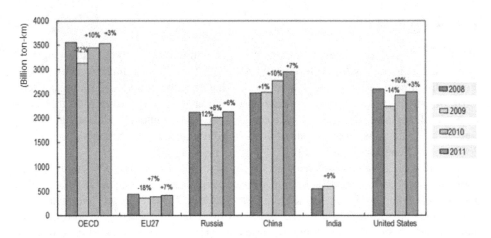

Fig. 1. Rail Freight Traffic in Key Countries [12]

As shown in Fig. 1, both OECD countries and the United States grew steadily, and nearly reached the 2008 pre-crisis level in 2011. Fig. 2 shows that Russia had already recovered the lost traffic in 2011, and had more rail freight pile up in 2013. During the same period, from 2008 to 2013, China showed a significant growth of 20% above the pre-crisis level in 2008, but EU and Japan remained below the pre-crisis peak, as shown in Fig. 2.

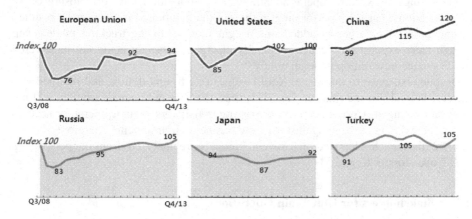

Fig. 2. Quarterly Percentage Change of Rail Freight [13]

In South Korea, which is smaller than Florida or Wisconsin in the United States, rail freight traffic follows a pattern similar to that of Japan, but at about half the level [11]. According to data from Korail [6], as of 2013, principal rail freight items include cement (37.3%), containers (29.8%), coal (11.5%), etc., as shown in Fig. 3. Cement, which has the highest percentage, is mostly carried for domestic consumption, and containers (containerized commodities), which have the second highest percentage, are transported for exporting/importing, in the area between Busan and Gyeonggi provinces.

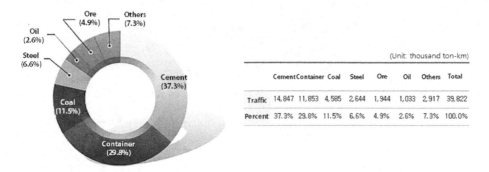

(Unit: thousand ton-km)

	Cement	Container	Coal	Steel	Ore	Oil	Others	Total
Traffic	14,847	11,853	4,585	2,644	1,944	1,033	2,917	39,822
Percent	37.3%	29.8%	11.5%	6.6%	4.9%	2.6%	7.3%	100.0%

Fig. 3. Rail Freight Traffic in Korea [14]

Up until 2013 in Korea, container rail transport was only undertaken by UTs; however, after Korail changed the previous rail operation policy to come into profitability in 2014, 100% of all commodities, including containers, was targeted to the key accounts of UTs. Thus, now, only precontracted orders (cars, schedule time, etc.) are fulfilled without any changes, unlike the previous policy that had additional car allocation available the day or week before departure [10].

The change in Korail's operating policy, in combination with the closure of 47 freight stations that were performing in the red (as mentioned in Section 1), resulted in uncontracted or urgently added rail freight moving to the trucking mode, even though more expensive, because seasonally changeable freight could not be covered by UTs in the new rail operational policy.

Despite resistance to this issue, Korail reduced its freight deficit, and returned to a surplus, in total, for the first time in 2014; thus, the new rail operating policy will not be easily changed. Consequently, the existing consignors or unit operators need to adjust the existing strategy against the new business environment. Therefore, in Section 2.2, the stakeholders and their revenue models are analyzed, from the perspective of UT operators in Korea.

2.2   Stakeholders for Unit Train Operations

For an advantageous strategy to improve the profitability of a UT operator, it is required to understand the stakeholders and their revenue models; however, they can be diversified situationally. Therefore, a UT operator (represented in Korea) in the area for exporting/importing between Busan (port) and Uiwang (ICD), is explained in terms of its physical operation, as shown in Fig. 4.

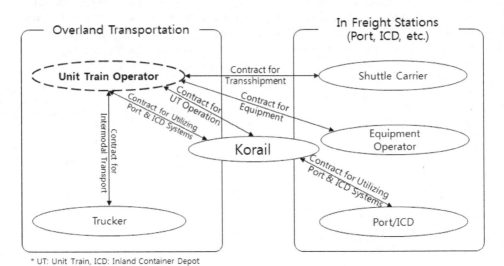

* UT: Unit Train, ICD: Inland Container Depot

Fig. 4. Stakeholders for Unit Train Operations

First, an operator has to enter into a contract for UT operations with Korail. In the contract, service section, train length (number of cars), number of operations, etc. are all incorporated into the costs. Second, if a UT operator is not an operator of freight stations (port, ICD, etc.), a contract to utilize port operating systems (information systems, facilities, etc.) is required with a port operating company. However, as Fig. 4 shows, it is also possible that, first, Korail makes a contract for utilizing port or ICD systems with a port or ICD operator, and re-rents to a UT operator with a margin. Then, a UT operator only has to calculate the charges collected, in port or in the ICD, with Korail. Third, a UT operator is necessary to load and unload the freight in each freight station; thus, an additional contract with an equipment operator is required. Fourth, in a port or an ICD, freight transported by UT needs to be stored in a yard to wait for its connecting mode of transportation (trucking or shipping), before and after the railroads; therefore, a fourth contract with a shuttle carrier is required. Fifth, for completeness of rail transportation, if a UT operator does not have a truck, a fifth contract with a trucking company is needed.

3   Strategic Analysis

3.1   Revenue and Expense Analysis

We analyzed the revenue and expense accounts of a UT operator in Korea. Table 1 shows the operating accounts of a railway yard office in Busan. Operating revenues (ⓐ) consist of landing charges, shuttle charges (internal and external), and other charges. Moreover, operating expenses (ⓑ) are composed of administrative personnel, insurance, etc., but only have to be settled up with Korail. Further, the operating results of UTs are sensitive to seasonality, operational strategy, trucking cost, etc.

Table 1. Revenue and Expense Accounts

Operating Revenues (ⓐ)	Operating Expenses (ⓑ)
Landing	Administrative Personnel
Internal Shuttle	Port Operating Labor
External Shuttle	Inspection
Others	Equipment Operating Labor
	Insurance
	Internal Shuttle
	External Shuttle
	Railway Yard Maintenance
	Depreciation
	Other Maintenance

In Table 2, which represented the 2013 history of the UT operating results in a company, we found that freight volumes, in January, April, and May of 2013, are similar, but not in the revenue and expense accounts. In the area between Busan and Uinwang, one twenty-foot equivalent unit (TEU) costs 345–390 US dollars, when one US dollar is equivalent to 1,000 Korean won. The posted cost for trucking is about 640 US dollars, but the market price is about 350 US dollars, which is nearly half the cost [15].

Table 2. Operating Results (Unit: Thousand US $)

	Freight (TEU)	Revenues	Expenses	Net Profit
2013-01	41,322	590	511	80
2013-02	38,134	540	491	49
2013-03	46,083	665	559	106
2013-04	41,547	591	525	66
2013-05	41,976	574	497	77
2013-06	37,001	513	477	36
2013-07	37,940	515	463	52
2013-08	38,393	535	501	34
2013-09	35,790	488	461	27
2013-10	39,087	564	520	44
2013-11	30,748	441	432	9
2013-12	25,207	383	375	8

3.2   Constraints and Considerations

In Korail's new rail operating policy, if a UT is contracted once, the operating capacity or supply (train length, schedule, etc.) is fixed for a half or full year. In this condition, an agile change of the supply capacity is impossible; thus, the minimization of the sum of gaps ($|c-f_t|$) between capacity (c) and freight (f), to load on a train at each time (t) (refer to equation (1)), is required.

At the same time, if there were an empty space on the UT at time (t), it would be better to sell the space at a giveaway price, even lower than the first leased price, to maximize revenue (r_t). In terms of operating a UT, the total expenses at each time (e_t) should to be minimized (refer to equation (2)).

$$\min_{x \in R} \left[\sum_{t=1}^{k} |c - f_t| \right] . \tag{1}$$

$$\max_{x \in R} \left[\sum_{t=1}^{k} (r_t - e_t) \right] . \tag{2}$$

In the past, apart from both of the conventional considerations above, an extra income or benefit, like a green point for reducing carbon emissions, could be expected; however, this is not the case now. In Korea, Certified Emission Reduction (CER) credits are not applicable in the realm of logistics now. Above all, the recent collapse of oil prices has become a driving force for trucking companies, to compete with UT operators over long distances.

3.3   Simulation Modeling and Analysis

For activating rail freight or modal shift, diverse research was performed. Recently, Jung and Kim (2012) introduced an educational simulator to facilitate the understanding of the operational strategies of UTs, based on more than 1,000 scenarios, which were composed of car type, train length, number of operations, freight volume, etc. [16]. The educational simulator was developed in Powersim, which was one of the simulation tools of system dynamics, as shown in Fig. 5. Our research refers to the simulation model.

In addition to the existing scenarios, a fuel scenario can be considered as a decision alternative, like in the case of Farrell et al. [17], but it has low practicality at the moment.

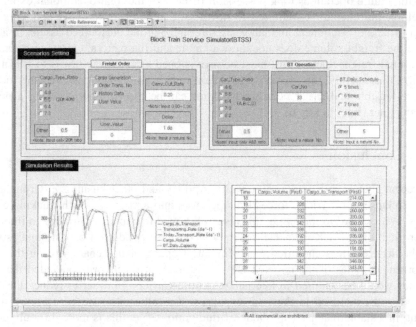

Fig. 5. A Unit Train Operational Simulator [16]

However, these simulation scenarios are available to understand a UT operating mechanism, but not for revenue and expense calculations. Therefore, the variables and parameters in Tables 1 and 2 are reflected in our simulation model, with equations (1) and (2).

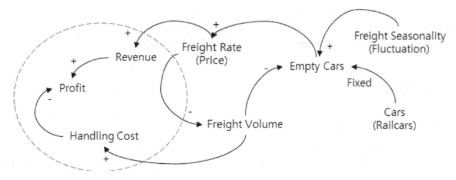

Fig. 6. Causation of Unit Train Operations

As a result, we found that the profitability of the UT operator would be difficult to significantly improve.

The case operator ran his or her UTs 384 times, in almost the only the section to have a gain, with about 453,000 TEUs in 2013, and recorded earning rates of 9%. As part of the expenses, which accounted for 91% of the total operational revenue, the costs for shuttle operations took possession of 64% of the total operational expenses. Therefore, the operator is required to develop a strategy, in terms of reducing the costs of trucking equipment and labor, by internalizing the outsourced assets. In addition, if the freight volume does not show a stable and flat pattern, empty cars, caused by seasonal freight fluctuations, must be provisioned at giveaway prices to fill in the deficit. Consequently, the two strategies seem to be meaningful; however, the former is more feasible and easier because the latter can be true without securing a major shipping liner (refer to Fig. 6).

4   Conclusion

The recent change in Korail's rail operating policy and other business environments, like a collapse in oil prices, forces UT operators into a price war with trucking companies. With regard to this issue, this research aimed to explore a strategy to improve the profitability of a Korean UT operator. Thus, we conducted a basic analysis of stakeholders and loss causation, and simulated operational scenarios using the cost and profit variables of the targeted UT operator. As a result, we concluded that the operator would be required to internalize shuttle carriers, in order to reduce the outsourced costs, included in the total expenses. The other strategy is to secure more freight, in a stable manner with the least fluctuations. Yet this is more difficult, compared to the former strategy, because shipping liners already have their own key accounts that are competitors of the analyzed company.

Our acceptable strategies, the internalization of shuttle carriers and the securing of more freight, were scientifically reinterpreted with the system dynamics methodology but their expectations were not evaluated in this research. Thus, their feasibility is required to evaluate in a large-scale experiment.

Acknowledgments. This work was supported by the National Research Foundation of Korea Grant funded by the Korean Government (NRF-2012S1A5B5A07036074).

References

1. Encyclopaedia Britannica. http://global.britannica.com/EBchecked/topic/615316/unit-train
2. Rail Freight Portal. http://www.railfreightportal.com/Block-Train-Full-Train
3. Sanchez Rodrigues, V., Pettit, S., Harris, I., Beresford, A., Piecyk, M., Yang, Z., Ng, A.: UK Supply Chain Carbon Mitigation Strategies using Alternative Ports and Multimodal Freight Transport Operations. Transportation Research Part E-Logistics and Transportation Review, In Press (2015)
4. Kim, K.: Block Train Reorganizing the Global Logistics Market. Monthly Maritime Korea, **416** (2008)

5. Korea Intellectual Property Rights Information Service. http://engdtj.kipris.or.kr/engdtj/searchLogina.do?method=loginTM&checkPot=Y
6. Korail: 2013 Statistical Yearbook of Railroad, vol. 51. Daejeon (2014)
7. Korea Shipping Gazette. http://www.ksg.co.kr/news/news_view.jsp?bbsID=news&pNum=94605&bbsCategory=KSG&categoryCode=FJY
8. ChosunBiz. http://biz.chosun.com/site/data/html_dir/2014/07/16/2014071602727.html
9. Business Post. http://www.businesspost.co.kr/news/articleView.html?idxno=9619
10. The Financial News. http://www.fnnews.com/news/201501111658302668
11. International Transport Forum: Key Transport Statistics-2013 Data, OECD (2014)
12. International Transport Forum: Global Transport Trends in Perspective. International Transport Forum Statistics Brief-Trends in the Transport Sector, OECD (2013)
13. International Transport Forum: Global Freight Data Show Diverging Trends for Developed and Developing Economies. International Transport Forum Statistics Brief-Global Trade and Transport, OECD (2014)
14. Korail. http://info.korail.com/mbs/www/subview.jsp?id=www_020302000000
15. Busan Development Institute and Busan Port Authority: 2014 Logistics Trends and Predictions, Busan (2013)
16. Jung, J.U., Kim, H.S.: An Educational Simulator for Operational Strategies of the Block Train Services. Journal of Digital Convergence 10(11), 197–202 (2012)
17. Farrell, A.E., Keith, D.W., Corbett, J.J.: A Strategy for Introducing Hydrogen into Transportation. Energy Policy 31, 1357–1367 (2003)
18. Tako, A.A., Robinson, S.: The Application of Discrete Event Simulation and System Dynamics in the Logistics and Supply Chain Context. Decision Support Systems 52(4), 802–815 (2012)

Other Applications

High Performance NetFPGA and Multi-fractal Wavelet Model Based Network Traffic Generator

Zhenxiang Chen$^{(\boxtimes)}$, Keke Wang, Lizhi Peng, Tao Sun, and Lei Zhang

University of Jinan, Nanxin Zhuang West Road No. 336, Jinan 250022, P.R. China
{czx,nic_wangkk,plz,ise_sunt,zhanglei}@ujn.edu.cn

Abstract. Internet traffic generator plays very important roles in network measurement and management field.A distributed digital filter is designed in this paper,which united with Multi-fractal Wavelet model(MWM) to generate network traffic data, then a network traffic generator was researched and implemented on the NetFPGA platform. Experiment result shows that the designed network traffic generator has good ability to generate traffic in accordance with the real network condition. More importantly, the generated traffic shows the characters of self similar and multi-fractal, which are two of the most important features of real Internet traffic. With high speed network packets process ability of the NetFPGA Platform, the designed network traffic generator can generate traffic on a highest speed of 10Giabit/s. The designed network traffic generator displays the distinct advantages in performance and real condition simulation ability contract to the invented works.

Keywords: Network traffic generation · Multi-fractal wavelet model · NetFPGA · Simulation

1 Introduction

New Internet application spring up constantly,which cause vast amount network traffic. Recently, traffic congestion has became one of the serious problems that received extensive attention. Network traffic measurement plays important roles in network management.Research shows that the characters of self similar and multi-fractal are two of the most important features in real network condition.Simulating and generating Internet traffic based on some reasonable model can effetely support the measurement and evaluation of network performance [1–4], which is helpful for improving and optimizing network condition.

The traffic generated based on some pre-created models, which accord with real network characters is useful for research on network equipment performance evaluation,traffic forecasting and traffic classification [5]. Self similarity in large time scale and multi-fractal nature in small time scale are two of the recognized important features in network measurement field. When designing and implementing a Internet traffic generator, create a reasonable traffic generator model which can satisfied the given characters is important. The main contribution of our work is that we improved the network data flow generate method

© Springer International Publishing Switzerland 2015
Y. Tan et al. (Eds.): ICSI-CCI 2015, Part III, LNCS 9142, pp. 287–296, 2015.
DOI: 10.1007/978-3-319-20469-7_31

which based on WMW, and then researched an FPGA based high performance generator structure. At last,we realized a NetFPGA based high speed Internet traffic generator, which shows distinct advantages in performance and real traffic simulation ability.

The rest of the paper is organized as follows. Related works are reviewed in the next section. Section 3 presents the Multi-fractal Wavelet Model theory and wavelet-based functions selecting method. In section 4, the NetFPGA based traffic generator simulation and implementation research is presented. The experiment and analysis work are given in section 5. Finally, Section 6 concludes the paper.

2   Related Works

Different kinds of Internet traffic generators were generated accompany with the development of computer network software and hardware technology. With the program speciality of FPGA, it was selected as one of the most common platform for traffic generator,since it's excellent characters of high performance,parallel computing and flexible programming ability [6]. Since 1996, Bernoulli model was applied to design and implement network traffic generator on FPGA platform [7]. However,the performance and traffic generate rate of the generator can't satisfied the real requirement owns to the restriction of gate array number on FPGA at that time. At the same time, the selected simple Internet traffic models can't fit the real network traffic features effectively. With the development of Internet technology and new applications, the similar and multi-fractal characters becomes two of the most obviously ones, which can portray the real Internet features roundly. Recent years, wavelet model was focused and using for Internet traffic simulation and generating [8–10]. But,most of the wavelet based Internet traffic generator are implemented on software structure, which only support much low traffic generate rate and more flexible configuration interface compared with hardware structure. The programable FPGA technology make it become reality to realize the wavelet model on a hardware structure. In this paper,we realized a NetFPGA and wavelet model based high speed Internet traffic generator, which shows distinct advantages in performance and real condition simulation ability at the same time.

3   Multi-fractal Wavelet Model(MWM)

Since 1990's, wavelet model was applied for the research of Internet model analysis [11]. In this section, the wavelet model was optimized and analysed firstly,then Internet traffic generated based on different wavelet functions were used to evaluate the feasibility of the selected wavelet model and basis function.

3.1   Multi-fractal Wavelet Model

For a continuously time output signal $X(t) \in L^2(R)$, it's continuous wavelet transform can be defined as:

$W_{ab} = <\ X, \psi_{a,b}\ > = \int X(t)\psi^*_{a,b}(t)dt = \frac{1}{\sqrt{a}}\int X(t)\psi^*(\frac{t-b}{a})dt, \psi_{a,b}(t) = \frac{1}{\sqrt{a}}\psi(\frac{t-b}{a}), a > 0, b \in R$. The W_{ab} is called as wavelet factor and $\psi(t)$ as wavelet generating function. Generally, wavelet functions include Harr wavelet, Morlet wavelet and Maar wavelet, and most of the wavelet model are achieved based on Haar wavelet. Commonly discrete wavelet transform for multi-dimension signal analysing can be presented as follows:

$$x(t) = \sum_k u_k 2^{-J_0/2}\phi(2^{-J_0}t - k) + \sum_{j=-\infty}^{J_0}\sum_k w_{j,k}2^{-j/2}\psi(2^{-j}t - k)j, k \in Z \quad (1)$$

In formula(1), J_0 is the number of iteration,u_k and $w_{j,k}$ respectively means scale factor and wavelet factor of wavelet transformation. The scale factor can be seemed as iteration approximate value and the wavelet can be seemed as high frequency detail information.

In the Haar wavelet transform procedure, the scale and wavelet factors can be computed as follows:

$$U_{j-1,k} = 2^{-\frac{1}{2}}(U_{j,2k} + U_{j,2k+1}) \quad (2)$$

$$W_{j-1,k} = 2^{-\frac{1}{2}}(U_{j,2k} - U_{j,2k+1}) \quad (3)$$

After analysing $U_{j,2k}$ and $U_{j,2k+1}$ we can get follows:

$$U_{j,2k} = 2^{-\frac{1}{2}}(U_{j+1,k} + W_{j+1,k}) \quad (4)$$

$$U_{j,2k+1} = 2^{-\frac{1}{2}}(U_{j+1,k} - W_{j+1,k}) \quad (5)$$

In order to get the nonnegative signal, for any value of j and k, it must make sure that $u_{j,k} >= 0$. That is make sure

$$|w_{j,k}| \le u_{j,k}, \forall j, k \quad (6)$$

In order to describe the fractal character of traffic, united with the restriction of formula (4)and (5), a wavelet energy decay scale function was defined to describe a procedure of nonnegative traffic generator. Suppose $A_{j,k}$ is a random variable in interval[-1,1], then the wavelet factor can be defined as formula(7).

$$W_{j,k} = A_{j,k}U_{j,k} \quad (7)$$

United with formula (4), formula (5) and formula (7), it can get the result of (8) and (9).

$$U_{j,k} = 2^{-\frac{1}{2}}(1 + A_{j+1,k}U_{j+1,k}) \quad (8)$$

$$U_{j,2k+1} = 2^{-\frac{1}{2}}(1 - A_{j+1,k}U_{j+1,k}) \quad (9)$$

At last, the wavelet transform procedure can be shown as a transform tree which present in Fig.1.

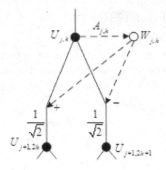

Fig. 1. The wavelet transform tree structure

3.2   The Selection of Wavelet Basis Function

In order to get the traffic generation character of different wavelet basis function, four kinds of representative wavelet basis Haar, Daubechies, Symlets and Coiflets were selected. The selected basis were applied to simulate and generate related traffic flows. Simulation result shows that all the generated traffics are similar to the real traffic on the whole. But it still some detail difference between the different wavelet basis based generation traffic, which can be evaluated by the Hurst value computed through R/S estimation method. The Hurst value of generated traffic data based on different wavelet basis shows in Table 1.

Table 1. The hurst value of generated traffic data based on different wavelet basis

name	Hurst parameter		
Real traffic 0.739	0.754	0.801
Haar traffic 0.732	0.751	0.806
db3 traffic 0.713	0.774	0.829
Sym3 traffic 0.721	0.31	0.738
Coifl traffic 0.757	0.817	0.835

Evaluation result from Table 1 shows that traffic generated based on the Haar wavelet basis function can get the most similar network data statistical feature to real traffic, which prove the feasibility of selecting the Haar as the wavelet basis function of our traffic generate model.

4   NetFPGA and Wavelet based Network Traffic Generator

Our Internet traffic generator was designed and implemented based on NetFPGA platform ,invented by Stanford University and specially used for high speed network information processing. In this platform, the Virtex2 pro is mainly used for logical development and the Xilinx Spartan FPGA chips are mainly used for communication between NetFPGA and host machine. The detail designing and implementing are as follows.

4.1   The Structure of Traffic Generator

As show in Fig.2, the designed traffic generator include two parts, one is the traffic packets generate module, and the other is the traffic data control module MWM control module which embed in NetFPGA platform. The packet generator model mainly for generator network packets which fit the standard network data format. The MWM control model are focus on controlling that the sending data can satisfied the rate,speed and interval character with real network condition.

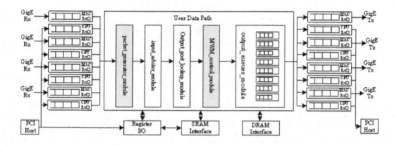

Fig. 2. The NetFPGA based traffic generator structure

4.2   The Implementation of Traffic Generator

Random Packets Generation Module The traffic generator module consisted from a state machine, which state variate from HEAD0 to HEAD6 controls the generated head of traffic frame must obey the TCP/IP protocols. It must satisfied the requirement of source IP address, target IP address, sources port, target port , protocol and packet size. As show in Fig.3, it can be generated by the Hdr_Add and Pkt_Gen module. The state of DATA control the generation of random traffic content by the Data_Gen module in Fig.3. The simulated date from Data_Gen and heads from Pkt_Gen can be organized into whole packets in the FiFO and then being sent out continuously.

The Implementation of MWM The general wavelet mainly composed the three parts:filter, random parameter and out put judgement.

(1)FPGA based distributed filter algorithm. As presented in Fig.4,when Load_x keep on the low level, the x_in will download the filter factor to the registers of reg0 and reg15. The Load_x will change to high level when all the filter factor have been downloaded, then the filter required signal input x_in will

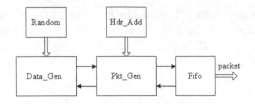

Fig. 3. The structure of traffic generator model

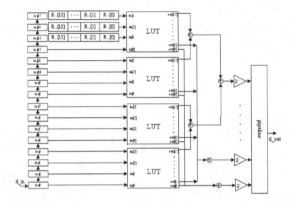

Fig. 4. FPGA based distributed algorithm

transform to output signal y_out after several clock cycle. For high level filter realization, as the same principle, only related filter factor need be changed.

(2)Random frictions. In order to make sure that traffics generated based on random A are nonnegative, a random function degree of freedom was used to control the long range dependence of the select model. As the requirement of the simply and flexibility for selecting random function, Beta distribution function $\beta(p,p)$ was selected to generate random variable [12] in interval[-1,1]:

$$f(a) = \frac{(1+a)^{p-1}(1-a)^{p-1}}{\beta(p,p)2^{2p-1}} \tag{10}$$

$f(a)$ will show different style when different p is selected, and it's variance can be presented as:

$$var(a) = E(A^2) = \frac{1}{2p+1} \tag{11}$$

(3)Judgement of Output. The Judgement of Output module was designed to detect if the digital smoothed traffic flow can fit the target requirement. The generated traffic will be sent if fit the requirement, otherwise will be kept on the continuous superposition process until reach the target. As a Output state machine rule controlling, the result traffic data was saved in FIFO, which prepared for reading by the next module.

5   Experiment and Analysis

MWM model was selected for Internet traffic simulation and generation, then a NetFPGA based traffic generator was implemented to generate traffic data that meets Internet traffic features in this paper. The traffic generate ability, self-similarity, multi-fractal nature and performance of the designed generator shows it's validity and accuracy.

5.1   Traffic Generation

The real traffics from the Shandong Provincial Key Lab of Network based intelligence computing were used to evaluate the generated traffic. As show in Fig.5 (X:time and Y:packe number at X time), the captured real and simulated traffic in time scale of 0.1s, 1s and 8s are used for evaluation. As in 1000s times, the simulated traffic shows the similar variation trends and general features as the real traffic.

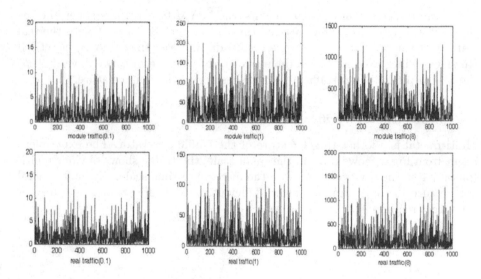

Fig. 5. The traffic generated in different scale

5.2   The Hurst of Generated Traffic

In the same network environment described in last section, three data sets were captured in 30 seconds. Experiment shows the average Hurst value on the three real traffic data sets is 0.742. In the same experiment scheme, three simulated traffic were generated in 30 seconds,the Hurst value of captured traffic from generator shows is 0.748. The generated traffic shows truly similar characters with real traffic in acceptable error rates of 0.01.

5.3   Traffic Generate in a Given Hurst Value

Generating traffic as some settled Hurst value is a effective method for evaluation the accuracy of the target system. The experiment settled the Hurst value from 0.55 to 0.95 increased by 0.05,the Hurst and error of generated traffic list in Table 2.

Table 2. The hurst value of generated traffic data

Set Hurst	MWM Hurst	error	Set Hurst	MWM Hurst	error
0.55	0.533	0.017	0.8	0.786	0.016
0.6	0.615	0.015	0.85	0.866	0.016
0.65	0.647	0.018	0.9	0.889	0.011
0.7	0.689	0.011	0.95	0.938	0.012
0.75	0.762	0.012	-	-	-

As can be seen from the experiments, the MWM based traffic generator has the ability to generate traffic satisfied the self-similarity. Contrast with the real traffic and generated traffic, the generated traffic increase the Hurst as the settled Hurst value increased in a acceptable error rate. All experiments can verify that the MWM model based traffic generator has good self-similar nature.

5.4   Multi-fractal Verification

Multi-fractal is another target feature of the traffic generator. The experiment result from Fig.6 shows that, in the small scale, the traffic shows obvious multi-fractal nature in any frictions. With the increase of time scale, the multi-fractal nature turn to Single fractal nature. Obviously, the traffic generator can meet the target requirement in some extent.

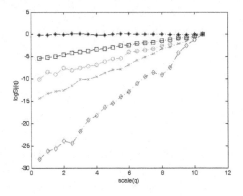

Fig. 6. The Multi-fractal feature verification

5.5 Performance Evaluation

The designed FPGA based traffic generator run on the frequency of 125 MHz, with the 72 bit data bit width and 64 efficient bit, the designed throughput is 8Gbps. As analysis from Xilinx ISE9.2 static time sequence report, the max time frequency can be 156.963MHz, and the max throughput can be 10.046Gbps, which can totaly meet the designed performance target. Contrast with the invented traffic generator in recent years, the designed traffic generator can generate traffic on a highest speed of 10Giabit/s. The designed network traffic generator displays the distinct advantages in performance and real condition simulation ability.

6 Conclusion

In this paper, some classic Internet traffic models are analyzed firstly. Then a distributed digital filter is designed in this paper,which united with Multi-fractal Wavelet model(MWM) to generate network traffic data, then a network traffic generator was researched and implemented on the NetFPGA platform. Evaluation by Hurst value of generated traffic, the experiments result show that the designed network traffic generator has the good ability to generate traffics accord with real network condition. More importantly, the generated traffic shows the characters of self similar and multi-fractal, which are two of the most important features of Internet traffic. With the high performance of high speed network packets process ability of the NetFPGA Platform, the designed traffic generator can generate traffic on a highest speed of 10Giabit/s. The designed network traffic generator displays the distinct advantages in performance and real condition simulation ability contract to related works. The designed network traffic generator displays some distinct advantages in performance and real condition simulation ability contract to related works. Although a effectively prototype is achieved, the complexity reduce and extendibility improve will be our timely future work.

Acknowledgments. This work was supported by the National Natural Science Foundation of China under Grants No.61472164 and No.61203105,the Natural Science Foundation of Shandong Province under Grants No.ZR2014JL042 and No.ZR2012FM010.

References

1. Riesi, R.H., Crouse, M.S., Rbeiro, V.J.: A multifractal wavelet model with application to network traffic. IEEE Trans. on Information Theory **45**(3), 992–1018 (1999)
2. Gilbert, A.C.: Multiscale analysis and data networks. Applied and Computational Harmonic Analysis **10**(3), 185–202 (2001)
3. Ma, S., Ji, C.: Modeling heterogeneous network traffic in wavelet domain. IEEE/ACM Trans. on Networking **9**(5), 634–649 (2001)

4. Zhang, B., Yang, J.H., Wu, J.P.: Survey and Analysis on the Internet Traffic Model. Journal of Software **22**(1), 115–131 (2011)

5. Takuo, N., Toshinori, S.: self-similar property for TCP traffic under the bottle-neck restrainment. In: Advanced Information Networking and Application Workshops, vol. 1, pp. 228-233 (2007)

6. Labrecque, M., Steffan, J.G.: NetTM: faster and easier synchronization for soft multicores via transactional memory. In: FPGA, pp. 29–32 (2011)

7. Watson, G., MxKeown, N., Casado, M.: Netfpga-a tool for network research and education. In: The 2nd Workshop on Architecture Research using FPGA Platforms (WARFP) (2006)

8. Huang, L., Wang, S.: Improved Multi-fractal Network Traffic Module and Its Performance Analysis. IEEE Transaction on Information Theory **18**(5), 102–107 (2011)

9. Richard, J., Harris, J.: Wavelet Spectrum for Investigating Statistical Characteristics of UDP-based Internet Traffic. International Journal of Computer Networks and communications **4**(5), 73–88 (2012)

10. Burros, C.S., Ramesh, A.: Introduction to Wavelets and Wavelet Transforms: A Primer (2006)

11. Flandrin, P.: Wavelet analysis and synthesis of fractional Brownian motion. IEEE Trans. on Information Theory **38**(2), 910–917 (1992)

12. Chen, S.Y., Wang, W., Gao, F.Q.: Combining wavelet transform and markov medole to forecast ttraffic. In: Proceedings of the Third Intemational Conference on Machine Learning and Cybernetics, pp. 26–29 (2004)

An Approach to Integrating Emotion in Dialogue Management

Xiaobu Yuan[✉]

University of Windsor, Windsor, Ontario, Canada
xyuan@uwindsor.ca

Abstract. Presented in this paper is a method for the construction of emotion-enabled embodied (conversational) agents. By using a modified POMDP model, this method allows dialogue management not only to include emotion as part of the observation of user's actions, but also to take system's response time into consideration when updating belief states. Consequently, a novel algorithm is created to direct conversation in different contextual control modes, whose dynamic changes further provide hints for emotion animation with facial expressions and voice tunes. Experiment results demonstrate that the integration of emotion in dialogue management makes embodied agents more appealing and yields much better performance in human/computer interaction.

Keywords: Embodied agents · Dialogue management · POMDP model · Emotion recognition · Emotion animation

1 Introduction

Due to their huge potential for providing online assistance over the Internet without time or location limits, embodied agents play a vital role to the wider acceptance of Web applications in the daily life of human beings [12]. Different projects have been conducted to explore the use of embodied agents in a variety of areas, including business, education, government, healthcare, entertainment, and tourism. Meanwhile, active research is being conducted to improve embodied agents with better appearance and higher level of intelligence. In particular, emotion has been identified as one of the most desirable capabilities that could lead to the creation of highly appealing experience in the interaction between human and embodied agents [3].

The typical structure of embodied agents employs four basic components to handle the input, fusion, fission, and output of information. It also includes a knowledge base to maintain the knowledge of an application domain, and relies on a dialogue manager to control system's overall operation [1]. Dedicated work has been devoted in the past a couple of decades to, especially, develop techniques for the critical task of dialogue management. Among them, techniques based upon the Partially Observable Markov Decision Process (POMDP) model overcame the shortcomings of other techniques, and had become the most advanced

© Springer International Publishing Switzerland 2015
Y. Tan et al. (Eds.): ICSI-CCI 2015, Part III, LNCS 9142, pp. 297–308, 2015.
DOI: 10.1007/978-3-319-20469-7_32

so far. Further efforts have been made to incorporated emotion in the POMDP model [2], but deficiency in performance is still noticeable [10].

This paper presents an integrated approach to emotion recognition and animation in embodied agents. By modifying the POMDP model with an added observation of user's emotion and a new component of system's response time, it enables dialogue management to integrate emotion in the two-way channel of communication. The rest of the paper is organized as follows. Prior work related to the subject of investigation is first highlighted in Section 2. Section 3 then presents the main focus of this paper, which includes emotion recognition with a customized decision tree, dialogue management with the modified POMDP model, and a new mechanism to use contextual control model. Experimental results and performance analysis of the proposed approach are presented in Section 4. Finally, Section 5 concludes the paper with discussions.

2   Related Work

2.1   Emotion Modeling

Emotions are a mix of physiological activation, expressive behaviors, and conscious experience. Although the study of emotions is a complex subject that involves with psychology and physiology, a few models have been developed. Among them, one of the oldest is Plutchik's wheel of emotion [14], which is well-known for its postulates about primary emotions and their combinations to generate new emotions. However, its lack of support to integrate with generic computational models has limited its applications. In comparison, the OCC (Ortony, Clore, and Collin) model classifies emotions with variables and their reactions to three factors [13]. For its ability to assist reasoning with rules about emotion activation, it has become the first model applicable to machine processing and an inspiration to other models [11].

Emotions may be recognized and/or animated in multiple modalities via text (e.g., emotional keywords [9]), audio (e.g., voice tunes [6]), and video (e.g, facial expressions [5]). In a multimodal approach to emotion recognition, Fourier spectra was used to extract multi-scale emotional descriptions from signals that capture the characteristics of face appearance, head movements, and voice [8]. For the rich information enclosed in visual appearance, facial animation has itself become a major research area in computer graphics. A recent survey categorizes the work on modeling into two groups as generic model individualization and example-based face modeling, and the work on animation into three as simulation and animation driven by performance or based upon blend shapes [4].

2.2   Dialogue Management

Along with the development of techniques for more realistic appearance, research working on dialogue management has gone through different approaches from the initials based upon finite state machines, to those with frames, Bayes networks,

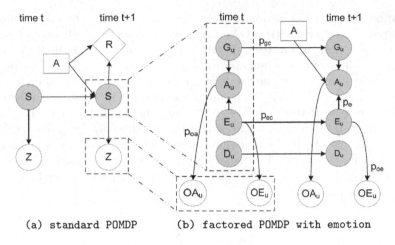

(a) standard POMDP   (b) factored POMDP with emotion

Fig. 1. Factored POMDP with Observation of Emotion [2]

or Markov decision processes, until the most advanced with POMDP [20]. As an extension to MDP, a standard POMDP (Fig. 1(a)) adds observation O and observation function Z to the original set of triples, producing a new definition as $\{S, A, T, R, O, Z\}$ [19]. Instead of relying on estimations of machine states $s \in S$ at time t, POMDP-based dialogue management selects an action $a \in A$ based upon the current belief state b, receives a reward r, and changes to a new unobserved state s'. Consequently, the system receives a new observation $o' \in O$ depending upon s' and a. Finally, $b(s)$ is updated using o' and a at the new time step t'. At the meantime, system's action $a_{m,t}$ at t results in a reward $r(b_t, a_{m,t})$.

While researchers working on dialogue management place their focus on accurate intention discovery under the influence of uncertainty, some others consider the process of decision making to be more dynamic and context sensitive [17]. Based upon cognitive modes, the Contextual Control Model (COCOM) was developed in [7]. This model argues that the operation of a system needs to decide what actions to take according to the context of situation. In reaction to both the environment and individual perspective of the user, the system may operate in one of the four modes, i.e., strategic, tactical, opportunistic, and scrambled modes, with each control mode associated with its characteristics and type of performance.

3   Integration of Emotion in Dialogue Management

3.1   An Integration of Emotion Recognition

The standard POMDP model has been extended by Bui et al to include user's emotion as part of the observation [2]. As shown in Fig. 1(b), the observation set is divided into observed action (OA_u) and observed emotional state (OE_u) of the user, and therefore $o' \in O$ in the POMDP model becomes a combined observation of both user action and user emotion. While OA_u maintains the original meaning of user action, OE_u may include a set of effects extracted from different modalities, such as text, speech, images, or a combination of them. The collection of all emotional observations produces a dataset, which can be used to train the system and produce a decision tree with emotional values.

Suppose $\{x_1, x_2, \cdots, x_n\}$ is an n-dimensional set of emotional values. For example, n is nine if there are seven features, including the strengths for six different emotion types (angry, sad, surprise, fear, disgust, joy) and an emotion intensity. The last two in the nine-dimensional set are for the positive and negative sentimental strength. The mean is x_m, which is the average of all values, i.e., $\sum x_i/n$; and the root mean square (rms) is x_{rms}, which can be obtained with the following equation.

$$x_{rms} = \sqrt{\frac{1}{n}\sum_{i=1}^{n} x_i^2} \tag{1}$$

By analyzing the entire dataset, the system produces two two-dimensional arrays, one for the calculated mean and the other for RMS. These two arrays can then be used as a metric in a decision tree to predicate the emotional values of new observations and hence to determine user's emotion state E_u. More details about emotion recognition with the customized decision tree are available in [15].

3.2   An Integration of Emotion Animation

For emotion animation, the time needed for the system to generate a response can be introduced as a new component to the POMDP model. Initially, the system at time t is normally in an unobserved state $s \in S$. When conversation is established between the user and the system, the state makes a transition from s to s' in a time increment. A decision of action selection depends on the time available for the particular context of the dialogue. Let T_A represent the available time for choosing the best action. It is a function $T_A(s_m, a_m, o', b)$ that relies on the machine state s_m, machine action a_m, the combined observation o', and belief state $b(s)$ of the machine at time t. As a result, the POMDP model can be modified as below.

$$\begin{aligned} b'(s') &= P(s'|o', a, b, t_r) \\ &= \frac{P(o'|s', a, b, t_r)P(s'|a, b)P(s'|a, t_r)}{P(o'|a, b, t_r)} \\ &= \frac{P(o'|s', a)\sum_{s \in S} P(s'|a, b, t_r, s)P(s|a, b)P(s|a, t_r)}{P(o'|a, b, t_r)} \\ &= \frac{P(o'|s', a)\sum_{s \in S} P(s'|a, s)P(t_r|a)b(s)}{P(o'|a, b, t_r)} \end{aligned} \tag{2}$$

The new component $P(t_r|a)$ in Eq. 2 makes it possible to introduce the concept of control modes in COCOM to POMDP-based dialogue management.

Reward calculation is also updated in such a way that the favorable strategic and tactical modes receive a positive value of +100, the opportunistic mode receives a negative value of -100, and the scrambled mode receives a 0 value. The following equation is used to calculate rewards, where λ is a discount factor at time t.

$$R = \sum_{t=0}^{\infty} \lambda^t r(b_t, a_{m,t}, t_r)$$

$$= \sum_{t=0}^{\infty} \lambda^t \sum_{s \in S} b_t(s) t_r(a_{m,t}) r(s, a_{m,t}) \tag{3}$$

As each action is determined by a policy π, dialogue management now uses the modified POMDP model to find an optimal policy π^* that maximizes the rewards for the given application.

$$\pi^*(s_t) = a[argmax_{a \in A}(b_t, t_r)] \tag{4}$$

In the equation, the confidence buckets represent how exactly the system understands the user's utterances. In the proposed approach, the original POMDP model operates as a special case when the dynamic control is always in the tactical mode.

3.3   An Algorithm to Integrate Emotion

Shown in Fig. 2 is an algorithm to integrate both emotion recognition and animation in dialogue management. The first five lines of codes initialize t, b, s_m, and m after the first machine action a_m. The algorithm then repeats operations to extract user emotion, update belief states, select modes with response time, decide actions in the context of control mode, and calculate rewards for mode adjustment until reaching a goal or terminating the dialogue due to failure. In each iteration, the algorithm sets the machine emotion in one of the mode-dependent types e_m when there is no mode change or one of the switch-dependent types e_k, $1 \leq k \leq 8$, when a mode change occurs from m_μ to m_ν, where $\mu, \nu \in \{\alpha, \beta, \gamma, \delta\}$ for the four modes and $\mu \neq \nu$. A synchronized rendering of facial expression and voice tune conveys the system's emotion.

By default, dialogue management operates in the strategy mode, aiming at maintaining control at the higher level. If a panic situation occurs, the system goes into the scrambled mode due to the loss of control in dialogue. In such a case, the system takes random actions depending upon the context of the situation and time availability. In a situation when the user fails to provide any useful information but throwing errors or repeating the same query, the system switches to opportunistic mode, and provides a list of options for the user to select. It helps the user to avoid the same mistake in the next round of dialogue, and reduces the time to achieve an optimal goal. By incorporating emotion in dialogue management, an embodied agent becomes more efficient to handle situations in which there are variations in user's domain knowledge and in the types of services.

Algorithm: DIALOGUE MANAGEMENT WITH INTEGRATED EMOTION ( )

$t \leftarrow time\ stamp\ at\ 0$
$s_m \leftarrow system\ state\ (unobserved)$
$b \leftarrow belief\ state\ b_0$
$m \leftarrow control\ mode\ 0$
$a_m \leftarrow system\ action$
comment: initialization completed

repeat
$\quad s_u \leftarrow user\ dialogue\ state$
$\quad a_u \leftarrow user\ dialogue\ acts\ \pi(s_u)$
$\quad$**comment:** in simulation of the user's action

$\quad extract\ user\ emotion\ e_u$
$\quad calculate\ belief\ state\ b(s)$
$\quad calculate\ confidence\ score\ c$
$\quad calculate\ system\ response\ time\ t_r$
$\quad m \leftarrow check\ current\ control\ mode$
$\quad a_m \leftarrow dialogue\ act\ in\ the\ context\ type$
$\quad calculate\ rewards\ r(s, a, t_r)$
$\quad m' \leftarrow calculate\ dialogue\ act\ type$
$\quad$**if** $m = m'$
$\quad\quad$**then** $emotion \leftarrow e_m$
$\quad\quad$**else** $\begin{cases} \textbf{switch}\ (m \Rightarrow m') \\ \textbf{case}\ (m_\delta \Rightarrow m_\gamma) : emotion \leftarrow e_1 \\ \textbf{case}\ (m_\delta \Rightarrow m_\beta) : emotion \leftarrow e_2 \\ \textbf{case}\ (m_\gamma \Rightarrow m_\beta) : emotion \leftarrow e_3 \\ \textbf{case}\ (m_\beta \Rightarrow m_\alpha) : emotion \leftarrow e_4 \\ \textbf{case}\ (m_\alpha \Rightarrow m_\beta) : emotion \leftarrow e_5 \\ \textbf{case}\ (m_\beta \Rightarrow m_\gamma) : emotion \leftarrow e_6 \\ \textbf{case}\ (m_\beta \Rightarrow m_\delta) : emotion \leftarrow e_7 \\ \textbf{case}\ (m_\gamma \Rightarrow m_\delta) : emotion \leftarrow e_8 \\ m \leftarrow m' \end{cases}$
$\quad prepare\ v_m\ with\ a_m\ and\ emotion$
$\quad prepare\ f_m\ with\ a_m\ and\ emotion$
$\quad animate\ with\ synchronized\ v_m\ and\ f_m$
$\quad t \leftarrow t + 1$
until the dialogue terminates

Fig. 2. Pseudocode of the Proposed Approach

4   Experiment Results and Discussions

Experiments follow the pizza-ordering example in [18], and are based upon the scenario that an agent provides assistance at a pizza restaurant to a human user for the purchase of a pizza.

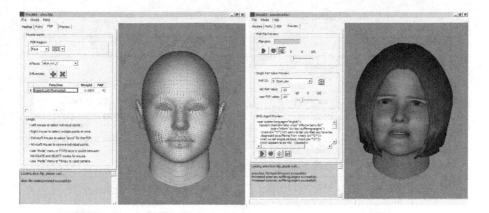

Fig. 3. Feature Setting and Fine Tuning for the 3D Model

4.1 Multimodal Modeling of Emotion

In experiments, five types of emotions are created with Xface editor in their morph targets respectively for `relief`, `hope`, `pleased`, `displeased`, and `fear`, plus the default `neutral`. Lip-audio synchronization further adds 22 morph targets for phonemes such as `aah`, `oh`, `ee`, etc. In addition, several modifiers are created for the embodied conversational agent to imitate human behaviors by, for example, blinking eyes continuously at regular time intervals. After setting up feature points (the left figure in Fig. 3), the 3D model for individual facial expressions are fine tuned with preview (the right figure in Fig. 3).

To match facial animation with audio playback, emotion animation in the experiments first uses a generator to produce scripts in the Synchronized Multimedia Integration Language, and then passes the scripts to Xface player together with the configuration file created during facial modeling. As Xface implements a client-server architecture, the dialogue system sends the scripts to Xface player by means of TCP/IP communication. Working as the server, Xface queues all the tasks of facial animation and audio playback as specified in the scripts for the player to start playing the animation one by one, producing a realistic embodied agent to communicate with software clients via conversation. In the implementation, acknowledgment of sending and receiving the tasks between the client and the server is done by sending out notification messages for each task.

4.2 Pizza-Ordering: Emotion Animation

In the pizza-ordering example, the dialogue starts with a greeting message from the system and continues in several time-stamps till an optimal goal is achieved. The experiment covers three scenarios to check the operation of the proposed method when there is or no uncertainty caused by speech/text errors or conflicts in user's requirements. Each column of Table 1 corresponds to one of the test cases, whose results are shown respectively in the tables in Fig 4–6 with the values

Table 1. Pizza-Ordering Scenarios

CASE 1	CASE 2	CASE 3
greeting U: *nonvegpizza*	greeting U: *vegpizza*	greeting U: *"request"*
ask type U: *crusttype*	ask type U: *crusttype*	wrong request; please order a pizza U: *vegpizza*
ask size U: *pizzasize*	ask size U: *pizzasize*	ask type U: *crusttype*
ask topping U: *pizzatopping*	ask topping U: *conflicttopping*	ask size U: *pizzasize*
end	topping conflicts type; ask topping U: *pizzatopping*	ask topping U: *nonvegtopping*
	end	conflicting request; please try option U: *pizzatopping*
		end

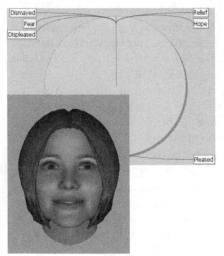

Round	Reward	Score	Conflict	B(s)	Mode
1	+100	71.42	N	0.10	Strategic
	+100	86.95	N	0.21	Tactical
	+100	53.84	N	0.35	Strategic
	+100	93.29	N	0.60	Strategic
2	+100	84.42	N	0.12	Strategic
	+100	76.75	N	0.12	Tactical
	+100	79.83	N	0.33	Strategic
	+100	91.16	N	0.58	Strategic
3	+100	63.29	N	0.14	Strategic
	+100	66.36	N	0.21	Tactical
	+100	85.51	N	0.25	Strategic
	+100	94.60	N	0.56	Strategic

Fig. 4. Results of A Normal Dialogue Process

of belief state, transition between the control modes, and indication of conflicts in simulation of user's domain knowledge, and actions taken by the system. The belief state values are calculated by considering four main attributes, including pizza type, crust type, pizza size, and pizza topping.

The first test case assumes a normal dialogue when the user consistently provides correct information. For each dialogue state, the dialogue management

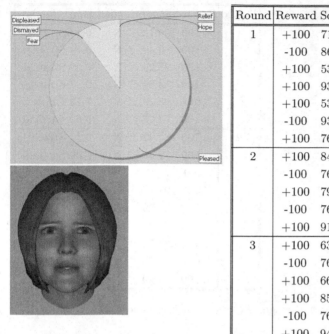

Round	Reward	Score	Conflict	B(s)	Mode
1	+100	71.42	N	0.10	Strategic
	-100	86.95	Y	0.21	Opport.
	+100	53.84	N	0.35	Strategic
	+100	93.29	N	0.60	Strategic
	+100	53.84	N	0.35	Strategic
	-100	93.29	N	0.60	Opport.
	+100	76.75	N	0.12	Tactical
2	+100	84.42	N	0.12	Strategic
	-100	76.75	Y	0.12	Opport.
	+100	79.83	N	0.33	Strategic
	-100	76.75	N	0.12	Opport.
	+100	91.16	N	0.58	Strategic
3	+100	63.29	N	0.14	Strategic
	-100	76.75	Y	0.12	Opport.
	+100	66.36	N	0.21	Tactical
	+100	85.51	N	0.25	Strategic
	-100	76.75	Y	0.12	Opport.
	+100	94.60	N	0.56	Strategic

Fig. 5. Results of A Dialogue Process with Conflict

updates its history, whose values are collected at the end of the dialogue. In the table in Fig 4, the values are arranged to show system's actions at each time steps, belief state values, and transition between control modes. In addition, the experiment tracks the rewards and transition mode for each dialogue state. The rewards are all positive as the system is always in control and therefore is either in the strategic or tactical mode. In comparison, a negative reward is triggered when the user provides a conflicting answer due to mistake or lack of domain knowledge as illustrated in the second column of Table 1. This is the second case of testing that checks reaction of the system as it changes to opportunistic mode for option selection (Fig. 5).

The third test case requires the system to handle combined speech error and conflict in simulation of both noisy environment and user's lack of domain knowledge. After receiving a negative reward due to speech error, the system in reaction changes its mode to opportunistic and provides options for selection. In the dialogue, a conflict appears when the user requests a vegetarian pizza with non-vegetarian topping. Once again, a negative reward is obtained. The system changes to the opportunistic mode, and provides a list of options for the user to choose. This test case provides the evidence that the dialogue system needs to maintain a dialogue history, and use it for making decisions. Shown in Fig. 6 is the experimental result that shows system's operation when handling the situation.

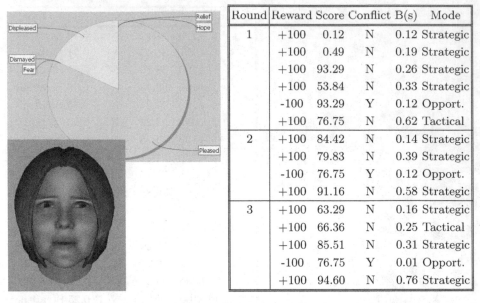

Round	Reward	Score	Conflict	B(s)	Mode
1	+100	0.12	N	0.12	Strategic
	+100	0.49	N	0.19	Strategic
	+100	93.29	N	0.26	Strategic
	+100	53.84	N	0.33	Strategic
	-100	93.29	Y	0.12	Opport.
	+100	76.75	N	0.62	Tactical
2	+100	84.42	N	0.14	Strategic
	+100	79.83	N	0.39	Strategic
	-100	76.75	Y	0.12	Opport.
	+100	91.16	N	0.58	Strategic
3	+100	63.29	N	0.16	Strategic
	+100	66.36	N	0.25	Tactical
	+100	85.51	N	0.31	Strategic
	-100	76.75	Y	0.01	Opport.
	+100	94.60	N	0.76	Strategic

Fig. 6. Results of A Dialogue Process with Error and Conflict

In the three scenarios, a pie chart is used to illustrate the distribution of different types of emotions. They cover the cases when the level of user's knowledge about the application domain varies from almost perfect to somewhat lost. Correspondingly, they result in increased efforts for the agent to respond with messages or explain with clarifications. In the second and third case, the agent is in a tactical mode or reaches to an opportunistic mode briefly but in the strategic mode all the other time. While the happy face in Fig. 4 is normal when everything goes well, the partially displeased facial expressions in Fig. 5 and Fig. 6 are the result of a transition from completely pleased emotion to a partially pleased emotion, with an increasing percentage of displeased emotion appearing in the two distributions.

4.3   Comparison Studies

In experiments, policy files are used to calculate belief states at every stage of the dialogue. To simulate users with different levels of knowledge, a user's goal is changed with random noise. Shown in Table 2 is the simulation results after a total of 1,000 runs, including the average number of dialogue turns, the number of users who successfully completed the process, the accuracy of intention discovery, and the stability of belief states in terms of deviations from the overall average during dialogue. The results demonstrate that the dialogue management with integrated emotion has the lowest standard deviation, and the proportion of number of turns per dialogue is more compact. In addition, the highest number of successful users shows that the proposed approach is more capable of handling uncertainties in belief states. Overall, the proposed

Table 2. Comparison Results for Dialogue Management with Emotion

Approach	No. of Turns	No. of Users	Accuracy	Deviation
no emotion	8	712	71	5.4
machine emotion	11	745	75	4.6
machine and user emotion	5	822	82	3.8

approach of integrating both emotion recognition and animation performs the best in comparison to the original POMDP-based approach without emotion [16] and an approach with incorporated machine emotion only.

5   Conclusions

This paper presents a method for dialogue management to add user's motion in observation, to include response time in a modified POMDP model, and to incorporate COCOM in emotion animation. In addition to the common advantages of POMDP-based dialogue management, it becomes possible to select actions dynamically according to different control modes, to handle uncertainty caused by errors in speech recognition or even user's lack of knowledge in a particular domain, and to engage embodied agents in lively interaction. In such a way, the proposed approach makes embodied agents more human-like in behavior and more accurate in goal identification, thus becoming more useful when employed to provide services/support in practice. Active investigation is ongoing to further refine the result of emotion recognition and animation, and to deal with issues such as high complexity of modeling, expressiveness of domain properties, and uncertainties in dynamic environments.

Acknowledgments. The authors of this paper acknowledge the funding support from NSERC of Canada, and experiments by Mr. Sivaraman Sriram and Mr. Rajkumar Vijayarangan.

References

1. Bui, T.: Multimodal dialogue management: State of the art. Technical Report TR-CTIT-06-01, Centre for Telematics and Information Technology, University of Twente, Enschede, Netherlands (2006)
2. Bui, T., Zwiers, J., Poel, M., Nijholt, A.: Toward affective dialogue modeling using partially observable markov decision processes. In: Proc. Workshop Emotion and Computing, 29th Annual German Conference on Artificial Intelligence (2006)
3. Creed, C., Beale, R., Cowan, B.: The impact of an embodied agent's emotional expressions over multiple interactions. Interacting with Computers **27**(2), 172–188 (2015)
4. Ersotelos, N., Dong, F.: Building highly realistic facial modelling and animation: a survey. The Visual Computer **24**(1), 13–30 (2008)

5. Fridlund, A.: Human Facial Expression: An Evolutionary View. Academic Press (2014)
6. Hashemian, M., Moradi, H., Mirian, M.: Determining mood via emotions observed in face by induction. In: Proc. 2nd RSI/ISM International Conference on Robotics and Mechatronics, pp. 717–722 (2014)
7. Hollnagel, E.: Coping, coupling and control: the modelling of muddling through. In: Proc. 2nd Interdisciplinary Workshop on Mental Models, pp. 61–73 (1992)
8. Jeremie, N., Vincent, R., Kevin, B., Lionel, P., Mohamed, C.: Audio-visual emotion recognition: A dynamic, multimodal approach. In: 26e conférence francophone sur l'Interaction Homme-Machine, pp. 44–51 (2014)
9. Kuber, R., Wright, F.: Augmenting the instant messaging experience through the use of brain-computer interface and gestural technologies. International Journal of Human-Computer Interaction **29**(3), 178–191 (2013)
10. López-Cózar, R., Griol, D., Espejo, G., Callejas, Z., Ábalos, N.: Towards fine-grain user-simulation for spoken dialogue systems. In: Minker, W., Lee, G., Nakamura, S., Mariani, J. (eds.) Spoken Dialogue Systems Technology and Design, pp. 53–81. Springer, New York (2011)
11. Marsella, S., Gratch, J., Petta, P.: Computational models of emotions. In: Scherer, K., Banzinger, T., Roesch, E. (eds.) A blueprint for an affectively competent Agent. Oxford University Press (2010)
12. Mavridis, N.: A review of verbal and non-verbal humanrobot interactive communication. Robotics and Autonomous Systems **63**(Part 1), 22–35 (2015)
13. Ortony, A., Clore, G., Collins, A.: The Cognitive Structure of Emotions. University Press, Cambridge (1988)
14. Plutchik, R.: The Emotions. University Press of America, revised edition edition (1991)
15. Sriram, S., Yuan, X.: An enhanced approach for classifying emotions using customized decision tree algorithm. In: Proc. IEEE 2012 Southeastcon, pp. 1–6 (2012)
16. Thomson, B.: Statistical Methods for Spoken Dialogue Management. Springer (2013)
17. Uithol, S., Burnston, D., Haselager, P.: Why we may not find intentions in the brain. Neuropsychologia **56**, 129–139 (2014)
18. Williams, J., Poupart, P., Young, S.: Partially observable markov decision processes with continuous observations for dialogue management. In: Recent Trends in Discourse and Dialogue of Text, Speech and Language Technology, vol. 39, pp. 191–217. Springer, Netherlands (2008)
19. Young, S.: Using POMDPs for dialog management. In: Proc. IEEE Spoken Language Technology Workshop, pp. 8–13 (2006)
20. Yuan, X., Bian, L.: A modified approach of POMDP-based dialogue management. In: Proc. 2010 IEEE International Conference on Robotics and Biomimetics (2010)

Mathematics Wall: Enriching Mathematics Education Through AI

Somnuk Phon-Amnuaisuk[✉], Saiful Omar, Thien-Wan Au, and Rudy Ramlie

Media Informatics Special Interest Group, Institut Teknologi Brunei,
Gadong, Brunei
{somnuk.phonamnuaisuk,saiful.omar,thienwan.au,rudy.ramlie}@itb.edu.bn

Abstract. We present the progress of our ongoing research titled *Mathematics Wall* which aims to create an interactive problem solving environment where the system intelligently interacts with users. In its full glory, the wall shall provide answers and useful explanations to the problem-solving process using artificial intelligence (AI) techniques. In this report, we discuss the following components: the digital ink segmentation task, the symbol recognition task, the structural analysis task, the mathematics expression recognition, the evaluation of the mathematics expressions and finally present the results. We then present and discuss the design decisions of the whole framework and subsequently, the implementation of the prototypes. Finally, future work on the explanation facility is discussed.

Keywords: Handwritten mathematics expression recognition · Automated reasoning · Enriching mathematics education with AI

1 Introduction

Enriching mathematics education can be accomplished via a number of approaches; for examples, computer guided content presentation, learning through games, intelligent tutoring system, etc. Artificial Intelligence plays a crucial role in this enriching process as it provides intelligent behaviours required by those applications. AI technology such as automated reasoning, intelligent interactive environment, students modelling, etc. are important enabling technologies.

In this paper, we present our ongoing research titled *Mathematics Wall*. This report discusses our current implementation, in particular, the recognition of handwritten digital ink and the processing of mathematical expressions. Readers can find a good summary of related works in this area in [1–5].

The term *digital ink* used in this report refers to our own representation of handwritten contents in a digital format. The handwriting information can be recorded as a sequence of coordinates in a 2D space. Each ink stroke (i.e., from pen down to pen up) contains, a sequence of discrete (x, y) coordinates which can be rendered as a continuous ink stroke. The recorded digital ink can be

© Springer International Publishing Switzerland 2015
Y. Tan et al. (Eds.): ICSI-CCI 2015, Part III, LNCS 9142, pp. 309–317, 2015.
DOI: 10.1007/978-3-319-20469-7_33

interpreted using handwritten recognition techniques. Since the interpretation of mathematical formula is dependent on the *spatial relationships* among symbols, this whole handwritten process encompasses the following sub-processes: *digital ink segmentation, symbol recognition, structural analysis, mathematics expression recognition, evaluation of the mathematics expressions and display of the result.* In this paper, we report our current progress on the mentioned sub-processes. The rest of the materials in the paper are organized into the following sections: Section 2: Related works; Section 3: Mathematics wall; Section 4: Discussion and Conclusion.

2   Related Works

In recent years, advances in touch/multi-touch technology have revived interests in pen-based computing, handwritten symbols recognition research and their applications in specialised domain e.g., mathematics [2,3,6,7], music [8,9], specific engineering applications [10]. In such applications, the system gathers ink strokes where each symbol may be constructed from one or more ink strokes. Challenges in these areas result from spatial relationships among symbols which allow multiple local ink segmentation and multiple local interpretations of ink strokes.

With advances in pattern recognition and machine learning techniques in recent decades, the accuracy of isolated symbol recognition has improved significantly. Online recognition of handwritten symbols using methods based on Hidden Markov Models was reported at the rates of 75 to 95% accuracy (depending on the number of symbols). The state of the art single digit classification system implemented using deep learning neural network, could recognize individual digit better than human i.e., more than 99% accuracy. However, the recognition of symbols in specific applications such as mathematical equations or music editing tasks are still an open research issue. In the observation made in [11], commercial OCR systems could obtain a recognition rate of 99% with a simple printed text. However, its accuracy fell below 10% or less once tried on perfectly formed characters in mathematical equations. This is due to the variations in font size, multiple baselines, and special characters in mathematical notation that cause useful heuristics for text processing fail to work well in a different domain. This observation is still true today for commercial OCR system. The state of the art handwritten equation entering system found in the market today is still not reliable and this justifies further research effort put into this area.

The recognition and evaluation of handwritten mathematical expressions involves four major tasks: digital ink segmentation, isolated symbol recognition, mathematical expressions interpretation and automated reasoning of those expressions.

The digital ink segmentation task aims to accurately group ink strokes of the same symbol together for the further isolated symbol recognition process. This may not be as simple as it might appear. Mathematical symbols are normally constructed from connected ink strokes, but symbols having isolated ink strokes are also not uncommon. Resolving spatial ambiguity among ink groups is crucial in the ink segmentation process and various tactics have been devised. Projection profile is one of the common tactics employed to detect boundary among symbols and a given expression can be recursively decomposed using X-Y cutting [12]. Spatial proximity and temporal proximity (in online handwritten) are two other popular heuristics employed to disambiguate ink strokes.

An accurate ink segmentation improves the accuracy of isolated symbol recognition, however, this does not guarantee an accurate interpretation of mathematical expressions. This is due to the fact that mathematical expressions are expressed on a 2D space and their interpretations are context dependent. A structural analysis of spatial relationships among symbols is often represented using tree [10,13,14]. A symbol layout tree or an operator tree are two examples in this line of approach. In a symbol layout tree, each node representing a handwritten symbol, spatial information such as superscript, subscript, etc., is captured in the link between nodes. If each individual symbol is accurately recognised, a layout tree can be parsed to form an operator tree which can be used to evaluate an expression by the symbolic computing facility in Matlab, Mathematica, or Python.

3   Mathematics Wall

Figure 1 summarises the main processes in the *Mathematics Wall* project: (i) mathematical expressions are entered using a pen as one would do using pen and paper; (ii) digital inks are segmented and grouped together; (iii) digital ink groups are recognised; (iv) their spatial structural relationships are evaluated, for example, x^y mean x to the power of y, and $x + y$ means the addition of x and y; (v) the mathematical expressions are recognised, evaluated and displayed in the author's own handwriting style.

3.1   Digital Ink Segmentation & Grouping

In our approach, mathematical expressions were entered using a standard pen-based device or using a touch technology. After a complete mathematical expression was entered, the digital ink strokes were segmented to form different symbols. In a more formal description, let $\mathbf{s}_i$ be a sequence of 2D coordinates $(x, y)_n$ where $|\mathbf{s}_i| = n$ and $x, y \in \mathcal{R}$. Each symbol $\mathbf{D}^{h \times w} \in \{0, 1\}$ was segmented from these ink strokes according to their connectivity and proximity. The extracted matrix $\mathbf{D}^{h \times w}$ was a binary matrix representing the digital ink pattern of a symbol. Each ink pattern image was a collection of ink strokes in a bounding box $h \times w$.

It was also our design choice that each symbol would be scaled to a standard dimension. Since each symbol was of a different height h and width w, each image $\mathbf{D}^{h \times w}$ was normalised to a fixed size of $\mathbf{D}_n^{28 \times 28}$ using the following function:

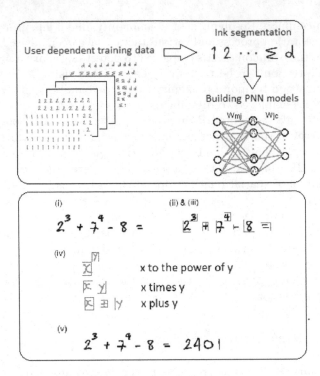

Fig. 1. An overview of the main processes of the *Mathematics Wall*: Top pane - offline PNN training, Bottom pane - online symbol recognition, mathematical expression evaluation and display of the calculation results in a user's own handwriting style

function NORMALISE($\mathbf{D}^{h \times w}$) **returns** $\mathbf{D}_n^{28 \times 28}$

$offset \leftarrow |\lceil \frac{w-h}{2} \rceil|$

if $h \geq w$

    $newD \leftarrow 0^{h \times h}$

    $newD(1\text{:}h, offset\text{:}offset+w) = \mathbf{D}^{h \times w}$

else

    $newD \leftarrow 0^{w \times w}$

    $newD(offset\text{:}offset+h, 1\text{:}w) = \mathbf{D}^{h \times w}$

$\mathbf{D}_n^{28 \times 28} \leftarrow$ SCALE($newD$)

return $\mathbf{D}_n^{28 \times 28}$

where $newD(1\text{:}h,1\text{:}w)$ denoted all the rows and columns in the matrix $newD$. The *SCALE* function linearly transformed the size of the matrix using bicubic interpolation technique. There were pros and cons to this decision. The standard dimensional approach allows a more flexible way to deal with different handwriting symbol sizes using just one model. On the other hand, the scaling distorted the stroke size i.e., scaling a digit to a bigger size would thin the stroke while scaling down would thicken the stroke.

3.2 Symbol Recognition

The normalised images $\mathbf{D}_n^{28 \times 28}$ were used for training the recognition models. Each entry j was recorded as a pair of training information $t_j = (\mathbf{D}_n, c)$. The training examples were employed to create a classification model using a probabilistic neural network technique (PNN). PNN was chosen since it had a good performance record with a limited number of training examples. To improve PNN efficiency, the dimensionality of the input data was reduced from 1568 to 100 using the principal component analysis technique (PCA).

Dimensionality Reduction PCA was employed here to reduce the size of the input data. We flattened all t_j, each image $\mathbf{D}_n^{28 \times 28}$ in the training dataset and this formed a training data matrix $\mathbf{T}^{1568 \times j}$. This was obtained by concatenating each row, row by row, and then each column, column by column. Each normalised symbol $\mathbf{D}_n$ was flattened to a vector $\mathbf{d}$ with the size of 1,568. PCA transformed the 1,568 dimensional vectors to a new coordinate system that maximised the variance of the projected data in a few principal components. In most dataset, this transformation dramatically reduced the dimension of the representation. In our implementation, the dimension was reduced from 1,568 to only 100 while still maintaining more than 95% of the total variance of the dataset [1].

> **function** REDUCE_DIM($\mathbf{T}$) **returns** $\mathbf{T}'$
>
> $\mu_i \leftarrow \frac{\sum_k^j d_{ik}}{j}$; d_{ik} is the element ik of the matrix $\mathbf{T}$
>
> $\forall_{r,c \in \{1,\dots,1568\}}$ $x_{rc} \leftarrow \sum_k^j (d_{rk} - \mu_r)(d_{ck} - \mu_c)$; $x_{rc} \in \mathbf{X}$
>
> $\mathbf{E}^t \Lambda \mathbf{E} \leftarrow \mathbf{X}$
>
> where Λ is the ordered diagonal Eigen value matrix, and
> $\mathbf{e}_m$ is a column Eigen vector where $\mathbf{e}_m \in \mathbf{E}$
>
> $\mathbf{T}'^{m \times j} \leftarrow \mathbf{E}^{m \times 1568} \mathbf{T}^{1568 \times j}$; if m principal components are desired.
>
> **return** $\mathbf{T}'$

Probabilistic Neural Network In this implementation, a two layer feedforward PNN network was constructed as a classifier [15]. The number of input nodes was the same as the number of principal components plus one bias node i.e., the number of reduced dimensionality input plus one. The number of node in the first layer was the same as the number of training examples and the number of nodes in the output layer was the same as the number of classes. The input node m was connected to the nodes j in the first layer with the weight w_{mj}. These nodes j in the first layer had a radial basis transfer function φ_{rb}. Hence, the output of the radial basis node j could be expressed as follow:

$$a_j = \varphi_{rb}(\|\mathbf{w}_j - \mathbf{d}'\| b_j) = e^{-(\|\mathbf{w}_j - \mathbf{d}'\| b_j)} \tag{1}$$

where b_j is the bias and $\|\mathbf{w}_j - \mathbf{d}'\|$ denotes a distance between the weight vector of node j and the input vector $\mathbf{d}'$. The radial basis nodes j are connected to the

---

[1] Let s_m be the variance of the principal component m, $\frac{\sum_{m=1}^{100} s_m}{\sum_{m=1}^{1568} s_m} \geq 0.95$.

output layer nodes c with the weight w_{jc}. The output nodes c have a competitive transfer function φ_c, here implemented as a dot product. Hence, classification result is obtained from:

$$argmax_c \quad \varphi_c(\mathbf{w}_c, \mathbf{a}) = argmax_c \quad \mathbf{w}_c\mathbf{a} \tag{2}$$

3.3 Layout Analysis

The interpretation of a mathematical expression requires more than correctly recognising symbols since its meaning is dependent on its contexts. Spatial relationships among symbols have been employed in many previous works [2,14]. In this implementation, spatial relationships were inferred from Euclidean distance and direction (see Figure 2). This allowed spatial relationships such as *above, below, adjacent, superscript, subscript,* etc. to be inferred.

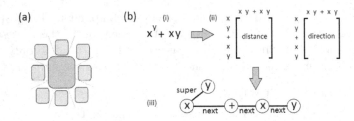

Fig. 2. Figure 2 (a) Spatial relationships are inferred from distance and direction between symbols, (b) Steps involved in a layout analysis: (i) each isolated symbol is recognised, (ii) distance matrix and directional matrix are calculated, and use to (iii) infer spatial relationships between symbols

Next, the expression underwent the mathematical reasoning process and the result was printed out in the user's own handwriting style. This was obtained by displaying the user's own handwriting symbols randomly chosen from the training data.

4 Results and Discussion

We decided to experiment with a limited set of symbols $\{0, 1, ..., 9, d, x, n, +, -, *, /, (,), \sum, \int\}$ since (i) they were sufficient for testing all common mathematical expressions and (ii) it was not the focus here to get the best classifier for all mathematical symbols (at least not at this stage) but to investigate the feasibility of the proposed framework. In this implementation, each user provided 100 examples of handwriting data for each symbol, 20 random samples were set aside as a testing data and 80 samples were used to train the user dependent

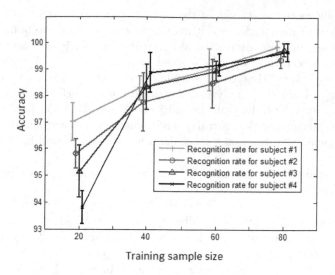

Fig. 3. Average recognition rate of handwritten symbols for 4 users, presented with error bars

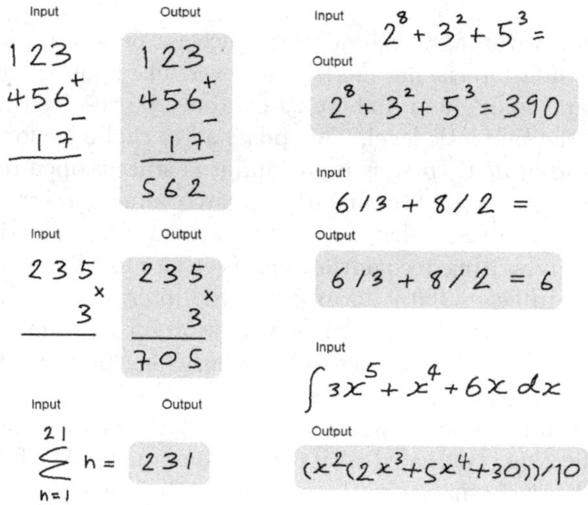

Fig. 4. Snap shots of various input/output from the *Mathematics Wall*

PNN models. Here the PNN models were constructed from 80, 60, 40, and 20 samples randomly selected from the pool of 80 samples. Each model was tested with 10 randomly selected testing samples. This whole process was repeated 10 times, i.e., all PNN models were reconstructed. Figure 3 summarises the average accuracy for each user.

Figure 4 shows the snapshots of the output from our system. At the current stage, the system can perform arithmetic operations and symbolic computation such as summation, derivative and integration. Various future directions can be pursued in future works - student modelling and explanation generation are the two areas that could seamlessly connect to the current research. A student model aims to analyze the student's own performance and reason about his/her understanding of the subject. An explanation facility aims to provide a better interactive learning environment by analyzing the student's answer and generating relevant explanation if errors are found in the student's answer.

In [16], the authors showed that it was possible to analyse student's answer and if the answer was incorrect, then the system could infer why the student has made such a mistake. For example, by comparing a student's answer tree and the system's answer tree, discrepancies can be located, then the system can examine its problem solving tree and further infer how and why the mistakes are made. This kind of reasoning is typical in expert systems and it is feasible for a well-scoped domain such as in a mathematical reasoning problem.

5   Conclusion

One of the major advantages behind the touch or pen-based interface is its universal and natural interfacing model. However, like spoken language, handwriting can be ambiguous and ambiguity can only be resolved using contexts which is beyond the syntactic level. This poses a big challenge for computers to interpret handwritten mathematics input and it is still an open research area.

In this paper, we report our ongoing research aiming to create an intelligent interactive environment that supports mathematics education. Although the research in handwriting recognition has been around for many decades, it is still impossible to get a 100% accuracy in the interpretation of mathematical expressions. This type of zero-error expectation for this kind of application is one of the factors that hinders research in enriching mathematics education using AI.

We have learned from our implementation that the error issue may be mitigated if the application is designed with proper constraints. The following tactics could improve the performance of the system: (i) implement a user-dependent handwriting recognition process, (ii) implement topic-dependent recognition models where each specific topic is associated with a different set of symbols' likelihood, and (iii) implement a backtracking mechanism that allows inconsistent interpretations at the layout analysis stage to be reinterpreted with a different set of plausible symbols.

Acknowledgments. We wish to thank anonymous reviewers for their comments, which help improve this paper. We would like to thank the GSR office for their financial support given to this research.

References

1. Chan, K.F., Yeung, D.Y.: Mathematical expression recognition: A survey. International Journal on Document Analysis snd Recognition $3(1)$, 3–15 (2000)
2. Anderson, R.H.: Two-dimensional mathematical notation. In: Fu, K.S. (ed.) Syntactic Pattern Recognition Applications, pp. 147–148. Springer Verlag, Berlin Heidelberg (1977)
3. LaViola, Jr. J.J., Zeleznik, R.: MathPad2: a system for the creation and exploration of mathematical sketches. In: Proceedings of the 31st International Conference on Computer Graphics and Interactive Techniques, pp. 432–440. ACM SIGGRAPH (2004)
4. Bainbridge, D., Carter, N.: Automatic reading of music notation. In: Bunke, H., Wang, P.S.P. (eds.) Handbook of Character Recognition and Document Image Analysis, pp. 583–603. World Scientific, Singapore (1997)
5. Fornés, A., Lladós, J., Sánchez, G.: Primitive segmentation in old handwritten music scores. In: Liu, W., Lladós, J. (eds.) GREC 2005. LNCS, vol. 3926, pp. 279–290. Springer, Heidelberg (2006)
6. Chang, S.K.: A method for the structural analysis of two-dimensional mathematical expressions. Information Sciences $2(3)$, 253–272 (1970). Elsevier
7. LaViola, Jr. J.J., Bott, J.N.: A pen-based tool for visualising vector mathematics. In: Proceedings of the 7th Sketch-based Interfaces and Modelling Symposium (SBIM 2010), pp. 103–110. Eurographics Association (2010)
8. Forsberg, A., Holden, L., Millrt, T., Zeleznik, R.: Music notepad. In: Proceedings of the 11th Annual ACM Symposium on User Interface Software and Technology, pp. 203–210. ACM (1998)
9. Lee, K.C., Phon-Amnuaisuk, S., Ting, C.Y.: Handwritten music notation recognition using HMM: a non-gestural approach. In: Proceedings of International Conference on Information Retrieval and Knowledge Management (CAMP 2010), March 16–18, 2010
10. Yang, R.U., Su, F., Lu, T.: Research of the structural-learning-based symbol recognition mechanism for engineering drawings. In: Proceedings of the 6th International Conference on Digital Content, Multimedia Technology and its Applications (IDC), Seoul, pp. 346–349, August 16–18, 2010
11. Berman, B.P., Fateman, R.J.: Optical character recognition for typeset mathematics. In: Proceedings of International Symposium on Symbolic and Algebraic Computation (ISSAC 1994), pp. 348–353. ACM Press (1994)
12. Ha, J., Haralick, R.M., Phillips, I.T.: Recursive X-Y cut using bounding boxes of connected components. In: Proceedings of International Conference on Document Analysis and Recognition, (ICDAR 1995), vol. II, pp. 952–955, August 1995
13. Kim, D.H., Kim, J.H.: Top-dowm search with bottom-up evidence for hand written mathematical expressions. In: Proceedings of the 12nd International Conference on Frontiers in Handwriting Recognition (ICFHR 2010), Kolkata, pp. 507–512 (2010)
14. Zanibbi, R., Blostein, D.: Recognition and retrieval of mathematical expressions. International Journal on Document Analysis snd Recognition 15, 331–357 (2012)
15. Specht, D.F.: Probabilistic neural networks. Neural Networks 3, 109–118 (1990)
16. Htaik, T.T., Phon-Amnuaisuk, S.: Intelligent tutoring system for mathematical problems: explanation generations for integration problem (EGIP). In: Proceedings to the third International Conference on Computer Application, Yangon, Myanmar 2005, March 9–10, 2005

Segmentation and Detection System

Spatial Subdivision of Gabriel Graph

M.Z. Hossain, M.A. Wahid, Mahady Hasan, and M. Ashraful Amin[✉]

Computer Vision and Cybernetics Group,
Department of Computer Science and Engineering,
Independent University Bangladesh, Dhaka 1229, Bangladesh
mzhossain@gmx.com, wahidrahman@gmail.com, mahadyh@yahoo.com,
aminmdashraful@iub.edu.bd

Abstract. Gabriel graph is one of the well-studied proximity graphs which has a wide range of applications in various research areas such as wireless sensor network, gene flow analysis, geographical variation analysis, facility location, cluster analysis, and so on. In numerous applications, an important query is to find a specific location in a Gabriel graph at where a given set of adjacent vertices can be obtained if a new point is inserted. To efficiently compute the answer of this query, our proposed solution is to subdivide the plane associated with the Gabriel graph into smaller subregions with the property that if a new point is inserted anywhere into a specific subregion then the set of adjacent vertices in the Gabriel graph remains constant for that point, regardless of the exact location inside the subregion. In this paper, we examine these planar subregions, named redundant cells, including some essential properties and sketch an algorithm of running time $\mathcal{O}(n^2)$ to construct the arrangement that yields these redundant cells.

1 Introduction

The notion of proximity graph is used to describe the neighboring relationship among a set of objects. Several proximity graphs have been extensively studied in literature, probably because of their theoretical interest and large number of application fields [1,14]. In geometric proximity graphs, vertices and edges can be denoted as points and line segments in metric space respectively. Two vertices of these proximity graphs are connected by an edge if the vertices satisfy some specific nearness properties in metric space. In this text, we study two closely related classes of proximity graphs known as Gabriel graph and Delaunay graph in Euclidean plane.

The Gabriel graph is one of the most common abstract representations to perform various analyses on different relationships among a set of spatially distributed real world objects such as wireless sensors, ethnic groups, points of services, taxonomic units etc., [13,12]. Because of this essential practical interest, other variations of the Gabriel graph including the higher order Gabriel graph [4] and the relaxed Gabriel graph [3] are also studied to extend the application fields of the conventional Gabriel graph.

The subdivision method of the plane producing the redundant cells provides a constant adjacency relationship of a candidate point in each redundant cell

© Springer International Publishing Switzerland 2015
Y. Tan et al. (Eds.): ICSI-CCI 2015, Part III, LNCS 9142, pp. 321–328, 2015.
DOI: 10.1007/978-3-319-20469-7_34

for a Gabriel graph. The term, candidate point, refers to a new point to be inserted into the corresponding Gabriel graph. Searching for a location or a region in the plane where a given set of adjacent vertices in a Gabriel graph can be found if a new point is inserted, is an infinite time computational problem since all regions including the plane contain infinite number of points [6]. However, number of the redundant cells is finite for a finite graph and thus the subdivision method reduces this infinite time computation to a finite time computation by considering the set of points denoting a redundant cell, at one time instead of considering each individual point at one time. Consequently, it becomes possible to solve different queries related to adjacency and degree of a candidate point when interconnections among the vertices are mapped according to the Gabriel graph. For instance, one may want to place a wireless sensor at any location in a specific region of a wireless sensor network such that the vertex degree of that sensor needs to be either maximum, minimum or a given number in the Gabriel graph representing the wireless sensor network.

Moreover, the spatial trends of genetic association and genetic variation can be explored in detail using the Gabriel graph [12] and the redundant cells. For example, a biologist may want to find a shared common region among a cluster of ethnic groups, which is in fact the redundant cells where a candidate point has the set of adjacent vertices referring to the ethnic groups, to study the spread pattern of some genetic diseases. Besides answering the point location queries, answering various region related queries also requires constructing the arrangement for the redundant cells, which enables analysis of different regional and morphological information of the cells, including area, shape, eccentricity, centroid, diameter etc.

2   Preliminaries

Let P be a finite set of points in the Euclidean plane $\mathbb{R}^2$ in general position and P contains at least three points. Hence, no three points in P are collinear and no four points in P are co-circular. The Gabriel graph GG of a set of points P in the plane $\mathbb{R}^2$ is a proximity graph where any two points $p_i, p_j \in P$ are connected by an edge if and only if there exists a circle through p_i and p_j with $p_i p_j$ diameter, contains no other point in its interior and on its boundary from P. The Delaunay graph DG of a set of points P in the plane $\mathbb{R}^2$ is another proximity graph where any three points $p_i, p_j, p_k \in P$ construct a triangle if and only if the circumcircle of the triangle $\triangle p_i p_j p_k$, contains no other point in its interior from P. Although the edges of both graphs are undirected (i.e., an edge $p_i p_j$ can be written as $p_j p_i$), we maintain a consistent order throughout the text which is essential for the correctness of our discussion. A circle through points p_i, p_j with diameter $p_i p_j$ is denoted as $\mathcal{C}(p_i, p_j)$ and a circle through points p_i, p_j, p_k is denoted as $\mathcal{C}(p_i, p_j, p_k)$. In this text, both the Gabriel graph GG and the Delaunay graph DG are obtained from the same set P of points.

Lemma 1. [7] *For every point $p_k \in \mathcal{V}(GG)$ except the points $p_i, p_j \in \mathcal{V}(GG)$, the angle $\angle p_i p_k p_j$ is acute iff the edge $p_i p_j \in \mathcal{E}(GG)$.*

Howe [10] showed in his Ph.D. thesis that the Gabriel graph GG is actually the sub-graph of the Delaunay graph DG and it is possible to efficiently construct GG from DG.

Lemma 2. [10] *The Gabriel graph on a vertex set V is a sub-graph of the Delaunay triangulation for V. Furthermore, the edge AB of the Delaunay triangulation is an edge of the Gabriel graph iff the straight line joining A to B intersects the boundary line segment common to the Thiessen polygons for A and B at a point other than the endpoints of that boundary line segment.*

3   Spatial Subdivision

From Lemma 1, it is clear that the right angle is the highest angle limit determines the valid edges of the Gabriel graph GG of a point set P. Thus the orthogonal lines through the incident vertices of an edge in a geometric graph divide the plane into regions with identical edge forming criteria in the Gabriel graph GG as shown in Figure 1.

Let p_i, p_j be two incident vertices of an edge $p_i p_j$ in a planar geometric graph and an orthogonal line $\mathcal{L}(\widehat{p_i}, p_j)$ on $p_i p_j$ passing through p_i divides the plane into two half planes. The right open half plane $ROH(\mathcal{L}(\widehat{p_i}, p_j))$ contains the part of the plane enclosed p_j and the left closed half plane $LCH(\mathcal{L}(\widehat{p_i}, p_j))$ contains the rest of the plane. Similarly, the left open half plane $LOH(\mathcal{L}(p_i, \widehat{p_j}))$ containing p_i and the right closed half plane $RCH(\mathcal{L}(p_i, \widehat{p_j}))$, can be found by considering the orthogonal line $\mathcal{L}(p_i, \widehat{p_j})$ passing through p_j. The left region $\mathcal{R}(\widehat{p_i}, p_j)$, the right region $\mathcal{R}(p_i, \widehat{p_j})$ and the middle region $\mathcal{R}(\widehat{p_i}, \widehat{p_j})$ constructed by partitioning the plane $\mathbb{R}^2$ based on $p_i p_j$ can be defined as

$$
\begin{aligned}
\mathcal{R}(\widehat{p_i}, p_j) &= LCH(\mathcal{L}(\widehat{p_i}, p_j)) \\
&= LOH(\mathcal{L}(p_i, \widehat{p_j})) \setminus ROH(\mathcal{L}(\widehat{p_i}, p_j)) \\
\mathcal{R}(p_i, \widehat{p_j}) &= RCH(\mathcal{L}(p_i, \widehat{p_j})) \\
&= ROH(\mathcal{L}(\widehat{p_i}, p_j)) \setminus LOH(\mathcal{L}(p_i, \widehat{p_j})) \\
\mathcal{R}(\widehat{p_i}, \widehat{p_j}) &= ROH(\mathcal{L}(\widehat{p_i}, p_j)) \cap LOH(\mathcal{L}(p_i, \widehat{p_j}))
\end{aligned}
$$

If $\mathcal{R}(p_i, p_j)$ is the set of these three regions whose union produces the $\mathbb{R}^2$ plane, then the set $\mathcal{R}(p_i, p_j)$ can be given as

$$
\mathcal{R}(p_i, p_j) = \{\mathcal{R}(\widehat{p_i}, p_j), \mathcal{R}(\widehat{p_i}, \widehat{p_j}), \mathcal{R}(p_i, \widehat{p_j})\}
$$

If p_m is a point in $\mathcal{R}(\widehat{p_i}, \widehat{p_j})$ and the angle $\angle p_i p_m p_j$ is acute, then the three edges $p_i p_j, p_i p_m, p_j p_m \in \mathcal{E}(GG)$ however if the $\angle p_i p_m p_j$ is not acute (denoted as p'_m in the figure), then only two edges $p_i p_m, p_j p_m \in \mathcal{E}(GG)$. Furthermore, any point, for instance p_l in $\mathcal{R}(\widehat{p_i}, p_j)$ cannot construct the edge $p_j p_l \in \mathcal{E}(GG)$ since the angle $\angle p_j p_i p_l$ is not acute. Similarly, any point, such as p_r in $\mathcal{R}(p_i, \widehat{p_j})$ cannot construct the edge $p_i p_r \in \mathcal{E}(GG)$ because the angle $\angle p_i p_j p_r$ is not acute.

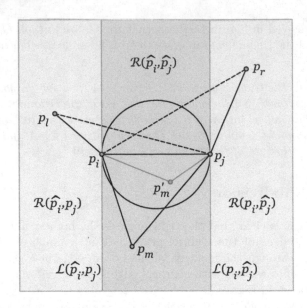

Fig. 1. The three regions: left region $\mathcal{R}(\widehat{p}_i, p_j)$, right region $\mathcal{R}(p_i, \widehat{p}_j)$ and middle region $\mathcal{R}(\widehat{p}_i, \widehat{p}_j)$

3.1   Redundant Cell

A redundant cell is the largest non-empty convex region $h \subset \mathbb{R}^2$ constructed from the common intersection of the each unique region r such that $r \in \mathcal{R}(p_i, p_j)$ and $h \subseteq r$ for every $p_i p_j \in \mathcal{E}(DG)$. Formally, redundant cell rc can be defined as

$$rc = \bigcap \{r : \forall p_i p_j (p_i p_j \in \mathcal{E}(DG) \implies \exists! r (r \in \mathcal{R}(p_i, p_j) \wedge h \subseteq r))\}$$

where rc cannot be empty since h cannot be empty. For an example, the convex region rc_1 in Figure 2 can be given as

$$rc_1 = \bigcap \{\mathcal{R}(\widehat{p}_i, \widehat{p}_j), \mathcal{R}(\widehat{p}_j, p_k), \mathcal{R}(\widehat{p}_k, p_l), \mathcal{R}(\widehat{p}_i, \widehat{p}_k), \mathcal{R}(\widehat{p}_i, p_l)\}$$

The convex region rc_1 is a redundant cell since it satisfies the conditions of redundant cell according to the definition. The following three lemmas simplify the formal proof of correctness for the subdivision method and also deduce some fundamental properties of redundant cell and Gabriel graph.

Lemma 3. [9] *For points $p_i, p_j \in \mathcal{V}(GG)$, if there exists no point inside the circle $\mathcal{C}(p_i, p_j)$, then $p_i p_j \in \mathcal{E}(DG)$.*

Lemma 4. [9] *For points $p_i, p_j, p_k \in \mathcal{V}(GG)$, iff the edge $p_i p_j \notin \mathcal{E}(GG)$, then there exists a point $p_k \in \mathcal{C}(p_i, p_j)$ such that $p_i p_k \in \mathcal{E}(DG)$.*

Lemma 5. [9] *For points $p_i, p_j, p_k \in \mathcal{V}(GG)$, iff the edge $p_i p_j \notin \mathcal{E}(GG)$, then there exists a point $p_k \in \mathcal{C}(p_i, p_j)$ such that $p_i p_k \in \mathcal{E}(DG)$ and $p_j \in \mathcal{R}(p_i, \widehat{p}_k)$.*

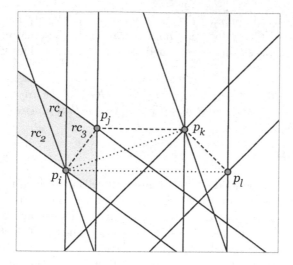

Fig. 2. An arrangement yielding redundant cells in the Delaunay graph of four vertices and five edges

Our goal is to develop a planar subdivision method that produces the set of subregions where the adjacent vertices in a Gabriel graph for a candidate point, remain unchanged. It is claimed that an edge $p_i p_j$ is not a Gabriel edge if and only if there exists a point $p_k \in \mathcal{V}(GG)$ in the circle $\mathcal{C}(p_i, p_j)$ with an edge $p_i p_k \in \mathcal{E}(DG)$ for which $p_j \in \mathcal{R}(p_i, \widehat{p_k})$. In other words, if we insert a point p_j at any location in the Gabriel graph, then p_j has edges in GG with all the point $p_i \in \mathcal{V}(GG)$ except those which have p_k as an adjacent in DG such that $p_j \in \mathcal{R}(p_i, \widehat{p_k})$. This is actually the basis of our proposed subdivision method of the plane for the Gabriel graph. With the help of Lemma 3, Lemma 4 and Lemma 5, we deduce the Theorem 1 to prove that such subdivision method and the resultant subregions or cells which are defined as redundant cells, exist for the Gabriel graph in $\mathbb{R}^2$ plane.

Theorem 1. [9] *If a point is inserted anywhere in a redundant cell, then the set of adjacent vertices of that point in the Gabriel graph remains the same.*

Proposition 1. [9] *The same set of adjacent vertices in the Gabriel graph can be found for a point inserted into different redundant cells.*

In a redundant manner, the redundant cells can share the same set of adjacent vertices in a Gabriel graph (which is the reason we call it redundant). Since we subdivide the plane based on the edges of the Delaunay graph, a question might arise, can the redundancy be reduced by considering a subset of the Delaunay edges. The edges of the Gabriel graph could be the appropriate subset of the Delaunay graph edges for constructing the cell complex to reduce or eliminate the redundancy. But the following proposition proves that it is not possible since the modified form of Lemma 4 for the Gabriel edges, that is for points $p_i, p_j,$

$p_k \in \mathcal{V}(GG)$, iff the edge $p_i p_j \notin \mathcal{E}(GG)$, then there exists a point $p_k \in \mathcal{C}(p_i, p_j)$ such that $p_i p_k \in \mathcal{E}(GG)$, does not hold for all cases. Therefore, the modified form of Theorem 1 based on the modified form of Lemma 4 does not hold for all cases as well.

Proposition 2. [9] *For points $p_i, p_j, p_k \in \mathcal{V}(GG)$, if the edge $p_i p_j \notin \mathcal{E}(GG)$, then there may not exist a point $p_k \in \mathcal{C}(p_i, p_j)$ such that $p_i p_k \in \mathcal{E}(GG)$.*

4 Algorithm

The bounded and unbounded convex redundant cells are joined together to form the cell complex of an arrangement RA which can be constructed by the boundary lines of the regions based on the edges in the Delaunay graph DG. The arrangement RA has at most $6n - 12$ boundary lines and $3n(6n - 23) + 67$ cells since the Delaunay graph DG with n vertices has at most $3n - 6$ edges [2]. In contrast to the regular arrangement, a cell region of this arrangement can include some edges while excluding other edges of the cell. To construct the redundant cell complex of RA, first we need to construct the Delaunay graph DG and then we need to construct the arrangement RA based on the edges of DG using appropriate algorithms from literature [11,8]. The Delaunay graph DG of n points in the plane can be computed in $\mathcal{O}(n \log n)$ time and the arrangement RA of $6n - 12$ lines can be computed in $\mathcal{O}(n^2)$ time.

Several data structures including the quad-edge and twin-edge with some modifications can be used to represent the arrangement RA [14]. The boundary lines of the arrangement RA require storing the information of the corresponding Delaunay edge and the incident vertex through which it passes. For every half edge, the redundant cells need to store the corresponding boundary line (pointer) including the half plane that contains it, so that we can compute whether a half edge is included in the cell region or not.

Maintaining the time bound, we can compute the set of adjacent vertices in the Gabriel graph for a candidate point in one redundant cell. Then the entire cell complex is traversed from that redundant cell of known adjacent vertex set, through neighboring cells in sequence. Trivially, it can be shown that at most one vertex in Gabriel graph can be added or deleted from the adjacent vertex set between two neighboring redundant cells and we can find that candidate vertex using the information stored in the half edges of the neighboring cells. However, to confirm a candidate vertex is added to the adjacent vertex set in the traversal, the algorithm can perform required additional test in constant time since the total degree of all the vertices in the Delaunay graph is linear. Hence, the total running time of the algorithm to construct the arrangement of redundant cells with their adjacent vertex sets, is $\mathcal{O}(n^2)$.[1]

The point location (point query) algorithms for conventional arrangement of lines [5,15] can be used for the arrangement of redundant cells with minor

---

[1] All our studies related to improve the running time of the algorithm and dynamic maintenance of the arrangement are not included due to the page limitation.

modifications. More precisely, the algorithms need to find the correct redundant cell containing the query point when it is on the boundary lines (i.e., on the cell edges) of the regions based on the Delaunay edges. Since the every boundary line and cell edge can retrieve the information of the corresponding Delaunay edge and the incident vertex through which it passes, thus it is possible to identify the desired cell by performing a constant time test on the retrieved Delaunay edge and vertex.

5   Conclusion

The subdivision method actually assists to find a desired location to insert a new point rather than provides an efficient technique to insert the point and update the Gabriel graph since it does not yield the deletion edges. In the Gabriel graph, the consecutive redundant cells with the same adjacent vertex set can be merged to construct a larger cell region without the redundancy. This reduces a large number of cells in the arrangement but produces both convex and non-convex cells. Currently, we are investigating different properties and dynamic maintenance of these convex and non-convex cells.

References

1. Baumgart, M.: Partitioning bispanning graphs into spanning trees. Mathematics in Computer Science **3**(1), 3–15 (2010)
2. Berg, M.D., Cheong, O., Kreveld, M.V., Overmars, M.: Computational Geometry: Algorithms and Applications, 3rd edn. Springer-Verlag TELOS, Santa (2008)
3. Bose, P., Cardinal, J., Collette, S., Demaine, E.D., Palop, B., Taslakian, P., Zeh, N.: Relaxed gabriel graphs. In: Proceedings of the 21st Annual Canadian Conference on Computational Geometry, CCCG, pp. 169–172, British Columbia, Canada (2009)
4. Bose, P., Collette, S., Hurtado, F., Korman, M., Langerman, S., Sacristan, V., Saumell, M.: Some properties of higher order delaunay and gabriel graphs. In: Proceedings of the 22nd Annual Canadian Conference on Computational Geometry, CCCG, pp. 13–16, Manitoba, Canada (2010)
5. Devillers, O., Teillaud, M., Yvinec, M.: Dynamic location in an arrangement of line segments in the plane. Algorithms Review **2**(3), 89–103 (1992)
6. Di Concilio, A.: Point-free geometries: Proximities and quasi-metrics. Mathematics in Computer Science **7**(1), 31–42 (2013)
7. Gabriel, K.R., Sokal, R.R.: A new statistical approach to geographic variation analysis. Systematic Zoology (Society of Systematic Biologists) **18**(3), 259–278 (1969)
8. Guibas, L., Stolfi, J.: Primitives for the manipulation of general subdivisions and the computation of voronoi. ACM Transactions on Graphics **4**(2), 74–123 (1985)
9. Hossain, M.Z., Wahid, M.A., Hasan, M., Amin, M.A.: Analysis of spatial subdivision on gabriel graph. http://www.cvcrbd.org/publications
10. Howe, S.E.: Estimating Regions and Clustering Spatial Data: Analysis and Implementation of Methods Using the Voronoi Diagram. Ph.D. thesis, Brown University, Providence, R.I. (1978)

11. Kao, T., Mount, D.M., Saalfeld, A.: Dynamic maintenance of delaunay triangulations. Tech. rep., College Park, MD, USA (1991)
12. Laloë, D., Moazami-Goudarzi, K., Lenstra, J.A., Marsan, P.A., Azor, P., Baumung, R., Bradley, D.G., Bruford, M.W., Cañón, J., Dolf, G., Dunner, S., Erhardt, G., Hewitt, G., Kantanen, J., Obexer-Ruff, G., Olsaker, I., Rodellar, C., Valentini, A., Wiener, P.: Spatial trends of genetic variation of domestic ruminants in europe. Diversity **2**, 932–945 (2010)
13. Matula, D.W., Sokal, R.R.: Properties of gabriel graphs relevant to geographic variation research and the clustering of points in the plane. Geographical Analysis **12**(3), 205–222 (1980)
14. O'Rourke, J.: Computational geometry in C, 2nd edn. Cambridge University Press, New York (1998)
15. Seidel, R.: A simple and fast incremental randomized algorithm for computing trapezoidal decompositions and for triangulating polygons. Computational Geometry: Theory and Applications **1**(1), 51–64 (1991)

Efficient Construction of UV-Diagram

M.Z. Hossain, Mahady Hasan, and M. Ashraful Amin[(⊠)]

Computer Vision and Cybernetics Group, Department of Computer Science and
Engineering, Independent University Bangladesh, Dhaka 1229, Bangladesh
mzhossain@gmx.com, mahadyh@yahoo.com, aminmdashraful@iub.edu.bd

Abstract. Construction of the Voronoi diagram for exploring various
proximity relations of a given dataset is one of the fundamental problems
in numerous application domains. Recent developments in this area allow
constructing Voronoi diagram based on the dataset of uncertain objects
which is known as Uncertain-Voronoi diagram (UV-diagram). In compare
to the conventional Voronoi diagram of point set, the most efficient algorithm known to date for the UV-diagram construction requires extremely
long running time because of its sophisticated geometric structure. This
text introduces several efficient algorithms and techniques to construct
the UV-diagram and compares the advantages and disadvantages with
previously known algorithms and techniques in literature.

Keywords: Uncertain-voronoi diagram · UV-diagram · Uncertain
object · Voronoi diagram · Delaunay triangulation · Computational
geometry

1 Introduction

A geometric proximity graph is a geometric graph where nodes are represented
using geometric objects and two nodes are adjacent in the graph if and only
if the nodes are satisfy some particular proximity properties. The concept of
the Voronoi diagram, the dual of the Delaunay triangulation, is based on the
proximity graph and frequently formulated in Euclidean metric space [9, 16]. The
Voronoi diagram is one of the extensively studied proximity structures because
of its numerous applications and theoretical interests [3]. The research on the
Voronoi diagram is extended recently to comprise the sites with imprecision and
uncertainty [18].

There are many different concrete data sources where data can be imprecise
and uncertain in nature [12]. For instance, environmental noises produce uncertainty in data obtained from sensing devices including the most often used image
sensing devices. Moreover, many areas of knowledge such as pattern recognition
and natural language processing have to deal with uncertain data for training the
statistical model and attaining high performance [14]. Additionally, limitation
of measuring equipment raises imprecision in data and for security reasons sometimes uncertain data are intentionally generated to ensure maximum protection.

© Springer International Publishing Switzerland 2015
Y. Tan et al. (Eds.): ICSI-CCI 2015, Part III, LNCS 9142, pp. 329–340, 2015.
DOI: 10.1007/978-3-319-20469-7_35

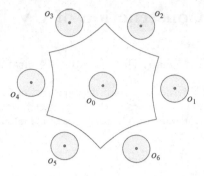

Fig. 1. A partial UV-diagram representing the UV-cell of o_0 with respect to $o_1, o_2, o_3, o_4, o_5, o_6$ uncertain objects

If a set of uncertain records denoted as uncertain objects is represented in Euclidean space by plotting associated attributes of the records to the axes of the space, the Uncertain-Voronoi diagram (UV-diagram) partitions the space based on the set of uncertain objects. This space partitioning technique subdivides the metric space into smaller region called Uncertain-Voronoi cell (UV-cell) as illustrated in Figure 1, to reduce the query time [7,22]. Consequently, answering a point query requires to determine the UV-cell where the point is located. Although a point query can be answered without constructing the exact UV-cell as describer in [6] but various region related queries requires to construct the exact UV-cell. Moreover, retrieving different regional and morphological information including area, shape, eccentricity, centroid, diameter etc., also needs the exact UV-cell construction. Additionally, if the number of queries on the UV-diagram is large, the exact UV-cell construction saves the accumulated running time for the queries.

The conventional Voronoi diagram is one of the eminent techniques for cluster analysis of point datasets used in many fields, including machine learning, pattern recognition, image analysis, and bioinformatics, however it fails to address the datasets with uncertain points [21]. Traditionally this limitation is handled by selecting a statistical method depending on application requirements to transform the dataset with uncertain points into a dataset of crisp points before the execution of clustering algorithms such as the Voronoi diagram [4]. The UV-diagram provides new perspectives and procedures to solve the clustering problems of the datasets with uncertain points without reducing the existing level of accuracy for the construction of the diagram.

2   Preliminaries

An uncertain object o_i is a d-dimensional region in Euclidean space $\mathbb{R}^d$ associated with a probability distribution function (pdf) $f_i(u)$ describes the distribution of position of the object inside the region such that,

$$\int_{o_i} f_i(u)du = 1, \quad u \in o_i$$

$$f_i(u) = 0, \quad u \notin o_i$$

To simplify our discussion, the uncertain object $o_i(c_i, r_i)$ is considered as a circular region with center c_i and radius r_i in Euclidean plane $\mathbb{R}^2$. Let O be a finite set of uncertain objects and D be the polygonal region that contains O in Euclidean plane $\mathbb{R}^2$. The region $D \subset \mathbb{R}^2$, named as domain, limits all the computation in finite range.

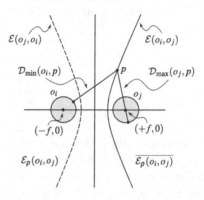

Fig. 2. The hyperbolic edge $\mathcal{E}(o_i, o_j)$

The minimum and the maximum distances of object $o_i(c_i, r_i)$ from p are denoted as $\mathcal{D}_{\min}(o_i, p)$ and $\mathcal{D}_{\max}(o_i, p)$ respectively. If $\mathcal{D}(c_i, p)$ denotes the distance between c_i and p then,

$$\mathcal{D}_{\min}(o_i, p) = \begin{cases} 0, & p \in o_i \\ \mathcal{D}(c_i, p) - r_i, & p \notin o_i \end{cases}$$

$$\mathcal{D}_{\max}(o_i, p) = \mathcal{D}(c_i, p) + r_i$$

The UV-edge $\mathcal{E}(o_i, o_j)$ of $o_i(c_i, r_i)$ with respect to $o_j(c_j, r_j)$ is a line such that the minimum distance $\mathcal{D}_{\min}(o_i, p)$ of o_i and the maximum distance $\mathcal{D}_{\max}(o_j, p)$ of o_j from every point p on the line is equal as shown in Figure 2. Symbolically,

$$\mathcal{D}_{\min}(o_i, p) = \mathcal{D}_{\max}(o_j, p)$$
$$\mathcal{D}(c_i, p) - \mathcal{D}(c_j, p) = r_i + r_j$$

No such line representing $\mathcal{E}(o_i, o_j)$ exists if $o_i \cap o_j \neq \emptyset$. If the coordinates of c_i and c_j are $(-f, 0)$ and $(+f, 0)$ respectively, then from the above equation, any point p of (x, y) on the edge $\mathcal{E}(o_i, o_j)$ satisfy the following equation,

$$\frac{x^2}{a^2} - \frac{y^2}{b^2} = 1$$

where $a = (r_i + r_j)/2$ and $b = \sqrt{f^2 - a^2}$. The equation represents a hyperbola with center at the origin of the Euclidean coordinate system. If the hyperbola is centered at (x_0, y_0) and rotated anticlockwise by an angle θ, then the above equation becomes,

$$\frac{x_\theta^2}{a^2} - \frac{y_\theta^2}{b^2} = 1$$

$$x_\theta = (x - x_0)\cos\theta + (y - y_0)\sin\theta$$

$$y_\theta = (x_0 - x)\sin\theta + (y - y_0)\cos\theta$$

Obviously, after the affine transformation, the point p of (x, y) is located at (x_θ, y_θ). Let the coordinates of c_i and c_j be (x_i, y_i) and (x_j, y_j) after the transformation. Then,

$$x_0 = \frac{x_i + x_j}{2}, \quad y_0 = \frac{y_i + y_j}{2}$$

$$\sin\theta = \frac{y_j - y_i}{d}, \quad \cos\theta = \frac{x_j - x_i}{d}$$

$$f = d/2, \quad d = \mathcal{D}(c_i, c_j)$$

The solid line branch of the hyperbola in Figure 2 is the UV-edge $\mathcal{E}(o_i, o_j)$ for the non-overlapping objects o_i and o_j. The interior region $\mathcal{E}_\rho(o_i, o_j)$ of UV-edge $\mathcal{E}(o_i, o_j)$ is the closed half-plane containing c_i and the complement open half-plane is the exterior region $\overline{\mathcal{E}_\rho(o_i, o_j)}$ of UV-edge $\mathcal{E}(o_i, o_j)$.[1] The hyperbola vertex $\mathcal{V}_\lambda(\mathcal{E}(o_i, o_j))$ of UV-edge $\mathcal{E}(o_i, o_j)$ is the intersection point of the major axis and the hyperbolic branch representing $\mathcal{E}(o_i, o_j)$. If (x_λ, y_λ) is the coordinate of hyperbola vertex $\mathcal{V}_\lambda(\mathcal{E}(o_i, o_j))$ then,

$$x_\lambda = \frac{1}{2d}\left(2a(x_j - x_i) + d(x_j + x_i)\right)$$

$$y_\lambda = \frac{1}{2d}\left(2a(y_j - y_i) + d(y_j + y_i)\right)$$

The UV-diagram $\mathcal{UV}(O)$ of a set $O = \{o_1, o_2, \ldots, o_n\}$ of uncertain objects in $\mathbb{R}^2$ is a subdivision of plane $\mathbb{R}^2$ into n UV-cells such that a point p lies in the UV-cell $\mathcal{U}(o_i)$ corresponding to an uncertain object o_i, denoted as the site of the UV-cell, iff $\mathcal{D}_{\min}(o_i, p) \leq \mathcal{D}_{\max}(o_j, p)$ for all $o_j \in O \setminus \{o_i\}$. Formally,

$$\mathcal{UV}(O) = \{\mathcal{U}(o_i) : o_i \in O\}$$

$$\mathcal{U}(o_i) = \{p \in \mathbb{R}^2 : \forall o_j \in O \setminus \{o_i\}$$

$$(\mathcal{D}_{\min}(o_i, p) \leq \mathcal{D}_{\max}(o_j, p))\}$$

A UV-cell $\mathcal{U}(o_i)$ can also be considered as the common intersections of the interior regions $\mathcal{E}_\rho(o_i, o_j)$ for all $o_j \in O \setminus \{o_i\}$ not overlapped with o_i. More precisely,

$$\mathcal{U}(o_i) = \bigcap\{\mathcal{E}_\rho(o_i, o_j) : o_j \in O \setminus \{o_i\}\}$$

---

[1] $\mathcal{E}_\rho(o_i, o_j)$ and $\overline{\mathcal{E}_\rho(o_i, o_j)}$ are also denoted as $\mathcal{E}_\rho(e)$ and $\overline{\mathcal{E}_\rho(e)}$ respectively letting e be $\mathcal{E}(o_i, o_j)$.

if no such o_j exists, then $\mathbb{R}^2$ is the $\mathcal{U}(o_i)$. The above alternative definition of UV-cell offers a new perspective to develop finite running time algorithms and thus our algorithms are based on this alternative definition. An approximate UV-cell $\mathcal{U}_\alpha(o_i)$ of o_i is the region which contains the UV-cell $\mathcal{U}(o_i)$. More precisely, $\mathcal{U}(o_i) \subseteq \mathcal{U}_\alpha(o_i)$. Therefore, the domain D is an approximate UV-cell for all uncertain object sites in our computational context.

3   Algorithms

The entire process of UV-cell construction can be divided into four steps, the seed selection, the approximate UV-cell construction, the execution of pruning methods, and the exact UV-cell construction. Some of the steps have alternative approaches to produce the similar results and thus any combination of these alternative approaches can be selected to achieve the resultant UV-cell based on the dataset configurations and implementation requirements. The first three steps remain same to build indexing of the UV-diagram based on candidate edges for the uncertain sites without constructing the exact UV-cells. However, there are different circumstances where the exact UV-cell construction is most appropriate as described in Section 1. In this text, we consider all domain edges are straight line and the number of edges $|\mathcal{E}(D)|$ enclosing the domain D is constant.

Lemma 1. [11] *The upper bound on the complexity of a UV-cell is quadratic.*

An exponential upper bound on the complexity of a UV-cell is proved in [7,22]. However, Lemma 1 suggests the existence of polynomial running time algorithms of the UV-cell construction and the UV-diagram construction.

3.1   Basic Approaches

The vertices of a UV-cell is the subset of the intersection points of the UV-edges. The intersection points which are located in the interior region of all the UV-edges are the actual vertices of the UV-cell. Since we only consider the vertices inside the domain D, some of the UV-cells are limited by the region D and formed vertices with the domain edges $\mathcal{E}(D)$.

BASIC-CONSTRUCT(E)
1 **begin**
2 $\quad$ $V \leftarrow \emptyset$
3 $\quad$ $E \leftarrow E \cup \mathcal{E}(D)$
4 $\quad$ **for all** pair $e, e' \in E$ **do**
5 $\quad\quad$ $V \leftarrow V \cup$ EDGE-VERTICES(E, e, e')
6 $\quad$ **return** V

The procedure BASIC-CONSTRUCT(E) computes vertices $\mathcal{V}(\mathcal{U}(o_i))$ of the UV-cell $\mathcal{U}(o_i)$ in $\mathcal{O}(|E|^3)$ running time. Since each vertex object $v \in \mathcal{V}(\mathcal{U}(o_i))$ contains the corresponding edges, a linear time post processing can construct the edge set $\mathcal{E}(\mathcal{U}(o_i))$ of the UV-cell from $\mathcal{V}(\mathcal{U}(o_i))$. To construct the edge or vertex chain, we need an additional quasilinear time sorting procedure. The operation EDGE-VERTICES(E, e, e') calculates the UV-cell vertices formed by e and e' edges in $\mathcal{O}(|E|)$ time [11]. The input, candidate edge set E, has to contain all the UV-cell edges of $\mathcal{U}(o_i)$ (i.e., $\mathcal{E}(\mathcal{U}(o_i)) \subseteq E$) to produce the correct output UV-cell.

3.2   Seed Selection

The average number of edges of a UV-cell is small but there is no simple method to find them ahead of the UV-cell construction. Thus we construct an initial approximate UV-cell $\mathcal{U}_\alpha(o_i)$ of $o_i(c_i, r_i)$ using a small number of selected uncertain objects, named as seeds, from the set $O \setminus \{o_i\}$ to obtain an efficient method of exact UV-cell $\mathcal{U}(o_i)$ construction. There are no strict rules for seed selection to construct the approximate UV-cell since the correctness of the algorithm does not depend on the selected seeds. However, an arbitrary set of uncertain objects can cause a large (area) approximate UV-cell that significantly reduce the efficiency of the UV-cell construction process. To produce an optimal approximate UV-cell $\mathcal{U}_\alpha(o_i)$, a k-Nearest-Neighbor query (k-NN) on the R-tree with c_i as the query point is used to retrieve the k nearest objects [7]. The R-tree is built by the set O of all uncertain objects. Then k_s objects are selected out of k objects by dividing the domain D into k_s sectors centered at c_i. For each division, the object closest to c_i is assigned as a seed.

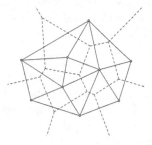

Fig. 3. The relation between the Delaunay triangulation (solid line) and its dual the Voronoi diagram (dashed line)

The Delaunay triangulation provides an alternative approach for seed selection. We construct the Delaunay triangulation using the point set $\{c_i : o_i(c_i, r_i) \in O\}$. Then, we select the uncertain objects associated with adjacent nodes of the point c_i, as seeds for producing approximate UV-cell $\mathcal{U}_\alpha(o_i)$. Since the Delaunay triangulation is the dual structure of the Voronoi diagram in $\mathbb{R}^2$, the adjacent nodes of a site in Delaunay graph are responsible for the edges of the conventional Voronoi cell of that site as shown in Figure 3. Similarly, the associated uncertain objects of the adjacent nodes of an uncertain site o_i in the Delaunay graph are the appropriate seeds for the UV-edges of the approximate UV-cell $\mathcal{U}_\alpha(o_i)$.

Moreover, $|\mathcal{E}(\mathcal{U}(o_i))|$ directly depends on the local configuration of the uncertain objects surrounding o_i but the former method always selects a constant number of seeds k_s ignoring the local configuration of o_i. Conversely, this method utilizes the information of local configuration in seed selection process. Furthermore, for each uncertain object retrieving k nearest neighbor requires $\mathcal{O}(k|O|)$ time where retrieving adjacent objects requires only $\mathcal{O}(1)$ time.

3.3   UV-Cell Approximation

The approximate UV-cell $\mathcal{U}_\alpha(o_i)$ of an uncertain site o_i can be constructed using the BASIC-CONSTRUCT procedure by the seeds obtained using appropriate seed selection method. Since BASIC-CONSTRUCT has an inefficient running time $\mathcal{O}(|E|^3)$, we develop an efficient alternative for the approximate UV-cell construction in this section.

A base edge $\mathcal{E}_\beta(o_i, o_j)$ is the UV-edge $\mathcal{E}(o_i, o_j)$ of $o_i(c_i, r_i)$ with respect to $o_j(c_j, r_j)$ such that $\mathcal{D}(c_i, c_j) + r_j$ is minimal for all $o_j \in O \setminus \{o_i\}$. There may exist more than one base edge of o_i since $\mathcal{D}(c_i, c_j) + r_j$ can be minimal for multiple $o_j(c_j, r_j)$.

Lemma 2. [11] *A base edge $\mathcal{E}_\beta(o_i, o_j)$ is an edge of the UV-cell $\mathcal{U}(o_i)$ of uncertain site o_i.*

The algorithm APPROXIMATE-CELL(E) constructs the approximate UV-cell $\mathcal{U}_\alpha(o_i)$ from the input of candidate edge set E of uncertain site o_i. The edges of E is needed to be sorted in anticlockwise order according to their coordinates of the o_j uncertain objects but no explicit sorting is required since both of the seed selection methods maintain an implicit sorting order. The procedure FIND-BASE (E) can be formulated based on Lemma 2 to finds a base edge $\mathcal{E}_\beta(o_i, o_j)$ from the input edge set E of o_i. The UV-cell edge next to v intersected with e in anticlockwise direction is searched in the procedure FIND-NEXT(E, e, v) which also updates the reference parameters e and v on the occurrence of the searching edge [11]. The average running time of FIND-NEXT would be $\mathcal{O}(|E|)$ although the worst case running time remains $\mathcal{O}(|E|^2)$. To improve further efficiency e of FIND-NEXT can be flagged and reordered edges E based on the flag.

In contrast to the BASIC-CONSTRUCT, this algorithm needs not to consider all $\mathcal{O}(|E|^2)$ pairs of edges to compute the vertices of the approximate UV-cell $\mathcal{U}_\alpha(o_i)$. The average-case behavior would require $\mathcal{O}(|E|)$ pairs to consider for the approximate UV-cell construction. Thus the average running time of this algorithm would be $\mathcal{O}(|E|^2)$. Both BASIC-CONSTRUCT and APPROXIMATE-CELL calculate the domain vertices included in the UV-cell.

3.4   UV-Cell Construction

The final candidate edge set E of o_i is prepared to ensure $\mathcal{E}(\mathcal{U}(o_i)) \subseteq E$ and $|E| \ll |O|$ using different pruning methods based on $\mathcal{U}_\alpha(o_i)$ constructed by a suitable algorithm described in the previous sections. An efficient algorithm is

Approximate-Cell(E)

```
1  begin
2  │  V ← ∅
3  │  e ← b ← Find-Base(E)
4  │  v ← V_λ(b)
5  │  E ← E ∪ E(D)
6  │  do
7  │  │  │  Find-Next(E, e, v)
8  │  │  │  V ← V ∪ {v}
9  │  while e ≠ b
10 │  return V
```

developed in this section which can construct the UV-cell $\mathcal{U}(o_i)$ from the approximate UV-cell $\mathcal{U}_\alpha(o_i)$ and the final candidate edge set E. The circle centered at c with radius r is denoted by $\mathcal{C}(c, r)$. A UV-edge can have multiple edge segments constituting a UV-cell.

Lemma 3. [11] *If $o_i(c_i, r_i)$, $o_j(c_j, r_j)$, $o_k(c_k, r_k)$ are three uncertain objects and the UV-edge $\mathcal{E}(o_i, o_k)$ intersects or replaces the edge segment $v_1 v_2$ associated with the UV-edge $\mathcal{E}(o_i, o_j)$ of $\mathcal{U}_\alpha(o_i)$, then $c_k \in \mathcal{C}(v_1, \mathcal{D}(c_i, v_1)) \cup \mathcal{C}(v_2, \mathcal{D}(c_i, v_2))$.*

Corollary 1. [11] *If $\mathcal{E}(o_i, o_k)$ is a UV-edge of $o_i(c_i, r_i)$ with respect to $o_k(c_k, r_k)$ and $v_1 \in \overline{\mathcal{E}_\rho(o_i, o_k)}$ for any vertex $v_1 \in \mathcal{V}(\mathcal{U}_\alpha(o_i))$, then $c_k \in \mathcal{C}(v_1, \mathcal{D}(c_i, v_1))$.*

Lemma 4. [7] *Given an uncertain object $O_i(c_i, r_i)$ and P_i's convex hull $CH(P_i)$, let $v_1, v_2, \ldots, v_n$ be $CH(P_i)$'s vertex. If another object O_j's center c_j is not in any of $\{Cir(v_m, dist(v_m, c_i))\}_{m=1}^n$, then $P_i = P_i - X_i(j)$.*

There is a major advantage of Lemma 3 over Lemma 4 that is, if $c_k \in \mathcal{C}(v, \mathcal{D}(c_i, v))$ for some vertex v of $\mathcal{U}_\alpha(o_i)$, then according to Lemma 4 the algorithm checks each edge of $\mathcal{U}_\alpha(o_i)$ for intersections which consumes a large fraction of the total time depending on the complexity of $\mathcal{U}_\alpha(o_i)$. On the contrary, it is sufficient to check the two incident edges of v for intersections according to Lemma 3.

Lemma 5. [11] *If a UV-edge $\mathcal{E}(o_i, o_k)$ intersects the approximate UV-cell $\mathcal{U}_\alpha(o_i)$, then either hyperbola vertex $\mathcal{V}_\lambda(\mathcal{E}(o_i, o_k)) \in \mathcal{U}_\alpha(o_i)$ or $v \in \overline{\mathcal{E}_\rho(o_i, o_k)}$ for some $v \in \mathcal{V}(\mathcal{U}_\alpha(o_i))$.*

Instead of directly calculating the intersections with the incident edges of a vertex v of $\mathcal{U}_\alpha(o_i)$, we can first check that the edge $\mathcal{E}(o_i, o_k)$ for effect on $\mathcal{U}_\alpha(o_i)$ using Lemma 5, which reduces a number of intersection calculation cost. Moreover, from Corollary 1 we know that if $v \in \overline{\mathcal{E}_\rho(o_i, o_k)}$, then $c_k \in \mathcal{C}(v, \mathcal{D}(c_i, v))$ which lead to develop the following efficient algorithm.

The fields site and reference of an UV-edge $\mathcal{E}(o_i, o_j)$ refer to o_i and o_j respectively. The Get-Edges(V) procedure lists all the UV-edges of the approximate UV-cell $\mathcal{U}_\alpha(o_i)$ constituted by V. The field edges of the vertex object v contains

CONSTRUCT-CELL(V, E)

```
 1  begin
 2  |   E' ← GET-EDGES(V)
 3  |   for each e ∈ E do
 4  |   |   {V', T} ← {∅, ∅}
 5  |   |   for each v ∈ V do
 6  |   |   |   if D(v, reference[e]) ≤ D(site[e], v) then
 7  |   |   |   |   if e ∈ edges[v] then continue
 8  |   |   |   |   V' ← V' ∪ {v}
 9  |   |   for each v' ∈ V' do
10  |   |   |   RESHAPE-CELL(V, E', v', e, T)
11  |   |   if count[e] > 0 then E' ← E' ∪ {e}
12  |   return V
```

the edges responsible for forming the vertex. The vertex number field count tracks the number of valid vertex for e. The approximate UV-cell of V is reshaped (calculated) in the procedure RESHAPE-CELL(V, E', v', e, T) considering the edge e. The set T stores the edges previously checked for UV-cell vertices to avoid producing duplicate vertices [11]. The average time complexity of this algorithm would be $\mathcal{O}(|E|)$ since the number of vertices of a UV-cell is a small constant number with respect to $|E|$ in practice. Alternatively, we can construct the UV-cell using BASIC-CONSTRUCT with candidate edges E as input in $\mathcal{O}(|E|^3)$ running time.

4   Experimental Results

In this section experimental results are reported on different datasets. The domain for each of the dataset used in these experiments is a square region having sides of length 30×10^3 centered at the origin. Additionally, each uncertain object in the datasets has a circular region of uncertainty with a diameter of 20 units and a Gaussian uncertainty pdf is associated with the region. A uniform random dataset of 60×10^3 objects is considered for the next two experiments which is a frequent real world scenario. We organize the same experimental setup discussed in [7] for R-tree based seed selection method that is the value of k is set to 300 for performing the k-NN search and the domain is divided into eight $45°$ sectors to obtain the seeds.

Guttman [10] illustrated the basic algorithms used in our implementations for R-tree construction and management. There are several variants of the R-tree, optimized for producing more well-clustered groups of entries in nodes for non-uniform dataset and reducing node coverage after deletion of entries [5]. However, our uniform random dataset requires only batch insertion of the complete dataset at the beginning. To ensure efficient k-NN query in R-tree, we follow the analyses and experiments in [13]. For computing the Delaunay triangulation, a divide and conquer algorithm [9] is used in these experiments although the algorithm by Dwyer [8] is best known for its high performance [20].

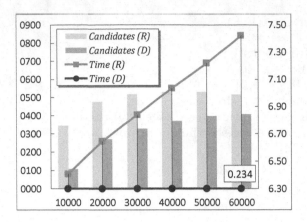

Fig. 4. The seed selection times (y-axis left) for the approximate UV-cells and the average numbers of candidate UV-edges (y-axis right) obtained after I-pruning and C-pruning are plotted against the number of uncertain objects $|O|$ (x-axis) from the synthetic dataset

There are two important criteria to measure the performance of a seed selection method. First criterion is the running time of the method where the Delaunay based seed selection is over a magnitude faster than the seed selection based on k-NN query in R-tree. Second criterion is the number of candidate edges after pruning since the exact UV-cell construction time directly depends on this number. Thus a smaller number of the candidate edges ensure less construction time of UV-cell for all known algorithms. The construction times of Delaunay graph and R-tree are shown in Figure 5 using right y-axis. It is clear from the two graphs that the construction times of both Delaunay graph and R-tree are much lower than the seed selection times.[2]

The former method as discussed in [7] uses k-NN query on R-tree for each uncertain object to retrieve the seeds and then BASIC-CONSTRUCT is applied to build the approximate UV-cell which is then used to execute the I-pruning and C-pruning to obtain the candidate edges of the UV-cell. Lastly, the exact UV-cell is computed by BASIC-CONSTRUCT using the candidate edges as parameter. On the other hand, the efficient method uses the Delaunay adjacent objects to acquire the seeds for each uncertain object and from these seeds, APPROXIMATE-CELL computes the approximate UV-cell which is at first used to execute the I-pruning to get the candidate edges. Finally, the exact UV-cell is constructed by CONSTRUCT-CELL with the approximate UV-cell and the candidate edges as parameters. The total construction times in Figure 5 also include the Delaunay triangulation time and R-tree construction time.

Answering different types of quires requires efficient detection of desired UV-cell. To find the desired UV-cell using linear search throughout all the UV-cells

---

[2] The suffixes (R) and (D) denote the results related to R-tree and Delaunay triangulation respectively.

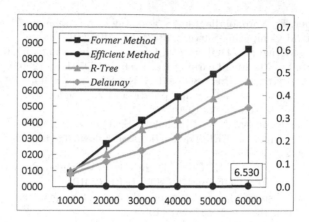

Fig. 5. The total construction times of the UV-cells (y-axis left), the Delaunay triangulation time (y-axis right), and the R-tree construction time (y-axis right) are plotted against the number of uncertain objects $|O|$ (x-axis) from the synthetic dataset

is an expensive process. Moreover, geometric structures of the UV-cells are non-primitive. Therefore, construction of a hierarchical index structure enables these quires to be answered quickly [1]. R-tree is one of the hierarchical tree structures used to index different geospatial arrangement including the conventional Voronoi diagram [19]. Quad-tree is another structure used in [7,22] to index the UV-diagram. In fact, if the index tree is distributed between main memory and disk storage, quad-tree shows high performance because of the reduction in expensive disk access.

5   Conclusion

The uniform random dataset has similar local configuration around all the uncertain sites but for other nonuniform datasets, frequently found in real word analysis such as the dataset followed normal distribution or arranged in clusters, the proposed UV-cell construction method would show higher performance since the Delaunay triangulation based seed selection method utilizes the local configuration around each uncertain site. The concept of the approximate UV-cell construction algorithm APPROXIMATE-CELL can be modified to construct the exact UV-cell efficiently. However, the modified algorithm would not be as efficient as the exact UV-cell construction algorithm introduced in this text.

References

1. Agarwal, P.K., Cheng, S.W., Tao, Y., Yi, K.: Indexing uncertain data. In: ACM SIGMOD-SIGACT-SIGART, pp. 137–146. ACM, New York, NY, USA (2009)
2. Akopyan, A.V., Zaslavskiĭ, A.A.: Geometry of Conics. Mathematical World, American Mathematical Society (2007)

3. Aurenhammer, F.: Voronoi diagrams – a survey of a fundamental geometric data structure. ACM Computing Surveys **23**(3), 345–405 (1991)
4. Banerjee, A., Merugu, S., Dhillon, I.S., Ghosh, J.: Clustering with bregman divergences. Journal of Machine Learning Research **6**, 1705–1749 (2005)
5. Beckmann, N., Kriegel, H.P., Schneider, R., Seeger, B.: The R*-tree: an efficient and robust access method for points and rectangles. In: ACM SIGMOD, pp. 322–331. ACM, New York (1990)
6. Cheng, R., Kalashnikov, D.V., Prabhakar, S.: Querying imprecise data in moving object environments. IEEE Transactions on Knowledge and Data Engineering **16**(9), 1112–1127 (2004)
7. Cheng, R., Xie, X., Yiu, M.L., Chen, J., Sun, L.: UV-diagram: A voronoi diagram for uncertain data. In: ICDE, pp. 796–807. IEEE (2010)
8. Dwyer, R.A.: A faster divide-and-conquer algorithm for constructing delaunay triangulations. Algorithmica **2**, 137–151 (1987)
9. Guibas, L., Stolfi, J.: Primitives for the manipulation of general subdivisions and the computation of voronoi. ACM Transactions on Graphics **4**(2), 74–123 (1985)
10. Guttman, A.: R-trees: A dynamic index structure for spatial searching. In: SIGMOD, pp. 47–57. ACM, New Yor (1984)
11. Hossain, M.Z., Hasan, M., Amin, M.A.: Analysis of efficient construction for uncertain voronoi diagram. http://www.cvcrbd.org/publications
12. Kao, B., Lee, S.D., Lee, F.K.F., Cheung, D.W., Ho, W.S.: Clustering uncertain data using voronoi diagrams and R-tree index. IEEE Transactions on Knowledge and Data Engineering **22**(9), 1219–1233 (2010)
13. Kuan, J.K.P., Lewis, P.H.: Fast k nearest neighbour search for R-tree family. In: International Conference on Information, Communications and Signal Processing, vol. 2, pp. 924–928. Singapore (1997)
14. Li, R., Bhanu, B., Ravishankar, C., Kurth, M., Ni, J.: Uncertain spatial data handling: Modeling, indexing and query. Computers & Geosciences **33**(1), 42–61 (2007)
15. Nagy, T.A.: Documentation for the Machine-Readable Version of the Smithsonian Astrophysical Observatory Catalog (EBCDIC Version). Systems and Applied Sciences Corporation R-SAW-7/79-34 (1979)
16. O'Rourke, J.: Computational geometry in C, 2nd edn. Cambridge University Press, New York (1998)
17. Richter-Gebert, J.: Perspectives on Projective Geometry: A Guided Tour Through Real and Complex Geometry, 1st edn. Springer Publishing Company, New York (2011)
18. Sember, J., Evans, W.: Guaranteed voronoi diagrams of uncertain sites. In: Canadian Conference on Computational Geometry, Montreal, Canada (2008)
19. Sharifzadeh, M., Shahabi, C.: VoR-Tree: R-trees with voronoi diagrams for efficient processing of spatial nearest neighbor queries. VLDB Endowment **3**(1–2), 1231–1242 (2010)
20. Su, P., Drysdale, R.L.S.: A comparison of sequential delaunay triangulation algorithms. Computational Geometry **7**, 361–385 (1997)
21. Vinh, N.X., Epps, J., Bailey, J.: Information theoretic measures for clusterings comparison: Variants, properties, normalization and correction for chance. Journal of Machine Learning Research **11**, 2837–2854 (2010)
22. Xie, X., Cheng, R., Yiu, M., Sun, L., Chen, J.: UV-diagram: a voronoi diagram for uncertain spatial databases. The VLDB Journal, 1–26 (2012)

Portrait Image Segmentation
Based on Improved Grabcut Algorithm

Shuai Li[1,2,3], Xiaohui Zheng[1,2,4], Xianjun Chen[4,5(✉)], and Yongsong Zhan[1,2,3,4]

[1] Key Laboratory of Cloud Computing and Complex System,
Guilin University of Electronic Technology, Guilin 541004, Guangxi,
People's Republic of China
[2] Guangxi Experiment Center of Information Science,
Guilin University of Electronic Technology, Guilin 541004, Guangxi,
People's Republic of China
[3] Key Laboratory of Intelligent Processing of Computer Image and Graphics,
Guilin University of Electronic Technology, Guilin 541004, Guangxi,
People's Republic of China
[4] Guangxi Key Laboratory of Trusted Software, Guilin University of Electronic Technology,
Guilin 541004, Guangxi, People's Republic of China
hingini@126.com
[5] Information Engineering School, Haikou College of Economics,
Haikou 571127, Hainan, People's Republic of China

Abstract. Traditional computer portrait caricature system mainly take the method that exaggerate and deform real images directly, that lead the facial image background also been deformed when exaggerate facial image. If in pretreatment stage, we segmented the characters and background of the input image, and then do the subsequent processing, the problem may be solved. But for better portrait caricature effects, we need an excellent segmentation algorithm. So, we propose an improved Grabcut image segmentation algorithm and use it to extract the prospect character image for exaggeration and deformation. In practical application, we separate deform and exaggerate the foreground characters image with TPS method, then fuse it with the original or new background picture, get the final image. Application proves, the method solves the background deformation problem well, and improves the quality and rate of image segmentation, caricature synthesis effect reality and natural.

Keywords: Image segmentation · Grabcut · Portrait caricature exaggeration

1 Introduction

In recent years, digital media industry develops rapidly, portrait caricature type`s applications gradually enter user`s view and subject to favorite. But studying traditional computer portrait caricature system find that it mainly adopts direct exaggeration and deformation method on real images and generate personalized caricature exaggeration. But at the same time, it lead to the deformation of facial image background, reduce the realistic effect of caricature scenes. If we separated character and background from the

© Springer International Publishing Switzerland 2015
Y. Tan et al. (Eds.): ICSI-CCI 2015, Part III, LNCS 9142, pp. 341–348, 2015.
DOI: 10.1007/978-3-319-20469-7_36

input image in the preprocessing stage, and then do subsequent exaggeration processing, final caricature effects will be better. But the speed and qualities of segmentation could do a directly effect on ultimately generated time of caricatures and quality of the caricature effect. Therefore, put forward a fast, effective and suitable image segmentation algorithm for computer portrait caricature system is critical.

Image segmentation is a basic computer vision technique, has many algorithms. In 2001, Boykov [1] et al proposed the Graph Cut algorithm based on energy minimization, it is considered to be one of the most widely used image segmentation method in present research and application. In 2004, Rother [2] proposed Grabcut algorithm based on Graphcut algorithm. Its extraction effect is better, and application in graph segmentation field is more mature, attracts many researchers to study and improve it. Such as Chen [3] improved GrabCut`s segmentation performance by optimized GMM; Zhou [4] used multi-scale watershed denoise smoothly, and did watershed operation to the gradient image again, finally used entropy penalty factor optimized segmentation energy function; Han [5] used a new method to replace GMM model. The above three methods all improved segmentation`s accuracy, but time-consuming is long. In order to reduce the time consuming, literature [6-8] proposed used the watershed algorithm to preprocess image; Zhang [9] proposed adopting binary tree based on color quantization algorithm to instead k-means algorithm for clustering; Ding [10] provide reduce the resolution of the image and narrow image to accelerate convergence speed.

Studies shown that the purpose of kinds of segmentation improved method aims to two points: improve segmentation speed and accuracy. So we put forward an improved Grabcut image segmentation method, combined watershed algorithm with image compression algorithm which based on wavelet transform together, and applied it into the pretreatment process of Grabcut. In practical application, we also supply it into portrait caricature creation which bases on TPS [11] method.

2   Facial Image Background Segmentation

For facial image background segmentation, we propose an improved Grabcut algorithm based on original Grabcut algorithm. It combined super-pixel and wavelet transform image compression algorithm together to preprocess the original image, aims to ensure segmentation quality, reduce the estimated time of GMM parameters, improve image segmentation`s execution efficiency, obtain favorable segmentation results.

2.1   The Original Grabcut Algorithm

Grabcut algorithm used iterative energy minimization to segment image. Its segmentation accuracy is well and can extracted prospect target from complex background effectively and has good robustness, but time-consuming is longer.

The Color Data Model of Grabcut Algorithm. Grabcut algorithm adopts RGB color space and introduces Gaussian Mixture Model (GMM). GMM model used the full covariance with K numbers of Gaussian component to model target foreground and background respectively. For a better use of GMM model, it introduces a vector $k =$ tor $k = \{k_1, \ldots, k_n, \ldots k_N\}$, $k_n \in \{1, \ldots, K\}$ and uses to specify the corresponding and the only GMM elements for each pixel. So the Gibbs energy of image segmentation is determined by the variable k of GMM, the formula is shown below:

$$E(\underline{\alpha}, k, \underline{\theta}, z) = U(\alpha, k, \theta, z) + V(\alpha, z). \tag{1}$$

$$U(\alpha, k, \theta, z) = \sum_n D(\alpha_n, k_n, \underline{\theta}, z_n). \tag{2}$$

Of which, $\alpha_n \in [0,1]$ is opacity, 0 denotes background, 1 represents foreground. Parameter $\underline{\theta}$ signifies image's gray histogram. $z = (z_1, \ldots, z_n, \ldots, z_N)$ means image gray value array. U is region item. After sorted and taken its negative logarithm, mixture Gaussian density formula can express as formula (3).

$$D(a_n, k_n, \underline{\theta}, z_n) = -\log \pi(\alpha_n, k_n) + \frac{1}{2} \log \det \sum (\alpha_n, k_n) \tag{3}$$
$$+ \frac{1}{2}[z_n - \mu(\alpha_n, k_n)]^T \sum (\alpha_n, k_n)^{-1}[z_n - \mu(\alpha_n, k_n)].$$

$$\underline{\theta} = \left\{\pi(\alpha, k), \mu(\alpha, k), \sum (\alpha, k), \alpha = 0,1, k = 1 \ldots K\right\}. \tag{4}$$

By the above reasoning, the model parameter could be determined as formula (4). Color image's smoothing item V could be expressed as formula (5).

$$V(\alpha, z)\gamma \sum_{(m,n)\in C} [\alpha_n \neq \alpha_m] \exp - \beta \|z_m - z_n\|^2. \tag{5}$$

Energy Minimization Segmentation Algorithm Iteratively. Grabcut is an iterative minimization, and its each iteration is an optimization process for GMM parameter. Its algorithm process could be described as below:

(1) User gets the initialized trimap T by interaction, pixels outside the box are background pixel T_B, pixels in the box are "possible target" pixel T_U. Initialized pixels in T_B as $\alpha_n = 0$ and pixels in T_U as $\alpha_n = 1$, and initialized the GMM model of foreground and background by K-means algorithm.

(2) Distributed matching GMM Gaussian component for each pixel in T_U.

$$k_n := \arg \min D_n(\alpha_n, k_n, \underline{\theta}, z_n). \tag{6}$$

(3) Learn optimizing parameter model of GMM from the given image data Z.

$$\theta := \arg \ min_{\underline{\theta}} U(\underline{\alpha}, k, \underline{\theta}, z). \tag{7}$$

(4) Though the min cut algorithm to segmentation estimate the image.

$$\min_{\{a_n : n \in T_U\}} \min_k E(\alpha, k, \theta, z). \tag{8}$$

(5) Repeat steps 1 to 3, until convergence.

2.2   The Improved Grabcut Algorithm and Result

Grabcut algorithm is a NP-hard problem substantially, so if image's node numbers are overmuch, problem solving is more time-consuming. In order to reduce node number, we use secondary watershed algorithm to segment image into super-pixels and compressed image with wavelet transform. Its specific steps describe as follows: (a) Demarcate target rectangular area manually. (b) Do preliminary segmentation by secondary watershed algorithm, generate super-pixels. (c) Initialize ternary diagram T by the marked rectangular. Set foreground is empty, T_U as the complement of background. Used $\alpha_n = 0$ and $\alpha_n = 1$ to initialize foreground and background's GMM model by K-means algorithm. (d) Used wavelet transform algorithm to compress the above image, get compressed images. (e) To sign the foreground GMM in T_U, and label the background GMM in the whole block diagram after watershed segmentation. (f) Regard super-pixels as nodes and the connection between adjacent nodes as edge structure s-t network graph model, then used the minimize segmentation algorithm to solve formula (3)'s data item, get initial segmentation. (g) Updated GMM parameter θ. (h) Do the segmentation process (e)-(g) iteratively on the compressed image. (i) Got each pixel's value in original image from convergence results by formula (9). (j) Do Grabcut algorithm iteration minimize segmentation process again on the original image, and used Border matting method to smooth intersected edges.

$$Q_\alpha(2x, 2y) = Q_\alpha(2(x + 1), 2y) = Q_\alpha(2x, 2(y + 1)) \tag{9}$$
$$= Q_\alpha(2(x + 1), 2(y + 1)) = R_\alpha(x, y).$$

In order to test the performance of the improved Grabcut algorithm, we compare our method with the original Grabcut algorithm, result shows as Fig.1. The first column is the original images; The second column is original Grabcut algorithm's results; The third column is our results. From comparative results, we could see them segmentation results are consistent relatively.

Fig. 1. Effect comparison between our algorithm and Grabcut algorithm

Table 1 compares the two methods` segmentation time-consuming, and lists super-pixels number. I represent our method, G denotes Grabcut algorithm. The first three iterations are on preprocessing image, the fourth iteration is on the original image. We find that segmentation achieve convergence roughly at the second or third iteration.

Table 1. Efficiency comparison between our and original algorithm

Image	flower		cat		fish		roo	
Resolu-tion/super-pixel no.	1024*736 （6025个）		1450*1140 （12915个）		800*600 （3692个）		1920*1080 （15360个）	
Time/method	I/s	G/s	I/s	G/s	I/s	G/s	I/s	G/s
1	4.015	18.300	6.821	32.304	2.132	9.890	14.587	72.852
2	3.548	10.064	6.016	19.630	2.027	6.161	9.316	27.625
3	3.519	10.063	5.595	19.530	1.923	6.101	8.524	24.781
4	7.018	10.062	12.525	19.551	3.523	6.093	16.119	24.663

3 Application of the Improved Algorithm in Portrait Caricature Creation

The above parts put forward a foreground extraction method for character image. This part applied the segmentation result portrait caricature`s creation process, exaggerates and deforms the extracted character image independently, and then synthesis segmentation images, gets final results. TPS algorithm is easy operation and exaggeration effect well, so we choose it as our experimental subject.

3.1 Thin Plate Spline Algorithm

Face deformation technology based on TPS algorithm is proposed by Bookstein [11] in 1989. It is a deformation method that using finite point set in an irregular space, its essence is a kind of RBF algorithm. Formula （10） is the radial basis function of TPS algorithm, and TPS method`s mapping function is defined as formula (11).

$$U(r) = -r^2 log r^2. \tag{10}$$

$$f(x, y) = a_1 + a_x x + a_y y + \sum_{i-1}^{n} w_i U(\|P_i - D_i\|). \tag{11}$$

a_1, a_x, a_y represent the key parameter, w_i is weight coefficient, P_i denotes original image`s feature points. $D_i = (x, y)$ is the new coordinate of corresponding feature points after exaggeration. Formula $\sum_{i-1}^{n} ...$ represents a linear combination of k numbers basis functions. Other parts are an affine transformation between deformation pair point sets. In order to solve $f(x, y)$, three matrixes K, P, P^T need to be defined. K is a matrix with (n, n), P is(n, 3), P^T is the transposed matrix of P, shows as below.

$$K = \begin{bmatrix} 0 & U(r_{12}) & \cdots & U(r_{1n}) \\ U(r_{21}) & 0 & \cdots & U(r_{2n}) \\ \vdots & \vdots & \ddots & \vdots \\ U(r_{n1}) & U(r_{n2}) & \cdots & 0 \end{bmatrix}; P = \begin{bmatrix} 1 & x_1 & y_1 \\ 1 & x_2 & y_2 \\ 1 & x_n & y_n \end{bmatrix}. \tag{12}$$

Construct partitioned matrix L by matrix K and P. Matrix O is a null matrix with 3*3. Besides, define matrix V is（2，n）and used to represent the pixel coordinate of target image, show as formula (13):

$$L = \begin{bmatrix} K & P \\ P^T & O \end{bmatrix}, (n + 3) \times (n + 3); V = \begin{bmatrix} x_1' & x_2' & \cdots x_n' \\ y_1' & y_2' & \cdots y_n' \end{bmatrix}. \tag{13}$$

Define matrix $C = L^{-1}V$ and used it to denote RBF's coefficient matrix. At last, obtained $f(x, y)$ and satisfy that the following integral deflection norm is minimal.

$$I_f = \iint_{R^2} \left(\left(\frac{\partial^2 f}{\partial x^2}\right)^2 + 2\left(\frac{\partial^2 f}{\partial x \partial y}\right)^2 + \left(\frac{\partial^2 f}{\partial y^2}\right)^2 \right) dxdy. \tag{14}$$

3.2   Portrait Caricature Synthesis

Caricature synthesis part aims at composite background and caricature which coped by image segmentation and TPS exaggeration. We choose the polygonal approximation algorithm to extract the exaggerated image's contour and treated pixels in the contour as foreground pixel, remains it unchanged. Other pixels are replaced by the user chooses image. The synthesis result shows as Fig.2, they are beautiful and practical. The first column is original image; the second column is synthesis images with same background; the third column is with new background.

Fig. 2. Image exaggeration and synthesis effects

4   Experiment Results

Compared our method with Huang hua`s [12] method, its effect is shown as Fig.3: the first line is the original image; the second line is the effect of the method proposed by Huang et al; the third line is the effect of our method. The first image in the third line shows the result that exaggerated on original image directly; the second image represents the synthesis effect with original image after image segmentation and exaggeration, it is obviously seen that the image coped by our method keep the background`s invariance well; the last image represents the synthesis effect with new background after image segmentation and exaggeration, it is more natural. Experiments show that used the improved method to extract image characters and do post caricature exaggeration processing could ensure the original background`s invariance well, and the synthetic deformation effect is also good. As a whole, caricature looks natural and reality.

Fig. 3. Effect comparison with other achievement

5   Conclusion

This paper introduces an improved Grabcut algorithm, and applies it into portrait caricature generation which based on TPS algorithm. As a whole, image segmentation and exaggeration`s effect is all well. It really improves the background`s deformation problem in the exaggerated facial portrait, and enhances the flexibility and diversity of caricature creation. But it still has some deficiency and improvement, shows as follows:(1) however, the improved Grabcut algorithm improves segmentation efficiency, but for poor light and complex background images, its segmentation effect is still not ideal; (2) Deformation and exaggeration templates could be further created in exaggeration phase in order to enhance interaction. Therefore, the above problem should be study further for a better segmentation and caricature effect in the future work.

Acknowledgments. This research work is supported by the grant of Guangxi science and technology development project (No: 1355011-5), the grant of Guangxi Experiment Center of Information Science of Guilin University of Electronic Technology, the grant of Key Laboratory of Cloud Computing & Complex System of Guilin University of Electronic Technology, the grant of Key Laboratory of Intelligent Processing of Computer Image and Graphics of Guilin University of Electronic Technology (No:GIIP201403), the grant of Guangxi Key Laboratory of Trusted Software of Guilin University of Electronic Technology, the grant of Hainan Natural Science Foundation (No: 610233), and the Universities and colleges Science Research Foundation of Hainan (No: Hjkj2013-48).

References

1. Boykov, Y., Jolly, M.-P.: Interactive graph cuts for optimal boundary and region segmentation of objects in N-D images. In: Proc. IEEE Int. Conf. on Computer Vision, CD–ROM (2001)
2. Rother, C., Kolmogorov, V., Blake, A.: " Grabcut"- Interactive Foreground Extraction using Iterated Graph Cuts
3. Chen, D., Chen, B., Mamic, G., et al.: Improved GrabCut segmentation via GMM optimization. In: Proc. of the 2008 International Conference on Digital Image Computing: Techniques and Applications. IEEE Computer Society, Washington, DC, pp. 39–45 (2008)
4. Liang-fen, Z., Jian-nong, H.: Improved image segmentation algorithm based on GrabCut. Journal of Computer Applications 33(1), 49–52 (2013)
5. Han, S.D., Tao, W.B., Wang, D.S., et al.: Image segmentation based on GrabCut framework integrating multi-scale nonlinear structure tensor. IEEE Trans. on Image Processing 18(10), 2289–2302 (2009)
6. Qiu-ping, X.: Target extraction method research based on graphcut theory. Shanxi Normal University (2009)
7. Yue-lan, X.: Superpixel-based Grabcut Color Image Segmentation. Computer Technology and Development 23(7), 48–51 (2013)
8. Jun-ming, W., Li-xin, G., Li, Z.: The research of Grab-Cut color image segmentation algorithm. TV Technology 32(6), 15–17 (2008)
9. Yu-jin, Z.: The transition zone and image segmentation. Chinese Journal of Electronics 24(1), 12–16 (1996)
10. Hong, D., Xiao-feng, Z.: Object abstraction algorithm with fast Grabcut. Computer Engineering and Design 33(4), 1477–1481 (2012)
11. Bookstein, F.L.: Principal warps: Thin-plate splines and the de-composition of deformation. Pattern Analysis and Machine Intelligence 11(6), 567–585 (1989)
12. Huang, H., Ma, X.W.: Frontal and Semi-Frontal Facial Caricature Synthesis Using Non-Negative Matrix Factorization. Journal of Computer Science and Technology 25(6), 1282–1292 (2010)

A Multiple Moving Targets Edge Detection Algorithm Based on Sparse Matrix Block Operation

Kun Zhang[✉], Cuirong Wang, and Yanxia Lv

Northeastern University at Qinhuangdao, Qinhuangdao 066004, Hebei, China
zkhbqhd@163.com

Abstract. A multiple moving targets edge detection algorithm based on sparse matrix block operation is proposed in this paper. The algorithm uses background subtraction algorithm to obtain the foreground image contains multiple moving targets. After getting the ideal foreground image, active contour model is used for edge detection. Here, we improved the active contour model by introducing the sparse matrix block operation. Through the quad-tree decomposition of the foreground image, the proposed algorithm uses the sparse matrix block operation to calculate the corresponding regional seed position of multiple moving targets. Finally, it executes the active contour model in parallel to complete the edge detection. Experimental results show that edge detection of the algorithm similar to the human visual judgment, and the introduction of sparse matrix block operation to calculate regional seed for active contour model reduces the time, improves the convergence of the profile curve and edge detection accuracy.

Keywords: Quad-tree · Sparse matrix block operation · Active contour model · Edge detection

1 Introduction

In order to deal with the massive recorded video content, it requires an intelligent video surveillance system to analyze the critical target automatically. Image edge detection is always a chief, classical and knotty problem in image analysis and pattern recognition field. It also plays an important role in object recognition system.

The edge detection of multiple moving targets from a video stream is a basic and fundamental problem of moving targets segment in intelligent monitoring system. Main edge detection operators include: Roberts operator [1]; Sobel operator [2]; Canny operator [3]; LOG operator[4] etc.. Other edge detection methods include: edge detection based on wavelet transform [5], Mathematical Morphology [6], Active contour model [7] etc.. People for the shortcomings of traditional active contour models made many improvements to the model, there are more classic model as : the balloon force Snake model[8], the distance potential model[9] which overcomes balloon model initial contour line not intersect with the target boundary and disadvantage of weak boundary leakage, the GVF (Gradient Vector Flow) snake model[10] which gradient vector flow by gradient diffusion equation to extend even farther region, make away

© Springer International Publishing Switzerland 2015
Y. Tan et al. (Eds.): ICSI-CCI 2015, Part III, LNCS 9142, pp. 349–356, 2015.
DOI: 10.1007/978-3-319-20469-7_37

from the target boundary initial contour correctly converge to the true boundary. Paper [11] integrated the global information of the image into the balloon force Snake model and relaxed restrictions on the initial contour position, to some extent overcome the shortcomings of weak boundary leakage and so on.

This paper is organized as follows. In Section 2, we explain the related work of our algorithm, including quad-tree decomposition and active contour model. Section 3 describes the proposed algorithm in detail. Section 4 provides the experimental results and the analysis. Finally, there are conclusion and references.

2   Related Work

In order to complete multiple moving targets edge detection, the first job we have to do is foreground image extraction based on background modeling. In this paper we use our background modeling algorithm based on Chebyshev Inequality [12] for foreground image getting. Then we use the quad-tree decomposition method to get the corresponding sparse matrix of foreground image which include multiple moving targets.

2.1   Quad-tree Decomposition

Quad-tree decomposition is an image decomposition method based on the uniformity of detection. The basic idea is that it put an image equally divided into four regions, if one sub-regional conformance standards, then that sub-region no longer split down. Quad-tree decomposition process is iterative repetition, until all regions of the image conformance standards. In general, the consistency of the standard refers to that the closeness of each pixel gray value in the region does not exceed a preset threshold value. Take use of quad-tree decomposition for foreground image can detect different moving targets belongs to different image areas. By quad-tree decomposition, the corresponding sparse matrix can be obtained. Through sparse matrix block operation regional seed point coordinates corresponding to different moving targets can be obtained.

2.2   Active Contour Model

In order to solve the problem of multiple moving targets edge detection, this paper considers the active contour model to determine the outline of different moving targets. People usually called the active contour model proposed by Kass [9] as the Traditional Snake Model. The profile curve of Traditional Snake Model can be represented by parametric equations:

$$v(s) = [x(s), y(s)], \ s \in [0,1] . \tag{1}$$

In here: $v(s)$ is a two-dimensional point coordinate on C, s is the normalized arc length. Energy E of profile curve C is defined as:

$$E_{snake} = \int_0^1 [E_{\text{int}}(v(s)) + E_{i\,mage}(v(s)) + E_{ext}(v(s))]ds = E_{\text{int}} + E_{i\,mage} + E_{ext} \, . \quad (2)$$

In the formula (2), E_{int} is the internal energy of C, $E_{i\,mage}$ is the force generating energy of image itself, and E_{ext} is the energy generated by the external restraining force. Internal energy controls the smoothness and continuity of C. According to the principle of variations, take variational calculation for the function defined by formula (1), obtained as follows Euler-Lagrange equation:

$$-\alpha(s)\frac{\partial^2 v}{\partial s^2} + \beta(s)\frac{\partial^4 v}{\partial s^4} + \nabla E_{image}(v) = 0 \, . \quad (3)$$

Contour convergence results of the active contour models are affected by setting initial contour position.

In this paper, for multiple moving targets edge detection, the sparse matrix block operation is used, seeds of different moving target region corresponding to the active contour model is introduced, the initial contour active contour models line is set to be divided in different sports around the target, get the convergence of different contours of moving targets, and ultimately achieve the purpose of the foreground image edge detection of multiple moving targets.

3   Multiple Objects Edge Detection Algorithm

In this section, we describe the proposed multiple moving targets edge detection algorithm based on sparse matrix block operation algorithm in detail. The proposed algorithm is as follows:

(1) Use background modeling algorithm based on Chebyshev Inequality, obtain the foreground image which contains multiple moving targets from the video image. Foreground image may be marked as $I(x, y)$, W is the total width of the image, H is the total length of the image, the location of the pixel in the image is marked as (x, y), $x \in [1, W]$, $y \in [1, H]$,

(2) Extend image $I(x, y)$ to get a new image $I_1(x, y)$, width and length of the new image to be expanded as $W_1 = H_1 = 2^m$, meet the requirements $W_1 = H_1 \geq \max(W, H)$. Expanded image pixel gray values of the whole set to zero. The upper left corner of the new image $I_1(x, y)$ is the original image $I(x, y)$.

(3) Conduct quad-tree decomposition of image $I_1(x, y)$, and get the corresponding sparse matrix marked as $D(x, y)$, the size of the sparse matrix $D(x, y)$ is the same size as the image matrix that is decomposed. The non-zero elements of sparse matrix can describe the size of region compliance with decomposition consistent standards. For example, if $D(1,1) = 16$, it means that in the image $I_1(x, y)$, pixels $I(1,1)$, $I(1,16)$, and $I(16,1)$, $I(16,16)$ are the vertex of a rectangular region, the region compliance with quad-tree decomposition consistent standards, the gradation of each pixel in the region does not exceed a difference threshold value, the size of the region is 16×16.

(4) Starting from the upper left corner of the matrix $D(x, y)$, with $2^t \times 2^t$ as the standard for the size, the sparse matrix $D(x, y)$ is divided into several small regional blocks. Conduct sparse matrix block operation of the foreground image (shown in Fig. 1).

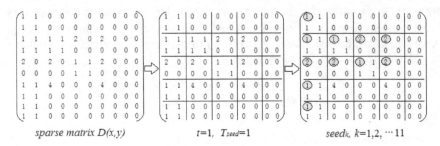

sparse matrix D(x,y) $t=1, \ T_{seed}=1$ $seed_k, \ k=1,2, \cdots 11$

Fig. 1. Sparse matrix block operation chart

In each $2^t \times 2^t$ small block area, except for boundary points, starting from the first matrix element that $D(x, y) = 1$ or $D(x, y) = 2$, looking up and down according to four directions, judge the following four value $D(x-1, y), D(x+1, y), D(x, y-1), D(x, y+1)$ whether it is one or two , If present, the matrix element is recorded as new starting point. Then discriminate from the new starting point, repeat the above judgment, until no additional point is reached.

(5) Record the coordinate collection $S_k, k = 1, 2, \cdots M$ and the total number n_k of points corresponding to the block of each small $2^t \times 2^t$ area to start judging in step (4), these recorded points meet the condition of non-repeating. Exist the circumstance that there has the same coordinate collection looking in adjacent areas, therefore $M \leq [m/t]$.

(6) Calculate region seed coordinates $seed_k (\overline{x}_k, \overline{y}_k)$ for each target area S_k :

$$\overline{x}_k = \left\lfloor \sum_{i=1}^{n_k} x_i \Big/ n_k \right\rfloor, \overline{y}_k = \left\lfloor \sum_{i=1}^{n_k} y_i \Big/ n_k \right\rfloor . \qquad (4)$$

Calculate $w_k = \max(|x_i - x_j|)$ and $h_k = \max(|y_i - y_j|)$, $i \neq j$, therein $(x_i, y_i), (x_j, y_j) \in S_k$, and. Get initial radius and center coordinates of the curve for potential model.

(7) Set threshold T_{seed}, when $w_k \leq T_{seed}$ or $h_k \leq T_{seed}$, cancel the region seed point, The threshold value T is set to reduce the effect of noise, it can be experience, can also use the formula (5) to calculate:

$$T_1 = \min(\sum_{i=1}^{M} w_i, \sum_{i=1}^{M} h_i) \Big/ M , \ T_{seed} = \min(\sum_{t=1}^{P_w} w_t \Big/ P_w, \sum_{t=1}^{P_h} h_t \Big/ P_h), (w_t, h_t \leq T_1) . \ (5)$$

(8) Start from different effective region seeds, take use of distance potential model and run in parallel, finally complete the multiple moving targets edge detection

The proposed algorithm takes advantage of the Quad-tree decomposition, sparse matrix block operation and Active contour model for multiple moving targets Edge Detection, so this algorithm may be named as QSA-ED algorithm.

4   Experimental Results

To demonstrate the feasibility and effectiveness of the QSA-ED algorithm, in this paper, we use the Intel (R) Core (TM)2 6300 CPUs, 1.86GHz, 1GB memory PC, use the Java language programming in the Eclipse development platform. The video images are captured from 336×448 traffic surveillance video.

Here we chose gray-scale images for example to carry out the experiments. If we define a random variable X to represent the value of each pixel for foreground segmentation, the probability of event $\{|X-\mu|<\varepsilon\}$ can be estimated by the Chebyshev inequality which reflects the changing circumstances about the corresponding value of video image pixels. The background modeling and foreground image extraction experimental results of the QSA-ED algorithm is shown in Fig. 2.

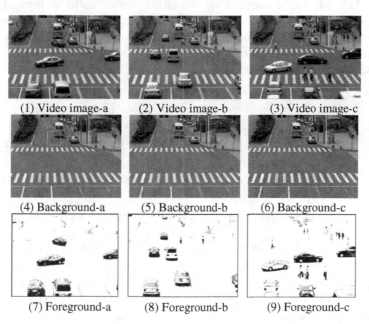

(1) Video image-a	(2) Video image-b	(3) Video image-c
(4) Background-a	(5) Background-b	(6) Background-c
(7) Foreground-a	(8) Foreground-b	(9) Foreground-c

Fig. 2. Experimental results of background and foreground

Through background modeling, we can get ideal foreground images concluding multiple moving targets. Quad-tree decomposition experimental results for these foreground images are shown in Fig. 3.

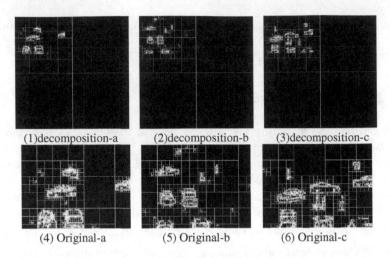

| (1)decomposition-a | (2)decomposition-b | (3)decomposition-c |
| (4) Original-a | (5) Original-b | (6) Original-c |

Fig. 3. Quad-tree decomposition experimental results

From Fig. 3 it can be seen the moving target characteristics of the corresponding sparse matrix are consistent with the foreground image, but the massive nature of the sparse matrix is unique. So for sparse matrix, we can perform block operations, extract location features of multiple moving targets. Experimental results of sparse matrix block operation are shown in Fig. 4.

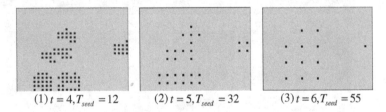

$(1) t = 4, T_{seed} = 12$ $(2) t = 5, T_{seed} = 32$ $(3) t = 6, T_{seed} = 55$

Fig. 4. Experimental results of sparse matrix block operation

From Fig. 4 it can be seen that in Fig. 4-(1), we set $t = 4, T_{seed} = 12$, the sparse matrix block size is 16×16, and we received intensive region seeds for each moving target. Adjusted the sparse matrix block operation density, made the size of $2^t \times 2^t$ larger, then we can see in Fig. 4-(2) and Fig. 4-(3), block size from 32×32 to 64×64, the number of region seeds obtained less and less. It can also be seen that the position relationship between region seeds and multiple moving targets not changed.

Sparse matrix block size $2^t \times 2^t$ is actually set up the density of the block operation, the threshold T_{seed} is set density threshold for each block matrix foreground object pixel. The size of $2^t \times 2^t$ and threshold T_{seed} moving are related with the number of moving targets in the foreground image. The more moving target, the greater the density calibration, the value of 2^t should be smaller, threshold T_{seed} should be decreased correspondingly.

This paper improved the active contour model by introducing the sparse matrix block operation. Experimental results of the QSA-ED algorithm are shown in Fig. 5

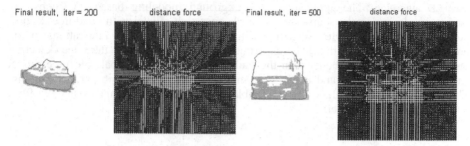

Fig. 5. Experimental results of the QSA-ED algorithm

In order to test the experimental results of the QSA-ED algorithm, we chose different active contour model: the traditional Snake model [9], the balloon force Snake model [10] and the GVF-Snake model [12] as comparison algorithms. For the same video frame images, Edge detection experimental results of different algorithms are shown in Fig. 6.

From Fig. 6 it can be seen that target boundary curves obtained by the proposed QSA-ED algorithm are better than the traditional snake model, balloon force model and GVF-Snake model. The traditional snake model does not converge, the balloon force model overflow, GVF-Snake model boundary inaccurate. The proposed QSA-ED algorithm can save computation time through parallel run of different region seeds. In summary, the QSA-ED algorithm not only can extract the desired multiple moving target edge detection, but also this algorithm is time savings which makes it more suitable for a real-time monitoring system.

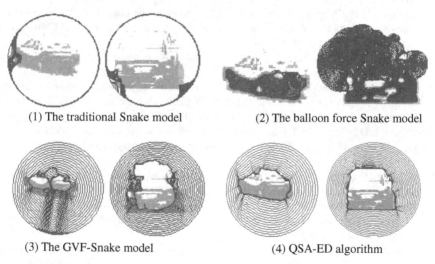

(1) The traditional Snake model   (2) The balloon force Snake model

(3) The GVF-Snake model   (4) QSA-ED algorithm

Fig. 6. Different Snake model edge detection comparison chart

5   Conclusion

In this paper, relating to the multiple moving objects edge detection we proposed a QSA-ED algorithm. The application of background modeling based on Chebyshev inequality, designing the sparse matrix block operation related with quad-tree decomposition, introducing region seed into active contour model are innovations of the proposed QSA-ED algorithm. The experimental results show that the edge detection results are significantly better than the traditional snake model, balloon force model and GVF-Snake model, and it can save more running time throng parallel running. Therefore, the QSA-ED algorithm is feasible, rapid and effective and is suitable for a real-time monitoring system.

References

1. JinWei, T., JingZhi, C.: Based on otsu thresholding roberts edge detection algorithm research. In: 2nd International Conference on Information, Electronics and Computer, pp. 121–124 (2014)
2. Hailong, X., Jingzhi, C., Xin, M., Mingxin, Z.: A novel edge detection algorithm based on improved sobel operator. In: BCGIN 2013 Proceedings of the 2013 International Conference on Business Computing and Global Informatization, pp. 837–840 (2013)
3. Yong, H., Mu, Q., Xue, W., Qing, C.: Application of image analysis based on canny operator edge detection algorithm in measuring railway out-of-gauge goods. In: Advanced Materials Research, vol. 912-914, pp. 1172–1176 (2014)
4. Amira, S.A., El-Sayed, A.M., Waheed, E.S., Abdel-Khalek, S.: New Method Based on Multi-Threshold of Edges Detection in Digital Images. International Journal of Advanced Computer Science & Applications 5, 90–99 (2014)
5. He, G.: The discussion and simulation for image edge detection techniques based on wavelet transform algorithm. Journal of Chemical & Pharmaceutical Research 6, 351–355 (2014)
6. JinYun, L., HaiPeng, P., YongMing, X.: The weld image edge-detection algorithm combined with canny operator and mathematical morphology. In: 2013 32nd Chinese Control Conference (CCC), pp. 4467–4470 (2013)
7. Kass, M., Witkin, A., Terzopolous, D.: Snakes: active contour models. International Journal of Computer Vision 1, 321–331 (1988)
8. HungTing, L., Sheu, W.H., HerngHua, C.: Automatic segmentation of brain MR images using an adaptive balloon snake model with fuzzy classification. Medical & Biological Engineering & Computing 51, 1091–1104 (2013)
9. Cohen, L.D., Cohen, I.: Finite- element methods for active contour models and balloons for 2-D and 3-D images. IEEE Transactions on Pattern Analysis and Machine Intelligence 15, 1131–1147 (1993)
10. Bohuang, W., Yan, Y., Jiwang, Y.: Boundary segmentation based on improved GVF snake model. In: Advanced Materials Research, vol. 756–759, pp. 3920–3923 (2013)
11. Xuanjing, S., Kaiye, W., Qingji, Q., Yingjie, L., Xiang, L.: Image segmentation using adaptive balloon force snake model. Journal of Jilin University (Engineering and Technology Edition) 41, 1394–1400 (2011)
12. Kun, Z., Cuirong, W.: A novel moving object segmentation algorithm based on chebyshev inequality. In: Proceedings of 2011 International Conference on Opto-Electronics Engineering and Information Science, pp. 1346–1350 (2011)

Robust Corner Detection Based on Bilateral Filter in Direct Curvature Scale Space

Bin Liao[1], Jungang Xu[2(✉)], Huiying Sun[3], and Hong Chen[2]

[1] School of Electrical and Electronic Engineering,
North China Electric Power University, Beijing 102206, China
nathan@ncepu.edu.cn
[2] School of Computer and Control Engineering,
University of Chinese Academy of Sciences, Beijing 101408, China
xujg@ucas.ac.cn, chenhong113@mails.ucas.ac.cn
[3] School of Information Engineering, Shandong Yingcai University, Jinan 250104, China
sunhuiying1988@sina.com

Abstract. In traditional Curvature Scale Space (CSS) corner detection algorithms, Gaussian filter is used to remove noise existing in canny edge detection results. Unfortunately, Gaussian filter will reduce the precision of corner detection. In this paper, a new method of robust corner detection based on bilateral filter in direct curvature scale space is proposed. In this method, bilateral filter is adopted to reduce image noise and keep image details. Instead of curvature scale space, direct curvature scale space is applied to reduce the computational complexity of the algorithm. Meanwhile, multi-scale curvature product with certain threshold is used to strengthen the corner detection. Experimental results show that our proposed method can improve the performance of corner detection in both accuracy and efficiency, and which can also gain more stable corners at the same time.

Keywords: Corner detection · Direct curvature scale space · Bilateral filtering · Threshold

1 Introduction

Corner detection has been widely used in many computer vision applications such as target recognition, image matching and camera calibration, since that the information about the object is always concentrated on corners, which prove to be most useful descriptors. There are lots of definitions of the corner, but generally, most people believe that the corner should be the brightness change point of a two-dimensional image or the local maxima of absolute curvature on an edge contour curve [1]. At present, the main methods for corner detection are divided into two main categories: intensity-based method and contour-based method. Both Harris algorithm and SUSAN algorithm belong to the intensity-based method, which detects corners by the gradient values of the pixels. Comparatively, a corner detection algorithm based on curvature scale space (CSS) [2] is typically contour-based method, which can detect

© Springer International Publishing Switzerland 2015
Y. Tan et al. (Eds.): ICSI-CCI 2015, Part III, LNCS 9142, pp. 357–365, 2015.
DOI: 10.1007/978-3-319-20469-7_38

corners from the contour curvature and achieve good results. However, this method has a few disadvantages, for example, some important and detailed information is easily lost because of adopting high scale factor, and the global threshold affect the selection of candidate corners. Mokhtaria and Mohanna [3] proposed an enhanced CSS algorithm, which focuses on dealing with some of these problems by using different scales for contours with different lengths. This method can reduce the missing corner and improve the detection rate of true corners in certain degree. He and Yung [4] proposed an algorithm which computes the absolute curvature value of each point on a contour at low scale and regards local maxima of absolute curvature as initial corner candidates. And an adaptive curvature threshold is choose to remove round corners from the initial list and the angles of the remaining corner candidates are evaluated to eliminate any false corners, arising from quantization noise and trivial details. Since the angle and curvature threshold can reflect the local characteristics of the candidate corners dynamically, this algorithm proves to achieve better detection results for multi-scale image.

Carefully studying the corner detection algorithms listed above, the following points can be drawn. Unlike the pipeline of the general intensity-based methods, which mainly exploit gradient values of the pixels, these CSS based methods do canny edge detection firstly to extract the contours of the image before detecting corners, and then remove noise by Gauss filter. Note that some weak corners are easily smoothed out, which results in lower accuracy of corner locating. What's more, curvature at all scales is calculated, which will definitely cause high computational complexity.

To solve the problems discussed above, a direct curvature scale space (DCSS) corner detection method based on bilateral filter is proposed. Compared with the CSS approach, the DCSS has less computational cost, because it only needs once convolution with the curvature function. In addition, the adoption of bilateral filter not only reduces the Gauss noise, but also reserves the edge details, which is helpful for corner detection. Then multi-scale curvature product with certain threshold is used to strengthen the corner detection.

The remainder of this paper is organized as follows. Section 2 introduces the bilateral filter. Section 3 describes the proposed corner detection method based on Bilateral Filter. Section 4 presents the experimental evaluation of our method. Finally, section 5 concludes the work.

2   Bilateral Filter

A bilateral filter [5] is non-linear, edge-preserving and noise-reducing smoothing filter, which has demonstrated great effectiveness for varied problems in image denoising and target identification [6][7][8][9][10]. The main idea of bilateral filter is to replace each pixel in an image with the weighted average of intensity values from nearby pixels. The weights depend not only the intensity distance but also on Euclidean distance. In this way, the edges can be preserved efficiently while the noise is smoothed out. Assuming that I denotes the image, at a pixel position x_c, the bilateral filtering function is defined as

$$N_{\rho,\sigma}(x_c) = \frac{1}{C}\Sigma_{\xi \in W} \exp\left(-\frac{\|\xi-x_c\|^2}{2\rho^2}\right) \cdot \exp\left(-\frac{|I(\xi)-I(x_c)|^2}{2\sigma^2}\right) \cdot I(\xi) . \tag{1}$$

Where ρ and σ are smoothing parameters, controlling the fall-off of weights in spatial and intensity respectively. W denotes the spatial neighborhood of the center pixel x_c. $\|\cdot\|$ operation denotes the Euclidean distance between ξ and x_c, and C is the normalization constant, given in Equation 2.

$$C = \Sigma_{\xi \in W} \exp\left(-\frac{\|\xi-x_c\|^2}{2\rho^2}\right) \cdot \exp\left(-\frac{|I(\xi)-I(x_c)|^2}{2\sigma^2}\right) . \tag{2}$$

3 The DCSS Corner Detection Algorithm Based on Bilateral Filter

3.1 Curvature Scale Space

The curvature scale space technology [2] is suitable for describing invariant geometric features (curvature zero-crossing points and/or extrema) of a planar curve at multi-scales. To compute it, the curve Γ is first parameterized by the arc length parameter u:

$$\Gamma(u) = (X(u), Y(u)) . \tag{3}$$

An evolved version Γ_σ of Γ can be computed, and Γ_σ is defined by

$$\Gamma_\sigma(u) = \left(X(u,\sigma), Y(u,\sigma)\right) . \tag{4}$$

where $X(u,\sigma) = X(u) \otimes g(u,\sigma)$, $Y(u,\sigma) = Y(u) \otimes g(u,\sigma)$, $\otimes$ is the convolution operator and $g(u,\sigma)$ denotes a Gaussian function with deviation σ. Note that σ is also referred to as the scale parameter.

The evolved versions of the curve Γ are generated as σ increases from 0 to infinity. This technique is suitable for removing noise and smoothing a planar curve as well as gradual simplification of its shape. Nowadays, many corner detection algorithms are based on curvature scale space technology.

In order to find curvature zero-crossings or extrema from evolved versions of the input curve, one needs to compute curvature accurately and directly on an evolved version Γ_σ. The curvature K [11] is given by

$$\kappa(u,\sigma) = \frac{X_u(u,\sigma)Y_{uu}(u,\sigma) - X_{uu}(u,\sigma)Y_u(u,\sigma)}{(X_u(u,\sigma)^2 + Y_u(u,\sigma)^2)^{\frac{3}{2}}} . \tag{5}$$

Where $X_u(u, \sigma) = X(u) \otimes g_u(u, \sigma)$, $X_{uu}(u, \sigma) = X(u) \otimes g_{uu}(u, \sigma)$, $Y_u(u, \sigma) = Y(u) \otimes g_u(u, \sigma), Y_{uu}(u, \sigma) = Y(u) \otimes g_{uu}(u, \sigma)$.

And $g_u(u, \sigma)$ and $g_{uu}(u, \sigma)$ denote the 1st and 2nd derivatives of $g_u(u, \sigma)$ respectively. Due to the CSS technology needs to calculate the evolved curve curvature at all scales, the computational complexity will be high.

3.2   Direct curvature Scale Space

The theory of the direct curvature scale space (DCSS) and its mathematical proof were given in Reference [3]. It is defined as CSS that results from convolving the curvature of a curve with a Gaussian kernel directly at different scales. Assuming that the curve parameter equation has the same definition as that is defined in Equation 3. The curvature equation of a curve Γ at the first scale $\sigma = 0$ is defined as Equation 6.

$$\kappa(u) = \dot{X}(u)\ddot{Y}(u) - \ddot{X}(u)\dot{Y}(u). \tag{6}$$

where $\dot{X}(u) = \frac{\partial X}{\partial u}$, $\dot{Y}(u) = \frac{\partial Y}{\partial u}$, $\ddot{X}(u) = \frac{\partial^2 X}{\partial u^2}$, $\ddot{Y}(u) = \frac{\partial^2 Y}{\partial u^2}$. Then the curvature of the evolved version Γ_σ can be expressed approximately by Equation 7.

$$\kappa(u, \sigma) = \kappa(u) \otimes g(u, \sigma). \tag{7}$$

Comparing with CSS, the DCSS technology has less computation cost, because it only needs to calculate the convolution with the curvature function.

3.3   Our Proposed Method

Both the original CSS algorithm and the improved algorithm [4][12] choose the canny operator for edge detection, which is very sensitive to the noise. Thus the accuracy of the corner detection will be reduced accordingly. Combining with bilateral filtering, a robust corner detection method based on the DCSS technology is proposed in this paper. The main steps are described as follows:

1. Apply the edge detection algorithm based on bilateral filter to extract the edge of original image:

   - Assuming I denotes the input image, filter the original image through Equation 1.
   - Calculate the gradient and direction of smooth image by using the first derivative of finite difference within 2×2 neighborhood.
   - Locate the edge accurately by means of the non-maximum suppression method [15] and preserve the largest local amplitude points.
   - Perform image segmentation on above result image by double thresholds algorithm, two edge images T_H and T_L are produced. The image T_H is obtained under higher threshold, which contains true edges, but it may have gap in the contours. Double threshold algorithm can link the edge endpoints of T_H to

reduce the edge discontinuity through checking the 8 neighborhood of those endpoints in T_L. In this way, the edge in T_L is searched recursively until all gaps in T_H are connected.

2. Extract the edge contours from the edge-map, and find the T-junctions.
3. Calculate curvatures at the different scales based on direct curvature scale space, and then calculate the curvature product at these scales. Regard local maximums as candidate corners whose curvature products exceed certain threshold.
4. Compare the T-junction corners with other corners, and remove one of any two corners that are close to each other.
5. Finally, consider the endpoints of open contours, and mark them as corners unless they are very close to another corner.

In the step of (1), bilateral filtering not only reduces the Gauss noise, but also reserves the edge details, preparing for the corner detection. In the step of (3), multi-scale curvature product is used to enhance the corner detection, and the curvature product is defined as follows.

$$P_N(u) = \prod_{j=1}^N \kappa(u, \sigma_j) \ . \tag{8}$$

In order to verify the importance of the curvature product, we choose the simple image corrupted by white Gaussian noise as our experimental subject, showed in Fig. 1. The curvatures of the contour at scales 1, 2, 6 are shown in Fig. 2(a)-(c). It is obvious that true corners present observable magnitudes across the scales, while magnitudes of the noise decrease rapidly. Moreover, the curvature orientation of corners propagates at different scales, while that of non-corner changes randomly. As depicted in Fig. 2(d)-(f), the small features and the noise are suppressed as the scale increases and the responses of the corners become more salient. Obviously, from (d)-(f) it is easy to know the multi-scale product enhances the corners peaks while suppressing the small features and noise.

Fig. 1. The simple image corrupted by Gaussian noise

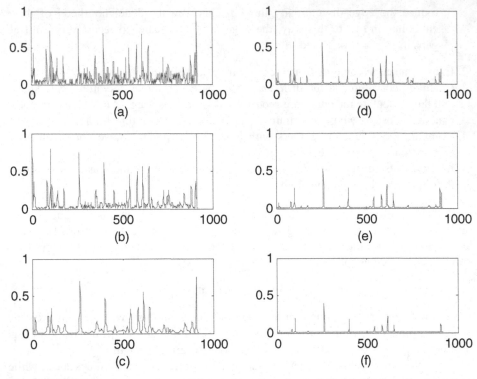

Fig. 2. (a)-(c) show the curvatures of the contour at scales 1, 2, 6; (d)-(f) show the curvature products at scales 1, 2/2, 6/1, 2, 6 respectively

4   Experimental Results

4.1   Evaluation Criteria

There are several metrics that can measure the corner detection accuracy, such as the number of the true detections, the number of false corners, the detection rate and the false alarm rate. However, a corner detector depends upon the threshold seriously. As a result, we use recall and precision, which can reveal the overall detector performances according to threshold variation. The recall value measures the number of correct corner detections out of the ground truth corners defined by Equation 9, while the value of the precision measures the correct corners out of all possible corner detections, as shown in Equation 10.

$$recall = \frac{correctcorners}{truecorners} . \qquad (9)$$

$$precision = \frac{correctcorners}{truecorners+falsecorners} . \qquad (10)$$

4.2   Performance Evaluation

In this section, the experimental results of the proposed corner detector are presented. Also the comparative experiments with Harris, SUSAN and CSS are performed. All the tests are carried out on the simple image in Fig. 3 and house image in Fig. 4.

Firstly, the outputs of all detectors given the simple artificial test image are shown in Fig. 3 and the squares denote the detected corners. Since we know the exact corner location, the ground truth location is prepared by human vision to evaluate the location error. Compared with Fig. 3(a)-(c), outputs from Harris, SUSAN and CSS detectors, the location accuracy and reliability of the proposed algorithm in Fig. 3(d) is apparently better. The proposed algorithm can detect all true corners and no false corner occurs, other algorithms don't achieve such result no matter how their parameters and thresholds are adjusted. The detection results of multiple-size features images are illustrated in Fig. 4(a)-(d). Subjective observation of Fig. 4(a)-(d) shows that the true corner numbers of proposed algorithm are more than other algorithms while the false corner numbers are less comparatively. The detection results are summarized in table 1. Moreover, the proposed algorithm achieves better accuracy of location described in Fig. 5.

(a) Harris   (b) SUSAN   (c) CSS   (d) proposed method

Fig. 3. Comparison of corner detection results on simple image

(a)Harris   (b) SUSAN   (c) CSS   (d) proposed method

Fig. 4. Comparison of corner detection results on house image

Table 1. Evaluation results

Test images	Detector	True corners	Missed corners	False corners	Recall	Precision
Simple image	Harris	24	1	14	96%	61.5%
	SUSAN	25	0	37	100%	40.3%
	CSS	23	2	7	92%	71.9%
	Proposed algorithm	25	0	0	100%	100%
House image	Harris	59	22	60	31%	41.8%
	SUSAN	63	18	55	78%	46.3%
	CSS	66	15	16	81%	68.0%
	Proposed algorithm	68	13	7	84%	77.3%

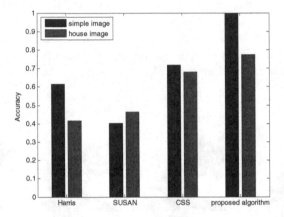

Fig. 5. Comparison of correct detection ratios

5   Conclusion

In this paper, we proposed one direct curvature scale space corner detection algorithm based on bilateral filtering, which aims to improve the accuracy of corner detection. In order to prepare for the corner detection, bilateral filter uses the spatial proximity and brightness similarity information of the neighborhood pixels to smooth image. The algorithm is proposed in the framework of direct curvature scale space (DCSS), then multi-scale curvature product with certain threshold is used to enhance the corner detection, the most significant property of the scale product is effectively enhancing curvature extreme peaks while suppressing noise and improving localization. The experimental results show that our method has the results of good quality and high location accuracy with low computational cost.

Acknowledgements. This work was supported by the National Natural Science Foundation of China [grant number 61372171]; the National Key Technology R&D Program of China [grant number 2012BAH23B03]; the Fundamental Research Funds for the Central Universities of China [grant number NCEPU2014MS02].

References

1. Sun, X.Y., Yao, L., Wan, Y.: An adaptive corner detection algorithm based on linear fitting. Journal of Shanghai Engineering and Technology University **23**(1), 46–50 (2009)
2. Mokhtarian, F., Suomela, R.: Robust image corner detection through curvature scale space. IEEE Trans. on Pattern Analysis and Machine Intelligence **20**(12), 1376–1381 (1998)
3. Mokhtarian, F., Mohanna, F.: Enhancing the curvature scale space corner detector. In: Scandinavian Conference on Image Analysis, pp. 145–152 (2001)
4. He, X.C., Yung, N.H.C.: Curvature scale space corner detector with adaptive threshold and dynamic region of support. In: 17th IEEE International Conference on Pattern Recognition, pp. 791–794. IEEE Press, New York (2004)
5. Tomasi, C., Manduchi, R.: Bilateral filtering for gray and colour images. In: IEEE International Conference on Computer Vision, pp. 839–846. IEEE Press, New York (1998)
6. Xie, Q.L.: Image denoising combining with bilateral filtering and more frames average filtering. Computer Engineering and Application **27**, 154–156 (2009)
7. Chen, W.J., Zhang, E.H.: Noise image magnification method keeping the edge character. Computer Engineering and Application **12**, 178–180 (2009)
8. Yang, X.Z., Xu, Y., Fang, J., Lu, J., Zuo, M.X.: A new image denoising algorithm combining with the regional segmentation and bilateral filtering. Chinese Journal of Image and Graphics **17**(1), 40–48 (2012)
9. Li, F., Liu, S.Q., Qin, H.L.: Dim infrared targets detection based on adaptive bilateral filtering. Acta Photonica Sinica **39**(6), 1129–1131 (2010)
10. Mokhtarian, F., Mackworth, A.K.: A theory of multi-scale, curvature based shape representation for planar curves. IEEE Trans. on Pattern Analysis and Machine Intelligence **14**(8), 789–805 (1992)
11. Zhong, B.J., Liao, W.H.: Direct curvature scale space: theory and corner detection. IEEE Trans. on Pattern Analysis and Machine Intelligence **29**(3), 508–512 (2007)
12. Neubeck, A., Gool, L.V.: Efficient non-maximum suppression. In: 18th International Conference on Pattern Recognition, pp. 850–855 (2006)

An Infrared Thermal Detection System

Zahira Ousaadi[(✉)] and Nadia Saadia

Faculty of Electronic and Computing (FEI),
Laboratory of Robotic Paralelism and Embeded Systems (LRPE),
University of Science and Technology Houari Boumediene (USTHB), Bab Ezzouar, Algeria
oussadizahira@gmail.com, Nadia_saadia@hotmail.fr

Abstract. Anybody whose temperature differs from absolute zero (0K) emits and absorbs electromagnetic radiation coming from, one hand, the physico-chemical nature and on the other hand, the action of intrinsic mechanisms of vibrational energies of the molecules. In this paper, we propose a robot that exploits this characteristic to detect and to track a thermal source in its environment. The system consists of a mobile platform having mounted a thermal detection device using thermopiles that is controlled by a computer via a graphical user interface.

Keywords: Thermal source · Infrared waves · Thermal detector · Mobile robot · Anti-fire safety · Thermal source detection and racking

1   Introduction

Infrared waves are part of the electromagnetic spectrum which lies between the microwave and visible light. The detection of this signal category requires the use of transcoders for its conversion into an exploitable signal. The research development in the infrared radiation led to the realization of the passive infrared detectors. Their working principle is based on the measurement of the emitted infrared flow according to the object's temperature. Moreover, two large families of passive infrared detectors are distinguished. The first one is the quantum detector called cooled; it is photoelectric-effect sensor that relies on the creation of mobile electrical charges by electromagnetic radiation of wavelength [1]. The second one is the thermal detector named uncooled, it is transducer in which the infrared radiation is directly converted into heat by absorption [2]. This basic structure allows the measurement of small changes through different effects [3]. The detection of thermal sources has a big importance in different fields. In the literature, several applications that uses thermal detectors have been proposed, in particular the tracking and detection field, as well as the anti-fire safety. For instance, we can cite the works on rescuing people [4], the anti-invasion monitoring [5], the monitoring of persons [6], the detection and counting of people [7], the representation of the trajectories of moving people [8], [9], [10] and also the work on the detection of fires and their extinguishing [11].

In this paper, we focus on the tasks of detection and tracking of the thermal sources in a closed environment. The remainder of the paper is organized as follows. First, the

© Springer International Publishing Switzerland 2015
Y. Tan et al. (Eds.): ICSI-CCI 2015, Part III, LNCS 9142, pp. 366–374, 2015.
DOI: 10.1007/978-3-319-20469-7_39

thermal system is described in section 2. Then, in section 3, we present the realized applications and we explain the adopted strategies to carry the desired tasks. Finally, we resume the obtained results and give some perspectives to improve the proposed thermal system.

2   Thermal System Description

In this work, we choose to design a thermal detection device using the TPA81 module for its advantages. This module is a passive monolithic system that can be similar to a low-resolution thermal camera and it's composed of a vertical row of eight thermopiles [12]. Its field of vision forms a cone of 41° horizontally and 6° vertically [7]. The thermal module TPA81 can provide power and control a servo motor [12]. Indeed, it includes a microcontroller which generates a PWM signal that allows the servomotor to have a sweeping angle of "180°" or "30°" in "32" steps, according to the command sequence sent by the circuit. To perform a 360° horizontal scanning of the environment and reduce the detection time, two thermal module are placed in opposition on the servo motor (Fig.1.a) and have chosen the sequence that allows to sweep an angle of 180° in 32 steps. This configuration divides the environment into a foreground and a background where each one will be swept by one of the two thermal modules. To increase the range of the detection device and ensure the movement towards the heat source or its follow up, we have set up the detection device on a robotic unicycle-type mobile platform. The control of the robot is performed by a computer via a graphical user interface that is implemented using language. This interface also allows viewing the detected thermal source on a low resolution thermal image, the different values of the measured temperatures, the value of the highest temperature and the motor's speeds. Data exchange between the control unit and robot is provided by a USB/I2C protocol. The thermal system realized is shown in the figure (Fig.1.b).

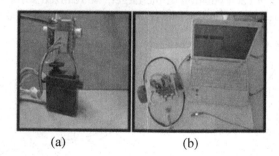

(a)   (b)

Fig. 1. (a) Thermal detection device (b) Thermal detection system

2.1   Evaluation of the Thermal Detection Device

To check the influence of the ambient temperature on the detection, we have positioned a candle at 50cm distance from the thermal detection device and performed a horizontal scanning of the environment for the detection of the candle in the case of two ambient temperatures (8°C and 29°C). Temperatures values from different thermopiles in the two scans are shown in the following figure.

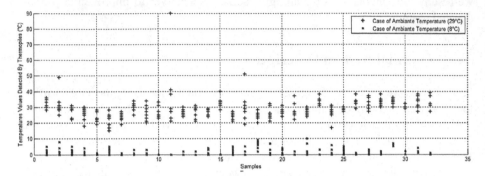

Fig. 2. Comparison results with two different ambient temperatures

The candle is seen as a bright spot on the thermal image with a temperature of 10°C for an ambient temperature of 8°C, and 90°C for an ambient temperature of 29 C.

To secure the robot against collisions with the thermal source, a relation has been established between target's temperature and the distance from the detection device to set a threshold temperature in the program. As soon as the detection device detects a higher temperature measurement or equal to the threshold, the system must stop. The first experiments were carried out based on a candle, but during testing we noticed that the thermal modules do not give the same values in the same experimental conditions. To check whether these measures differences are due to non-robust thermal modules, we set up each of the two thermal modules on a servomotor and built triangulation system. Then we proceeded to the detection of a candle which is successively moved along a graduated ruler for a distance of 120cm in the case of an ambient temperature of 17°C. For each position of the target, we realized several measures of the highest point and retained average. The results obtained are shown below:

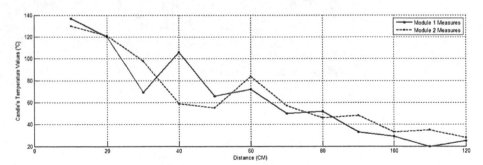

Fig. 3. Highest temperatures values detected using candle

The figure shows that the measured temperatures allow us to detect target. However, these values decrease each time the distance between the candle and the thermal detection device increases and the detection becomes poor when the distance exceeds

120 cm, and they are not identical for the same distance from the same target. In addition, they have significant differences that can sometimes reach 47°C. The results thus confirm that the measures identified by the TPA81 module are not robust and limit its use to detection tasks. Therefore, we can say that these errors may be due to the non-stability of the candle flame. To clarify the situation, the experiment should be repeated with a more or less stable thermal source using a filament lamp for the same ambient temperature. The obtained results are shown in the Fig.4.

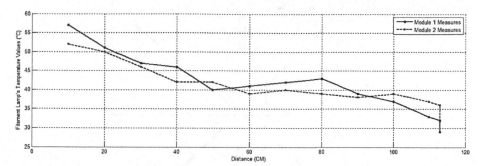

Fig. 4. Highest temperatures values detected using lamp

The measurements given by the two thermal modules are roughly similar. Indeed, the maximum difference obtained in this experiment is 5°C.

To confirm that the estimated position is correct, we performed a test bench. It consists of a range of 64 zones. The upper part of the range is restricted to the first module of front sensing. It is numbered from right to back left. Each zone of the range covers an angle equal to 5.625. This angle corresponds to one of the servomotor as it travels 180 by performing 32 steps. The tests have shown that the robot manages to locate the candle in the environment in a single scan and the precision of the estimated position given by the program depends on the position of highest temperature detected in the area bounded by two servomotor positions. Indeed, if it is in the middle of the field of view of the thermal module, the thermal detection device is positioned exactly in front of the thermal source. Otherwise, if it's between two successive positions of the servomotor, or on its middle, the estimated position given by the program is the servomotor position when it detects the thermal source. Therefore, the final position of the robot after moving will be shifted from the thermal source.

3 Applications

The first application consists to detect and localize the thermal source situated in the robot's environment, to move the robot towards the thermal source and stop it up when the temperature threshold set in the program has been detected. The algorithm of this application can be used in detection tasks and fire extinguishing. The second application consists to follow a mobile thermal source. For this, a tracking algorithm is implemented.

3.1   Strategy

To position the robot in front of a thermal source, we divided the plan into four planes (Fig. 5.). Each of the thermal modules scans a part of the plan; the first scans the upper half plane which consists of sub-planes 1 and 2 while the second half is scanning the lower half plane which consists of sub-planes 3 and 4.

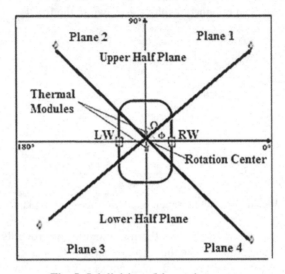

Fig. 5. Subdivision of the environment

From the realized robot, we get the angle "Φ" which corresponds to the servomotor position up on detection of the source. This position is a multiple by the step number "P" of the angle "α" which is the movement step of the servomotor (Equation 1).

$$\Phi = \alpha.P \tag{1}$$

To position the robot in front of the source, we make the robot rotate around the center axis of the driving wheels by considering two cases.

1. If thermal source is in the sub plane 1 or 3, we make a forward rotation of the left wheel (LW) and a rear rotation of the right wheel (RW). The robot deviates by an angle Ω:

$$\Omega = 90° - \Phi \tag{2}$$

2. If thermal source is in the sub plan 2 or 4, a rear rotation of the left wheel (LW) and a forward rotation of the right wheel (RW) are done. The robot has to deviate by an angle of Ω:

$$\Omega = \Phi - 90° \tag{3}$$

3.2   Application 1

To explain the adopted strategy in the first application, we present curves obtained when the target in in the plane 1. The following figures illustrate the highest temperature values detected by the thermal modules used and the maximum temperatures selected by the program such as target and the speed of the motors.

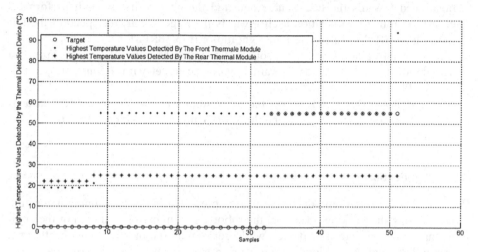

Fig. 6. Highest temperature values detected by the thermal detection device

The target has been detected by the front thermal module. At the end of the scanning phase, the selected value for the temperature of the target is 55 ° C. It corresponds to the maximum value detected by the system during the scanning phase which lasts from the first sample to the thirty-second (32nd) sample. This value is maintained from the sample 33. It should be noted that during the scanning phase, the value of the heat source is considered as zero. At the fifty-one sample, the robot detected a temperature value equal at 94°C. It exceeds the threshold temperature which is 90°C.

The robot behavior during this task is visualized through the curve illustrating the speed of the motors (Fig. 7).

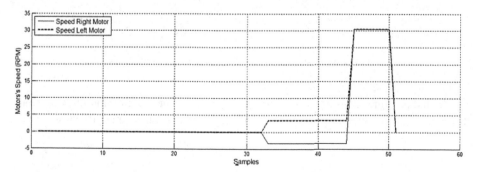

Fig. 7. Motor' speeds during different phases

We distinguish in figure four (4) phases:

- The detection phase (from sample 1 to sample 32): In this phase, the detection device performs the scanning (in 32 steps of the servomotor) of the environment of the robot. It remains stationary so its speed is zero.
- The positioning phase (from sample 33 to sample 44): In this phase, the robot is positioned towards the heat source detected during the first scan. It performs a clockwise rotation to position itself towards the target. Thus, the speed of the left motor is 3.38RPM while that of the right motor is -3.38RPM.
- The robot shift phase towards the source (from sample 45 to 49 sample): From the sample 45, the robot moves forward toward the target with a constant speed of 30.46 RPM.
- The stop phase: During this phase, the robot stops when the temperature measured by the detection device exceeds the threshold temperature which is 90 ° C (Fig. 6), speed becomes therefore zero.

3.3   Application 2

We note that the robot reproduces the phases presented in the positioning algorithm. Indeed, during the tracking operation, the robotic system performs a scan of the environment using the detection device while remaining stationary. Once the thermal source is localized, it positions in its direction and moves toward it, covering a distance of 15cm and then stops. These steps are repeated until the stop condition is verified or the system is manually terminated as in our case. We present in the following some results obtained when tracking the mobile thermal source.

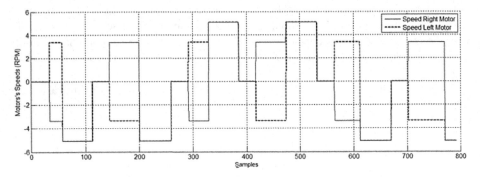

Fig. 8. Motor's speeds obtained in the case of tracking

The temperature measurements from thermopile of the thermal modules used and the curve of the temperatures maintained as target during tracking are illustrated in the figure (Fig. 9).

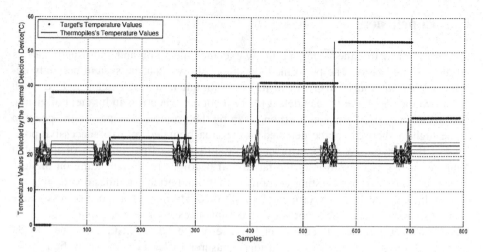

Fig. 9. Target's temperature values during tracking

During the realization of the first scan of the environment by the detection system, the value of the target's temperature is maintained at 0. At the thirty-third sample, a value of 38°C is assigned to the value of the target. This value corresponds to the maximum temperature detected in the first scan. It is maintained until the end of the second scan where it will be replaced by the maximum value of the second scan. This value will be maintained until the end of next scan. Thus, the last temperature measurements detected at the end of each scan are maintained until the beginning of the next scan. This procedure should be repeated until the tracking is stopped. Moreover, the trajectory followed by the robot is illustrated in the following figure:

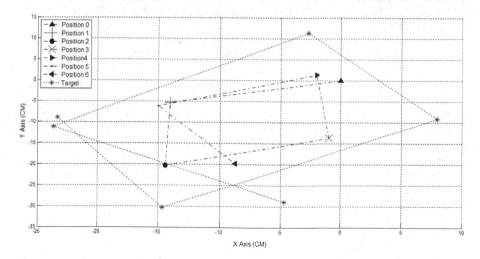

Fig. 10. Trajectory obtained during tracking of target

4   Conclusion

The results show that the system perfectly executes the command and usually carries out the assigned tasks. The performed tests have shown that the system perfectly detects the thermal source disposed in its environment in a single scan. Indeed, each of the two thermal modules can detect the heat source when it is in his field of vision. The precision of the estimated position given by the program depends on the position of the highest thermal source detected in the area bounded by two successive positions of the servomotor. This error causes a shift in the final position of the robot compared to the real position of the target and becomes large depending on the distance robot-target. During testing we noticed that the thermal modules do not give the same values in the same experimental conditions. The use of a more or less stable thermal source therefore does not allow to completely eliminate the gap measurement. Indeed, measurement errors can occur because of the non-robustness of the TPA81 modules. Moreover, the tests showed that measures differences are amplified in the case where the heat source is unstable (case of a candle). So we will not be able to consider this module for applications requiring precise measurement of temperature.

References

1. Haffar, M.: Etude et Réalisation de Matrices de Microcapteurs Infrarouges en Technologie Silicium pour Imagerie Basse Resolution. Diss, Univ. des Sci et Tech de Lille-I (2007)
2. Dereniak, E.L., et al.: Infrared Detectors and Systems. Wiley (1996)
3. Belconde, A.: Modélisation de la Détection de Présence Humaine. Diss, Université d'Orléans (2010)
4. Pellenz, J.: Rescue robot sensor design: an active sensing approach. In: SRMED: 4th International Workshop on Synthetic Simulation and Robotics to Mitigate Earthquake Disaster (2007)
5. Kastek, M., et al.: Passive Infrared Detector Used for Detection of Very Slowly Moving of Crawling People. Opto-Electronics, Review 16(3), 328–335 (2008)
6. Lv, X., Yongxin, L.: Design of human motion detection and tracking robot based on pyroelectric infrared sensor. In: 8th World Congress on Intelligent Control and Automation (WCICA). IEEE (2010)
7. Honorato, J., Luis, I., et al.: Human detection using thermopiles. In: Robotic Symposium, LARS 2008. IEEE Latin American (2008)
8. Shankar, M., et al.: Human-Tracking Systems Using Pyroelectric Infrared Detectors. Optical Engineering 45(10), 106401–106401 (2006)
9. Hosokawa, T., Kudo, M.: Person tracking with infrared sensors. In: Khosla, R., Howlett, R.J., Jain, L.C. (eds.) KES 2005. LNCS (LNAI), vol. 3684, pp. 682–688. Springer, Heidelberg (2005)
10. Kuki, M., et al.: Human Movement Trajectory Recording for Home Alone by Thermopile Array Sensor. In: IEEE International Conference on Systems, Man, and Cybernetics (SMC). IEEE (2012)
11. Rehman, A., et al.: Autonomous fire extinguishing eystem. In: International Conference on Robotics and Artificial Intelligence (icrai). IEEE (2012)
12. TPA81ThermopileArray. www.robot-electronics.co.uk/htm/tpa81tech.htm

FDM-MC: A Fault Diagnosis Model of the Minimum Diagnosis Cost in the System Area Network

Jie Huang[✉] and Lin Chen

School of Computer, National University of Defense Technology, Changsha, China
hj_nudt@qq.com, chenlin@nudt.edu.cn

Abstract. There are many problems in the fault diagnosis of system area network which facing enormous challenges. These problems are mainly due to the fact that network fault is often conditionally dependent on many factors, which are usually dependent on complex association relationship. Non-linear mapping may exists between symptoms and causes of network fault, and the same network fault often has different symptoms at different time, while one symptom can be the result of several network faults. Because there is a lot of correlative information in the network, how to construct the model of fault diagnosis is a challenging topic. In this paper, we firstly provided the description of the diagnosis costs, and then we proposed the model based on the condition of dependent diagnosis actions and the model based on the condition of dependent faults. Through a series of theoretical support, we have seen that our diagnostic model produces expected cost of diagnosis that are close to the optimal result and the lower than the simple planners for a domain of troubleshooting.

Keywords: Expected cost of diagnosis · System area network · Diagnosis model

1 Introduction

Nowadays system area network is widely used in supercomputers and data centers. Unlike the Internet, system area network such as InfiniBand [1] provides an integrated service environment by connecting compute nodes, storage node, and other network devices. The specific characteristics of system area network are described as following.

- Increasingly network scale

The scale of system area network becomes increasingly larger. For example, the number of compute nodes in Tianhe-2 [2] is around 16,000 and its switch nodes are close to the 1,000 magnitude.

- Complex flow behavior

In high-performance computers or data centers, the traditional MPI(Message Passing Interface) applications or the emerging big data applications are working with a very large amount of burst traffic. For example, the many-to-one communication pattern,

© Springer International Publishing Switzerland 2015
Y. Tan et al. (Eds.): ICSI-CCI 2015, Part III, LNCS 9142, pp. 375–382, 2015.
DOI: 10.1007/978-3-319-20469-7_40

e.g., Incast in Hadoop, produces much burst traffic. In addition, lots of data transmission induced by applications like virtual machine migration can also bring heavy loads on the network resulting in more and more complex traffic behavior.

- exponential increase of network failure rate

Long-term practice shows that the network failure rate increases rapidly with the number of nodes in a system area network. Provided N denotes the number of nodes and α denotes the average reliability ratio of a single node, the number of network failure points would reach $p = 1 - \alpha N$. Considering the influence made by every failure point on its neighbors, the number of the real failure points are much larger than p.

- Requirements of fine-grained network measurement and timely diagnosis

Applications running on system area network are often highly sensitive to network performance. To make a controllable and reliable network, managers need to know the status of the whole network and the end-to-end communication performance more timely and accurately. When the network performance decreases, managers should come up with a troubleshooting solution quickly by discovering network faults and congestion, and locating the specific related points [3].

To summarize, we argue that obtaining the running status of a system area network in a real-time manner, and identifying the performance failure points (once we found degraded performance) is dramatically important to guarantee the performance of network applications.

The rest of the paper is organized as follows. We briefly review the related works in Section 2 and present the cost of diagnosis in Section 3. In Section 4 and 5, through a series of theoretical support, we present the model based on the condition of dependent diagnosis actions and the model based on the condition of dependent faults. Finally, we conclude the paper and point out potential future works in Section 6.

2   Related Work

In this section, we review the related works in this area. Classical approaches to fault diagnosis in network management are mostly based on passive event correlation [4]. The typical limitation of these techniques is that they require heavy instrumentation since each managed entity needs to have the ability to send out the appropriate alarms. But in many real applications, only network devices are able to report status information initiated, for example, through SNMP traps. Many managed entities, such as applications developed by third parties, middleware and hosts, do not even provide management interfaces, let alone event notifications that reflect their status. In these cases, the management system has to actively inquire about their health. In recent years, active probing which provides end-to-end monitoring capability has received significant attention [5]. By measuring probing responses, fault diagnosis can be performed without instrumentation on managed entities. Moreover, fault diagnosis can be accomplished more efficiently if we can actively observe specific data points.

Expert system technology has been widely used to solve the network fault diagnostic problem. Reference [6] has used real-time expert system for monitoring of data packet networks. Reference [7] has used expert system for fault location in paired cables. Reference [8] has used expert system for telephone cable maintenance. And, intelligent knowledge based systems for network management has been studied [9]. Most of the existing expert systems for fault diagnosis are based on the deductive reasoning [10]. Though deductive inference mechanism generates only appropriate explanations, which will not generate those required explanations, if the missing information were to be present [11].

3   The Model Based on the Condition of Dependent Diagnosis Actions

We make the fault diagnosis model and introduce the theory of exact cover by 3-sets (abbreviated as X3C) in the conditions of dependent diagnosis operations.

Theorem 1. Exact Cover by 3-sets (X3C) that give a finite set X with $|X| \models 3q, q \in Z^{+}$, for some integer q and a collection C of 3-element sets of X, is there a set $C' \subseteq C$ such that every element of X occurs in exactly one element of C'. This problem is NP-complete problem which proved in reference [12]

Assumption 1. In the conditions of dependent diagnosis actions, we make the following assumptions to the problem of troubleshooting.

(1) Single-fault assumption that means the failure is caused by the single or multiple non-related fault reasons.

(2) For the possible faults set F, we assume that there is $|F| \models 3m, m \in Z^{+}$.

(3) For any failure f_i, we assume that there is $P_{f_i} = p(f_i \; happens \mid E_c) = 1/3m$, which means fault happen in the same probability.

(4) For any diagnosis action a_i, there is the fault set T_i can be diagnosed by the action a_i, and $|T_i|=3$. We can diagnose all faults in the set T_i with the probability one and outside the set T_i with the probability zero, that is $p(a_i \; succeeds \mid f_j \in T_i) = 1$, and the action cost is a constant.

You can conclude that the simplified fault diagnosis is the subset of the fault diagnosis described above. The same probability of occurrence of failure and the operation cost is a constant, so we can randomly select an action a_k. If the reason of the problem cannot be found by the action a_k, we try to select the next action. Because there is the correlation between the two actions, for any two actions a_i and a_j, if $T_i \cap T_j \neq \varnothing$, there is the correlation between the actions a_i and a_j. Therefore if there are the subsequent actions associated with the action a_k, which is clearly not optimal, and we can choose the action a_l with $T_k \cap T_l = \varnothing$. So we finally try to get an array of actions named s, which with the highest efficiency and the least cost of diagnosis.

3.1   Single-Fault Assumption

Firstly we assume that only a fault occurs in the each diagnostic problem and introduce the following three lemmas.

Lemma 1. In the single-fault assumption, if there is the array of actions $s_o = \{a_{j1},...,a_{jm}\}$, each failure happens only in the subset of three elements that associated to a action in the set s_o, that is three elements exact cover problem and $ECD_o(s) = (m+1)/2$. Here ECD_o is expected cost of diagnosis in the condition of single-fault assumption, and O means only one fault. Because of single-fault assumption, the following formula establishes.

$$p(f_1 \ \ happens \vee ... \vee f_r \ \ happens) = \sum_{i=1}^{r} p(f_i \ \ happens) = r/3m \qquad (1)$$

By the formula (1), the following formula establishes.

$$p(f_1 \ \ unhappens \vee ... \vee f_r \ \ unhappens) = 1 - \sum_{i=1}^{r} p(f_i \ \ happens) = (3m-r)/3m$$

By $p(a_i \mid f_j \in T_i) = 1$ and the formula (1), the following formula establishes.

$$p(a_1 \ \ fails,...,a_l \ \ fails) = 1 - p(a_1 \ \ succeeds \vee ... \vee a_l \ \ succeeds)$$
$$= 1 - \sum_{f \in \cup_{i=1}^{l} T_i} p(f \ \ happens) = 1 - |\bigcup_{i=1}^{l} T_i| / 3m \qquad (2)$$

According to the agreement of each action corresponds to only three failures in the assumption 1, we can get $|\bigcup_{i=1}^{l} T_i|/3m = 3l/3m$, and formula (2) further reduced to the formula $p(a_1 \ \ fails,...,a_l \ \ fails) = 1 - 3l/3m = (m-l)/m$.

We can get the *ECD* of the array of diagnostic actions according to the agreement in assumption 1.

$$ECD_o(s_o) = \sum_{i=1}^{m} p(a_{j1} \ \ fails,...,a_{j(i-1)} \ \ fails) \cdot c_{a_{ji}} = \sum_{i=1}^{m} (\frac{m-(i-1)}{m}) \cdot 1 = \frac{m+1}{2}$$

Lemma 2. In the context of the single-fault assumption, for any diagnostic operation array $s_o = (a_1,...,a_r)$, the expected cost of diagnosis in solving the problem of $3m$ fault diagnosis is $ECD_o(s_o) \geq (m+1)/2$.

For any array of diagnosis actions s_o, its preferred action performs diagnostics with probability 1. According to assumption 1, the action up to solve the three failures, if the action is not able to determine the reason of the problem, the second action in the operations array s_o diagnose fault in at least the probability of $1 - 3/3m = 1 - 1/m$, and so on, the i th action in the operations array s_o diagnose fault in at least the probability of $1 - (i-1)/m$. We can get the ECD of the array of actions s_o as follow.

$$ECD_O(s_O) = \sum_{i=1}^{r} p(a_1 \quad fails,\ldots,a_{i-1} \quad fails) \cdot c_{a_i} \geq \sum_{i=1}^{m}(1 - \frac{i-1}{m}) = \frac{m+1}{2} \qquad (3)$$

Lemma 3. In the context of the single-fault assumption, there is an action a_x in an array of actions $s_o = (a_1,\ldots,a_x,\ldots,a_r)$ can only diagnose the remaining two or fewer failures (the faults associated with the action a_x may be completed firstly by the other action), the expected cost of the action array meets $ECD_o(s_o) > (m+1)/2$.

Because there is an action a_x in the array s_o can only diagnose the remaining two or fewer failures. Therefore, the diagnosis actions in the array s_o at least is $m+1$, that is $r > m$, the i th ($i > x$) action in the array s_O in the process of the diagnosis, the corresponding diagnostic probability is greater than or equal to $1 - 2 + 3 \cdot (i-2)/3m$, So we can get that the ECD_O of the array s_o as follow.

$$ECD_o(s_o) \geq \sum_{i=1}^{x}(1 - \frac{i-1}{m}) + \sum_{i=x+1}^{m+1}(1 - \frac{2 + 3(i-2)}{3m}) = \frac{m+1}{2} + \frac{1}{3m} > \frac{m+1}{2} \qquad (4)$$

3.2 Multiple Independent Failure Assumption

According to the hypothesis 1, we further present the fault diagnosis model based on the multiple faults problem.

We assume that there is a multiple fault problem and no correlation between faults. The case of the fault correlation will be presented in the following sections.

According to assumption 1 we can see that the probability of any failure is $p(f) = 1/3m$, and the probability of failure did not occur is $p(\neg f) = (3m-1)/3m$. ECD_M is the expected cost of diagnosis in the case of multiple faults, and M means multiple faults.

Lemma 6. In the case of multiple independent faults, if there are an array of actions $s_M = \{a_{j1},\ldots,a_{jm}\}$, and each fault of F is only the 3-subset associated by an action in s_M, which is the problem of the exact cover by 3-sets, and there is $ECD_M(s_M) =$

$\dfrac{m}{1 - p(\neg f)^{3m}} - \dfrac{1}{p(\neg f)^3 - 1}$.

When the network fails, we try to use the diagnostic array of actions s_M to discovery potential failures. It is the preferred action performs diagnostics with probability 1, if the action a_{j1} does not discovery the problem, the next action will be selected to perform. When we perform the $k+1$ th action in the array s_M, apparently the previous k actions did not find real failure. As the array s_M meets the problem of the exact cover by 3-sets, that is, no two operations can solve the same fault, there are still $3m-3k$ faults left, and no correlation between failures. So when the $k+1$ th action is performed, the probability of the remaining $3m-3k$ faults are $(1-p(\neg f)^{3m-3k})/p_0$. Then we can obtain the $ECD_M(s_M)$ of the array s_M as follow.

$$
\begin{aligned}
ECD_M(s_M) &= \sum_{k=0}^{m-1} p(a_{j1} \quad fails,\ldots,a_{jk} \quad fails) \cdot c_{j(k+1)} = \sum_{k=0}^{m-1} \frac{1-p(\neg f)^{3m-3k}}{p_0} \cdot 1 \\
&= \frac{m}{p_0} - \frac{p(\neg f)^{3m}}{p_0} \cdot \sum_{k=0}^{m-1} p(\neg f)^{-3k} = \frac{m}{p_0} - \frac{p(\neg f)^{3m}}{p_0} \cdot \frac{(p(\neg f)^{-3})^m - 1}{p(\neg f)^{-3} - 1} \\
&= \frac{m}{1-p(\neg f)^{3m}} - \frac{1}{p(\neg f)^{-3} - 1}
\end{aligned}
\tag{5}
$$

Lemma 7. In the context of multiple independent failure assumption, for any actions array $s_M = (a_1,\ldots,a_r)$, the expected cost of diagnosis in solving the problem of $3m$ is $ECD_M(s_M) \geq m/(1-p(\neg f)^{3m}) - 1/(p(\neg f)^3 - 1)$.

For any actions array s_M, whose preferred action perform diagnostics with probability 1, according to assumption 1, the action up to solve the three faults. If the first action is not able to determine the failures, the diagnosis probability of the second action in the array s_M is at least $(1-p(\neg f)^{3m-3})/p_0$. And so on, the diagnosis probability of the i th action in the array s_M is at least $(1-p(\neg f)^{3m-3(i-1)})/p_0$. According to the analysis, we can get the $ECD_M(s_M)$ of the array s_M as follow.

$$
\begin{aligned}
ECD_M(s_M) &= \sum_{i=1}^{r} p(a_1 \quad fails,\ldots,a_{i-1} \quad fails) \cdot c_{a_i} \\
&\geq \frac{1}{p_0} \sum_{i=1}^{m} (1-p(\neg f)^{3m-3(i-1)}) \cdot 1 = \frac{m}{1-p(\neg f)^{3m}} - \frac{1}{p(\neg f)^{-3} - 1}
\end{aligned}
\tag{6}
$$

Lemma 8. In the context of multiple independent failure assumption, if there is an action a_x in a array $s_M = (a_1,\ldots,a_x,\ldots,a_r)$, that can only diagnose the remaining two or fewer failures (the part faults in the set may in advance to be diagnosed by the other actions), the expected cost of diagnosis of the array is $ECD_M(s_M) > m/(1-p(\neg f)^{3m}) - 1/(p(\neg f)^3 - 1)$.

Because there is an action a_x in the array s_M that can only diagnose the remaining two or fewer faults, the number of actions in the array s_M is at least $m+1$, that is $r > m$. When the i th action of the array s_M performs, the corresponding probability is greater than or equal to $(1 - p(\neg f)^{3m-3(i-2)-2})/p_0$, so we can get the expected cost of diagnosis $ECD_M(s_M)$ of the array s_M as follow.

$$ECD_M(s_M) \geq \frac{1}{p_0}\left[\sum_{i=1}^{x}(1 - p(\neg f)^{3m-3(i-1)}) + \sum_{i=x+1}^{m+1}(1 - p(\neg f)^{3m-3(i-2)-2})\right]$$

$$> \frac{m}{1 - p(\neg f)^{3m}} - \frac{1}{p(\neg f)^{-3} - 1}$$

3.3   The Model Based on the Condition of Dependent Faults

When there are correlations between faults, whose characteristic cannot be described by the assumption 1, we further describe the problem of fault diagnosis.

Assumption 2. For the characteristics of the correlation between faults, we present the following simplifying assumptions.

(1) Suppose F is the faults set, and there is $|F| = m, m \in \mathbb{N}$. For a $f_i \in F$, there is as, that is occurrence of fault has the same probability.

(2) All factors set related the fault set F is named U, and there is $|U| = 3n, n \in \mathbb{N}$.

(3) For a fault f_i, the factors set related the fault f_i is named U_i, suppose there is $|U_i| = 3$.

(4) The action a_i only correspond the fault f_i, that is the action a_i diagnose the fault f_i with probability one, diagnose the non-fault f_i with probability zero, and there is $p(a_i \; succeeds \,|\, f_j \; happens) = 1$ and $p(a_i \; succeeds \,|\, f_j \; unhappens) = 0$ and the action cost $c_{a_i} = 1$ is a constant.

We can see that the simplified fault diagnosis is the subset of fault diagnosis described above. If the simplified problem of fault diagnosis is NP-complete problem, then the problem described above is obviously NP-complete problem.

The failure occurs with the same probability and the action cost is a constant, so we can randomly select an action a_k to diagnose the corresponding fault f_k. If the fault f_k is not a true failure reason, we try to select the next operation. Because there is the correlation between the two faults, for any two faults f_i and f_j, if $U_i \cap U_j \neq \varnothing$, there is the correlation between the faults f_i and f_j. Therefore if there are the subsequent faults associated with the faults f_k, which is clearly not optimal, and we can choose the action a_l with $U_k \cap U_l = \varnothing$. So we finally try to get an array of actions named s, which with the highest efficiency and the least cost of diagnosis.

4   Conclusions

We have developed the model based on the condition of dependent diagnosis actions and the model based on the condition of dependent faults. Through a series of theoretical support, we have seen that our diagnostic model produces expected cost of diagnosis that are close to the optimal result and the lower than the simple planners for a domain of troubleshooting. In future work, we shall relax the assumptions used for approximating the expected costs of repair for various scenarios, specifically examining extensions for dependent costs of repair.

Acknowledgments. This work is supported by the Natural Science Foundation of China (No.61379148).

References

1. Infiniband Trade Association. Infiniband Arichitecture Specification Volume 1. Release 1.2.1 (November 2007)
2. Dongarra, J.: Visit to the National University for Defense Technology Changsha, China (June 3, 2013). http://www.netlib.org/utk/people/JackDongarra/PAPERS/tianhe-2-dongarra-report.pdf
3. Gill, P., Jain, N., Nagappan, N.: Understanding network failures in data centers: Measurement, analysis, and implications. In: Proceedings of ACM SIGCOMM (August 2011)
4. Song, H., Qiu, L., Zhang, Y.: NetQuest: A flexible framework for large-scale network measurement. IEEE/ACM Transactions on Networking **17**(1), 106–119 (2009)
5. Brodie, M., Rish, I., Ma, S., Odintsova, N., Beygelzimer, A.: Active probing strategies for problem diagnosis in distributed systems. In: Proc. IJCAI 2003 (2003)
6. Rish, I., Brodie, M., Ma, S., Odintsova, N., Beygelzimer, A., Grabainik, G., Hernandez, K.: Adaptive Diagnosis in Distributed Systems. IEEE Trans. Neural Networks **16**(5), 1088–1109 (2005)
7. Tang, Y., Al-Shaer, E., Boutaba, R.: Active integrated fault localization in communication networks. In: Proc. IM 2005, France (2005)
8. Tang, Y., Al-Shaer, E., Boutaba, R.: Efficient fault diagnosis using incremental alarm correlation and active investigation for internet and overlay networks. IEEE Trans. Network and Service Management **5**(1), 36–49 (2008)
9. Natu, M., Sethi, A.S.: Probabilistic Fault diagnosis using adaptive probing. In: Clemm, A., Granville, L.Z., Stadler, R. (eds.) DSOM 2007. LNCS, vol. 4785, pp. 38–49. Springer, Heidelberg (2007)
10. Natu, M., Sethi, A.S.: Application of adaptive probing for fault diagnosis in computer networks. In: NOMS 2008, pp. 1055–1060 (2008)
11. Cheng, L., Qiu, X.S., Meng, L.M., Qiao, Y., Li, Z.Q.: Probabilistic fault diagnosis for IT services in noisy and dynamic environments. In: Proc. IM 2009, New York (June 2009)
12. VomlelovÁ, M.: Decision Theoretic Troubleshooting. PhD Thesis, Faculty of Informatics and Statistics, University of Economics, Prague, Czech Republic (2001)

A Robust Point Sets Matching Method

Xiao Liu, Congying Han[(✉)], and Tiande Guo

School of Mathematical Science, University of Chinese Academy of Sciences,
Beijing, China
hancy@ucas.ac.cn

Abstract. Point sets matching method is very important in computer
vision, feature extraction, fingerprint matching, motion estimation and
so on. This paper proposes a robust point sets matching method. We
present an iterative algorithm that is robust to noise case. Firstly, we
calculate all transformations between two points. Then similarity matrix
are computed to measure the possibility that two transformation are
both true. We iteratively update the matching score matrix by using the
similarity matrix. By using matching algorithm on graph, we obtain the
matching result. Experimental results obtained by our approach show
robustness to outlier and jitter.

Keywords: Point sets matching · Similarity matrix · Robust matching
method · Point to point

1 Introduction

The point sets matching problem is to find the optimum or suboptimal spatial
mapping between the two point sets. In its general form, the point sets matching
problem always consists of point sets with noise, outliers and geometric transformations. A general point sets matching algorithm needs to solve all these
problems. It should be able to find the correspondences between two point sets,
to reject outliers and to determine a transformation that can map one point set
to the other.

Some popular methods solve the problem of affine point sets matching. This
basic model searches interest points and tries to search the affine invariant points
in an affine Gaussian scale space [1]. A developed algorithm with the thin-plate
spline as the parameterizations of the non-rigid spatial mapping and the soft
assign for the correspondence was suggested [2]. [3] proposed a novel approach
for affine point pattern matching. When one point sets is rotated to match the
other point set, the value of a cost function which is invariant under special affine
transformations would be calculated. When the value reached the minimum, the
affine transformations were got. Using partial Hausdorff distance, a genetic algorithm (GA) based method to solve the incomplete unlabeled matching problem
was presented [4].

[5] proposed a sparse graphical model for solving the rigid point pattern
matching. Obviously, the point sets is larger, the matching time is much longer.

© Springer International Publishing Switzerland 2015
Y. Tan et al. (Eds.): ICSI-CCI 2015, Part III, LNCS 9142, pp. 383–390, 2015.
DOI: 10.1007/978-3-319-20469-7_41

To speed up the matching, it's better to match the subsets firstly to get the initial transformation parameters. [6] proposed an asymptotically faster algorithm under rigid transformations, and provided new algorithms for homothetic and similarity transformations. The point sets matching problem arises in the domains of computer vision. A common application is image registration [7] [8] [9] [10].

Another common application is super resolution which is the process of combining a sequence of low-resolution noisy blurred images to produce a higher resolution image or sequence. [11] proposed a method for matching feature points robustly across widely separated images.

In this paper, we propose a novel solution to the rigid point sets matching. Although we assume rigid motion, jitter and noise are allowed. This property is very important because noise and jitter are inevitable from the process of image acquisition and feature extraction. The rest of this paper is organized as follows. Section 2 introduces the assumption between point sets. Section 3 describes the main contribution of this paper, which is a robust matching algorithm to the rigid point sets. Simulations on synthetic data are presented in Section 4, and Section 5 concludes this paper.

2   Statement of the Problem

Suppose two point sets A and B in two dimensions are given. That is A $= \{p_1, p_2 \cdots p_m\}$ and B $= \{q_1, q_2 \cdots q_n\}$, where the p_i and q_j are points in R^2. We want to find a global similarity transformation T_{θ, t_x, t_y}, such that T(A) "matches" some subset of B, where matching will be made precise below. In the transformation T, θ is a rotation angle, t_x and t_y are the x and y translations, respectively. That is, for $(x, y) \in R^2$, we have

$$T \begin{pmatrix} x \\ y \end{pmatrix} = \begin{pmatrix} t_x \\ t_y \end{pmatrix} + \begin{pmatrix} \cos\theta & -\sin\theta \\ \sin\theta & \cos\theta \end{pmatrix} \begin{pmatrix} x \\ y \end{pmatrix} \tag{1}$$

The first condition dictates that there exists a rigid transformation which should match most of points between A and B. In other word, only a few points in set are noise points which are random and disordered.

The second condition states the optimal transformation cause similarity between neighbors. Suppose p_i and p_j are neighbors in A, the transformation T matches p_i and p_j to q_k and q_l, respectively. If T is optimal, q_k and q_l are like to be neighbors.

In this paper, we assume the point sets in R^2 is directed. The point feature, represented by feature location, is the simplest form of feature. However, in many applications, such as motion estimation and fingerprint matching, the point features have both location and direction.

3   Robust Directed Point Sets Matching

We now briefly describe our proposed methodology.

To simplify the description, we use P and Q to represent an arbitrarily point in A and B, respectively. The coordinates of P and Q are (P_x, P_y) and (Q_x, Q_y), respectively. The direction of P and Q is P_θ and Q_θ, respectively.

The transformation T_{θ,t_x,t_y} from P to Q, which means $T(P) = Q$, are firstly calculated. Since points are directed, the rotate angle θ is $Q_\theta - P_\theta$. Then the transformation T_{θ,t_x,t_y} is got, and

$$\begin{cases} t_x = Q_x - (P_x cos\theta - P_y sin\theta) \\ t_y = Q_y - (P_x sin\theta + P_y cos\theta) \end{cases} \tag{2}$$

The similarity between two transformations is considered. The correct matching should be similar to many other correct matching, while the incorrect matching not because of its randomness. If we randomly choose two matching and compute the similarity between them, the similarity between both true matching is significantly bigger than the similarity between both random matching.

Given two transformation T_{θ,t_x,t_y} and T'_{θ',t'_x,t'_y}, we have the similarity as:

$$similarity(T, T') = \begin{cases} 0, if \max\{\frac{|t_x - t'_x|}{\alpha}, \frac{|t_y - t'_y|}{\beta}, \frac{|\theta - \theta'|}{\delta}\} > 1 \\ 1 - \{\frac{|t_x - t'_x|}{\alpha} + \frac{|t_y - t'_y|}{\beta} + \frac{|\theta - \theta'|}{\delta}\}/3, otherwise \end{cases} \tag{3}$$

where α and β are threshold of x and y, respectively, δ is threshold of angle.

Let us define T_{ij} as the transformation $T_{ij}(p_i) = q_j$. Take similarity between neighbors into consideration, we iteratively update matching score matrix using

$$W^{t+1}(i,j) = W^t(i,j) + \sum_{k \in SMN_{A,i}, l \in SMN_{B,j}} W^t(k,l) similarity(T_{ij}, T_{kl}) \tag{4}$$

where W^t represents matching score matrix in the tth iteration, $SMN_{A,i}$ means the neighbor points of point i in A, $SMN_{B,j}$ means the neighbor points of point j in B.

There are several ways to define the neighbor relationship in point sets. In this paper, we use the K nearest points in Euclidean distance to define the neighbor relationship. The magnitude of K will be discussed in experiments.

It is obvious that the matching score matrix always increase during the update. To transform the matching score matrix into the ideal score interval, matching score matrix will be linearly normalized after each iterative.

While the iterations stop, we model point sets matching as a weighted graph matching problem, where weights correspond to the matching score matrix. Then we can find a maximum score by Kuhn-Munkres algorithm.

4 Experiments

The synthetic point sets are generated as follows. First, a random point sets, named A, whose number is given as N, is generated. Then, another point sets B is generated from rigid rotation and translation on A. It is obvious that each point in A has a unique corresponding point in B. Randomly choose some point matching pairs from A to B, use the random noise points to replace them. The ratio of number of noise points to number of whole points is called outlier ratio. The rest matching pairs are added by a jitter value, which is random chose. The ratio of jitter value's range to whole point sets' range is called jitter ratio.

We use the average correct point pair ratio (ACPPR) to measure the robustness. The correct point pair ratio is defined as follows:

$$Average\ Correct\ Point\ Pair\ Ratio = \frac{correct\ matching\ pairs'\ number}{all\ the\ correct\ matching\ pairs'\ number} \tag{5}$$

To reduce the uncertainty on initializing point sets, we repeated experiments and then calculated the average correct point pair ratio. In this paper, all ACCPR values are calculated from at least 20 simulation experiments.

Fig. 1 reports the correlation between ACPPR and K. The horizontal axis (K) is the number of neighbors to one point, dependent on different N which is the number of points of A. From the Fig. 1, it is well evident that ACPPR increases obviously and reaches to 95%, when K increases. An important result is that, when K is bigger than $[\frac{N}{4}]$, ACPPR converges to a limit, no matter how many N is. If K is bigger enough to N, the magnitude of N is independent of ACPPR. The further discussion is shown below.

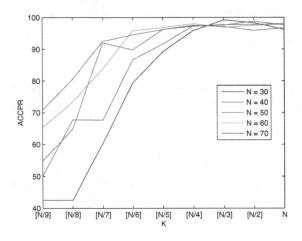

Fig. 1. The correlation between ACPPR and K. Outlier ratio is 20%, jitter ratio is 8%. And the (α, β, δ), which are thresholds to measure the similarity between transformations in Eq.3, are $(10, 10, \pi/6)$.

Fig. 2 reports the correlation between ACPPR and outlier ratio. While outlier ratio increases, noise points increase and true matching point pairs decrease. Take K equals 12 as an example, while outlier ratio increases from 10% to 50%,

ACPPR slowly decreases from 98% to 87%. This phenomenon indicates that our method is robust when true pairs are more than noises. However, while outlier ratio increases from 50% to 80% , ACPPR rapidly decreases from 87% to 16%. This is an inevitable trend. When the outlier ratio is more than 50%, noise points are more than true matching points, so that the correct and meaningful information has been covered up by the wrong and meaningless information.

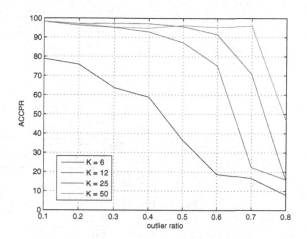

Fig. 2. The correlation between ACPPR and outlier ratio on different K. N is 50, jitter ratio is 8%, (α, β, δ) is $(10, 10, \pi/6)$. The options of K (6,12,25,50) represent $([\frac{N}{8}], [\frac{N}{4}], [\frac{N}{2}], N)$, respectively.

Fig. 3 reports the correlation between ACPPR and jitter ratio on different K. When K is no less than $[\frac{N}{4}]$, ACPPR decreases slowly. While jitter ratio reaches 12%, ACPPR is more than 85%. This graph illustrates our method is robust on jitter ratio when K is not too much less than N.

We tested our method on the condition that both outlier ratio and jitter ratio are variable. Table.1 reports the trend of ACPPR change caused by outlier ratio and jitter ratio. In Table.1 we show the performance of our algorithm for outlier ratio from 0% to 60%, different jitter ratio from 0% to 12%, (a),(b),(c),(d) are different of K, which are (6,12,25,50), respectively. N = 50, $(\alpha, \beta, \delta) = (10, 10, \pi/6)$.

Take Table.1(c) as an example. While outlier ratio increases quickly from 0% to 50%, ACPPR decreases slowly and is 70.6% even in the worst situation (outlier ratio = 60%, jitter ratio = 12%). While jitter ratio increases from 0% to 12%, ACPPR decreases from 100% to 87% in average ACCPR of all outlier ratios. We compare average ACPPR among different sub tables. Our method performs badly when K is small as 6. This is because the size of K decides the number of neighbor points. The more neighbor points, the more information is got. Small K results in lack of robustness.

Table 1. Comparisons of ACPPR for different K, outlier ratios and jitter ratios

(a) K=6

ACPPR(%)		Jitter ratio (%)							Average (%)
		0	2	4	6	8	10	12	
Outlier Ratio (%)	0	100	100	90.4	90.4	74	50	36.8	77.4
	10	97.2	98.9	94.7	87.7	85.9	69.7	38.3	81.8
	20	92.3	93.9	91.8	88.9	72.4	55.9	28.4	74.8
	30	93.8	92.6	82.2	81.5	57.6	52.1	32.1	70.3
	40	81.3	77.1	77.5	75.6	60.7	33.5	22.7	61.2
	50	70.7	71.2	69.8	65.1	39.8	32.3	14.2	51.9
	60	60	55.4	49.2	48.8	21.2	11	4.4	35.7
Average (%)		85	84.2	79.4	76.9	58.8	43.5	25.3	64.7

(b) K=12

ACPPR(%)		Jitter ratio (%)							Average (%)
		0	2	4	6	8	10	12	
Outlier Ratio (%)	0	100	100	100	100	98.4	96	92	98.1
	10	100	100	99.5	98.4	98.5	93.7	88	96.9
	20	100	100	99.8	99	95.3	92.7	89.4	96.6
	30	99.9	100	99.7	98.7	92.4	88.5	66.7	92.3
	40	99.7	99.7	99.3	96.1	93.7	82.5	59.6	90.1
	50	98.9	98.1	99.2	95.4	87.7	74.4	53.6	86.8
	60	92	96.2	95.2	79.8	69.4	63.6	23.2	74.2
Average (%)		98.6	99.1	99	95.3	90.8	84.5	67.5	90.7

(c) K=25

ACPPR(%)		Jitter ratio (%)							Average (%)
		0	2	4	6	8	10	12	
Outlier Ratio (%)	0	100	100	100	100	98.4	97.6	92.8	98.4
	10	100	100	100	99.8	98.9	95.9	91.8	98.1
	20	100	100	100	99.4	98.6	95.8	93.2	98.1
	30	100	100	100	99.4	97.1	94.6	88.3	97.1
	40	100	99.7	99.5	98.8	96.8	92	88.3	96.4
	50	100	99.4	98.9	98.2	96.6	94.4	84.2	96
	60	100	99.4	98.2	98	90.4	75.6	70.6	90.3
Average (%)		100	99.8	99.5	99.1	96.7	92.3	87	96.3

(d) K=50

ACPPR(%)		Jitter ratio (%)							Average (%)
		0	2	4	6	8	10	12	
Outlier Ratio (%)	0	100	100	100	100	99.2	98.4	98.4	99.4
	10	100	100	100	99.1	97.9	96.8	92.2	98
	20	100	100	100	99.1	98.2	96.6	93.9	98.3
	30	100	100	99.9	98.6	96.9	93.8	89.9	97
	40	100	100	99.7	98.8	98.1	94.1	88.8	97.1
	50	100	100	98.7	98.2	97.1	93.1	89.8	96.7
	60	99.6	99.8	99.6	98.6	96.8	89.4	87.4	95.9
Average (%)		99.9	100	99.7	98.9	97.7	94.6	91.5	97.5

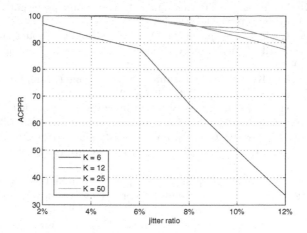

Fig. 3. The correlation between ACPPR and jitter ratio on different K. N is 50, outlier ratio is 20%, (α, β, δ) is $(10, 10, \pi/6)$. The options of K $(6,12,25,50)$ represent $([\frac{N}{8}], [\frac{N}{4}], [\frac{N}{2}], N)$, respectively.

5   Conclusion

In this paper, we have proposed a novel method for point sets matching. While our method resembles other point sets matching methods in calculating transformation, our method is novel in that it uses the similarity between transformations and the influence of neighbors. More specifically, our method does not depend on only point location, but also the point direction. An important implication of this property is that we can expect our method to work on some special situation, such as motion estimation in super resolution.

Besides, we believe the use of direction and similarity of transformation can better matching result.

Acknowledgments. This work was funded by the Chinese National Natural Science Foundation(11331012, 71271204 and 11101420).

References

1. Mikolajczyk, K., Schmid, C.: An affine invariant interest point detector. In: Heyden, A., Sparr, G., Nielsen, M., Johansen, P. (eds.) ECCV 2002, Part I. LNCS, vol. 2350, pp. 128–142. Springer, Heidelberg (2002)
2. Chui, H., Rangarajan, A.: A new point matching algorithm for non-rigid registration. Computer Vision and Image Understanding **89**(2), 114–141 (2003)
3. Suesse, H., Ortmann, W., Voss, K.: A novel approach for affine point pattern matching. In: Campilho, A., Kamel, M. (eds.) ICIAR 2006. LNCS, vol. 4142, pp. 434–444. Springer, Heidelberg (2006)

4. Zhang, L., Xu, W., Chang, C.: Genetic algorithm for affine point pattern matching. Pattern Recognition·Letters, 9–19(2003)

5. Caetano, T.S., Caelli, T., Schuurmans, D., et al.: Graphical models and point pattern matching. IEEE Transactions on Pattern Analysis and Machine Intelligence, 1646–1663 (2006)

6. Aiger, D., Kedem, K.: Approximate input sensitive algorithms for point pattern matching. Pattern Recognition, 153–159 (2010)

7. Fischer, B., Modersitzki, J.: Intensity based image registration with a guaranteed one-to-one point match. Methods of information in medicine, 327–330 (2004)

8. Modersitzki, J., Fischer, B.: Optimal image registration with a guaranteed one-to-one point match. In: Bildverarbeitung fr die Medizin, pp. 1–5. Springer, Heidelberg (2003)

9. Wang, H., Fei, B.: A robust B-Splines-based point match method for non-rigid surface registration. The 2nd International Conference on Bioinformatics and Biomedical Engineering, ICBBE 2008, pp. 2353–2356. IEEE (2008)

10. Dubuisson, M.P., Jain, A.K.: A modified hausdorff distance for object matching. In: Proceedings of the 12th IAPR International Conference on Pattern Recognition. -Conference A: Computer Vision and Image Processing, pp. 566–568. IEEE (1994)

11. Choi, O., Kweon, I.S.: Robust feature point matching by preserving local geometric consistency. Computer Vision and Image Understanding, 726–742 (2009)

Machine Translation

An Enhanced Rule Based Arabic Morphological Analyzer Based on Proposed Assessment Criteria

Abdelmawgoud Mohamed Maabid[(⊠)], Tarek Elghazaly, and Mervat Ghaith

Department of Computer and Information Sciences,
Institute of Statistical studies and Research, Cairo University, Giza, Egypt
amawgoud@pg.cu.ed.eg, tarek.elghazaly@cu.edu.eg,
mervat_gheith@yahoo.com

Abstract. Morphological analysis is a vital part of natural language processing applications, there are no definitive standards for evaluating and benchmarking Arabic morphological systems. This paper proposes assessment criteria for evaluating Arabic morphological systems by scrutinizing the input, output and architectural design to enables researchers to evaluate and fairly compare Arabic morphology systems. By scoring some state of the art Arabic morphological analyzers based on the proposed criteria; the accuracy scores showed that the best algorithm failed to achieve a reliable rate. Hence, this paper introduced an enhanced algorithm for resolving the inflected Arabic word, identifies its root, finds its pattern and POS tagging that will reduce the search time considerably and to free up the deficiencies identified by this assessment criteria. The proposed model uses semantic rules of the Arabic language on top of a hybrid submodel based on two existing algorithms (Al-Khalil and An Improved Arabic morphology analyzer IAMA rules).

Keywords: Morphology · Arabic morphology · NLP · Morphology assessment criteria · Stemmer · Morphology benchmarking

1 Introduction

Morphology in linguistics concerns with the study of the structure of words [1]. In other words, morphology is simply a term for that branch of linguistics concerned with the forms words take in their different uses and constructions [2].Arabic is one of the languages having the characteristics that from one root the derivational and inflectional systems are able to produce a large number of words (lexical forms) each having specific patterns and semantics [3]. The root is a semantic abstraction consisting of two, three, or (less commonly) four consonants from which words are derived through the superimposition of template based patterns [4]. Unfortunately if understanding is considered, non-diacritic words may make problems of meaning; where many words when they appears in non-diacritic text can have more than one meaning; these different meanings rises problems of ambiguity[5].

In Arabic, like other Semitic languages, word surface forms may include affixes, concatenated to inflected stems. In nouns, prefixes include conjunctions ("و" "and",

© Springer International Publishing Switzerland 2015
Y. Tan et al. (Eds.): ICSI-CCI 2015, Part III, LNCS 9142, pp. 393–400, 2015.
DOI: 10.1007/978-3-319-20469-7_42

ف "and, so"), prepositions ("بـ" "by, with", "كـ" "like, such as", "لـ" "for, to") and a determiner, and suffixes include possessive pronouns. Verbal affixes include conjunction prefixes and negation, and suffixes include object pronouns. Either object or possessive pronouns can be captured by an indicator function for its presence or absence, as well as by the features that indicate their person, number and gender [6]. A large number of surface inflected forms can be generated by the combination of these features, making the morphological generation of these languages a non-trivial task [7].

Morphological analysis can be performed by applying language specific rules. These may include a full-scale morphological analysis, or, when such resources are not available, simple heuristic rules, such as regarding the last few characters of a word as its morphological suffix. In this work, we will adapt some major assessment criteria for measuring advantage or drawback of any Arabic morphological system [8].

2   Background and Previous Work

We believe that this is the first proposed work to sum up assessment criteria for Arabic morphological analyzers and Generators. Several researches talked about building powerful stemmers for the Arabic language with accuracies normally exceeding 90% but none of these stemmers offer the source code and/or the datasets used. It is therefore difficult to verify such claims or make a comparison between different stemmers without having the full description of the proposed method or the source code for the implementation of the algorithm [9].

Mohammed N. Al-Kabi and Qasem A. Al-Radaideh [9] proposed analysis of the accuracy and strength of four stemmers for the Arabic language using one metric for accuracy and four other metrics for strength.

Al-Khalil Arabic Morphological System [10]; is java compiled application published on April 2010. The system can analyze a word or sentences typed in the text area.   The system can analyzes up to 10 words per second in small text and up to 35 words per second in larger documents.

An Improved Arabic morphology analyzer (IAMA) [11] is an algorithm has been developed by using some new semantic rules of the Arabic language to reduce the searching time in the ATN.   Also, this research introduces an algorithm for root identification that will reduce the search time considerably.

3   Proposed Assessment Criteria of Arabic Morphologies

Assessing and evaluating Arabic morphological systems depends on the input words and resulted output [12] according to a predefined criteria to measure and analyze given system in order to study its weakness and strength, trying to find an Arabic morphological analyzer free from all mistakes.

Assessments are carried out by executing some of the available Arabic morphological systems on a randomly selected Arabic political news article, an Arabic Sport News article "from Al-Ahram newsletter" and chapter number 36 of the Holy Qur'an "Surah Yassin". We then manually extracted the roots of the test documents' words to compare results from different stemming systems, thus creating our baseline. Roots extracted were then checked manually in an Arabic dictionary. Voting weights are assigned to each assessment item in order to accurately make comparisons between these algorithms. Here is the step by step procedure of executing the assessment criteria:

1. Manually extract the roots of the test documents' words.
2. Assign voting mark for each assessment item.
3. Manually check the extracted roots against Arabic dictionary.
4. Apply each assessment item separately on each of Arabic Morphological system.
5. For the output results, check them manually against Arabic dictionary.

Finally, we applied the assessment items separately on each of Arabic Morphological system and all items have been assigned a maximum value of 100 marks. Each assessment item has been applied and calculated as per system result of applying the analysis of the sample document words.

Table 1. Assessment Criteria Items

	Factor No.	Assessment Criteria	Wight
Input	1	The possibility of analyzing the modern standard texts	100
	2	The possibility of analyzing the common error words	100
	3	The possibility of analyzing new words	100
	4	Processing of Arabized and transliterated words	100
	5	Processing of non- tripartite verbs.	100
Output	6	Covering analysis of all input words	100
	7	Meet all possible cases for analysis	100
	8	Express grammatical function of the affixes	100
	9	Ambiguity and Overlapping of syntactic cases	100
	10	Identifying the root of the word and determining all possible roots	100
	11	Grammatical errors and misspellings in the context of the results of the analysis	100
	12	Cover all possible cases of syntactic word analyst	100
	13	Consistency between analyzed word and its patterns	100
	14	The result has to be coming from Arabic dictionary	100
System Architecture	15	Percentage of non-reliance on predefined knowledgebase of affixes	100
	16	Percentage of non-reliance on common words	100
	17	Processing Speed	100
	18	Ease of use and integration with larger applications	100
	19	Availability, documentation and customization	100
	20	User Interface (English - Arabic)	100
	21	Encoding and word representation	100
Sum			2200

4   Proposed Arabic Morphological Analyzer System

The proposed system will be a redesign, reuses and enhancement of Al-Khalil Arabic Morphological System [10] based on some of the Arabic morphology analyzer rules designed by Saad (IAMA) [11]. Al-Khalil Arabic morphological system is an open source application written using Java language and published on April 2010. The proposed system uses the modified version of Al-Khalil's knowledgebase. This modification proposed here uses linguistic algorithms and approach used for stem and root identification based on predefined linguistic rules and knowledgebase. The proposed system consists of five components, an Arabic Tokenizer, a Common word& non-derivational nouns extractor, a Stem identifier, a Stem refiner and a Root identifier. The first component, Arabic Tokenizer, receives as input the Arabic text and then converts it into a stream of tokens. The Arabic word may be a verb, noun or common word [13]. Each type may be inflected or not. The first step in the morphology analysis is to identify the common words and the non-derivational nouns. The remaining tokens are assumed to be inflected words.

4.1   Arabic Word Tokenizer

Word Tokenizer is first step in the proposed system which takes the input text and splits it into basic components where each component represents one Arabic word with no spaces or punctuation marks. The Arabic word may be a verb, noun or common word. Algorithm 1 illustrates the idea.

— Find the input text and i
— Split i into basic words ws
— For each word w in ws find the type of w.
— Return a token of w and its non-diacritic form

4.2   Word Analyzer

The analyzer receives as input a stream of tokens one by one and applies some linguistic rules to determine the inflection result of given token. Each token type may be inflected or not. The first step in the morphology analysis is to identify the exceptional words, common words (حروف الجر, حروف النصب, أدوات النداء ... الخ) and the non-derivational nouns (علم, اسم جنس ...الخ). The remaining tokens are assumed to be inflected words. The inflection means that, either there are affixes attached to the word or there are some inflectional modifications [14].

Word Segmentation. If the input word is not exceptional word then the analyzer will process the input word as segments. The segmentor is stem identification based component responsible for extract all possible valid Arabic components of the input word where the differences of these components are in affixes. For example the word 'بسم' "Besm" can be dealt as two components, 'بسم' "In the name of" or 'سم' "poison" with 'بـ' "with" as a prefix attached with it.

Stem Identification. To find the stem, the word inflection must be resolved. The Arabic word has to satisfy the following formula[11]:

$$[Prefix1|prefix2]+Stem+suffix1+suffix2+suffix3$$

Where affixes are list of prefixes and suffixes attached at beginning and the end of any Arabic word respectively, the affixes serve for a special linguistic purpose. Based on ordering and linking of the affixes to each other and to the stem, the inflection was resolved.

The system in hand applies new semantic and linguistic rules to reduce the number of search possibilities. Studying the Arabic affixes carefully will show that there is a semantic contradiction between these affixes determined with affix classes stated in the affix list. For example the prefixes for the present verbs cannot be joined in the same word with the suffixes which can be attached to past verbs only, (e.g. "ﺕ" "t" cannot be joined, as a suffix, to present verbs, "ﻓﺲ" "fas" cannot be joined, as a prefix, to the past verbs or the imperative).The same idea applies to the affixes that can be attached to nouns only and not to verbs. The division of these affixes into contradicted groups, then building separate methods for the non-contradicted groups only, will reduce the searching time considerably. Now, the following constraints can be formulated which can be applied during the process of the stem identification to determine a validity of stem; where the valid stem has to satisfy the following rules.

6. Stem length between 2 and 9 characters
7. Stem with prefix of class N cannot be suffixed by V class
8. Stem with prefix of class V cannot be suffixed by N class
9. Stem with prefix of class N1 , N2, N3 or N5 has to suffixed with any valid suffix

4.3   Tool Word Analysis

Tool word can be extracted from a predefined list of tool words where each are defined with type of suffixes and prefixes can be attached with and diacritic and non-diacritic form of the word.

The tool word extractor method uses the exact matches to identify all possible tool words form given segment where the segment has to satisfy the following conditions:

— The diacritic form of stem equals non- diacritic form property in tool words.
— The prefix class and suffix class of the stem in the allowed prefixes and suffixes of the tool word respectively.

4.4   Arabic Nouns Analysis

Arabic nouns classified into two types; proper noun and nominal nouns:

- **Proper nouns**: can be identified of any word only if non- diacritic form of input segment matches the non- diacritic form of predefined list of proper nouns and the segment's prefix is not in class of 'C'.

- **Nominal nouns**: using root identification algorithm to find the patterns of the input stem, nominal nouns are assumed to satisfy the following rules:

— The word with Fathatan, Kasratan or Dammatan must has any valid suffix
— The word with Fathatan, Kasratan or Dammatan cannot be prefixed with any prefix in class of N1, N2,N3, or N5
— Prefixes of N2, C2 and C3 are not valid with noun categories NCG (13 to 18)
— Prefixes of N4 and N5 are not valid with noun categories NCG (13 to 18)
— If Hamza appears in the diacritized form of the segment then the Hamza letter must be valid against hamza rule
— The diacritized form has to satisfy valid contaminative characters.

4.5   Arabic Verbs Analysis

Arabic verbs are assumed to satisfy the following formula:

$$P1+Stem+S1+S2$$

Where, P1 is the set of conjugates and special articles, S1 is the subject, and S2 is the first object. The second object is ignored, since one word that contains a verb plus a subject plus the first object plus the second object, appears rarely in the modern Arabic texts [11]. The proposed system uses the same algorithm of matching the nominal patterns to identify all possible roots of the verb.

4.6   Stem Refiner

The proposed system serves for the purpose of omitting the affixes. To deal with the stem omitted or converted characters a new component has been added. An important fact will be explained which results in the need of such component. The fact is that, Arabic verbs and derivational nouns have finite number of patterns for those roots with no weak characters (ا, و, ى) or hamza (ء). If the roots have weak characters or hamza, these basic patterns will have variants. One solution to find the root is to write down all these variants and then matching the word with them [13]. Applying this approach, 101 pattern categories to account for all pattern variants were used in; then matching the stem with these patterns to find a root. The same approach can be also applied to Arabic nouns. The omitted characters are manipulated in the next section. The proposed system stores the available conversion rules with all patterns.

4.7   Root Identification

This section is concerned with the root identifier component. The root identifier component is responsible for matching the stems with the stored patterns. Then the root can be easily identified. Another approach were to develop an algorithm that examines every combination of characters from the stem and check if this combination is a valid root [15], then it produces the pattern. This approach will be superior to that used in [13], but it still needs to store the patterns to check if the legality of produced pattern.

The approach in hand needs only the basic patterns set as mentioned previously since, the stem refiner component restores any converted characters in the stem due to the presence of the weak characters or the hamza in the root.

The final point is to deal with the omitted characters. The omitted characters in Arabic verbs are the weak characters or the hamza (if found in the root). This omission occurs under certain circumstances. This will cause the basic patterns to have variants. One solution to this problem is again to write these variants. Another one depends on a new matching Algorithm. The algorithm is built on the fact that the omitted characters must be elements from the set of weak characters and the hamza. The algorithm is further illustrated in the following points:

— Find the stem length L.
— Get undiacritized patterns whose lengths are L with related rules.
— Get all the pattern excess characters (all characters except for ف ع ل)
— Build result pattern list R
— Remove these characters form the corresponding locations in the stem.
— For each rule in pattern rules replace the corresponding numeric characters with stem character to build new root r.
— For new roots r replaces (أ ، و ، ى) with (ء)
— Match r in non-diacritized patterns
— Repeat for all patterns in step 2.

5   Conclusion and Future Research

The proposed assessment criteria are adapted to measure Arabic Morphological Analyzers with some features intended for integration with lager applications in natural language processing. Many other criteria can be added to the proposed items and may vary in weight and phase of testing; similar to the source code related metrics used for measuring the system as a product.

The stemming algorithms involved in the experiments agreed and generate analysis for simple roots that do not require detailed analysis. So, more detailed analysis and enhancements are recommended as future work.

The new algorithms used in the proposed morphological analyzer are minimal in searching time as explored in the previous sections. The proposed system is directed to the standard modern Arabic that covers non- diacritized, partially diacritized and fully diacritized words. It can resolve the inflected Arabic word, identifies its root, finds its pattern and POS tagging.

By rating the proposed system with this baseline standard measurement showed that it achieved better word analysis improvement, and minimized the searching time which yielded a better performance with processing speed of up to 1500 words per second in small text up to 3000 words per second in larger documents.

References

1. Kiraz, G.A.: Computational nonlinear morphology; with emphasis on semitic languages. In: Branimir Boguraev, I.(eds.) T.J. Studies in Natural Language Processing.Watson Research Center and L.D.C. Steven Bird, University of Pennsylvania, The Edinburgh Building, Cambridge CB2 2RU, UK: The press syndicate of the University of Cambridge, the Pitt building, Trumpington street, Cambridge, United Kingdom (2004)
2. Beesley, K.R.: Arabic morphological analysis on the internet. In: 6th International Conference and Exhibition on Multi-lingual Computing, Cambridge (1998a)
3. Buckwalter, T.: Buckwalter Arabic Morphological Analyzer Version 1.0. Linguistic Data Consortium (2002)
4. Watson, J.C.E.: The Phonology and morphology of srabic. In: Durand, J.(ed.) The Phonology of the World's Languages. Oxford University Press. New York (2007)
5. Mohammed, A.A.: An Ambiguity-Controlled Morphological Analyzer for Modern Standard Arabic Modelling Finite State Networks. School of Informatics, The University of Manchester (2006)
6. Darwish, K.: Building a shallow morphological analyzer in one day. In: 40th Annual Meeting of the Association for Computational Linguistics (ACL 2002), Philadelphia, PA, USA (2002
7. Soudi, A., Cavalli-Sforza, V., Jamari, A.:. A Computational lexeme-based treatment of arabic morphology. In: Arabic Natural Language Processing Workshop, Conference of the Association for Computational Linguistics (ACL 2001), Toulouse, France (2001)
8. Roark, B., Sproat, R.: Computational Approaches to Morphology and Syntax. Oxford University Press, United States (2007)
9. Al-Kabi, M.N., Al-Radaideh, Q.A., Akkawi, K.W: Benchmarking and assessing the performance of Arabic stemmers. Journal of Information Science 37(111) (2011)
10. Khawaja, A., Mazrui, A., Boodlal, A.R.: AL-Khalil Arabic Morphological System. Mohammed Al-Aoual University - Jeddah - Laboratory Research in informatics: The Arab League Educational, Cultural and Scientific Organisation (ALECSO) (2010)
11. Saad, E.-S.M., et al.: An Improved Arabic morphology analyzer, faculty of engineering. Helwan University. (2005)
12. Mazrui, A., et al.: Morphological analysis system specifications. In: Meeting of experts in Computational Morphological Analyzers for the Arabic language, Damascus (2010)
13. Ahmed, H.: Developing Text Retrieval System Using NL. Institute of Statistical Studies and Research. Computer Science and Information. Cairo University, Cairo (2000)
14. Farouk, A.: Developing an Arabic Parser in a Multilingual Machine Translation system. Faculty of Computers and Information, Cairo (1999)
15. Al-Fedaghi, S.S., Al-Anzi, F.S: A new Algorithm to generate the Arabic Root-Pattern forms (1988)

Morphological Rules of Bangla Repetitive Words for UNL Based Machine Translation

Md. Nawab Yousuf Ali[1(✉)], Golam Sorwar[2], Ashok Toru[1],
Md. Anik Islam[1], and Md. Shamsujjoha[1]

[1] Department of Computer Science and Engineering, East West University,
Dhaka 1212, Bangladesh
nawab@ewubd.edu, {ashoktoru,anik1400}@gmail.com,
dishacse@yahoo.com
[2] School of Business and Tourism, Southern Cross University, Lismore, Australia
golam.sorwar@scu.edu.au

Abstract. This paper develops new morphological rules suitable for Bangla repetition words to be incorporated into an interlingua representation called
Universal Networking Language (UNL). The proposed rules are to be used to
combine verb roots and their inflexions to produce words which are then combined with other similar types of words to generate repetition words. This paper
outlines the format of morphological rules for different types of repetition
words that come from verb roots based on the framework of UNL provided by
the UNL centre of the Universal Networking Digital Language (UNDL)
Foundation.

Keywords: Morphological rules · Repetition words · UNL · Machine translation

1   Introduction

The UNL is an artificial language replicating the function of natural languages for
human communications [1]. The motivation behind UNL is to develop an interlingua
representation in which semantically equivalent sentences of all languages should
have the same interlingua representation [2]. Rules play an important role in machine
translation process. Bangla language processing research communities have been
working on the development of morphological rules for morphological analyses of
words, verbs and parsing methodology of Bangla sentences [3-7]. To the best of our
knowledge, no attempts have been made to develop rules for Bangla repetition words
in a concrete computational approach. To address the above limitation, this paper
classifies different types of Bangla repetition words and developed formats of morphological rules for those words based on the structure of UNL. The core structure of
UNL is based on the *Universal Words*, *Attributes* and *Relation* [3]. *Universal Words*
(UW) are words that constitute the vocabulary of UNL. *Attributes* are for the purpose
to describe the subjectivity information of sentences and also used to express range of

© Springer International Publishing Switzerland 2015
Y. Tan et al. (Eds.): ICSI-CCI 2015, Part III, LNCS 9142, pp. 401–408, 2015.
DOI: 10.1007/978-3-319-20469-7_43

concepts and logical expressions in strengthening the express ability of the UNL. The relation between UWs is binary with different labels according to the different roles [1].

2   Bangla Word Dictionary and Enconverter

A word dictionary is a collection of word entries. Each entry in a word dictionary is composed of three kinds of elements: the *Headword* (HW), the *Universal Word* (UW) and the *Grammatical Attributes* (GA) [1]. A headword is a notation of a word in a natural language composing the input sentence and to be used as a trigger for obtaining equivalent UWs from the word dictionary. An UW expresses the meaning of word and is to be used in creating UNL expression. Grammatical attributes are the information on how word behaves in a sentence and are to be used in making analysis rules for conversions. Each dictionary entry has the following format for any native language word [8-10].

Data format:
[HW]{ID} "UW"(Attributes)<FLG,FRE,PRI>

where, HW stands for head word, ID for identification of head word which is omitable, UW for universal word, FLG for language flag, FRE for frequency of head word, and PRI for priority of head word. The attributes part divided into three sub groups: *grammatical attributes*, *semantic attributes* and *UNL attributes*.

The EnConverter (EnCo) [1] converts a native language sentence/word into UNL expressions. It is a language independent parser provided by the UNL project, a multi-headed Turing Machine [11] providing synchronously a framework for morphological, syntactic and semantic analysis. The machine has two types of windows namely *Analysis Windows* (*AW*) and *Condition Windows* (*CW*).

The machine traverses input sentence back and forth, retrieves the relevant dictionary entry (UW) from the Word Dictionary (Lexicon), depending on the *attributes* of the nodes under the AWs and those under the surrounding CWs. It then generates the semantic relationships between the UWs and /or attaches speech act attributes to them. As a result, a set of UNL expressions are made equivalent of UNL graph. EnCo is driven by a set of analysis rules to analyzing a sentence using Word Dictionary and Knowledge Base. The enconversion rule has been described in [12].

Morphological analyses are performed by the left and right composition rules. The basic type to this group is "+". This type of rule is used primarily for creating a syntactic tree with two nodes on the Analysis Windows [12].

3   Format of Morphological Rules of Bangla Repetition Words

We have rigorously analysed the Bangla grammar [13-15] and found the following types of Bangla repetition words that come from verb roots. We have also outlined the format of morphological rules for constructing these types of repetition words.

3.1   Rules for the Repetition Words from Vowel and Consonant Ended Roots

The words that come from vowel and consonant ended roots used with present tense. Two types of morphological rules for creating these types of repetition words are:

3.1.1   Rules of Repetition Words from Consonant Ended Roots

These types of rules are usually used to make the repetition words of the present tense for second person singular number. For example, 'ধরধর' (pronounce as dhoro dhoro) means catch, etc. Example of dictionary entry of the root is given below.

[ধর]{ }"catch(icl>hook>occur,obj>thing)"(ROOT,CEND,DITT,R011,R02,R051,R052, #OBJ)<B,0,0>

where, attributes ROOT denotes verb root, CEND for consonant ended root, semantic attribute, DITT for repetition words, R stands for rule, R01, R02, R05 are types of rules and R011 is the sub type of R01. We use attributes R011, R02, R051 and R052 with the above template because the repetition words under this template are fall into the above stated rule-groups. UNL attributes, #AGT indicates that an action is initiated by someone, #PLC denotes the place, where an action performs and #OBJ means which action is performed. FLG for language flag, here B for Bangla, FRE means how frequently a word used in a sentence and PRI denotes the priority of the word. <B,0,0> is appended with each of the avoid it for the next UWs in this paper.

- Rule for combining two roots for creating repetition words:

+{ROOT,CEND,DITT,R011:+cmp1::}{ROOT,DITT,CEND,R011:::}(BLK)

Description: The rule describes if one root is in LAW and another root is in RAW, two headwords of the left and right analyses windows are combined with each other to make a composite node. The new composite node is the desired repetition word that will be placed in the RAW.

3.1.2   Rules for Repetition Words from Vowel Ended Roots

These repetition words are usually used with present tense. For example, 'খাই খাই' (pronounce as khai khai) meaning eat. Example of dictionary entry is shown below.

[খা]{ }"eat(icl>consume>do,agt>living_thing,obj>concrete_thing,ins>thing)"(ROOT, VEND,DITT,R012,#ART,#OBJ,#INS)

[ই]{ }" "(INF,VINF,VEND,DITT,R012,#AGT,#PLF,#PLT,#OBJ,#INS,#EQU,#SRC)

- Rules for combining verb root and verbal inflexion:
+{ROOT,VEND,DITT,R012:+cmp::}{[ই]:::}(ROOT,DITT,VEND,R012:::)([ই])
- Rule for combining two verbs for making repetition words:
  +{DITT,VEND,R012,cmp1:cmp2::}{DITT,VEND,R012,cmp1:::}(BLK)

3.2 Rules for the Repetition Words from Consonant Ended Roots with 'আ-কার' (akar) and 'ই-কার' (ikar)

In this case, 'আ-কার' (akar) is added with consonant ended root with the first word and 'ই-কার' (ikar) is added with the same root for the second word to create these kinds of repetition words. For example, 'বলাবলি' (pronounce as bola boli) means *tell*. An entry is as follows.

[বল]{ }"tell(icl>say>do,cob>uw,agt>person,obj>uw,rec>person)"(ROOT,CEND, DITT, R011, R02, R051, R052, #COB,#ART,#OBJ,#REC)

[আ]{ }" " (INF, VINF,CEND, DITT,R011,R02,R051,R052, #COB, #AGT,#REC, #OBJ,#AOJ)

[ই]{ }" " (INF, VINF,CEND, DITT, R011,R02, R051,R052, #COB, #AGT,#REC, #OBJ,#AOJ)

- Rule for combining consonant ended root and verbal inflexion 'আ' to make verbal nouns:

  +{ROOT,DITT,R02,CEND:+cmp1::}{[আ:::]}(ROOT,DITT,R02,CEND:::)([ই])
- Rule for combining consonant ended root and verbal inflexion 'ই':
  +{DITT,R02,CEND,cmp1:cmp2::}{ROOT,CEND,DITT,R02:::}([ই])(BLK)

- Rule for combining two verbs to make repetition words:

  +{CEND,DITT,R02,cmp1,cmp2,ditt:-DITT,-CEND,-R02,-cmp1,-cmp2,+V::}
  {CEND, DITT, R02, cmp1,cmp2,ditt:-DITT,-CEND,-R02,-cmp1,-cmp2, +V::}

3.3 Rule for Repetition Words from Verbal Noun

Two different types of words with comparatively similar meaning form this kind of repetition words. For examples, 'লেখাপড়া' (pronounce as lekha pora) means *reading and writing*. Dictionary entry is:

[লেখ]{ }"write(icl>communicate>do,agt>person,obj>information,cao> thing, ins> thing,rec>person)"(ROOT,CEND,DITT, R02,R03,R051,R052 #ART, #OBJ, #CAO)

[পড়]{ }"read(icl>see>do,agt>person,obj>information)"(ROOT,CEND,DITT,R02,

R03, R051, R052,#AGT, #OBJ)

[আ]{ }""(INF,VINF,CEND,DITT,R02,R03,R051, #AGT,#OBJ,#CAO)

Morphological analyses of this kind of words are not possible for present enconverter. They can be combined with each other by applying semantic rules after semantic analyses.

3.4   Rule for Repetition Words with Sound উ(U), ই (I) and ও (O) in Initial Roots

This type of repetition words are formed with primary sounds উ (pronounce as U), ই (pronounce as I) and ও (pronounce as O), where the last parts of the repetition words have no meanings. As a consequence, it is not possible to analyse these kinds of words on present enconverter. To solve the translation problems in the current work, we can create direct dictionary entries of these repetition words. Three types of such words are given below.

3.4.1   Repetition Words with Primary Sound উ (U)

If the primary sound of the root is উ (U) then we can construct these types of repetition words. Template of the words:

[HW] { } "UW" (ROOT, VEND, DITT, R041, #EQU, #OBJ)

A dictionary entry is as follows:

[চুপচাপ] { }"silent(icl>adj,equ>mum)"(DITT,CEND,R041, #EQU)

3.4.2   Repetition Words with Primary Sound ই(I)

If the primary sound of the root is ই (I) then we can construct following types of repetition words. Template of the words:

[HW]{ }"UW"(ROOT,VEND,DITT,R011,R041 #EQU, #OBJ)

A dictionary entry is as follows:

[মিটমাট] { } "reconciliation (icl>cooperation>thing)" (DITT, CEND, R041, #OBJ)

3.4.3   Repetition Words with Primary Sound ও (O)

If the primary sound of the root is ও (O) then we can construct the following kinds of repetition words.

[HW]{ }"UW"(ROOT,VEND,DITT,R043,#AGT,#OBJ,#PLC,#EQU,#AOJ, #OPL)

An entry is as follows:
[কোলাকুলি]{ }"hug(icl>embrace>do,agt>person,obj>person,ins>thing)"(DITT, VEND, R043, #AGT, OBJ, #INS)

3.5   Repetition Words with Inflexion তে(te) and এ (e)

Inflexions তে (te) and এ (e) are added with both of the roots in creating such kind of repetition words. Two types of words are:

3.5.1   Repetition Words with Inflexion ত (te)

This kind of repetition words are formed by adding inflexion 'ত'(te) with both words.
Construction of repetition words:
repetition word= root + ত + root + ত
For example, হাসতেহাসতে   (pronounce as haste haste) means *laugh* etc.
Template of the words:

[HW] { } "UW" (ROOT, VEND, DITT, R02, R051, #AGT, #OBJ,#INS,#COM)

Some dictionary entries are as follows:

[হাস] { } "laugh (icl>utterance>thing)"(ROOT, CEND, DITT, R02, R051, #OBJ)

[ত]{ }""(INF,VINF,CEND,DITT,R02,R051,#COM,   #AGT,#INS, #OBJ)

- Rule for combining root and verbal inflexion 'ত':
  +{ROOT,DITT,R051,CEND:+cmp1::}{[ত]:::}(ROOT,DITT,R051,CEND:::)
  ([ত])
- Rule for combining two verbs to construct repetition word:
+{DITT,R051,CEND,cmp1:cmp2::}{DITT,CEND,R051,cmp1:::}(BLK)
Repetition Words with Inflexion এ (e)
Here, inflexion এ (e) is added with the first root and repeat it to form repetition words.
Construction of repetition words: *repetition word=root +   এ+ root + এ*
For example, নেচেনেচে (neche, neche) etc.
Some dictionary entries are as follows:

[নাচ]{ }"dance(icl>move>do,com>grace,agt>person,obj>thing)"(ROOT,CEND, DITT,
R052, #AGT, #OBJ)

[এ]{ }""(INF,VINF,CEND,DITT,R011,R02,R051,R052,#AGT,#PLC,#OBJ, #COB)

- Rule for combining root and verbal inflexion 'এ' to create infinite verb:
  +{ROOT,DITT,R052,CEND:+cmp1::}{[এ]:::}(ROOT,DITT,R052,CEND:::)([এ])
- Rule for combining two infinite verb to create repetition word.
  +{DITT,R052,CEND,cmp1:cmp2::}{DITT,CEND,R052,cmp1:::}(BLK)

3.6   Repetition Words from Consonant Ended Roots with Inflexion এ (e) with the Second Word

This type of repetition words are formed by adding inflexion এ(e) with the second root
while first same root remains unchanged .
Construction of the repetition words: *repetition word = root + root + এ*
For example, খিটখিটে (pronounce as khit khite) means *angry*, etc. An entry is:
[খিট]{ }"annoy(icl>displease>do,agt>thing,obj>volitional_thin,met>thing)"(ROOT,
CEND, DITT, R06,#AGT,#OBJ,#MET)

[এ] { }" " (INF, VINF ,CEND, DITT,R06,#AGT, #MET, #OBJ)

- Rule for combing second root and verbal inflexion 'এ' (e) to make a noun:
  +{DITT,CEND,R06,cmp1:cmp2::}{[এ]:::}

- Rule for combining first root and noun made by the previous rule to make repetition word:
+{ROOT,DITT,R06,CEND:+cmp1::}{N,DITT,R06,CEND:::}([অ])(BLK)

3.7 Repetition Words from Consonant Ended Roots with Inflexion আ (aa) with the First Word

This category of repetition words are formed by adding inflexion আ (aa) with the first root while second root remains unchanged.

Construction of the repetition words: *repetition word = root + আ + root*

For example, চলাচল (pronounce as chola chol) means *moving* etc. An entry is:

[চল]{ }"move(icl>occur,equ>displace,plt>thing,plf>thing,obj>thing)"(ROOT,CEN
D,DITT,R011,R051,R07,#EQU,#PLT #PLF,#OBJ)

[আ]{ }""(INF,VINF,CEND,DITT,R011,R051,R07,#EQU,#PLT,#PLF,#OBJ)

- Rule for combining first root and verbal inflexion আ (aa) to make noun:
+{ROOT,DITT,R07,CEND:+cmp1::}{[আ:::]}(ROOT,DITT,R07,CEND:::)(BLK)
- Rule for combining noun formed by the previous rule and second root to make repetition word:
+{N,DITT,VEND,R07,cmp1:cmp2::}{ROOT,DITT,R07,CEND:::}(BLK)

3.8 Repetition Words from Consonant Ended Roots with Inflexion উ (U) U with Both Words

This type of repetition words are formed by adding inflexion উ(U) U with both first and second roots.

Construction of the repetition words: *repetition word = root + উ +root + উ*

For example, তুলুতুল (pronounce as tulu tulu) means *softness* etc. Some entries are:
[তুল]{ }"softness(icl>consistency>thing,ant>hardness)(ROOT, DITT, CEND,#ANT)
[উ]{ }""(INF,VINF,CEND,DITT,R051,R052,R08,#ANT,EQU)
- Rule for combining first root and verbal inflexion উ (U) to make noun:
+{ROOT,DITT,R08,CEND:+cmp1::}{[উ]:::}(ROOT,DITT,R08,CEND:::)([উ])

Same rule will be applied to combine second root and verbal inflexion উ (U) to make noun.

- Rule for combining two nouns made by the previous rule to make repetition words.
+{N,DITT,R08,VEND,cmp1:cmp2::}{N,VEND,DITT,R08,cmp1:::} (BLK)

4 Conclusions

This paper has outlined the templates of morphological rules for different types of Bangla repetition words that are derived from verb roots. The developed templates are useful for morphological analyses of thousands of Bangla repetition words for Bangla-UNL language server. Theoretical analysis has proved that the proposed rules perform well in morphological analyses of Bangla repetition words. The proposed rules can be equally applicable to repetitive words in other languages.

References

1. Uchida, H., Zhu, M., Senta T.D.: Universal Networking Language. UNDL Foundation, International environment house, 2nd edn. UNDL Foundation, Geneva(2006)
2. Bhatia, P., Sharma, R.K.: Role of punjabi morphology in designing punjabi-UNL enconverter. In: Int. Conference on Advances in Computing Communication and Control, pp. 562–566 (2009)
3. Asaduzzaman, M.M., Ali, M.M.: Morphological analysis of bangla words for automatic machine translation. In: Int. Conference on Computer, and Information Technology, Dhaka, pp. 271–276 (2003)
4. Ali, N.Y., Sarker M.Z., Farook, G.A., Das, J.K.: Rules for morphological analysis of bangla verbs for universal networking language. In: Int. Conference on Asian Language Processing (IALP), Harbin, China, pp. 31–34 (2010)
5. Ali, N.Y., Noor, S.A., Zakir, M.H., Das, J.K.: Development of analysis rules for bangla root and primary suffix for universal Networking language. In: Int. Conference on Asian Language Processing (IALP), Harbin, China, pp. 15–18 (2010)
6. Hoque, M.M., Ali, M.M.: A Parsing methodology for bangla natural language sentences. In: Int. Conference on Computer and Information Technology 2003, vol. I, Dhaka, Bangladesh, pp. 277–282 (2003)
7. Ali, N.Y., Das, J.K., Al-Mamun, S.M.A., Nurannabi, A.M.: Morphological analysis of bangla words for universal networking language. In: Third Int. Conference on Digital Information Management, London, England, pp. 532–537 (2008)
8. Ali, N.Y., Yousuf, K.B., Shaheed, F.I.: Development of Bangla Word Dictionary through Universal Networking Language Structure. Gobal Science and Technology Journal 1(1), 53–62 (2013)
9. Ali, N.Y., Das, J.K., Al-Mamun, S.M.A., Choudhury, E.H.: Specific features of a converter of web documents from bengli to universal networking language. In: Proceedings of International Conference on Computer and Communication Engineering, Kuala Lumpur, Malaysia, pp.726–735 (2008)
10. Sarker, Z.H., Ali, N.Y., Das, J.K.: Dictionary Entries for Bangla Consonant Ended Roots in Universal Networking Language. International Journal of Computational Linguistics, 79-87 (2012)
11. Parikh, J., Khot, J., Dave, S., Bhattacharyya, P.: Predicate Preserving Parsing. Department of Computer Science and Engineering, Indian Institute of Technology, Bombay
12. Uchida, H., Zhu, M., Senta T.D.: EnConverter Specification, Version 3.3, UNL Center/UNDL Foundation, Tokyo Japan (2002)
13. Shohidullah, D.M.: Bangla Bayakaron. Ahmed Mahmudul Haque, Mowla Brothers Prokashoni, Dhaka, Bangladesh, (2003)
14. Shniti, K.C.: Vasha-Prokash Bangla Vyakaran. R and C. Prokashoni, Calcutta (1999)
15. Bangla, A.: Bengali-English Dictionary, Dhaka, Bangladesh (2004)

River Network Optimization Using Machine Learning

M. Saravanan[1(✉)], Aarthi Sridhar[2], K. Nikhil Bharadwaj[2],
S. Mohanavalli[2], and V. Srividhya[2]

[1] Ericsson Research India, Ericsson India Global Services Pvt. Ltd., Chennai, India
m.saravanan@ericsson.com
[2] Department of Information Technology, SSN College of Engineering, Chennai, India
{aarthisridhar14i,nikil.bharadwaj}@gmail.com,
{mohanas,srividhyav}@ssn.edu.in

Abstract. Lack of potable water is a perennial problem in the day-to-day life of mankind around the world. The demand-supply variations have been on an increasing trend for so many years in different developing countries. To address this prevailing issue is the need of the hour for the society and the relevant government agencies. In this paper, as an explorative approach, we address this predominant issue in the form of an alternate solution which re-routes the course of the natural water sources, like rivers, through those areas, where the water supply is minimal in comparison with the demand, in a cost-effective and highly beneficial manner. Our analysis and discussions are more prone to Indian scenario where India is one of the worst affected fast developing countries for the water crisis. This involves the consideration of the physical, ecological and social characteristics of the lands on the route that fall under the course of the river and also the regions dependent on its flow. In order to understand and predict the optimized new flow paths to divert the water sources, we have employed Machine Learning algorithms like Multiple Regression and Multi-Swarm Optimization techniques. For selecting the most needed re-route, we have also considered the areas that are prone to droughts, and unite the re-routed water with the original course of the river, finally, draining into the sea, for the sustainable development. The proposed methodology is experimented by analyzing the flow areas (river basins) of river Mahanadi in India, one of the considerably important projects cited many times without any real implementation. The results are validated with the help of a study conducted earlier by the National Water Development Agency (NWDA), Government of India, in 2012.

Keywords: River networks · Machine learning · Multi-swarm optimization · Multiple regression · Fitness criterion

1 Introduction

Water resources are one of the limiting factors of development around the world. According to the UN, more than one in every six people in the world are water stressed, meaning that they do not have access to potable water [1]. These people make up about 1.1 billion people in the world and are living in developing countries like India, China, etc. Water shortages may be caused by climate change,

© Springer International Publishing Switzerland 2015
Y. Tan et al. (Eds.): ICSI-CCI 2015, Part III, LNCS 9142, pp. 409–420, 2015.
DOI: 10.1007/978-3-319-20469-7_44

including droughts or floods, increased pollution, and human demand and over usage. Developed countries like Australia and the United States were prey to such water problems that was largely mitigated with the development of inter-basin water transfers [2]. Israel, Canada, Iraq and China followed the trend set by the developed countries in inter-basin water transfers. In this paper one of the regions in India has been considered as the area of study with respect to the river network optimization and selection of re-route in the river basin.

Water scarcity has been a serious issue especially in the Indian sub-continent, due to the lack of utilization of continuous supply of water from natural sources such as rivers and rainfall throughout the year. India receives well distributed rainfall for about 5-6 months in a year and has a good amount of water resources like rivers; however, growing human population, poor storage infrastructure, low per capita availability of utilizable water, variability of rainfall and the associated drought and floods are major factors contributing to Indian water crisis [1]. Given that the potential for increasing the utilization of water is only about 5-10%, India is bound to face severe scarcity of water in the near future. This calls for immediate action to make a sustainable use of the available water resources by optimizing the river water utilization, which is notably represented in Fig. 1. It is clearly visible that drained water to sea from rivers is significantly higher than the utilized water resources. River network optimization techniques discussed in this paper aim to solve these major issues by exploring different set of algorithms in Machine Learning. These algorithms were implemented by considering various river basin characteristics and its causal relationships with relevant parameters.

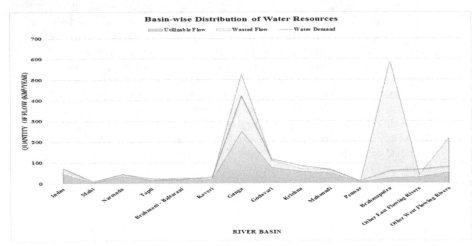

Fig. 1. Basin-wise distribution of utilizable and wasted water resources in Indian rivers

Optimization is a process generally employed to figure out the most suitable solution from a set of candidate solutions. In river networks, it involves the identification of the most suitable re-routing path for the river water flow. Optimization is achieved through the use of Machine Learning techniques by involving features/parameters relevant to the river network. Machine Learning techniques [3] involve the study of algorithms that can build a model using the available inputs and make predictions or decisions, and are applicable to specific domains. Swarm intelligence is a collection of Machine Learning based optimization techniques that uses the intelligence of a collection of simple agents and their interactions [4]. Particle Swarm Optimization (PSO) is one of the Swarm Intelligence based techniques that provides an optimal solution by iteratively improving a candidate solution with respect to relevant quality measures [4]. In this paper, Multi-Swarm Optimization (MSO) [5], an extended version of PSO, is employed as one of the methods to estimate the optimal re-route of the river by considering different locations, which involves facts about a number of parameters associated with the surrounding regions.

Our idea of river network optimization involves collection of various kinds of data associated with river spanning and surrounding regions, for each major river basin in and around the sub-continent. Based on the features arrived, the study is undertaken to predict a possible re-routing path of the river flow to utilize the surplus water. In addition, the possibility of interconnecting the river with other nearby rivers that are characterized by extremely irregular water flows. To accomplish this task, machine learning techniques are employed, which predicts the possible combination of the river features. This will help to narrow down the route to redirect the existing river water for the up-liftment of the society. Various other factors like precipitation, vegetation and forest cover are also considered. This study has been evaluated with the report on Inter-basin Water Transfers of Mahanadi River Basin published by NWDA [6].

2   Related Work

The idea of linking rivers for various purposes in the Indian sub-continent is a much researched area. The Ganga-Cauvery Link was proposed by Dr. K.L. Rao, former Union Minister for Irrigation, in 1975, which was eventually dropped, owing to budget and energy constraints [7]. The National Water Development Agency (NWDA) was established in 1982 to study basin-wide surplus and deficit regions and explore the possibilities for storage and transfer [8]. There have been extensive studies on the Mahanadi River Basin. A case study by IIT Delhi on Mahanadi River Basin explains the climatic changes impacting the course and surrounding regions [9]. It models the water flow of Mahanadi River using hydrological models like Soil and Water Assessment Tool (SWAT) and explains the various parameters to be considered for the formulation of the model. Another study of the Mahanadi River Basin [10] considers the vegetation and soil effects of the river flow and describes the effects that it possesses on the periodic change in the flow of rivers. Telci and Ilker [11] have worked based on the above study, providing clear insight into the different factors for modelling the river flow. Karunanithi and Nachimuthu [12] have applied Artificial Neural Networks (ANN) to predict the river flow, and results are found to be in close coordination with actual data. The features that they have extracted clearly predict the river flow model. Another study compares two popular and significant approaches to the application of ANN namely, Multi-layer Perceptron (MLP) network and Radial Basis Function (RBF) network for the modelling of the Apure river basins in Venezuela [13]. However, these studies have not considered the re-routing of river flow or for any societal benefits.

The importance of applying Swarm Intelligence techniques for optimization, in particular, Particle Swarm Optimization (PSO) algorithm [14], has tremendously helped to program the fuzzy logic that a river network involves. The application of PSO to the task of optimization has been covered in the works of Eberhart et al [14]. The classical PSO has since undergone various hybridizations according to the domain, nature and properties of the networks to which it is applied, proving to be highly cost-effective. Multi-Swarm Optimization extends PSO by using several swarms of simulated particles rather than a single swarm [15]. It is also used for estimating the weights and bias values for an artificial neural network or for weak learners in ensemble classification and prediction. Hence in this paper we have applied MSO for predicting the re-routing river path by calculating the fitness values of different locations with the help of relevant parameters.

3   Need of River Network Optimization in Indian Scenario

India has a potential to store about 200 billion cubic meters of water [16] that can be utilized for irrigation and hydropower generation. However, with growing population, increasing demand and over usage, India is on the verge of facing severe water crisis in the near future. India has approximately 16 per cent of the world's total population as compared to only 4 per cent of water resource availability. And the per capita water availability is around 1,588 cubic meters per year [16, 17]. This is very low compared to developed nations such as the USA where it is close to 5,600 cubic meters, which is more than four times the availability in India. Recurrent droughts affect 19 percent of the country's regions, 68 percent of the cropped area and 12 percent of the Indian population [18]. From Fig 2 and 3, it is clear that the demand for water resources is increasing over the years for different sectors.

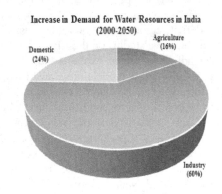

Fig. 2. Demand and Per Capita Availability of Water Resources in India

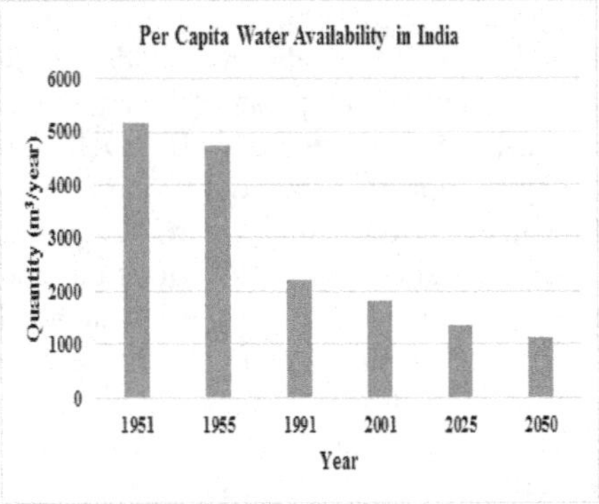

Fig. 3. Per Capita Water Availability in India influenced by the population

The total available water amounts to 4000 billion m^3 per annum, out of which the usable water is only 1123 billion m^3[17]. Against the above supply, the water consumed during the year 2006 in India was 829 billion m^3 which is likely to increase to 1093 billion m^3 in 2025 and 1147 billion m^3 in 2050, as estimated by the Government of India [17]. Most of the Indian States will be expected to reach the water stress condition by 2020 and water scarcity condition by 2025. This would further put at risk the food security, as the scarcity of water will directly affect agricultural production.

Although India is not a water deficit country, due to lack of planning, monitoring of water resources and without considering the optimization of river networks, it experiences water stress from time to time. It is therefore necessary to prevent this crisis by making best use of the available technologies in optimizing the river networks. In this paper, we have attempted to trial out an experiment related to this with the help of Machine Learning techniques.

4   Overview of Optimization and Prediction Techniques

Optimization is considered to be one of the essential techniques in Machine Learning. Network optimization has been an important area of research and application to enable decision making process. A collection of optimization techniques fall under the category of Swarm Intelligence, which use the intelligence of groups of simple agents, interacting with one another and with their environment [4]. Examples in natural systems include ant colonies, bird flocking, animal herding, fish schooling and microbial intelligence. Combining the optimization techniques with prediction is considered to be one of the most efficient methods in real time applications. Following algorithms are explored to find the suitability of use in the river network optimization scenario.

4.1   Multiple Linear Regression

It is used for modelling the relationship between a scalar, dependent variable y and one or more independent variables, denoted as x [19]. The model takes the form

$$y_i = \beta_i x_{i1} + \cdots + \beta_p x_{ip} + \varepsilon_i, i = 1, \ldots, n \qquad (1)$$

where β refers to the slope of the line and ε_i is an error variable. This technique is used in the estimation of the parameter values (y_i) at specific locations (i) on the river basin, characterized by latitude (L_1) and longitude (L_2) values as x_i's. Although the technique used here is Multiple Regression, the underlying concept is essentially applying linear regression for each parameter in different locations. The estimated parameter values are then applied in the fitness function to determine the optimal re-routing path in the river network.

4.2   Particle Swarm Optimization

Particle Swarm Optimization (PSO) is an optimization technique [14] used in problems where a best solution is a point or surface in an n-dimensional space. The advantage of PSO is its ability to handle optimization problems with multiple local optima reasonably well owing to the presence of large number of particles that make up the swarm and its simplistic implementation, especially in comparison with other optimization techniques. Since in this study, our intention is to arrive at a best fitness value for different locations based on the usage of various parameters and optimize to get a best route, we extend the PSO into Multi-Swarm condition.

4.3 Multi Swarm Optimization

Multi-Swarm Optimization (MSO) is a variant of PSO based on the use of multiple sub-swarms instead of the standard one swarm procedure [5]. The general approach in multi-swarm optimization is that each sub-swarm focuses on a specific region that is decided by a specific criterion. The algorithm proceeds as follows:

Let $f: \mathbb{R}^n \to \mathbb{R}$ be the fitness function to be minimized. Here S is the number of sub-swarms and P is the number of particles in each sub-swarm, each having a position (L_1, L_2 as l_{id}) in the search-space and a velocity v_i. Both location and velocity values of each particle are initialized to random values within the dimensions of the search space. The best known position of particle i is $pbest_i$, the best known position of each sub-swarm j is $gbest_j$ and the best known position of all swarms put together is $gbest$. A basic MSO algorithm is formulated related to river network optimization-

```
Do
        For each sub-swarm, sⱼ ∈ S, with dimensions, d
                For each particle, pᵢ ∈ sⱼ, sub-swarm with dimensions, d
                        INITIALIZE location of the particle, lᵢ𝒹 ∀ d
                        INITIALIZE velocity of the particle, vᵢ𝒹∀ d
                End
        End
        Do until completion of path
                For each sⱼ ∈ S
                        For each pᵢ ∈ sⱼ
                                CALCULATE fitness value f(yᵢ) = Σⁿₖ₌₁ wᵢₖyᵢₖ²,
                                where k refers to the parameters associated with
                                        each particle i, yᵢₖ are the regression
                                        parameter values and wᵢₖ are the corresponding
                                        weights (constants)
                                        If f(yᵢ) < personal best value, pbestᵢ then
                                        UPDATE pbestᵢ
                                        CALCULATE generation best value, gbestⱼ,
                                        where gbestⱼ = minᵢ pbestᵢ
                        End
                        CALCULATE generation best value, sbest, where
                        sbest = minⱼ gbestⱼ
                        For each pᵢ ∈ sⱼ
                                UPDATE velocity, vᵢ𝒹 = w * vᵢ𝒹 + c1 * r1 *
                                        (pbestᵢ − lᵢ𝒹) + c2 * r2 * (gbestⱼ − lᵢ𝒹) + c3 *
                                r3 * (sbest − lᵢ𝒹) ∀ d, where w=0.729,
                                c1=c2=1.49445, c3=0.3645 and r1, r2, r3 are
                                random numbers
```

UPDATE location, $l_{id} = l_{id} + v_{id} \; \forall \, d$

        End

      End

   End

cEnd

The values for constants w, c_1, c_2 are assigned the values based on various research studies [20]. It is clear that MSO is better suited to our optimizing procedure which involves quite a large of parameters and different locations. The use of large number of agents and simplistic calculations aids MSO in exploring diverse solutions using relevant features for river network optimization.

5 Extracting Relevant Features for River Networks Optimization

The following inherent features, relevant for understanding the flow of rivers combined with factors influencing redirection of river water which are identified based on various reports available [8-13] are discussed in this section. After arriving at the list of necessary features, they have been applied into the formulated algorithm for the task of river network optimization.

1. **Precipitation (y_1).** It is an important measure that influences the flow of rivers through the land mass, and also significantly impacts agriculture, cultivation and biodiversity. Precipitation levels characterize the water availability in the concerned regions.
2. **Cultivable land (y_2).** Cultivable land is a type of land capable of being ploughed and used to grow crops. Greater percentage of available cultivable lands helps more food production, and thus improving self-sufficiency. Hence, the existence of cultivable lands in the regions under consideration immensely shapes the re-route in river flow.
3. **Irrigated land (y_3).** Irrigation is heavily influenced by the availability and proximity of water resources especially natural sources like streams, lakes, ponds, etc. Better irrigational facilities imply better agricultural production. Hence, it is necessary to prioritize regions with lesser irrigational facilities.
4. **Sown land (y_4).** Sown land denotes the expanse of earth that is used for farming and cultivation of crops. Greater availability of water resources could improve the utilization of available fertile land for cultivation, thus improving agricultural economy.
5. **Forest land (y_5).** There should be minimum destruction of forest lands during the redirection of rivers, in order to preserve the ecosystem and biodiversity in the regions and ensure sustainability of environmental resources.
6. **Population density (y_6).** Rerouting of rivers might involve the displacement of a healthy amount of small settlements, communities and tribes. So, the displacement

and hence rehabilitation of these communities are serious causes for concern while considering the proposal for a new route.

6   Algorithmic Implementation of River Network Optimization for Mahanadi River Basin

In this section, Machine Learning techniques have been applied in the task of identifying and understanding the best route to redirect the existing Mahanadi River water. The selection of Mahanadi River Basin is substantiated by various reports [9, 10, 21] and the fact that the water demand closely coincides with the available water resources, as evident from Fig 1, thereby establishing the need for increasing the utilization of the Mahanadi River water.

As described in the previous section, the various features of the river that are essential for the application of our model are extracted from different studies and reports. These studies describe the significance of the chosen features and their influence on river flow and redirection. 13 regions along the proposed re-routing path have been identified and these routes belong to the southern districts of Jharkhand and northern districts of Odisha in India. The features discussed in Section 5 are gathered from various sources of governmental agencies [7, 21-25], a sample of which is shown in Table 1.

Table 1. Sample District-wise Statistics of mentioned features for sample regions of Mahanadi River basin in the states of Jharkhand and Odisha in India

Regions	West Singhbum	Saraikela	Simdega	East Singhbum	Jharsugudha	Sundergarh
State			Jharkhand		Orissa	
Precipitation *mm*	1422	1350	1478.3	1007.4	1527	1657.1
Cultivable Land *000 ha*	285	120	285	392	88	313
Irrigated Land *000 ha*	6	6	5	12	15.77	81.95
Net Sown Area *000 ha*	148	62	112	84	78	300
Forest Land *000 ha*	228	96	56	123	20	496
Total Land Area *000 ha*	563	237	372	557	208.1	971.2
Population Density */km^2*	209	390	160	648	274	214

With the extracted feature sets being district averages (Table 1), it is essential to establish a relationship between the chosen parameters and the locations. This enables the estimation of parameters for every particular location among the considered regions. This is achieved with the help of multiple linear regression technique, the results of which are presented in Table 2.

Table 2. Regression Equations for River parameters

Parameter (y_i)	Regression Equation (x_1: latitude x_2:longitude)
Precipitation	$y_1 = 12.43114489\ x_1 - 30.41487728\ x_2 + 45.07417168$
Cultivable Land %	$y_2 = -17.25100229\ x_1 + 45.10757844\ x_2 - 62.12951393$
Irrigated Land %	$y_3 = -34.03455462\ x_1 - 17.29225093\ x_2 + 79.82325286$
Sown Land %	$y_4 = 10.20654652\ x_1 - 39.43118445\ x_2 + 63.74998508$
Forest Land %	$y_5 = 84.47768184\ x_1 - 101.1913671\ x_2 + 82.84103224$
Population Density	$y_6 = -42.85820115\ x_1 + 84.66020393\ x_2 - 103.1191207$

The above equations are applied into the fitness criterion of MSO algorithm and the optimized re-routing path for the Mahanadi River is estimated from different locations. The objective fitness function is designed by assigning weights for each parameter and the weights are captured in the variations of parameter values of each location.

$$f(y) = (0.006139)y_1^2 + (0.01595)y_2^2 + (0.135484)y_3^2 + (0.01849)y_4^2 \\ + (0.063319)y_5^2 + (0.039358)y_6^2 \qquad (2)$$

The algorithm proceeds by calculating the fitness values of each considered location, according to equation (2). In each iteration, one optimal location with best fitness value is identified along the re-routing path. This is depicted with an example in Fig 4 and 5.

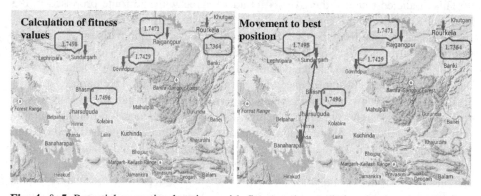

Fig. 4. & 5. Potential re-routing locations with fitness values and choosing location with best fitness value

The algorithm is run repeatedly until it reaches specified boundary regions by getting minimal values from the fitness function so as to obtain an optimized path. A local minimum is obtained at the 13th location, which represents the geographical location of Ghatshila in Jharkhand. The predicted re-rerouting path along with the existing river flow path is depicted in Fig 6.

Fig. 6. Re-routing path estimated using MSO with existing river routes in the considered regions

7   Discussion

The MSO technique is employed on our dataset (Table 1) concerning the Mahanadi River Basin to predict the re-routing path for the river and the results are represented in Figure 6. The proposed route originates from the Hirakud Dam in Orissa and extends north-east through the cities of Jharsugudha, Rourkela towards Jharkhand. It further traverses the southern districts of Jharkhand such as Simdega, West Singhbum, Saraikela and East Singhbum finally uniting with River Subarnarekha, near Ghatshila town. The redirected waters then flow through the course of the Subarnarekha River and finally drain into the Bay of Bengal.

This predicted route of the river is found to agree with the route proposed in the National River Linking Project (NRLP) [6] as estimated by NWDA. This link underlines a possible route from the river South Koel in West Singhbum district, Jharkhand, through the tributary river Kharkai eastwards and finally uniting with river Subarnarekha. This link forms the latter part of our predicted route, which essentially validates the efficiency and significance of the proposed optimization technique.

We have faced few interesting challenges in this study including the selection of regions as well as choosing the Machine Learning techniques for finding suitable optimization criterion. Significant amount of research on the Mahanadi River Basin was available to analyze the influence and effect of the selected features and their relevance to the optimization problem. Thereafter, a number of reports published by Indian governmental organizations had to be studied to determine the feature values with respect to each district under consideration. Moreover, the application of regression analysis involved evaluating its suitability for the proposed model. Further, the selection of the optimization algorithm involved extensive study on many algorithms such as Ant Colony Optimization (ACO), Artificial Bee Colony (ABC) algorithm and so on. The applied MSO algorithm was arrived at after huge focus on Particle Swarm Optimization (PSO) algorithm to justify the use of multiple swarms for our problem at hand. Since the MSO algorithm relies largely on the fitness criterion for optimization, large numbers of trials were performed on the data to finally determine and arrive at a suitable optimization criterion.

Wide range of advantages can be realized with the implementation of the proposal. Vast amount of land areas would become fertile, thereby increasing the agricultural production, hydroelectric power generation and domestic water supply. The project also provides benefits such as inland navigation, fisheries and groundwater recharge. When extended to all river basins, it will increase India's utilizable water resources by 25%, and it would ensure adequate supply of water in every part of the country.

8   Conclusion

The advancements in Machine Learning techniques, especially Optimization methods have enabled to apply them in real-time contexts concerning the worldwide issue of water scarcity. The application of Swarm Intelligence techniques to the field of river network optimization has allowed addressing the deficiency in availability and access to water in regions, by proposing a redirection route of a nearby river through them. This is done through the extensive study of characteristics of the river basin and formulation of models that explain the impact of various parameters. This approach is highly scalable and could be applied to address any given river basin in the world. The results of the study, although looks workable, allows room for the implementation of other optimization and deep learning techniques in addition to applying diverse mathematical and statistical models. The feature set of the river networks considered in this study can be further expanded to include water quality parameters and other geo-spatial features. This could help in devising accurate models to further optimize the river networks and prove highly beneficial to a larger population in the developing countries.

References

1. United Nations Development Programme Human Development Report 2006: Beyond Scarcity–Power, Poverty and the Global Water Crisis. Palgrave Macmillan, Basingstoke (2006)
2. Ghassemi, F., White, I.: Inter-basin water transfer: case studies from Australia, United States, Canada, China and India. Cambridge University Press (2007)
3. Anderson, J.R.: Machine learning: An artificial intelligence approach. In: Michalski, R.S., Carbonell, J.G., Mitchell, T.M. (Eds.), vol. 2. Morgan Kaufmann (1986)
4. Beni, G., Wang, J.: Swarm intelligence in cellular robotic systems. In: Proceed. NATO Advanced Workshop on Robots and Biological Systems, Tuscany, Italy, June 26–30, 1989
5. Blackwell, T., Branke, J.: Multi-swarm optimization in dynamic environments. In: Raidl, G.R., et al. (eds.) EvoWorkshops 2004. LNCS, vol. 3005, pp. 489–500. Springer, Heidelberg (2004)
6. Amarasinghe, U.: Water Policy Research: The National River Linking Project in India-some contentious issues. IWMI-TATA Water Policy Program (2012). www.iwmi.org/iwmi-tata/apm2012
7. Shukla, A.C., Vandana A.: Anatomy of interlinking rivers in india: a decision in doubt. Program in Arms Control, Disarmament, and International Security, University of Illinois at Urbana-Champaign (2005)

8. Amarasinghe, U., Shah, T., Malik, R.: Strategic analyses of the National River Linking Project (NRLP) of India. Series1. India's Water Future: Scenarios and Issues. International Water Management Institute, New Delhi (2008)
9. Gosain, A.K., Aggarwal, P.K., Rao, S.: Linking water and agriculture in river basins: Impacts of climate change – Impact Assessment on Water Resources of Mahanadi River Basin. Civil Engineering Department, Indian Institute of Technology, Delhi (2011)
10. Mishra, N.: Macroscale Hydrological Modelling and Impact of land cover change on stream flows of the Mahanadi River Basin. Indian Institute of Remote Sensing, Diss (2008)
11. Telci, I.T., et al.: Real time optimal monitoring network design in river networks. World environmental & water resources congress (2008)
12. Karunanithi, N., et al.: Neural networks for river flow prediction. Journal of Computing in Civil Engineering **8**(2) 201–220 (1994)
13. Dibike, Y.B., Solomatine, D.P.: River flow forecasting using artificial neural networks. Physics and Chemistry of the Earth, Part B: Hydrology, Oceans and Atmosphere **26**(1), 1–7 (2001)
14. Eberhart, R.C., Kennedy, J.: A new optimizer using particle swarm theory. In: Proceedings of the Sixth International Symposium on Micro Machine and Human Science, vol. 1 (1995)
15. McCaffrey, J.: Test Run - Multi-Swarm Optimization. MSDN Magazine, September 2013
16. Narayan, G.H.: Water Scarcity and Security in India. India Water Portal (2012). http://www.indiawaterportal.org/articles/water-scarcity-and-security-india
17. Govt. of India. Background note for consultation meeting with Policy makers on review of National Water Policy. Ministry of Water Resources (2009)
18. Amarasinghe, U.A., Shah, T., Turral, H., Anand, B.K.: India's water future to 2025-2050: Business as usual scenario and deviations. Research Report 123, IWMI (2007)
19. Freedman, D.A.: Statistical models: theory and practice. Cambridge university press (2009)
20. Eberhart, R.C., Yuhui, S.: Comparing inertia weights and constriction factors in particle swarm optimization. In: Proceedings of the 2000 Congress on Evolutionary Computation, vol. 1. IEEE (2000)
21. Orissa Watershed Development Mission, Agriculture Department, Orissa. Perspective and Strategic Plan for Watershed Development Projects, Orissa (2010-2025)
22. Joshi, N.M.: National River Linking Project of India. Hydro Nepal: Journal of Water, Energy and Environment **12**, 13–19 (2013)
23. Mehta, D., Mehta, N.K.: Interlinking of Rivers in India: Issues & Challenges. GeoEcoMarina 19 (2013)
24. Central Water Commission. Water Data Complete Book. Central Water Commission, Ministry of Water Resources, Govt. of India, New Delhi, India (2005)
25. Country Policy Support Programme. Water Resources Assessment of Brahmani River Basin, India. International Commission on Irrigation and Drainage (ICID), New Delhi, India (2005)

Virtual Management
and Disaster Analysis

Gauging the Politeness in Virtual Commercial Contexts Based on Patrons' Collective Perceptions

I-Ching Chen[1(✉)] and Shueh-Cheng Hu[2]

[1] Department of Information Management,
Chung Chou University of Science and Technology, Yuanlin, Taiwan, ROC
jine@dragon.ccut.edu.tw
[2] Department of Computer Science and Communication Engineering,
Providence University, Taichung, Taiwan, ROC
shuehcheng@gmail.com

Abstract. Politeness constantly plays a significant role in commercial contexts. In contrast to its importance, politeness-related issues in fast-growing virtual commercial contexts received rare attention. This article reports a work developing an instrument for gauging degree of politeness in online storefronts. The instrument's reliability and validity were confirmed through analyzing empirical data, which distilled collective perceptions of 282 sampled patrons. A second-order confirmatory factor analysis revealed that online consumers' tendency in paying relative more attention to their rights being respected and gaining useful information while they are assessing online retailers' politeness. Using the instrument, people can measure the degree of politeness in online retailers.

Keywords: E-commerce · Online storefronts · Politeness · Instrument · Collective perceptions · Confirmatory factor analysis

1 Introduction

1.1 Politeness and Successful Business

Impoliteness in commercial contexts often hurts patrons' feelings and faces, thus leave customers negative impression and words-of-mouth. Gradually, a merchant will lose its customers if it cannot treat them politely; even it has many other merits such as competitive pricing, plentiful product choices, advanced facilities, convenient layout, etc. Many prior studies [1-4] support this argument directly or indirectly. Among others, Berry [5], Reynolds and Beatty [6] found that rapport consisting of enjoyable interactions and personal connections, is a major determinant affecting customers' satisfaction and loyalty, which contribute to a successful business. Kim and Davis [7] further pointed out that politeness plays a key role in early stage of nourishing rapport between sales representative and customers. Accordingly, merchants not likely to build a satisfying and loyal customer base without paying attention to the politeness in their commercial contexts.

Y. Tan et al. (Eds.): ICSI-CCI 2015, Part III, LNCS 9142, pp. 423–430, 2015.
DOI: 10.1007/978-3-319-20469-7_45

1.2   Awareness of Politeness in Computerized Environments

Prior study found that people reciprocally expect politeness from computers, just like they treat their computers with politeness [8]. The findings indicate that people do care about the politeness of computers with which they interact. When waves of computer and telecommunication keep on permeating into various aspects of our daily life, customers eventually will well recognize the politeness issue in online storefronts, just like they do in physical commercial contexts, and they will pay considerable amount of attention to it accordingly.

1.3   Motivation and Goals

In the age of computers and Internet, theoretical works about politeness focusing on verbal communications among persons become inadequate to interpret, assess, and manage the interactions between human and computers. In light of this inadequacy, Brian Whitworth established a "polite computing" framework [9] that comprises five principles for judging whether computer-initiated actions in five different facets are polite or not, based on users' perceptions. Obviously, the framework is a proper basis for assessing the extent to which an online storefront treats its patrons with politeness.

Given that politeness in e-commerce contexts are well worth notice and consideration, but it is still vague about how to manage it. Due to the lack of an assessment mechanism, the present work aims to develop an instrument for gauging the degree of politeness in online storefronts, based on patrons' collective perceptions.

2   Relevant Studies

2.1   Politeness in Computing Environments

While computers are continuously penetrating people's work, life, education, and other activities, it is rational that people will pay increasing attention to the politeness of computers, mostly via software and Web sites, with which they interact often. In consequence, Brian Whitworth and his colleagues responded to this need; they gave a 5-item definition of software politeness based on theories about sociology and socio-technical interactions [10]:

"1. Respect the user. Polite software respects user rights, does not act pre-emptively, and does not act on information without the permission of its owner.

2.   be visible. Polite software does not sneak around changing things in secret, but openly declares what it is doing and who it represents.

3.   be helpful. Polite software helps users make informed choices by giving information that is useful and understandable.

4.   Remember you. Polite software remembers its past interactions and so carries forward your past choices to future interactions.

5.   Respond to you. Polite software responds to user directions rather than trying to pursue its own agenda."

Obviously, the politeness in Web-enabled contexts including online storefronts can be assessed according to the operationalized form of this definition. Nevertheless, there is no reported work that investigated how to assess it quantitatively yet.

2.2 Measurement of Politeness

Intuitively, the politeness is an abstract concept and thus hard to measure it directly. However, some researchers found ways to measure it due to the necessity of embedding this concept into people's behavioral model. The conventional "politeness theory" introduced by Brown and Levinson [11] in the 80s' of the last century has been operationalized to build instruments for measuring politeness in different physical contexts. Among others, Dawn Lerman [12] built a scale for measuring politeness in order to examine the relationship between consumer politeness and their propensity to engage in various forms of complaining behavior.

However, the conventional politeness theory focusing on contents of verbal communication among persons obviously is not a proper basis for this study. Instead, the 5-principle definition proposed by Brian Whitworth is a proper way to measure the politeness in online storefronts, where computer-initiated actions to human users affect patrons' feelings and perceptions.

3 Research Method

To operationalize the polite computing framework presented by Brian Whitworth [10], the present work drew 20 questionnaire items according to the 5 principles of polite computing, and examined the reliabilities of the instrument and its factors. All following analysis works rely on collecting and analyzing perceptions from swarm of patrons who possess insights into visited e-commerce contexts.

3.1 Instrument Development

Based on 5 principles of the polite computing framework, a group of 32 college students with seasoned (> 5 years) online shopping experience were invited to draw observable action items, which they thought were able to assess to what extent visited online storefronts conforming to latent principles of the polite computing concept. Each item aims to judge to what extent an online storefront treats patrons politely while it is performing a particular function. Both face and content validity of the instrument for gauging degree of politeness in online storefronts were confirmed.

Table 1 summarizes the 20 items in the instrument. Each item was assessed by a 7-point Likert continuum, with higher scores representing the high end of the impoliteness scale; 1 indicates "disagree strongly" while 7 means "agree strongly".

Table 1. Descriptions of items in the politeness instrument

Construct (latent var)	Observable Variable	descriptions
Respect Right of Users	RR1	online storefronts (OS) played noisy but hard-to-stoppable background music
	RR2	OS popped-up disturbing and irrelevant messages from time to time
	RR3	OS exploited membership information to send SPAM advertisement
	RR4	OS changed the default setting of homepage in my browser

Table 1. (*Continued*)

Behave Transparent	BT1	OS installed software or change configuration on my devices stealthily
	BT2	OS asked me to fill questionnaires without disclosing purposes honestly
	BT3	OS twisted images in product catalogues to make products look more attractive
	BT4	OS added me (member) to some online communities/groups without notification
Useful Information	UI1	OS did not categorized their products/services well
	UI2	OS showed me broken links or guided me to wrong destination via misleading link description
	UI3	OS did not provide inventory (in-stock) or available time of products
	UI4	OS provided me recommendation after buying a product, but there is low or even no relevance (to my purchased product) in their recommendation.
Familiar With Habits	FH1	OS asked me to input my username every time when I tried to enter it
	FH2	OS did not record what I shopped in the past
	FH3	OS cannot keep track of my periodical purchasing behaviour
	FH4	OS asked me to provide payment/contact information every time when I checked-out
Fidelity Response	FR1	OS placed other products' advertisements within a checking-out page
	FR2	OS ignored or changed my requests; such as payment method choice.
	FR3	OS popped out a window, but directed me to somewhere when I clicking its "close" button/icon
	FR4	OS delivered a product that is not fully similar to the one I saw in their catalogue

3.2 Empirical Data Collection - Participants and Procedure

An open online questionnaire was used to collect participants' opinions. 379 participants filled the online survey during the 2014 spring semester, 282 completed the survey effectively; 147 (52.1%) of them are male, while 135 (47.9%) are female.

4 Data Analysis

4.1 Reliability Checking

The Cronbach`s α values measure the internal consistency of the 5 constructs and the instrument. The data indicate that the 5 constructs have good reliabilities. The Cronbach`s α value of the overall instrument is 0.910, which indicates that the overall instrument has a very good internal consistency. In addition, deletion of two items: RR1 and FH3, resulted in higher construct reliability.

4.2 Item Adjustment

To check whether the 20-item instrument could be improved further, confirmatory factor analysis (CFA) was used to examine fitness of the models with 4 different item formulations. Because the models were derived based on the prior polite computing framework, CFA was a preferable method for assessing how well they fit the data collected by this research. As the indices suggested, the 18-item model has the best goodness-of-fit according to the indices collectively.

4.3   Model Selection

According to the polite computing theoretical framework and the approach for checking plausible alternative models presented by Doll and Torkzadeh [13], the present
study compared 3 different models' fitness to the sampled data. As figure 1 shows,
the 3 examined models are (A) the first-order, 5-factor uncorrelated model; (B) first-
order, 5-factor correlated model; and (C) second-order 1-factor, first-order 5-factor
model. According to the models' goodness-of-fit index values, the model B is much
better than its uncorrelated counterpart, model A. Model B and C generated close and
both good model-data fits according to their relative and absolute indices [14].

   Furthermore, in order to measure the ability of the second-order factor (politeness)
to explain the covariation among the five first-order factors, target coefficient [15],
which is equal to the ratio of the chi-square of model B to the chi-square of model C,
was 0.982, an obvious indication of the second-order factor (politeness) can explain
the covariation among the five first-order factors.

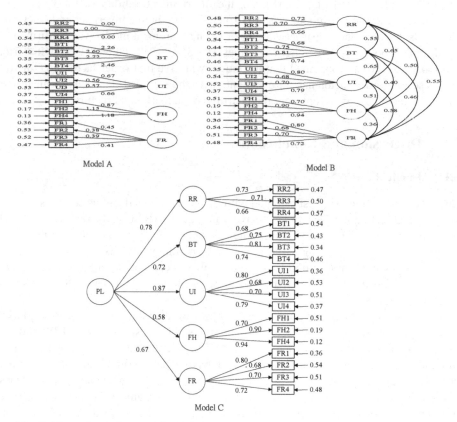

Fig. 1. Three alternative models (A,B,C) with factor loadings and structural coefficients

4.4   Measurement and Structural Model Analysis

According to the suggestions of Bagozzi and Yi [16], this work applied maximum likelihood estimation to test the measurement model. The criteria include factor loadings and indicator reliabilities, i.e., square multiple correlation (SMC) of the 18 observed items, composite reliabilities (CR) and variance extracted (VE) of the five first-order factors. Factor loadings above 0.32 represent substantial coefficient and structural equivalence [17], so all items in the instrument were considered meaningful and retained for their loaded factor. The SMC values indicated that the reliabilities of individual observed items are higher or very close to the recommended level of 0.5 [16], except the RR4 item. According to the recommended cut-off values of CR and VE, which are 0.6 and 0.5, respectively [18], the analysis results showed the measurement model has good reliability and convergent validity.

The square root of the average variance extracted (AVE) of each construct was much larger than all other inter-factor correlations, and greater than the recommended acceptable cut-off level of 0.7 [18]. So, the discriminant validity of the five latent constructs in the present model was confirmed. Taking both convergent and discriminant parts into account, construct validity of the measurement model was confirmed.

Multiple goodness-of-fit indexes' values (RMSEA=0.048; GFI=0.92; SRMR=0.046; PNFI=0.82; NFI=0.96) confirmed that the model with five first-order factors loading on single second-order factor has an excellent fit to the sampled data, which mean the structural model can well represent the instrument's underlying structure.

5   Discussion and Conclusions

5.1   Implications

The research findings suggest sampled patrons thought visited online retailers tend to be impolite. Among the 5 factors, "behave transparently (BT)" is the only one with all items being graded above 5.0. That indicates online shoppers thought virtual storefronts behave without enough transparency. Combining with its high loading ($\lambda = 0.72$) on the second-order politeness factor, online merchant should pay more attention to improve their behavioral transparency in order to gain better assessment in terms of politeness.

The second-order confirmatory factor analysis revealed that online shoppers placed relatively more weight on the factor of their rights being respected ($\lambda = 0.78$) and the factor of obtaining useful information ($\lambda = 0.87$), while they are interpreting the impoliteness in online storefronts. Patrons prefer having more control in their shopping contexts and during their shopping processes [19-22], which explain why patrons dislike that online storefronts preempt their rights. To many online shoppers, time efficiency is critical while they are going through a purchasing process, which comprises several steps and often is time-consuming. In consequence, patrons dislike any useless information distracting them during the course of online shopping, which is consistent with a prior study [22] saying that informativeness motive online shopping.

5.2   Contributions and Limits

This work developed an instrument that can gauge the degree of politeness in e-tailers' storefronts. After developing the new instrument, this study confirmed the psychometric properties of the instrument and its underlying model with a sample of 282 subjects. Among other properties, the factor structure was confirmed through testing a hierarchical model with five first-order factors loading on a second-order politeness construct by using confirmatory factor analysis. The research findings indicate that there is substantial room for improving online retailers' politeness.

Regarding the limitation of this work, because many aspects including society class, education, occupation, income, prior shopping experience, types of examined online business, and others collectively shape people's feelings, perceptions, and preferential items. Therefore, further research works and subsequent meta-analytic structural equation modeling [23] are necessary to generalize the findings of similar works.

5.3   Future Directions

This work sets stage for further research on two facets; one is the politeness management issue in other service-oriented online industries or e-government. Another direction is the influence of politeness on other constructs in various theoretical models. These constructs might include but not limit to rapport, customer affinity, trust, loyalty, revenue, profitability, and many others that interest business administrator.

Acknowledgements. This research work has being funded by the grant from the Ministry of Science and Technology, Taiwan, ROC, under Grant No. MOST 103-2410-H-235-001-. We deeply appreciate their financial support and encouragement.

References

1. Cooper, A.: The Inmates Are Running the Asylum: Why High Tech Products Drive Us Crazy and How to Restore the Sanity. Sams, Indianapolis (1999)
2. Parasuraman, R., Miller, C.A.: Trust and etiquette in high-criticality automated systems. Communications of the ACM **47**, 51–55 (2004)
3. Preece, J.: Etiquette online: from nice to necessary. Communications of the ACM **47**, 56–61 (2004)
4. Skogan, W.G.: Citizen satisfaction with police encounters. Police Quarterly **8**, 298–321 (2005)
5. Berry, L.L.: Relationship marketing of services—growing interest, emerging perspectives. Journal of the academy of marketing science **23**, 236–245 (1995)
6. Reynolds, K.E., Beatty, S.E.: A relationship customer typology. Journal of retailing **75**, 509–523 (1999)
7. Campbell, K.S., Davis, L.: The Sociolinguistic Basis Of Managing Rapport When Overcoming Buying Objections. Journal of Business Communication **43**, 43–66 (2006)
8. Nass, C.: Etiquette equality: exhibitions and expectations of computer politeness. Communications of the ACM **47**, 35–37 (2004)

9. Whitworth, B.: Polite computing. Behaviour & Information Technology **24**, 353–363 (2005)
10. Whitworth, B., Ahmad, A.: Polite Computing. The Social Design of Technical Systems: Building technologies for communities, pp. 77–105. The Interaction Design Foundation (2013)
11. Brown, P., Levinson, S.: Politeness: Some universals in language. Cambridge University, Cambridge (1987)
12. Lerman, D.: Consumer politeness and complaining behavior. Journal of Services Marketing **20**, 92–100 (2006)
13. Doll, W.J., Torkzadeh, G.: The measurement of end-user computing satisfaction. MIS quarterly, 259–274 (1988)
14. Kline, R.B.: Principles and practice of structural equation modeling. Guilford press (2011)
15. Marsh, H.W., Hocevar, D.: Application of confirmatory factor analysis to the study of self-concept: First-and higher order factor models and their invariance across groups. Psychological bulletin **97**, 562 (1985)
16. Bagozzi, R.P., Yi, Y.: On the evaluation of structural equation models. Journal of the academy of marketing science **16**, 74–94 (1988)
17. Tabachnick, B.G., Fidell, L.S.: Using multivariate statistics (2008)
18. Fornell, C., Larcker, D.F.: Structural equation models with unobservable variables and measurement error: Algebra and statistics. Journal of marketing research, 382–388 (1981)
19. Ganesh, J., Reynolds, K.E., Luckett, M., Pomirleanu, N.: Online shopper motivations, and e-store attributes: an examination of online patronage behavior and shopper typologies. Journal of retailing **86**, 106–115 (2010)
20. Eroglu, S.A., Machleit, K.A., Davis, L.M.: Atmospheric qualities of online retailing: a conceptual model and implications. Journal of Business research **54**, 177–184 (2001)
21. Lee, P.-M.: Behavioral model of online purchasers in e-commerce environment. Electronic Commerce Research **2**, 75–85 (2002)
22. Sorce, P., Perotti, V., Widrick, S.: Attitude and age differences in online buying. International Journal of Retail & Distribution Management **33**, 122–132 (2005)
23. Cheung, M.W.-L., Chan, W.: Meta-analytic structural equation modeling: a two-stage approach. Psychological methods **10**, 40 (2005)

Implementation and Theoretical Analysis of Virtualized Resource Management System Based on Cloud Computing

Yong Li[(✉)] and Qi Xu

China Tobacco Zhejiang Industrial Co., LTD, Hangzhou, China
{liy,xuq}@zjtobacco.com

Abstract. With the continuous and rapid development of computational science and data engineering related techniques, the transmission and protection of data are crucial in the computer science community. Cloud computing is becoming increasingly important for provision of services and storage of data in the Internet. Cloud computing as newly emergent computing environment offers dynamic flexible infrastructures and QoS guaranteed services in pay-as-you-go manner to the public. System virtualization technology providing a flexible and extensible system service is the foundation of cloud computing. How to provide the infrastructure for a self – management and independent cloud computing through virtualization has become one of the most important challenges. In this paper, using feedback control theory, we present VM-based architecture for adaptive management of virtualized resources in cloud computing and model an adaptive controller that dynamically adjusts multiple virtualized resources utilization to achieve application Service Level Objective (SLO) in cloud computing. Through evaluating the proposed methodology, it is shown that the model could allocate resources reasonably in response to the dynamically changing resource requirements of different applications which execute on different VMs in the virtual resource pool to achieve applications SLOs. Further optimization and n-depth discussion are also taken into consideration in the end.

Keywords: Cloud computing · Resource management system · Virtualized structure

1 Introduction

As IT industry infrastructure has become more complicated and Internet services have expanded dramatically today, offering reliable, scalable and inexpensive dynamic computing environments for end-users are the key challenge to service providers. Cloud computing [1-5] as a new computing environment provides dynamic flexible infrastructure and QoS guarantee service to the public in the form of pay-as-you-go. Cloud computing is actually a virtualized computing resources pool in a very large data centre, to adapt to a variety of applications. Virtualization in the cloud computing environment, resources can be dynamically allocated according to need a variety of end users and the calculation of each application running in a separate execution environment. System

© Springer International Publishing Switzerland 2015
Y. Tan et al. (Eds.): ICSI-CCI 2015, Part III, LNCS 9142, pp. 431–439, 2015.
DOI: 10.1007/978-3-319-20469-7_46

virtualization technology [6] which renders flexible and scalable system services is the base of the cloud computing. Through system virtualization technology which is a single physical machine logically is divided into one virtual machine monitor (VMM) and multiple virtual machines (VMs). Each VM (virtual machine) can run some applications perform an independent system. The VMM executive function such as isolated VM, resolve the conflict between virtual machine resource allocation and management of virtual machine. One nontrivial problem concerns the resource allocations on-demand of virtual machines in the cloud computing environment. The opportunity to overcome this problem is to offer efficient resource scheduling and manager architecture and model for Cloud Computing. How to provide a self-managing and autonomic infrastructure for cloud computing through virtualization becomes an important challenge.

Feedback control has developed a fairly sound theory and effective practices in many engineering systems over the last forty years. In the past ten years, the application areas of feedback control theory focuses on performance control in computer systems, such as a database server, multimedia system, Web server, etc. [7]. Recently, researchers have focused on the dynamic resource management virtualization server [8]. However, before the work is in a virtual server or multiple - tier application in multiple virtual servers and limits the scalability and flexibility. Not suitable for cloud computing environments. In this paper, using feedback control theory, we present VM-based architecture for adaptive management of virtualized resources in cloud computing. Feedback control provides a conceptual tool to solve the dynamic management of resource, especially the workload and configuration and unpredictable changes and interference is becoming a feasible method of self-management and autonomic computing system design.

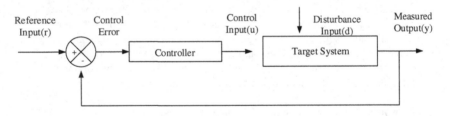

Fig. 1. The Standard Feedback Control System

Fig. 1 illustrates the essential architecture of a standard feedback control system. The target system is the computing system controlled by the controller. Desired goal (called the reference input) of the target system, the controller dynamic adjusting control input value according to the control error is the difference between the reference input and the output measurement. Because the interference such as workload control-error will frequently change, the measured output is a measurable index of the target system. We adopt KVM [9], the Kernel-based Virtual Machine, as the infrastructure of the cloud computing system instead of Xen [10]. Through analysing architectures of KVM and Xen, we believe KVM has more advantages than Xen. In the following sections we will discuss our methodology in detail.

2   Principles of Cloud Computing Architecture

In this Cloud Computing Open Architecture, we propose an integrated co-innovation and co-production framework to get cloud vendors, cloud partners, and cloud clients to work together based on seven principles, which are used to define ten major architectural modules and their relationships shown in Figure 2. The crucial principles are discussed as follows.

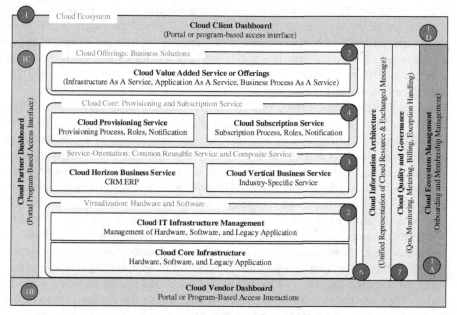

Fig. 2. The Cloud Computing Open Architecture Overview Diagram

1. Integrated Ecosystem Management for Cloud. The architecture must support the management of the ecosystem of Cloud Computing. This ecosystem includes all involved services and solutions vendors, partners, and end users to provide or consumer shared resources in the Cloud Computing environment. Cloud vendors expose its interaction interfaces of its internal operations and product development capability to the cloud. Cloud partners provide components to cloud vendors or serve as agents to provide value-added services to the cloud clients. Cloud clients are users of the cloud services that offer business goal driven resource sharing.
2. Virtualization for Cloud Infrastructure. There are two basic approaches for enabling virtualization in the Cloud Computing environment. The first method is the hardware virtualization, management plugin hardware equipment and the play mode. Hardware can be added or removed without affecting the normal operation of other equipment in the system. Of course, performance or storage space may be due to the dynamic changes that add and delete operations. The second approach is software virtualization, i.e., to use software image management or software code virtualization technology to enable software sharing.

3. Service-Orientation for Common Reusable Services. Introduced before, in addition to the characteristics of the service oriented virtualization, is another driving force of cloud computing for the further implementation of assets from the reusability, the commercial value of the composite application, and mix of services. Two main types of common reusable services: cloud horizontal and vertical business services.

4. Extensible Provisioning and Subscription for Cloud. Extensible service provisioning is the unique feature of a Cloud Computing system. No scalability, the configuration part of the cloud computing architecture to support only a specific type of resource sharing. This means that the architecture of free use service users and subscribers are all the same. Both types of users can be service providers or consumers from time to time. From service consumers' perspective, they are interested in how to easily access services based on their own business logics and goals. From service providers' perspective, three levels of service provisioning described previously will be the target offerings.

5. Configurable Enablement for Cloud Offerings. Cloud offerings are the final products or services that are provisioned by the Cloud Computing platform. In the Cloud Computing environment, business process as a service is a new model for sharing best practices and business processes among cloud clients and partners in the value chain. An example of the business process as the cloud computing environment provides a service is software testing, which is a software development procedure. It is very important for the business process. Because the software system or application need to deploy and run on different operating system, middleware environment, or different versions or configuration, it is very difficult for individual developers or companies to create a test "factory" solve the problem of testing in the application service area. Upgrading or maintaining testing environment involves the migration of hardware, software, and testing knowledge repositories. They are the two major expenses in addition to the testing engineers who manipulate the environment to conduct the testing.

6. Unified Information Representation and Exchange Framework. Information representation and message exchange of Cloud Computing resources are very important to enable the collaborative and effective features of Cloud Computing.

7. Cloud Quality and Governance. The last and most important module in out platform is the Cloud Quality and Governance shown in Figure 2. This module is responsible for the identification and definition of quality indicators for Cloud Computing environment and a set of normative guidance to govern the design, deployment, operation, and management of the cloud offerings.

3   The System Virtualization Model

System virtualization was first introduced in System/360 by IBM in the 1960s [8] and has a rapid resurgence in recent years as a way to server consolidation, security isolation and hardware independent with performance improvement of microcomputer hardware. It has become more attractive, research community and industry. System virtualization makes a computer seems to be one or more virtual computer system. System virtualization is the basis of the virtual machine monitor (VMM) or hypervisor is a software layer, it provides the illusion of a real hardware for multiple virtual machine (VM). The VMM support multiple virtual machine creation and execution in a physical machine, a single physical resource and the multiplexes in multiple virtual resources. Each virtual machine operating system (called a guest operating system) and the application is running on a virtual machine based on virtual resources, such as virtual CPU, virtual memory, virtual disks and virtual network adapter. There are two basic types of system virtualization according to implement structure of VMM. Hypervisor Model: VMM directly runs on the system hardware and own all physical resources. Xen is an example of this type of VMM. Host Model: VMM runs as a kernel module or a normal program inside a normal operating system (Host OS). Host OS is responsible for managing physical resources and VMM only provides virtualization functions. KVM and VMWare are examples of this type of VMM. In the next subsections, we describe and compare the virtualization architecture used in Xen and KVM respectively.

3.1   The Xen

Xen is a popular open-source VMM on a single x86-based physical platform. Xen supports para-virtualization [9] which requires the operating system changes to interact with the VMM. The lowest level and running of the Xen hypervisor is one of the most privileged processor. There can be many domain (vm) running at the same time a hypervisor. The driver domain (also known as dom0), as a privileged VM, using local device driver directly access I/O devices and perform I/O operations on behalf of other poor VM (also known as customer domain). Guest domains use virtual I/O devices controlled by a para-virtualized driver (called afront-end driver) to communicate with the corresponding back-end driver in the driver domain, which in turn, through software Ethernet bridge, multiplexes data for each guest onto the physical NIC device driver. Front-end driver and the data transmission between the back-end drivers is by using a zero copy page remapping mechanism. All I/O access hypervisor leads to longer within the territory and drive the I/O latency and higher CPU overhead because the context switch between the domain and the hypervisor. Driver areas become an I/O intensive workload performance bottlenecks. In the page remapping mechanism, the paging and grant table management are higher overhead functions.

3.2   The KVM

KVM (for Kernel-based Virtual Machine) is a full virtualization solution for Linux on x86 hardware containing. KVM is an open source software projects have been developed [8]. And Xen management execution scheduling, memory management, and on behalf of the I/O function privileges guests, KVM transmission field, using the Linux kernel as a hypervisor virtualization capabilities by adding standard Linux kernel. In this mode, each virtual machine is a regular Linux process processing standard Linux scheduler by adding a guest model execution, has its own kernel and user mode. Its memory allocation is based on Linux memory allocator. The kernel component of KVM, the device driver (KVM Driver) for managing the virtualization hardware, is included in the Linux 2.6.20. The other component that comprises KVM is the user-space program for emulating PC hardware. This is a modified version of QEMU [11]. The I/O model is directly derived from QEMU's. The inclusion of KVM driver in Linux2.6.20 has dramatically improved Linux's ability to act as a hypervisor. KVM runs in the Linux kernel as a kernel module and become a lightweight VMM. Compared with Xen and KVM consume fewer resources, reduces the service disruption, and eliminate the driver domain and the context switch between the program. When performing the dense network I/O, KVM has a performance advantage. Hence we will construct our architecture for cloud computing environments based on KVM.

4   The Implementation of the Model

4.1   The Motivation

Today's enterprise data centers host many different types of business critical applications on shared cluster infrastructures. It is responsible for ensuring fair use and maximum system workload. If there is an uneven distribution of the different applications on the whole or over time, this leads to unnecessary idle time, since resources used by one application cannot easily be moved to the other application. It's difficult to determine the real resource needs of each, and to provide additional resources as needed if multiple applications share hardware platform. It will limit the flexibility, functionality and performance of data centers. The next generation of enterprise data center is used to build the cloud computing infrastructure should provide reliable, "fee" and QoS guaranteed dynamic computing environments for end-users. A cost-effective way to achieve this, all of the hardware resources together into a common Shared cloud computing infrastructure, hosting application sharing these resources on demand response time varying load and resource allocation. As a result, the cloud computing infrastructure is faced with challenges to provide manageable and flexible resource management model and methods that enable dynamic resource allocation along with unpredictable time-varying workloads of applications and interactions across many applications in order to meet Service Level Objectives (SLOs) of applications.

4.2   Virtual Resource Management Architecture

This paper introduces the virtual resource management architecture, system virtualization and feedback control to deal with these challenges. The complete system structure is shown in figure 3. KVM through virtualization technology, physical nodes of cluster logically divided into a VMM (including the host operating system) and multiple VM. The VMM for multiplexing whose physical resources is used to the VM. Each VM host operating system can perform some application or service. The VMM executive functions such as isolation VM and solve the conflict between the virtual machine. As a result, many virtual machines of the cluster nodes to form a virtual resource pool as a cloud computing infrastructure. KVM as a kernel module is tightly integrated into Linux kernel and VM exists as a Linux process. We can reuse existing the scheduler, the memory manager, NUMA support as actuators for setting CPU shares, memory allocations and I/O allocations for all the VMs. Sensors periodically collect the application performance and the resource usage statistics. We also reuse the existing Linux process management methods, such as *top* and *proc* file system to measure CPU and memory consumption. Dynamic control system structure is an important contribution to reallocate resources to the VM migration they live in the network. The virtual machine live migration is the ability to transport from one host to another without interrupting the execution of the application. KVM VMM can manage VMs to create, terminate, suspend and resume services at any point of their execution. KVM provides a dirty page log facility [12-14] to speed up VM live migration.

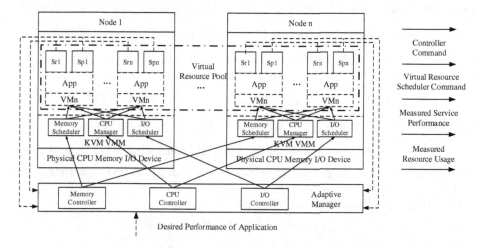

Fig. 3. The VM-based Architecture Adaptive Management

4.3   The Adaptive Manager

The adaptive manager is multi-input, multi-output (MIMO) resource controller in which the goal is to regulate multiple virtualized resources utilization to achieve application SLOs using the control inputs per-VM CPU, memory and I/O allocation. We use the dynamic state-space feedback control [11] to model the adaptive manager.

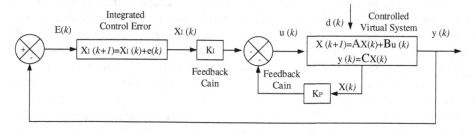

Fig. 4. The Adaptive Management Model

As showed in Fig. 4, in the controller system, reference input $r = [r_1, ..., r_n]^T$ which is desired performance of applications inside VMs, where r_i denotes desired performance of application i, namely application i' SLO value. The state-space model representing the input-output relationship in the controlled virtualization system is:

$$x(k+1) = Ax(k) + Bu(k) \tag{1}$$

$x(k) = [x_1(k), ..., x_n(k)]^T$, where $x_i(k) = S_i(k) - \overline{S}_i$ is offset value of application i's SLO during the kth time interval. $\overline{S}_i$ is the operating point of application i' SLO. The control input is $u(k) = [u_1(k), ..., u_n(k)]^T$ where $u_i(k) = R_i(k) - \overline{R}_i$ is offset value of application i's resource allocation during the kth time interval. $\overline{R}_i$ is the operating point of resource allocated of application i.

The control error vector is represented as: $e(k) = r - y(k) = r - Cx(k)$. The integrated control error $X_I(k+1) = X_I(k) + e(k)$, where $X_I(k)$ describes the accumulated control error. A disturbance input $d(k)$ that affects the control input denotes unpredictable changes and disturbances in workloads and configuration at the kth time interval.

This system uses the control law:

$$u(k) = -K[X(k), X_I(k)]^T = -[K_p, K_I][X(k), X_I(k)]^T \tag{2}$$

The adaptive manager model provides a way to allocate resources reasonably in response to the dynamically changing resource requirements of different applications which execute on different VMs in the virtual resource pool to achieve applications SLOs.

5   Conclusions and Summary

In this paper, we propose a VM-based architecture for adaptive management of virtualized resources in cloudcomputing environments. This architecture provides strong isolation between applications which simultaneously run in the virtual resource pool, and allows the dynamic allocation of resources to applications and applications to grow and shrink based on resource demand to achieve SLOs. A main problem under the architecture is how to allocate resources to every application on-demand and in response to time-varying workloads. We have designed an adaptive manager which includes three controllers: CPU controller, memory controller and I/O controller. The adaptive manager is modelled by the dynamic state-space feedback control method. Through analysing the virtualization architecture used in Xen and KVM respectively, we adopt KVM as the infrastructure of virtual system instead of Xen. Currently, we are verifying the validity and implementing the function of our proposed architecture and model. Also our evaluations exhibit that network I/O performance is still much room for further improvement. Furthermore, we are investigating direct I/O assignment for the issues and better modelling of network sharing.

References

1. Meinhardt, M., Rahn, J.: Empirical qualitative analysis of the current market situation in the context of cloud computing for logistics. In: Cloud Computing for Logistics, Lecture Notes in Logistics, Springer (2015)
2. Ahmed, E., Akhunzada, A., Whaiduzzaman, M., Gani, A., Hamid, S.H.A., Buyya, R.: Network-centric performance analysis of runtime application migration in mobile cloud computing. Simulation Modelling Practice and Theory **50**, 42–56 (2015)
3. Bahrami, M., Singhal, M.: The role of cloud computing architecture in big data. In: Information Granularity, Big Data, and Computational Intelligence, pp. 275-295. Springer (2015)
4. Garg, S.K., Buyya, R.: Green cloud computing and environmental sustainability. In: Harnessing Green IT: Principles and Practices, pp. 315-340 (2012)
5. Boru, D., Kliazovich, D., Granelli, F., Bouvry, P., Zomaya, A.Y.: Models for efficient data replication in cloud computing datacenters. In: IEEE International Conference on Communications (ICC). (2015)
6. Abolfazli, S., Sanaei, Z., Sanaei, M., Shojafar, M., Gani, A.: Mobile cloud computing: The-state-of-the-art, challenges, and future research (2015)
7. Zhao, S., Lu, X., Li, X.: Quality of service-based particle swarm optimization scheduling in cloud computing. In: Proceedings of the 4th International Conference on Computer Engineering and Networks, pp. 235-242. Springer (2015)
8. Nayar, K.B., Kumar, V.: Benefits of cloud computing in education during disaster. In: Proceedings of the International Conference on Transformations in Engineering Education, pp. 191-201. Springer (2015)
9. Hung, C.-L., Chen, W.-P., Hua, G.-J., Zheng, H., Tsai, S.-J.J., Lin, Y.-L.: Cloud Computing-Based TagSNP Selection Algorithm for Human Genome Data. International Journal of Molecular Sciences **16**, 1096–1110 (2015)
10. Ross, P.K., Blumenstein, M.: Cloud computing as a facilitator of SME entrepreneurship. Technology Analysis & Strategic Management **27**, 87–101 (2015)
11. Chan, C.-C.H., Lo, Y., Chen, C.Y., Tsai, P.C.: Using cloud computing to support customer service in the automobile industry: an exploratory study. In: New Ergonomics Perspective: Selected Papers of the 10th Pan-Pacific Conference on Ergonomics, p. 319. CRC Press, Tokyo, Japan, August 25-28, 2014
12. Fernandez-Llatas, C., Pileggi, S.F., Ibañez, G., Valero, Z., Sala, P.: Cloud computing for context-aware enhanced m-health services. In: Data Mining in Clinical Medicine, pp. 147-155. Springer (2015)
13. Puthal, D., Sahoo, B., Mishra, S., Swain, S.: Cloud Computing Features, Issues and Challenges: A Big Picture. Computational Intelligence (2015)
14. Boopathy, D., Sundaresan, M.: Enhanced Encryption and Decryption Gateway Model for Cloud Data Security in Cloud Storage. In: Satapathy, S.C., Govardhan, A., Raju, K.S., Mandal, J.K. (eds.) Emerging ICT for Bridging the Future - Volume 2. AISC, vol. 338, pp. 415–421. Springer, Heidelberg (2015)

Integration of a Mathematical Model Within Reference Task Model at the Procurement Process Using BPMN for Disasters Events

Fabiana Santos Lima[1(✉)], Mirian Buss Gonçalves[1],
Márcia Marcondes Altimari Samed[2], and Bernd Hellingrath[3]

[1] Department of Post Graduate Production and Systems Engineering,
Federal University of Santa Catarina, Florianópolis, Brazil
fabiana.lima@posgrad.ufsc.br
[2] Department of Production Engineering, State University of Maringá, Maringá, Brazil
[3] Department of Information Systems and Supply Chain Management,
Westfälische Wilhelms-Universität Münster, Münster, Germany

Abstract. The presented approach on this article is related to the task streams activities according to the reference model from humanitarian logistics tasks proposed by Blecken (2009) using the Business Process Modeling Notation (BPMN). Steps were presented initially for the mathematical model insertion, emphasizing on the function of supplies acquisition during the phase of response in case of disasters. The outcome from this model consists on determining which suppliers have the capability to deliver emergency supply items on the requested date. Illustrating the usage of this model, we will show the development of a small example. The first approach contributions is the BPMN specification for assessment tasks and procurement, according to the reference model, including new tasks. The second contribution is represented by a mathematical model insertion, specifically on the task "select supplier", from the reference model. Therefore, there was a graph model proposed for the Network Flow Model, which was adapted to the Humanitarian Logistics context.

Keywords: Humanitarian logistics · Disasters events · Reference task model · Business Process Modeling Notation · Network Flow

1   Introduction

In the last decade, we have noticed the concepts evolution about humanitarian logistics (HL) and supply chain management (SCM). According to [2], humanitarian logistics is a special branch of logistics where the challenges are the demand surges, uncertain supplies, critical time-window in face of infrastructure vulnerabilities and vast scope and size of the operations. [3] were one of the first to define the humanitarian supply chain characteristics, like: ambiguous objectives, limited resource, high uncertainty, urgency and politicized environment. Currently, these concepts are well-known, but in practice, it is still necessary to improve the actions efficiency from the

© Springer International Publishing Switzerland 2015
Y. Tan et al. (Eds.): ICSI-CCI 2015, Part III, LNCS 9142, pp. 440–452, 2015.
DOI: 10.1007/978-3-319-20469-7_47

humanitarian logistics and supply chain management. The speed of response in an emergency is essential and it depends on the logistics performance to transport and deliver supplies to the affected area. A large part of the execution from the decision made for the relief procurement is qualitative experience, which is developed by expert's experience in conducting relief operations. This means that it is difficult to allocate tasks to those who are less experienced and there are a few tools to support the decision makers in these high-pressure environments.

The usage of concepts that focus on behavioral aspects of an organization adapted to the humanitarian specificities assistance and management chain can be the differential to minimize improvisation actions, maximize efficiency and response time to emergencies. According to some researchers ([4], [5], [6]) the used process models offer organizations a set of benefits, for example, how to improve transparency and support relations harmonization between various sectors. It also serves to fortify the institutional capacity in order to manage its growth.

[1] developed a methodology denominated Reference Task Model (RTM) that aims to standardize the tasks of humanitarian logistics according to the time horizon of: strategic, tactical and operational planning. RTM offers a number of potential benefits to humanitarian organizations: "First, it provides a tool to humanitarian organizations enabling them to quickly visualize the tasks carried out by the organization and its supply chain partners. The RTM supports the SCM tasks standardization by clarifying roles, responsibilities and definitions. It enables humanitarian organizations to communicate tasks and processes to their supply chain partners. Thus, transparency and accountability to donors can be improved. The RTM is reusable and facilitates the processes and supply chain performance benchmarking"[1].

We have decided to expand the scope of the tasks "assessment and procurement" in the humanitarian logistics and supply chain operation with the advantages of RTM mentioned above. Under these circumstances, the research question that motivates this paper is: How can the process standardization and its activities be improved to support the preparation and response phases from humanitarian logistics? Taking it into consideration, we define the specific objectives from this paper in: Usage of BPMN to organize the process of activities and tasks standardization and in response phase, introduce a network flow model to determine the suppliers that are able to find the requested demand in capability function, price and time.

This paper is structured as described below. section 2 is focused on literature that addresses the process models in humanitarian logistics and addresses on some quantitative methods. The methodology is presented in section 3. The section 4 its objective is to demonstrate the BPMN development, the introduction of new tasks and the quantitative model with a small example, just like illustration. Concluding remarks are described in section 5.

2   Related Literature

2.1   Process Models in Humanitarian Logistics

There are few publications considering process modeling in the literature about LH. Moreover, three important contributions have been selected: [7], [8] and [1]. In sequence, it is presented by a brief description of these publications: [7] describes modeling processes in logistics disasters by developing a Logistic model Process in International Disaster Relief in considering the logistical disaster management under two foci: in one hand, non-specific (generic), which the activities that are included occur before there is a crisis situation, it is not targeted to a specific disaster. The generic phase ends as soon as the specific disaster operation starts. The specific phase consists of three levels, regarding support to operations, supply of specific disasters, the process control and integrated planning. The saving time has a highest priority and the fast transportation modes are chosen to delivery the material assistance to the beneficiaries as quickly as it is needed. One of the points highlighted by Tufinkgi refers to the network structure. The author states that more efficient and effective network structure settings could improve the   current system deficiencies. [8] presents a Management Supply Chain model for Health Products with the intention from assisting humanitarian organizations to improve the effectiveness and operations efficiency.

A framework was developed to manage the supply of health products provided as humanitarian assistance in complex emergencies chain. The structure is a decision support system for logistics and supply chain management that can be applied to specific health programs at a given humanitarian organization in a specific context.

The table shows factors that favor certain decisions, but it does not use mathematical deductions for the best solutions. In order to answer the main question of how humanitarian organizations can manage the supply chain of health goods in complex emergencies, the research is placed on the structure of hierarchical planning context of [9], which is modified and extended before adding the time, distance and criticality aspects. [1] presented the reference task model (RTM), consisting of a tool that supports the tasks standardization performed by the organization and its humanitarian partners in the supply chain. The RTM has involved 39 major international humanitarian non-profit organizations that operate around the world and, therefore reflects the activities of the real world in humanitarian supply chain management. It was developed for emergency tasks and post-emergency humanitarian organizations.

The model distinguishes between the tasks a humanitarian supply chain along two axes: the hierarchical horizon decomposition of the planning of strategic, tactical and operational tasks and a split in the functional axis on the evaluation, acquisition, storage, transport, and reporting support operations were separated. The reference task model (RTM) is a tool that supports the standardization tasks performed by the organization and its humanitarian partners in the supply chain. The reference processes of the RTM can be used as an adequate modeling framework for the network flow problem of the quantitative decision support tool.

2.2   Reference Task Model

RTM was developed for emergency and post-emergency supply chain management activity for humanitarian organizations [1]. "The RTM distinguishes approximately one hundred twenty activities in a humanitarian supply chain along the hierarchical decomposition of the planning horizon (strategic, tactical and operational) and a division according to the function (assessment, procurement, warehousing and transport)" [6]. Figure 1 shows the framework of RTM for humanitarian logistics.

[6] has used the RTM for mapping processes at two different relief organizations in order to support the optimization of humanitarian logistics processes. The authors also state that decisions on the supply chain structure (e.g. plan emergency preparedness, plan emergency supply strategy, plan transport network, plan warehouse capacities etc.) are made on a strategic level, in which the time horizon can exceed two years. On the tactical level, which covers a time horizon between six months and two years, the entire supply chain is planned and optimized (e.g. decisions regarding plan demand, plan supply of operations regional or local, pre-qualify suppliers, set inventory control policy etc.). An optimized use of resources, which are allocated on the tactical level, is the main aim at the operational level. Each task at this level extends up to six months (e.g. demand forecasts, mobilization of supplies (ad-hoc), obtaining quotations, qualification of suppliers etc.). In RTM each activity is formed by tasks, for example, the "Plan Programme Strategy" activity is composed of tasks as allocation of required logistical resources and capacities. Only a few quantitative approaches have been proposed in the literature for the procurement in humanitarian relief operations. Emphasized is the work of [10] which describes a tool for decision support MIP (Mixed-Integer Program) in order to improve the maritime carrier and the strategy of bidding prices from the food suppliers, and further the study of [11] which developed decision purchasing models for humanitarian relief. Both works aim at specific situations, in which general guidance for the usage of quantitative methods in relief to assist the procurement of items for relief operations process was not found in the literature. I.e., there are no studies that offer a methodology to support decision making that considers the time of service in the delivery of supplies during the response to natural disasters.

By the analysis of the activities proposed in the RTM, it was verified that the tasks related with the functions of assessment and procurement can be adapted and extended to be applied in quantitative tools to decision support in humanitarian logistics.

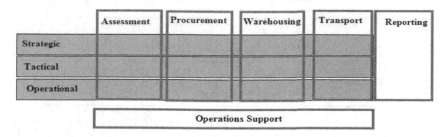

Fig. 1. Framework of the Reference Model for Humanitarian Logistics (Source: Blecken,2009, p.116)

2.3   Quantitative Methods

Based on a supply chain perspective, quantitative models existing in the literature to assist decision making in humanitarian operations can be classified into three main categories: (1) inventory management applications, (2) location of facilities, (3) distributing aid and transport of victims. It is proposed to integrate a quantitative method with the RTM for humanitarian logistics. Specifically, it is suggested to use the quantitative method 'network flow'. It is used to determine the suppliers that are able to attend the relief supplies demand in the shortest time as possible.

Network flow is a concept that is well employed in the context of humanitarian logistics. We can find important contributions in this area; here we describe some of these. [12] used a multi-commodity, multi-modal network flow in disaster relief operations. [13] applied the network flow in an interactive approach for hierarchical analysis of helicopter logistics in disaster relief operations. [14] suggested an emergency logistic planning in natural disasters using a network flow to define how to dispatch commodities (e.g., medical materials and personnel, specialized rescue equipment and rescue teams, food, etc.) to distribution centers in affected areas. [17] applied the concepts of the network flow in a multi-criteria optimization model for humanitarian aid distribution.

3   Methodology

We propose in this paper the usage of a methodology within the preparedness and response phases of disaster assistance by integrating tools into the disaster relief processes. Based on the structure demanded determined beforehand, suppliers, who are able to deliver the corresponding items in a short period of time, it should be selected by using an network flow model within the disaster response phase. The tool integration into the processes was achieved by utilizing and extending a reference model for humanitarian logistics. The new integrated tasks in the RTM are described in BPMN language for process standardization. Under this circumstance, it will be possible to analyze supply chain process in the procurement phase. The quantitative models integration with the RTM is divided into three phases. In the first phase, it will be identified in which activities the RTM can be supported by a proposed quantitative model. In the second phase, a new task will be proposed for the RTM extension in order to make the RTM integration possible and the quantitative model. In the third phase, the quantitative model itself will be formulated and adapted to the needs of humanitarian relief organizations. Figure 2 shows the process of integration.

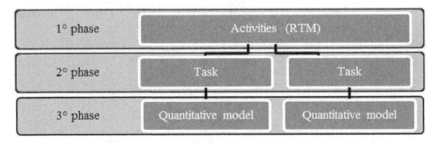

Fig. 2. Process of Integration (Source: authors)

The RTM was analysed and verified if that the procurement task is related to quantitative decision supporting tools on the operational level. Based on these findings, the integration process consists in analysing the RTM activities, integrating new tasks and describing the tasks in BPMN. Specifically, it was proposed an adaptation to a network flow model for the supplier selection in the context of humanitarian logistics. Figure 3 illustrates this process.

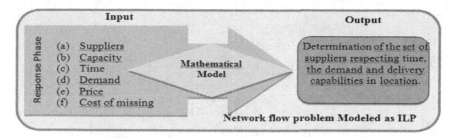

Fig. 3. Inputs e Outputs of the Problem of the Network Flow (Source: authors)

4   Extended Business Process Modeling Notation for Humanitarian Logistics

In this section, we presented the contributions to the RTM, considering not just the assessment task at the strategic, tactical and operational levels but also the procurement task in the operating level. Figure 4 presents the Extended Business Process Modeling Notation for humanitarian logistics.

The processes designs (Figure 4) are arranged in a column that indicates the tasks for assessment and procurement functions. The pools represents the Strategic Center Logistics (strategic level), Regional Logistics Center (tactical level) and Local Logistics center (operational level), hierarchically. In the upper left corner we indicate the name of the task used by [4], in the lower right corner we declare if the task is generic or specific and in the lower left corner we indicate the corresponding function (assessment or procurement).

In the BPMN development each new proposal subtask is highlighted by color. The new subtasks integration in the "Select Supplier" task is given quantitatively by the network flow problem. These approaches are made to assist the preparation phase in the type definition of goods by a type of disaster. The process continues until it receives the information that determines the set of suppliers respecting the demand and delivering capabilities at the affected location. In the next items are specifically described the integration of the referent subtask to the select suppliers.

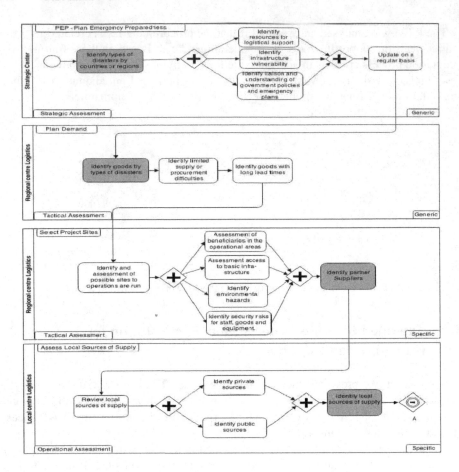

Fig. 4. Extended Business Process Modeling Notation for Humanitarian Logistics (Source: authors)

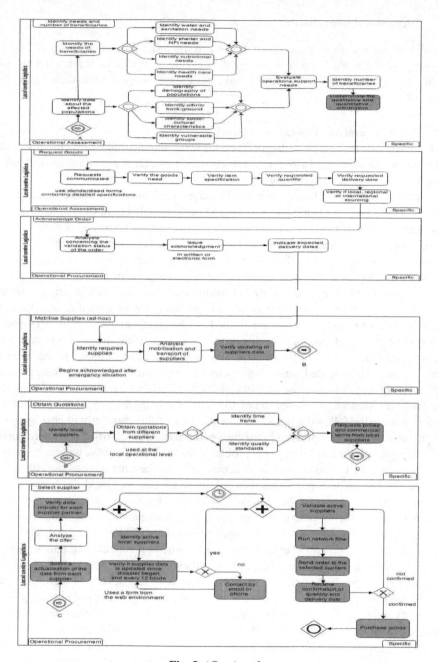

Fig. 5. (*Continued*)

4.1   Proposed Integration

In order to introduce the quantitative method in the tasks framework proposed by [1] we integrate the subtasks as it is proposed in Figure 2. We show the related tasks in RTM and we explain the additional task, which we suggest with their respective quantitative model.

4.1.1.   Procurement - Operational Level

In the procurement function at the operational level, the quantitative method is developed for select supplier. In order to present this process, we distinguish the Select Supplier task extracted from Figure 4

1° **phase Activity: Select Supplier-** [1] states "[...] suppliers should be easily accessible and flexible in responding to changing demand, reliable in case of high-priority orders and offer low total cost of ownership of goods and services. This is reflected in the RTM process elements." 2° **phase Task** - Define sets of suitable suppliers: Due to the intrinsic complexity of the factors involved in humanitarian logistics operations, it is necessary to adopt some considerations before presenting the modeling of the problem: (a) the existence of partnership between private companies and humanitarian organizations in which, somehow, commit to inform and update your identification, location, supply capacity and price; (b) the use of the Cost Insurance and Freight (CIF) system - transport on behalf of the supplier. It is considered that while the region is not in the normal situation, any kind of action helps generates a cost. For example, people who are homeless need to be hosted somewhere as tents, gymnasiums, schools, hotels etc., and the longer the time to reach the normal condition, the higher the total cost of the disaster. To better illustrate the definition of the cost of missing by day of the product can seek to answer the following question: how the lack of this product will cost by day to the affected region? 3° **phase Quantitative model**: After clarified the prerequisites needed to prepare the modeling of the problem, we try to define what will be the suppliers to be triggered by the humanitarian organization at the time of response, being primordial that the supplies are required in the areas affected as quickly as possible. The goal at this point is to act in the immediate response phase of a humanitarian operation and considering even in practice the immediate term means meet the demands of the affected areas in the shortest possible time, under penalty of lives being lost. Therefore, the modeling must be flexible enough to consider time as a variable and thus negatively weigh the possible delays. Thus, the solution will evolve to select suppliers that have the capacity to deliver relief supplies in the shortest possible time, even if a more expensive cost because the point of view of lives at risk, is the best option against a cheaper supplier, but cannot meet the demand on the requested date. Thus, the model should contain the cost of missing per day not attended, so that the resolving procedure seeks for those suppliers of lower cost, with product availability at the requested time, or as close to it.

Therefore, we propose a model graph that T is the time horizon for discrete periods, 1,2,...,T for example a period is equal to one day, K is the number of suppliers, M is the number of locations, F_k is the vector k supplier's capacity in the horizon planning T, D_m is the vector with the locale demand m in the planning horizon T. In this model, that are two types of nodes: nodes of types 1 is the ordered pair (supplier's capacity, time)=

(F, t), nodes of type 2 is the ordered pair (demand of location, time)= (D, t) and there are arcs of three types representing the flows: arch type 1, flows associated with inventory, $((F_k, t), (F_k, t+1))$; arch type 2, flows associated with the lack, $((D_m, t), (D_m, t-1))$ and arch type 3, related to the transport streams, $((F_k, t), (D_m, t+\text{time}(k, m)))$. Figure 5 presents the graph developed as well as the corresponding network flow problem. In the graph, the node label (supplier's capacity, time) represents the amount of product available for prompt delivery at that time. The node label (demand of location, time) represents the required amount of product at the disaster in the requested period. The goal is to determine which suppliers will be able to attend the event in such a way that the total cost of shipping the product (supply) available via the network to be minimized in order to satisfy the given demand .

The specialized simplex method for solving the problem of network flow based on the classic structure [17] was implemented (Figure 5). The objective function minimizes the total transportation cost of provision available through the network to meet demand given, subject to the constraint of conservation of flow. Where: c_{ij} is the unit cost for flow through the arc $i \to j$; x_{ij} represents the flow through the arc $i \to j$; r_k define the net flow defendant at node k; u_{ij} is the capacity given in arc $i \to j$. It is further considered that $r_k > 0$, node k is a demand node with demand equal to r_k; if $r_k < 0$, node k is offered, with an offer equal to $|r_k|$ and finally, if $r_k = 0$, k is a transshipment point. To facilitate understanding of this step, we present a small example of this model. Figure 6 shows the input data, with m_1, m_2 and m_3 municipalities. The suppliers were considered k_1, k_2 and k_3.

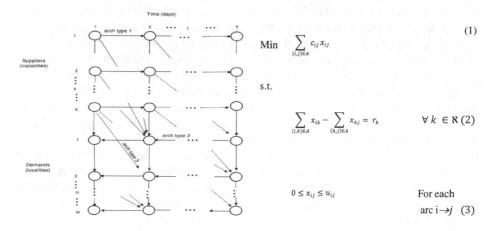

$$\text{Min} \sum_{(i,j) \in A} c_{ij} x_{ij} \tag{1}$$

s.t.

$$\sum_{(i,k) \in A} x_{ik} - \sum_{(k,j) \in A} x_{kj} = r_k \qquad \forall k \in \aleph \tag{2}$$

$$0 \leq x_{ij} \leq u_{ij} \qquad \text{For each arc } i \to j \tag{3}$$

Fig. 6. Model Proposed (Source: authors)

In order to trial the model, it is refereed that the lack of each product per day in the region represents a cost of 2 c.u.. This value is added to the price of each item of the day when the product is not in the specific location. We use the CIF system, i.e., the cost of shipping is not being built because the organization responsible decision making considers that this cost is already inserted in price from the supplier. In the

elaboration of the problem, the cost of lack of product per day not met assists in determining the supplier, aids in choice of supplier in the sense that, when the supplier
does not have the requested amount for that date, but for later date, the miss cost is
attributed in price supplier represented in the arc of the transport stream. Thus, the
model defines those suppliers with the nearest time the requested product and finds
the lowest total cost. The specialized simplex method to solve the networks
flow problems, based on [17], was implemented using the Object Pascal programming
language. This way, the graph resulting from the model we obtained 20 nodes and
58 arches and the simplex method developed solution arrived at after a total of
20 iterations in less than 1 second in a computer processor Intel Celeron 2.2 GHz,
2 GB memory and system operating Windows 7 32-bit. The total amount given in
the final solution was 60,880 $c.u.$, being that 7,400.$c.u.$ refers to the cost of missing
and 53,480 $c.u.$ the product cost (cost-effectively paid). In the graphical representation
of the result of the network flow problem, shown in Figure 6, data relating to the
quantity demanded are represented by nodes of demand at the specific date and the
daily capacity of each supplier is represented in the nodes of the supplier in respect
date.

The values next to the arcs represent how much of the capacity of their supplier
was used on the day. For example, the demand for m_3 of 5000 units of the product
ordered on the 13th was attended by k_1, k_2 and k_3 suppliers, with 1800 units of the
supplier k_3 were used on day 13. Since this amount was sent on the date requested,
not been submitted to any cost of missing. On the same day were obtained from the
supplier k_2 800 units and 2400 units were acquired on 14 k_1 suppliers, who suffered
the cost of missing for one day. To meet the demand m_1, 2000 units of product on the
day 14, it was provided on the same day, 1500 units of the product vendor k_3 and k_1
supplier of 500 units in 15 days, with the cost of embedded missing for a day. Have to
meet demand m_2, 800 units requested on the day 13, were acquired on the day 14, 500
units of the product vendor k_3 and k_1 supplier of 300 units, we considered the cost of
missing for one day. Thus, the needs of each municipality were met and due to cost of
missing, the program assisted in the nearest time the date of demand, using for this
purpose, the supplier with enough to meet on the date requested capacity. This small
example addresses the problem relating to the procurement process in Humanitarian
Logistics. To do so, it was set up the best set of suppliers to respect the time demand
and the ability to request delivery at developing the network flow problem. It was
found that the condition of the product be as fast as possible at the point of demand is
satisfied, and the capacity of the suppliers closer to the event dates with the lowest
cost is used. Overall, the method calculates the total cost is important to note that this
cost is generated; also incorporating the value of the cost of failure, but the final
amount to be actually paid this amount for the cost of failure is disregarded. This lack
cost may be an index that depicts how much the lack of this product costs per day to
the affected region. This shows the importance of the network flow problem adapted
in the context of Humanitarian Logistics in natural disasters.

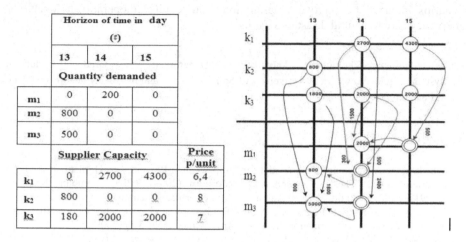

Horizon of time in day (r)			
13	14	15	
Quantity demanded			
m₁	0	200	0
m₂	800	0	0
m₃	500	0	0
Supplier Capacity			Price p/unit
k₁	0	2700	4300	6,4
k₂	800	0	0	8
k₃	180	2000	2000	7

Fig. 7. Inputs and graph model from example(Source: authors)

5   Concluding Remarks

Structures of a reference model for humanitarian supply chains were shown with the purpose of developing the integration from a decision support tool, which uses appropriated quantitative model reflecting the specific area of humanitarian logistics.

The tool integration into the processes was being done by utilizing and extending a reference model for humanitarian logistics. A methodology to operationalize the action strategy was developed with the implementation of new activities to be inserted into the RTM concept. It was done to support the responsible organizations for disaster response, in order to improve the agility, adaptability and flexibility in the supply chain. Using this methodology, decision makers are able to efficiently manage the material flow along the supply chain at humanitarian operations. We use BPMN as an enough flexible modeling tool to the reality of humanitarian logistics. The tasks described in the RTM were analyzed, aiming the integration of variables used in the quantitative decision support tools with the activities of the reference model.

Through the RTM combination between the quantitative decision support tool and standardized modelling technique (BPMN), it is possible to analyse an overall concept for a decision support system for procurement in disaster relief operations, which can be adapted for different needs of situations and relief organizations. The network flow problem can contribute to identify the suppliers that supply the demand in the shortest time, considering the cost of missing per day. The cost consideration of missing per day can be standardized to the usage in different disaster situations. For future researches, further variables can be introduced to this tool, such as modeling the choice of the shortest route or the appropriate modal types. The introduction of new variables, on the other hand, should influence the process for modeling for humanitarian logistics. In order to assist decision-making to managers during the emergency and using the previously described processes, for future works we have been developing a model for discrete event simulation. Thus, we intend to evaluate realistic scenarios intended to improve programming and results at operational levels for emergency care to natural disasters victims.

Acknowledgements. The authors acknowledge the Department of Post Graduate Production and Systems Engineering Federal University of Santa Catarina and the financial support of the CAPES and CNPQ.

References

1. Blecken, A: A Reference Task Model for Supply Chain Processes of Humanitarian Organisations. Ph.D. Thesis, University of Paderborn (2009). URN: urn:nbn:de:hbz:466-20091203022, Link: http://ubdok.unipaderborn.de/servlets/DocumentServlet?id=11585
2. Apte, A: Humanitarian Logistics : A new field of research and action. Foundation and trends in technology,information and operations management **3**(1) (2009)
3. Tomasini, R., Van Wassenhove, L.: Humanitarian logistics. Insead Business Press (2009)
4. Blecken, A.: Supply chain process modelling for humanitarian organizations. Journal of Physical Distribution & Logistics Management **40**(8/9), 675–692 (2010)
5. Becker, J., Rosemann, M., Uthmann, C.V.: Guidelines of Business Process Modeling. In: van der Aalst, W., Desel, J., Oberweis, A. (eds.) Business Process Management. LNCS, vol. 1806, pp. 30–49. Springer, Heidelberg (2000)
6. Widera, A.; Hellingrath, B.: Improving humanitarian logistics - towards a tool-based process modeling approach. In: Proceedings of the Logistik Management, Bamberg, pp. 273-295 (2011)
7. Tufinkgi, P.: Logistik im kontext internationaler katastrophenhilfe: Entwicklung eines logistischen referenzmodells für katastrophenfälle.Ed. Haupt Verlag. Bern. Stuttgart Wien (2006)
8. Mcguire, G.: A Development of a supply chain management framework for health care goods provided as humanitarian assistance in complex political emergencies. Ph.D. Thesis, Wirtschaftsuniversität Wien, Vienna (2006)
9. Thorn, J.: Taktisches supply chain planning. Planungsunterstützung durch deterministische und stochastische Optimierungsmodelle. Lang. Frankfurt am Main, Wien, Berlin, Bern (2002)
10. Trestrail, J., Paul, J., Maloni, M.: Improving bid pricing for humanitarian logistics. International Journal of Physical Distribution & Logistics Management **39**(5), 428–441 (2009)
11. Falasca, M., Zobel, C.: A two-stage procurement model for humanitarian relief supply chains. Journal of Humanitarian Logistics and Supply Chain Management **1**(2), 151–169 (2011)
12. Haghani, A., Sei-Chang, O.: Formulation and solution of a multi-commodity, multi-modal network flow model for disaster relief operations. Transportation Research Part A: Policy and Practive **30**(3), 231–250 (1996)
13. Barbarosoglu, G., Özdamar, L., Çevik, A.: An interactive approach for hierarchical analysis of helicopter logistics in disaster relief operations. European Journal of Operational Research **140**(1), 118–133 (2002)
14. Özdamar, L., Ekinci, E., Küçükyazici, B.: Emergency logistics planning in natural disasters. Annals of Operations Research **129**, 217–245 (2004)
15. Vitoriano, B., Ortuño, M.T., Tirado, G., Montero, J.: A multi-criteria optimization model for humanitarian aid distribution. Journal of Global Optimization **51**, 189–208 (2011)
16. Kennington, J.L., Helgason, R.V.: Algorithmfor Network Programming. John Willey&Sons, New York (1980)

Other Applications

An Efficient Design of a Reversible Fault Tolerant n-to-2^n Sequence Counter Using Nano Meter MOS Transistors

Md. Shamsujjoha[1]([✉]), Sirin Nahar Sathi[1], Golam Sorwar[3], Fahmida Hossain[2], Md. Nawab Yousuf Ali[1], and Hafiz Md. Hasan Babu[2]

[1] Department of Computer Science and Engineering,
East West University, Dhaka 1212, Bangladesh
dishacse@yahoo.com, snsathi82@gmail.com, nawab@ewubd.edu
[2] Department of Computer Science and Engineering,
University of Dhaka, Dhaka 1000, Bangladesh
fahmidacsedu@gmail.com, hafizbabu@hotmail.com
[3] School of Business and Tourism, Southern Cross University, Lismore, Australia
golam.sorwar@scu.edu.au

Abstract. This paper proposes an efficient reversible synthesis for the n-to-2^n sequence counter, where $n \geq 2$ and $n \epsilon$ N. The proposed circuits are designed using only reversible fault tolerant gates. Thus, the entire circuit inherently becomes fault tolerant. In addition, an algorithm to design the n-to-2^n reversible fault tolerant sequence counter based on fault tolerant J-K flip-flops has been presented. The functional verification of the proposed circuit is completed through the simulation results. Moreover, the comparative results show that the proposed method performs much better and is much more scalable than the existing approaches.

Keywords: Counter · Fault tolerant · Reversible logic · Transistor

1 Introduction

In logic computation, every bit of information loss generates $KTln2$ joules of heat [1]. These heat dissipation can be reduced to zero joule, if the circuit is constructed only with reversible gates as it maintains one-to-one mapping between inputs-outputs and thus recovers from bit loss [2]. Moreover, reversible circuit is also viewed as a special type of quantum circuit because quantum evaluation must be reversible [3]. On the other hand, fault tolerant reversible gate can detect faulty signal in its primary outputs through parity checking. Researchers showed that an entire circuit can preserve parity if its individual gate is parity preserving [4–6]. In addition, reversible and fault tolerant computing gained remarkable interests in the development of highly efficient algorithms [7,8], optimal architecture [4], simulation and testing [5], DNA and nano-computing [9], quantum dot cellular automata [10] etc. In these consequences, this paper investigates the design methodologies for the reversible fault tolerant sequence counter based on low power MOS transistor. The counter is used in the control unit of a

© Springer International Publishing Switzerland 2015
Y. Tan et al. (Eds.): ICSI-CCI 2015, Part III, LNCS 9142, pp. 455–462, 2015.
DOI: 10.1007/978-3-319-20469-7_48

microprocessor [11]. It has also been used in memory and I/O [12]. The transistor implementation of the reversible circuits is considered in this paper because of its scalability and easier fabrication process [4–6].

2 Basic Definitions and Literature Review

Some basic definitions, notations and background study of the existing works are presented in this section.

2.1 Reversible and Fault Tolerant Gates

An n bit **reversible gate** is a data block that uniquely maps between input vector $I_v = (I_0, I_1, ..., I_{n-1})$ and output vector $O_v = (O_0, O_1, ..., O_{n-1})$ [6]. The **Fault tolerant gate** is a reversible gate that preserves parity between inputs and outputs [5]. In other words, an n-bit fault tolerant gate maintains the following property among the inputs and outputs:

$$I_0 \oplus I_1 \oplus ... \oplus I_{n-1} = O_0 \oplus O_1 \oplus ... \oplus O_{n-1} \tag{1}$$

2.2 Garbage Output, Hardware Complexity and Quantum Cost

The output of a reversible gate that neither used as primary output nor as input to another gate is the **garbage output** [6], *i.e.*, the output which are needed only to maintain the reversibility is known as garbage output [7]−[12]. The number of basic operations (AND, OR, NOT, Ex-OR, etc.) needed to realize a gate/circuit is referred as its **hardware complexity** [4,5]. The **quantum cost** of a reversible gate is the total number of 2×2 quantum gate used in it, since quantum cost of the all 1×1 and 2×2 reversible gates is considered as 0 and 1, respectively [7]−[12]. The quantum computer with many qubits is difficult to realize. Thus, the reversible circuit with the fewer quantum cost is considered as beneficial.

2.3 Popular Reversible Fault Tolerant Gates

Feynman double gate ($F2G$) and Fredkin gate (FRG) are the most popular reversible fault tolerant gates. Following subsections present these gates along with their all necessary properties. This section also presents the transistor representations of these gates. We already proposed these representations in [5,6] using 901 and 920 MOS models. But in this paper we use advanced p-MOS 901(ax4) and n-MOS 902(bx5) for more efficient realization.

Feynman Double Gate. Inputs and outputs for a 3×3 $F2G$ is define as follows: $I_v = (a, b, c)$ and $O_v = (a, a \oplus b, a \oplus c)$. Block diagram of $F2G$ is shown in Fig. 1(a), whereas Fig. 1(b) represents $F2G$'s quantum equivalent realization. From Fig. 1(b), we find that an $F2G$ is realized with two 2×2 Ex-OR gates and thus its quantum cost is two. Figs. 1(c) and (d) represent the transistors realization and the corresponding timing diagram of $F2G^1$.

---

[1] Throughout the paper we consider the signal less than $0.01ns$ stability are glitches and thus omitted from the simulation results. Moreover, considering these gate representation as schema all the proposed circuit of Sec. 3 is simulated.

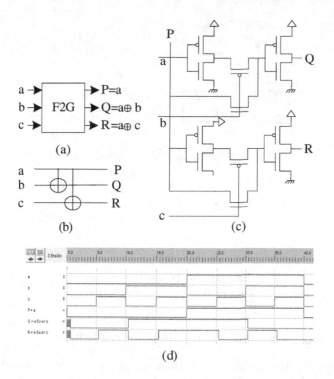

a →| F2G |→ P=a
b →| |→ Q=a⊕ b
c →| |→ R=a⊕ c

(a)

(b)

(c)

(d)

Fig. 1. Reversible 3 × 3 Feynman double gate (a) Block diagram (b) Quantum equivalent realization (c) Transistor equivalent realization (d) Simulation result

Fredkin Gate. Input-output vectors of a 3 × 3 FRG is define as follows: $I_v = (a, b, c)$ and $O_v = (a, a'b{\oplus}ac, a'c{\oplus}ab)$. The block diagram of an FRG is shown in Fig. 2(a). Fig. 2(b) represents the quantum equivalent realization of the FRG. To realize the FRG, four transistors are needed as shown in Fig. 2(c). The corresponding timing diagram of Fig. 2(c) is shown in Fig. 2(d), which proves the functional correctness of the proposed realization. It has been shown in [4–6] that both $F2G$ and FRG maintain the parity preserving property of Eq. (1).

3 Proposed Reversible Fault Tolerant Sequence Counter

This section illustrates the design methodologies of the proposed reversible fault tolerant sequence counter. The proposed counter for a reversible fault tolerant microprocessor based on J-K flip-flops is designed. The working procedure of the proposed J-K flip-flop is dependent both on its previous state and clock. There are several designs of reversible sequence counters in the literature among which the design from [13] is considered to be the most compact and efficient. In this section, initially we propose a reversible fault tolerant J-K flip-flop. Then, the proposed flip-flop is used to design the proposed fault tolerant counter circuit.

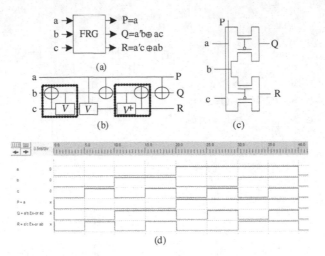

Fig. 2. Reversible 3×3 Fredkin gate (a) Block diagram (b) Quantum equivalent realization (c) Transistor equivalent realization (d) Simulation result

3.1   Proposed Reversible Fault Tolerant J-K Flip-Flop

Fig. 3(a) presents the architectural block diagram of the proposed reversible fault tolerant J-K flip-flop. The corresponding quantum equivalent realization is shown in Fig. 3(b). From the quantum representation of the proposed flip-flop, we find that the proposed J-K flip-flop is constructed with total of nine 2×2 quantum equivalent gates. Thus its total quantum cost should be nine. But, in this quantum realization, there are two consecutive Ex-OR gate, one from FRG and the other from $F2G$ (dashed area), which have the identical level from its I/O. As shown by the researchers [5,6], if there are two consecutive Ex-OR gates with identical I/O level, then it only represents a quantum wire and is realized without any quantum cost. Thus, the total quantum cost of the proposed fault tolerant J-K flip-flop is 7 rather than 9.

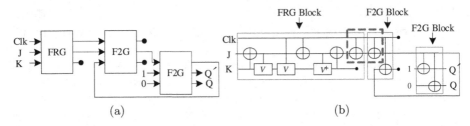

Fig. 3. Proposed reversible fault tolerant J-K flip-flop (a) Architectural block diagram (b) Quantum equivalent realization

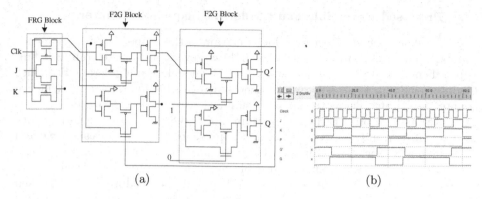

Fig. 4. Proposed J-K flip-flop (a) Transistor realization (b) Simulation result

Transistor equivalent realization of the proposed reversible fault tolerant J-K flip-flop is shown in Fig. 4(a). The corresponding simulation result is shown in Fig. 4(b), which proves the functional correctness of the proposed circuit. From Fig. 4(b), we find that the output Q depends on both the clock and the P. The P represents the value of previous state, which is initially set to high. When clock is low, the value of J or K has no effect on Q (0 to 2ns). However, when clock is high, Q doesn't change its state till 5 ns, as both J and K are low during this time. If the clock changes from low to high, J is set to high, K is set to low and Q becomes high (10 to 25 ns). In opposite case, Q also stays at high (25 to 35 ns). But, if both J and K are set to high, then Q is equal to P'. Table. 1 shows the comparison of the proposed fault tolerant J-K flip-flop with the existing non-fault tolerant flip-flops. Generally, a fault tolerant design is much more expensive than the non-fault tolerant design. However, Table. 1 shows that the proposed design performs much better than the existing non-fault tolerant designs which is also much scalable[2].

Table 1. Comparison of reversible J-K flip-flops

Evaluation Criteria	QC	HC	TR	P	CD	CL	DS
Existing Circuit [11]	14	$9\alpha + 8\beta + 2\gamma$	39	58.5	1.8901	1755	0.0325
Existing Circuit [13]	12	$4\alpha + 5\beta + 4\gamma$	33	49.5	2.1357	1485	0.0275
Proposed Circuit	7	$4\alpha + 4\beta + 2\gamma$	28	21.0	0.3082	1260	0.0233

---

[2] In this table and all the following tables, we consider QC=Quantum cost, HC=Hardware Complexity, TR=No. of Transistors, P=Average Power Consumption in nW, CD=Critical Path Delay in ns, CL=Average Channel Length and DS=Average Density per $kgates/mm^2$. We also consider that α, β, γ are the hardware complexities of a two-input Ex-OR, an AND and a NOT calculations respectively.

3.2 Proposed Reversible Fault Tolerant Sequence Counter

Fig. 5(a) shows the architectural block diagram of the proposed 2-to-4 sequence counter[3]. The combination of one $F2G$ and two reversible fault tolerant J-K flip-flops schema can work together as a 2-to-4 **R**eversible **F**ault tolerant **S**equence **C**ounter (RFSC). From now on, we denote a reversible fault tolerant sequence counter as RFSC. The corresponding simulation result is shown in Fig. 5(b). Figs. 6(a) and (b) present the architecture of a 3-to-8 RFSC and its corresponding simulation result, respectively. From the presented simulation results, we find that the initial clock transition occurs at $1ns$ and the output becomes available in $1.637ns$. Thus, the maximum possible delay of the proposed circuit is less than $0.65ns$. Algorithm 1 presents the detailed design procedure of the proposed n-to-2^n RFSC. Primary inputs of this proposed algorithm are clock pulse, $F2G$ and FRG. Initially, the algorithm builds an architectural block of the proposed reversible fault tolerant J-K flip-flop (RFJKFF), which is shown in lines 5 to 10 of the Algorithm 1. Then, n numbers of RFJKFF blocks and $(n$-$1)$ numbers of $F2G$s are used to create the entire structure of the n-to-2^n RFSC.

Algorithm 1. Algorithm for the proposed reversible fault tolerant n-to-2^n sequence counter, $\textbf{\textit{RFSC(clk, F2G, FRG)}}$

Input   : Clock pulse clk, $F2G$ and FRG
Output: n-to-2^n reversible fault tolerant sequence counter circuit

1 **begin**
2      $i = input, o = output$
3   **begin procedure**
4          $\textbf{\textit{RFJKFF(clk, F2G, FRG)}}$
5          $clk \rightarrow first.i.FRG,$
6          $J \rightarrow second.i.FRG,$
7           $K \rightarrow third.i.FRG$
8          $first.o.FRG \rightarrow first.i.F2G_1, second.o.FRG \rightarrow second.i.F2G_1,$
            $third.o.FRG \rightarrow third.i.F2G_1$
9          $second.o.F2G_1 \rightarrow first.i.F2G_2, 1 \rightarrow second.i.F2G_2, 0 \rightarrow third.i.F2G_2$
10         $second.o.F2G_2 \rightarrow Q', third.o.F2G_2 \rightarrow Q,$
11   **end procedure**
12   **for** $j \leftarrow 2$ **to** n **do**
13   **call** $\textbf{\textit{RFJKFF (clk, F2G, FRG)}}$ $first.o.RFJKFF \rightarrow$
            $first.i.F2G_{j-1}, 1 \rightarrow second.i.F2G_{j-1}, 0 \rightarrow third.i.F2G_{j-1},$
14         $second.o.F2G_{j-1} \rightarrow second.i.RFJKFF_j, remaining.i.RFJKFFR \leftarrow 1$
15         $third.o.F2G_j \rightarrow Q_{j-1}$
16         $first_j.o.RFJKFF \rightarrow Q_j$
17   **end for**
18   **return** $remaining\ F2G.o,\ FRG.o\ and\ RFJKFF. \rightarrow garbage\ outputs.$
19 **end**

---

[3] This is trivial case $i.e., n=2$.

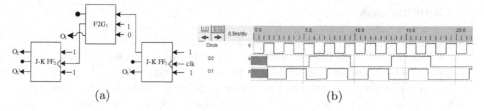

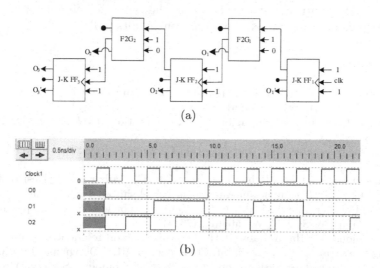

Fig. 5. Proposed 2-to-4 RFSC (a) Architecture (b) Simulation result

Fig. 6. Proposed 3-to-8 RFSC (a) Architectural block diagram (b) Simulation result

Table 2. Comparison of Reversible Sequence Counters

n	Proposed Method			Existing Method [11]			Existing Method [13]		
	QC	TR	HC	QC	TR	HC	QC	TR	HC
2	16	68	$10\alpha + 8\beta + 4\gamma$	35	95	$27\alpha + 24\beta + 6\gamma$	25	72	$7\alpha + 10\beta + 8\gamma$
3	25	108	$16\alpha + 12\beta + 6\gamma$	56	151	$45\alpha + 40\beta + 10\gamma$	38	111	$24\alpha + 15\beta + 12\gamma$
4	34	148	$22\alpha + 16\beta + 8\gamma$	59	207	$63\alpha + 56\beta + 14\gamma$	51	150	$34\alpha + 20\beta + 16\gamma$

4 Evaluation of the Proposed Methods

Table 2 compares the performance of the proposed scheme with the existing methods [11,13]. From this table, we find that the performance of the proposed method is much better and has significantly better scalability than the existing approaches with respect to all the evaluation parameters.

5 Conclusions

The paper has presented a compact design of the reversible fault tolerant n-to-2^n counter based on J-K flip flops. The representation of each gate is shown using transistors, which are further used in the design of the proposed circuit. The corresponding simulation results have also been presented. All the simulation results evidenced that the proposed fault tolerant circuits work correctly. The comparative results show that the proposed design is much more scalable than its counterparts [11,13]. In addition, this paper has addressed the problem with the gate level quantum cost calculation in the reversible circuits, which is an interesting research work for future quantum circuits [3,6].

References

1. Landauer, R.: Irreversibility and Heat Generation in the Computing Process. IBM J. Res. Dev. **5**(3), 183–191 (1961)
2. Bennett, C.H.: Logical Reversibility of Computation. IBM J. Res. Dev. **17**(6), 525–532 (1973)
3. Peres, A.: Reversible Logic and Quantum Computers. Phys. Rev. A. **32**(6), 66–76 (1985)
4. Shamsujjoha, M., Hasan Babu, H.M., Jamal, L., Chowdhury, A.R.: Design of a fault tolerant reversible compact unidirectional barrel shifter. In: 26th Int. Conference on VLSI Design and 12th Int. Conference on Embedded Systems, pp. 103–108. IEEE Computer Society, Washington (2013)
5. Shamsujjoha, M., Hasan Babu, H.M.: A low power fault tolerant reversible decoder using MOS transistors. In: 26th Int. Conference on VLSI Design and 12th Int. Conference on Embedded Systems, pp. 368–373. IEEE Computer Society, Washington (2013)
6. Shamsujjoha, M., Hasan Babu, H.M., Jamal, L.: Design of a Compact Reversible Fault Tolerant Field Programmable Gate Array: A Novel Approach in Reversible Logic Synthesis. Microelectronics J. **44**(6), 519–537 (2013)
7. Sharmin, F., Polash, M.M.A., Shamsujjoha, M., Jamal, L., Hasan Babu, H.M.: Design of a compact reversible random access memory. In: 4th IEEE Int. Conference on Computer Science and Information Technology, pp. 103–107 (2011)
8. Maslov, D., Dueck, G.W., Scott, N.: Reversible Logic Synthesis Benchmarks Page. http://webhome.cs.uvic.ca/~dmaslov
9. Jamal, L., Shamsujjoha, M., Hasan Babu, H.M.: Design of Optimal Reversible Carry Look-Ahead Adder with Optimal Garbage and Quantum Cost. Int. J. of Eng. and Tech. **2**, 44–50 (2012)
10. Morita, K.: Reversible Computing and Cellular Automata. Theor. Comput. Sci. **395**(1), 101–131 (2008)
11. Sayem, A.S.M., Ueda, M.: Optimization of reversible sequential circuits. J. of Computing **2**(6), 208–214 (2010)
12. Mahammad, S.N., Veezhinathan, K.: Constructing Online Testable Circuits Using Reversible Logic. IEEE Tran. on Instrumentation and Measurement **59**, 101–109 (2010)
13. Jamal, L., Alam, M.M., Hasan Babu, H.M.: An Efficient Approach to Design a Reversible Control Unit of a Processor. Sustainable Computing: Informatics and Systems **3**(4), 286–294 (2013)

Transfer of Large Volume Data over Internet with Parallel Data Links and SDN

S.E. Khoruzhnikov[1], V.A. Grudinin[1], O.L. Sadov[1], A.Y. Shevel[1,2(✉)], and A.B. Kairkanov[1]

[1] ITMO University, St.Petersburg, Russian Federation
shevel_a_y@niuitmo.ru
[2] National Research Centre, "Kurchatov Institute" B.P. Konstantinov, Petersburg Nuclear Physics Institute, Moscow, Russian Federation

Abstract. The transfer of large volume data over computer network is important and unavoidable operation in the past, now and in any feasible future. There are a number of methods/tools to transfer the data over computer global network (Internet). In this paper the transfer of data over Internet is discussed. Several free of charge utilities to transfer the data are analyzed here. The most important architecture features are emphasized and suggested idea to add SDN Openflow protocol technique for fine tuning the data transfer over several parallel data links.

Keywords: Big Data · Linux · Transfer · SDN · Openflow · Network

1 Introduction

Instead of large volume data the people quite often use the term "Big Data" [1] which is known for many years. Keeping in mind Big Data description - "triple V": Velocity, Volume, Variety we can pay attention that all those features are relative to current state of the technology. For example in 1970-s the volume of 1 TB was considered as huge volume. There is a range of aspects of the problem: to store, to analyze, to transfer, to visualize. In this paper we discuss one of important aspects of the Big Data – the transfer over global computer network.

1.1 The Sources of the Big Data

It is known the long list of human activities (scientific and business) which are the generators of large volume of data [2-10].

In according [2] total volume of business mails in the World in year 2012 is around 3000 PB (3×10^{18}). The consensus estimation for the total volume of stored data is growing 1.5-2.0 times each year starting from 2000. In this paper (and for our tests) we will assume that volume of data around 100 TB (10^{14}) and more could be labeled as Big Data.

Y. Tan et al. (Eds.): ICSI-CCI 2015, Part III, LNCS 9142, pp. 463–471, 2015.
DOI: 10.1007/978-3-319-20469-7_49

Another source of Big Data – the preservation of the data for long periods of time: several tens or more years. Many aspects of our personal, society, technical, and business life are now held in digital form. Large volume of those data needs to be stored and preserved. For example, results of medicine tests, data generated by important engines of various kinds (airplane engines, power station generators, etc) and other data have to be archived for long time. The preserved data will be kept in distributed (locally and globally) storage. It is assumed that replicas of preserved data have to be stored in several places (continents) to avoid data loss due to technical, nature or social disasters.

Historically one of the first field where Big Data came into reality was experiments in High Energy Physics (HEP). As the result a number of aspects for data transfer were analyzed and a range of problems was solved. Now more and more scientific and business sectors are dealing (or plan to) with the "Big data" [2-7, 33]. The interest to data transfer of increasing volumes is growing [11, 12].

1.2   Big Data Transfer over the Network

The time to transfer over global computer network (Internet) depends on the real data link bandwidth and volume of the data. Taking into account that we talk about volume 100TB and more we can estimate minimum required time for data copy over the data link with 1 Gbit capacity. It will give us about 100MB/sec, hence 100TB/100MB = 1000000 secs = 277.8 hours = 11.6 days. During this time the parameters of the data link might be changed. For example percent of dropped network packages and other data link parameters can be varied significantly. The data link might be suffered of operation interruptions for different periods: secs, hours, days. Also a number of Linux kernel network parameters might affect the data transfer speed. Among most important of them are TCP Window size, MTU, congestion control algorithm, etc. Finally in each data transfer of large volume we need to be able to tune the number of data transfer streams, the size of TCP Window, etc. Now it is time to observe popular freely available data transfer tools/utilities.

2   Freely Available Utilities/Tools for Data Transfer over the Network

2.1   Low Level Data Transfer Protocols

We could mention several tools:

one of low level protocols to transfer the data over the network is UDT [13]. UDT is library which implements data transfer protocol which permit to use *udp*. In some cases the library can help to improve data link usage, i.e. to reduce the data transfer time.

the protocol RDMA over Converged Ethernet (RoCE) has been studied in [33] and it was found that in many cases RoCE shows better results than UDP and conventional TCP.

MPTCP [14] is interesting protocol which permits to use several data links in parallel for one data transfer. The protocol is implemented as Linux kernel driver.

2.2   Experimenting with MPTCP

Among other things we tested available version driver MP TCP. In our scheme we used our own server with Linux kernel 3.14.15-303.mptcp.el7.centos.x86_64 and two data links with two different routing tables as it explained in MP TCP docs. Opposite server was mutipath-tcp.org. To test the speed of data transfer we used the command: `wget -c ftp://multipath-tcp.org/500MB`.

First of all we used mode when MPTCP is switched OFF (`systcl net.mptcp.mptcp_enabled=0`). Here we got the speed 1.84 MB/sec. Next test when MPTCP is switched ON and the speed was 3.33 MB/sec – reasonably higher. Another test measurement was dealing with MPTCP is ON however one of the data link was set down (i.e. one data link became broken). In this test we got the speed 911 KB/sec – reasonably less then with two data links and even less than MPTCP is OFF (with one data link).

Above results were signs for us that for long time data transfer we need to watch regular the real status of data links when we plan to speed up the data transfer.

2.3   Popular Programs for Data Transfer

openssh family [15] – well known data transfer utilities deliver strong authentication and a number of data encryption algorithms. Data compression before encryption to reduce the data volume to be transferred is possible as well. There are two well known openSSH flavors: patched SSH version [16] which can use increased size of buffers and SSH with Globus GSI authentication. No parallel data transfer streams.

bbcp [17] — utility for bulk data transfer. It is assumed that bbcp is running on both sides, i.e. transmitter, as client, and receiver as server. Utility bbcp has many features including the setting:

— TCP Window size;
— multi-stream transfer;
— I/O buffer size;
— resuming failed data transfer;
— authentication with ssh;
— other options dealing with many practical details.

bbftp [18] – utility for bulk data transfer. It implements its own transfer protocol, which is optimized for large files (larger than 2GB) and secure as it does not read the password in a file and encrypts the connection information. bbftp main features are:

— SSH and Grid Certificate authentication modules;
— multi-stream transfer;
— ability to tune I/O buffer size;

— restart failed data transfer;
— other useful practical features.

fdt [20] – Java utility for multi-stream data transfer and support I/O buffer size tuning.
gridFTP [21] is advanced redesign of well known utility *ftp* for globus security infrastructure (GSI) environment. The utility has many features:

— two security flavors: Globus GSI and SSH;
— the file with host aliases: each next data transfer stream will use next host aliases (useful for computer cluster);
— multi-stream transfer;
— ability to tune I/O buffer size;
— restart failed data transfer;
— other useful practical features.

Mentioned utilities are quite effective for data transfer from point of view of data link capacity usage. The more or less typical data transfer over real shared (the traceroute consists of 9 hops) 1 Gbit long distance data link is shown on Fig. 1. The program bbcp was used for real measurements. In this measurement the directory with test data files located in main memory was transferred. Total volume of the directory was 25 GB. The content of files was specially randomly generated binaries. The sizes of files in test directory was randomly distributed with average 100MB and standard deviation was 50 MB. Horizontal axis shows TCP Window sizes in bytes. It can be seen on the picture that with small number (1-2) of TCP streams the maximum speed is achieved with the size of TCP Window around 2097KB and with numbers of streams 3 and 4 the maximum is around 524KB. In contrast with relatively large number (18 and more) of TCP streams the maximum is achieved even with TCP Window size around 131KB. Existing fluctuations in data transfer speed curves are explained by the fact that shared data link is used for those measurements. The transfer of large volume data assumes significant transmission time (may be many hours, days or more). The transmission speed might change very seriously over shared data links and over so long time.

2.4   Middle Level File Transfer Service

The FTS3 [22] is relatively new and advanced tool for data transfer of large volume of the data over the network. It has most features already mentioned above and more. In particular FTS3 uses utility gridFTP to perform data transfer. There is advanced data transfer tracking (log) feature, ability to use http, restful, and CLI interfaces to control the process of the data transfer.

Another interesting development is SHIFT [23] which is dedicated to do reliable data transfer in LAN and WAN. There was paid much attention to the reliability, advanced tracking, performance of the data transfer and the usage of parallel data transfer between so called equivalent hosts (between computer clusters).

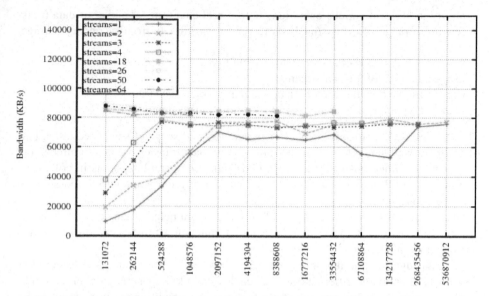

Fig. 1. Data transfer speed depends on the number of streams and size of TCP Window

2.5   High Level Data Management Service: PhEDEx

PhEDEx - Physics Experiment Data Export is used (and developed) in collaboration around Compact Muon Solenoid (CMS) experiment [24-27] at CERN [7]. The experiment does produce a lot of experimental data (in 2013 it was written around 130 PB). Data analysis requires to copy (to distribute) the data in a range of large computing clusters (about 10 locations in different countries and continents) for analysis and data archiving. Later on the fractions of the data might be copied to smaller computing facilities (more than 60 locations). Total data transfer per day is achieved 350 TB/day [25]. It is possible that in nearest future the volume per day will be increased. Because in between several sites there are more than one data link in PhEDEx there were developed routing technique which permit to try alternative route when default route is not available.

Finally the system PhEDEx is quite complicated and the management service depends on the physics experiment collaboration environment. It is unlikely that PhE-DEx is possible to use without redesign in different environment.

3   Consideration

Mentioned utilities have several common useful features for data transfer. Among them:

client-server architecture;
ability to set the buffer size, TCP Window size, number of streams, etc;

ability to perform various operations before real data transfer and after data transfer, use a range of drivers/methods to read/write files to/from secondary storage, etc;

usage more than one of authentication techniques;

usage a number of transfer streams;

usage in some conditions more than one data link for data transfer;

usage of a number of techniques to make data transfer more reliable.

The utilities are not equal in number of parameters and scope of suggested tasks. Part of them are well suited to be used as independent data transfer utilities in almost any environment. Others, like PhEDEx (in CMS) and comparable systems in collaboration ATLAS [29] are dedicated to be used as part of more complicated and specific computing environment.

In other words there is a stack of toolkit which might help in many cases to transfer the Big Data over networks. Also it is seen that quite a few utilities can use more than one data link in parallel.

No tool suggests fine tuning with parallel data links. Fine tuning is considered as possibility to apply the different policy to each data link. In general parallel data links might be completely different in nature, features, and conditions of use. In particular it is assumed individual QoS for each data link to be used in data transfer and ability to change the policy on the fly. All that give the idea that special application is required which might watch the data links status and change the parameters of data transfer accordingly to real situation in the data links. QoS is planned to be set with protocol Openflow [30-31]. The special tool PerfSonar [32] can be used to watch the data links status.

4   Proposal and Testing

Let us image two parallel data links between two points in Internet. Both data links might be different in its nature and independent of each other. The data transfer might be performed with use both data links. Usually any data transfer program creates several data streams. On both sides (independently on source and destination) a control program might know the port pairs on source and destination hosts. With such the knowledge the control program might build up the appropriate routing table for network switch with ability of SDN Openflow. To be synchronous both control program have to use exactly same algorithm to arrange data streams among the data links. If we have 20 data streams in our example for each data links 10 streams would be arranged. If the status of data links is changed, for example, one data link experienced interruption, control programs on both sides have to get information about the problem from perfsonar: each control program from its own perfsonar. The schematic picture can be seen at Fig.2.

In our local testing (on one laboratory table) everything is working as expected. Now we are preparing the larger series of long distance data transfer. To perform the testing the special long distant testbed has been developed and deployed.

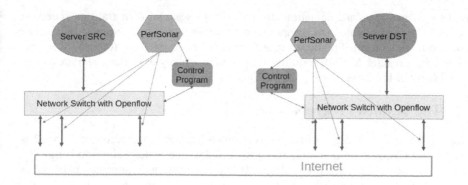

Fig. 2. Proposed scheme of the data transfer

5 Long Distance Testbed: Goals and Measurement Procedures

There is special aspect in the procedure of the comparison of the utilities to transfer the Big Data over real computer network. Not trivial scheme for data transfer demands the customized testbed which is able to simulate at least main network problems, e.g. changing RTT, delays, package drop percent, and so on. Such the testbed development has been started at the network laboratory [28]. The need for testbed is becoming obvious for us thanks to measurement results performed by authors [11]. The authors did the comparative measurements for one data transfer stream and many data streams. The data transfers were performed with special servers so called Data Transfer Nodes (DTN). DTNs have several specific techniques to transfer the data from LAN to WAN destinations. A number of utilities: rsync, scp, bbcp, GridFTP were discussed and measured just for concrete transferred file sizes (11 KB, 3.5 MB, 158 MB, 2.8 GB, and 32 GB) to transfer 6 files in each case. It was discovered that no change in the transfer speed after number of streams more than 8. At the same time in [11] there is no information about the Linux kernel parameters and other details of measurement.

In developed testbed it is planned to get precise answers on those questions. Also in the testbed we are taking into account the ideas described in [12].

The testbed is intended to be platform to compare a number of data transfer methods in particular with SDN Openflow approach.

As the first step it was planned to perform comparative measurements under variety of the measurement conditions. To do that we need detailed measurement tracking: writing all available details of the measurements (hardware types, buffer sizes, kernel parameters, generated messages, program parameters, etc). Everything has to be written in special log directory with date stamps. Presumably this log directory needs to be kept available long time: may be years. Obviously the writing of all conditions must be done with special script or program to perform saving all test conditions automatically.

Another question is dealing with test data which intended to be transferred over data links. Here we assume the transfer of the test directory consisting of files. It is known that sizes of files might affect the data transfer speed. That fact requires the special procedure to generate test directory with files of random sizes with defined average size and standard deviation. In addition to see real data transfer speed we need to eliminate possible data compression. In other words it is required to generate the random file content. Finally a range of scripts/programs have to be used to produce graphics results. Most of required scripts were already developed and are available in https://github.com/itmo-infocom/BigData.

6   The Testbed Progress

Currently the testbed consists of two servers: each has CPU Intel Xeon E5-2650v2 @ 2.6GHz, 96 GB of main memory, 100 TB of disk space under Scientific Linux 6.6. The distance in between servers is about 40 Km and they interconnected over public Internet (1 Gbit is maximum bandwidth). Because it is planned to test everything in virtual environment for each mentioned data transfer systems two virtual machines are used. One VM as transmitter and another VM as receiver. In other words we have around ten Vms for different tests. The cloud infrastructure Openstack (version Icehouse) has been deployed to organize VMs. PerfSonar has been deployed as well on both sides.

The special procedure has been developed to generate test directory with files of random length, the total volume of test directory is defined by the parameter of the procedure. During generation of test data it is possible to set mean value for file size and standard deviation of the file size. The data inside each file in test directory is intentionally prepared to eliminate possible effect of the data compression (if any) during data transfer.

As it was mentioned earlier in the paper many parameter values in the directory /proc might affect the speed of the data transfer. That means the requirement to write automatically whole directory /proc into "log directory". In addition there is need to write all the parameters used when data transfer starts. Also it is required to write all messages from data transfer engine/utility. Finally the data links status is also to be written as well. All those features have been implemented in the scripts dedicated to do measurements. Total volume of the log data for one measurement might exceed 5 MB.

Acknowledgements. The research has been carried out with the financial support of the Ministry of Education and Science of the Russian Federation under grant agreement #14.575.21.0058.

References

1. Big Data. http://en.wikipedia.org/wiki/Big_data
2. Information Revolution: Big Data Has Arrived at an Almost Unimaginable Scale. http://www.wired.com/magazine/2013/04/bigdata/
3. Square Kilometer Array. http://skatelescope.org/
4. Large Synoptic Survey Telescope. http://www.lsst.org/lsst/
5. Facility for Antiproton and Ion Research. http://www.fair-center.eu/
6. International Thermonuclear Experimental Reactor. http://www.iter.org/
7. CERN. http://www.cern.ch/
8. Borovick, L., Villars, R.L.: White paper. The Critical Role of the Network in Big Data Applications. http://www.cisco.com/en/US/solutions/collateral/ns340/ns517/ns224/ns944/critical_big_data_applications.pdf

9. The Center for Large-scale Data Systems Research at the San Diego Supercomputer Center. http://clds.sdsc.edu/
10. Johnston, W. E., Dart, E., Ernst, M., Tierney, B.: Enabling high throughput in widely distributed data management and analysis systems: Lessons from the LHC. https://tnc2013.terena.org/getfile/402
11. Ah Nam, H. et al: The Practical Obstacles of Data Transfer: Why researchers still love scp. http://dl.acm.org/citation.cfm?id=2534695.2534703&coll=DL&dl=ACM&CFID=5634854 33&CFTOKEN=25267057
12. Gunter, D., et al.: Exploiting Network Parallelism for Improving Data Transfer Performance. http://ieeexplore.ieee.org/xpls/abs_all.jsp?arnumber=6496123
13. UDT: Breaking the Data Transfer Bottleneck. http://udt.sourceforge.net/
14. MutiPath TCP – Linux Kernel Implementation. http://multipath-tcp.org/
15. OpenSSH. http://openssh.org/
16. Patched OpenSSH. http://sourceforge.net/projects/hpnssh/
17. BBCP – utility to trasnfer the data over network. http://www.slac.stanford.edu/~abh/bbcp/
18. BBFTP – utility for bulk data transfer. http://doc.in2p3.fr/bbftp/
19. Hodson, S.W., Poole, S.W., Ruwart, T. M., Settlemyer, B. W.: Moving Large Data Sets Over High-Performance Long Distance Networks. http://info.ornl.gov/sites/publications/files/Pub28508.pdf
20. Fast Data Transfer. http://monalisa.cern.ch/FDT/
21. Grid/Globus data transfer tool. Client part is known as globus-url-copy. http://toolkit.globus.org/toolkit/data/gridftp/
22. File Transfer Service. http://www.eu-emi.eu/products/-/asset_publisher/1gkD/content/fts3
23. Data Transfer Tools. http://fasterdata.es.net/data-transfer-tools/
24. The CMS Collaboration: The CMS experiment at the CERN LHC. http://iop science.iop.org/1748-0221/3/08/S08004/
25. Kaselis, R., Piperov, S., Magini, N., Flix, J., Gutsche, O., Kreuzer, P., Yang, M., Liu, S., Ratnikova, N., Sartirana, A., Bonacorsi, D., Letts, J.: CMS data transfer operations after the first years of LHC collisions. In: International Conference on Computing in High Energy and Nuclear Physics 2012. IOP Publishing Journal of Physics: Conference Series, vol. 396 (2012)
26. PhEDEx, CMS Data Transfers. https://cmsweb.cern.ch/phedex
27. PHEDEX data transfer system. https://twiki.cern.ch/twiki/bin/view/CMSPublic/Phedex AdminDocsInstallation
28. Laboratory of the Network Technology. http://sdn.ifmo.ru/
29. The Rucio project is the new version of ATLAS Distributed Data Management (DDM) system services. http://rucio.cern.ch/
30. Open Networking Foundation White Paper Software-Defined Networking: The New Norm for Networks. https://www.opennetworking.org/images/stories/downloads/white-papers/wp-sdn-newnorm.pdf
31. Nunes, B. A., Mendonca, M., Nguyen, X., Obraczka, K., Turletti, T.: A Survey of Software-Defined Networking: Past, Present, and Future of Programmable Networks. http://hal.inria.fr/index.php?halsid=ig58511e1q1ekqq75uud43dn66&view_this_doc=hal-00825087&version=5
32. Zurawski, J., Balasubramanian, S., Brown, A., Kissel, E., Lake, A., Swany, M., Tierney, B., Zekauskas, M.: perfSONAR: On-board Diagnostics for Big Data. https://www.es.net/assets/pubs_presos/20130910-IEEE-BigData-perfSONAR2.pdf
33. Tierney, B., Kissel, E., Swany, M., Pouyoul, E.: Efficient Data Transfer Protocol for Big-Data. http://www.es.net/assets/pubs_presos/eScience-networks.pdf

Program of Educational Intervention
for Deaf-Blind Students

Maria Zeza[1(✉)] and Pilios-D. Stavrou[2]

[1] Laboratory of Special and Curative Education, (LABESPEC),
University of Ioannina, Epirus, Greece
marialzeza@gmail.com
[2] Laboratoire Psychologie Clinique, Psychopathologie,
Psychanalyse (PCPP), Universitè Paris Descartes-Sorbonne, Paris, France
pstavrou@otenet.gr

Abstract. The purpose of the study was to develop an interventional educational plan for a deaf-blind student with screened difficulties in body schema awareness. This was part of a more extensive research in developing the screening inventory for the deaf-blind students' cognitive and communicative profile. The study used a qualitative research methodology and adopted an interpretative position. The aim of the inquiry was descriptive. We followed the case study methodology. The application of the interventional program aimed to help the deaf-blind student in promoting early concept development (body schema). The student was offered multisensory and concrete experiences in order to promote the body schema awareness.

After the intervention, it was observed that the student became aware of having a body with a center (midline) and two sides. The student succeeded in naming her body parts and matching them to others.

Keywords: Deaf-blind · Program of intervention · Body schema

1 Introduction

The multisensory impaired students receive limited and distorted sensory information, through which they define their relation to the word and construct their conceptual background. The concept development may be impeded, since multisensory deprivation imposes limitations on communication and cognitive development. Students with deaf- blindness may miss or misinterpret natural cues and incidental information due to their sensory loss [1, 2, 3].

1.1 Towards a Definition

Deaf-blindness is defined "as a concomitant hearing and visual impairment, the combination of which creates such severe communication and other developmental and educational needs that cannot be supported and accommodated in special educational programs aiming solely at children with deafness or blindness" [4]. The basic concept of heterogeneity is especially apparent among the deaf-blind population. Each child has a particular degree of visual and auditory loss, ranging from moderate to total. The sensory loss may be gradual or immediate, it may occur before birth or at any age and may be lost at a different or at the same time [5].

2   Concept Development

The child's level of cognitive functioning and its ability to establish and elaborate on meaningful concepts relies to a large extent on its ability to receive and assimilate input from the world around it. Therefore, the ability to assimilate sensory input influences the communication and the concept development [2]. The combined loss of both vision and hearing imposes limitations on the communication and the perception of primary concepts (e.g. time, space, body schema), which are important for the concept formation. [3].

Deaf-blind children cannot learn from the interaction with their environment as easily as their non-handicapped peers due to their multi-sensory deprivation. The information perceived is not always clear, simultaneous or consistent. The environment is often limited to what is approachable by the deaf-blind children's hands, or by means of their sensory potential. Therefore, deaf-blind children usually express limited motivation to explore their environment [1,2,6,7].

Children who cannot rely on their distance senses to accumulate information about their world may learn less effectively, because the received information is unreliable, distorted or inadequate. Hence, learning and experience are restricted, and the child's relation to the world and conceptual background is difficult to be defined [8,9, 10, 11].

2.1   The Body Schema

The body schema concept refers to the body unity perception. It refers to the representation that each person has for its own body. Simultaneously, it involves the basic point of reference in the person's relationship and interaction with the wider social and natural environment [12, 13, 14].

The body schema concept is important for the development of the child's personality. Through the gradual perception of the body and its potential in action, the child becomes aware of itself and the way he/she can influence the environment. [15, 16].

For the deaf-blind child the body schema perception and body awareness may not occur naturally. The deaf-blind child should be introduced to the fact that he/she constitutes a separate individual, different from others and his/her environment, who can also influence others and his/her environment. In this way, he/she is encouraged to develop his/her self-awareness.

Deaf-blindness impedes also the child's opportunity to model and imitate the way other people move and act, which is an important factor in building the body schema. Therefore, forming an accurate body image becomes an interventional goal, which is further analyzed in daily routine activities and games. Every movement should be demonstrated by physical manipulation and appropriate movement of the child's body parts.

For the deaf-blind child, its body becomes the basic means of communication and information perception. The visual and auditory communication is replaced by tactual communication. The deaf-blind child understands that tactile cognitive signs and cues may be perceived and transmitted through the body [1,2,17,18].

Gradually, the deaf-blind child becomes aware through experience that he/she has a body with different body parts performing different functions, and in turn learns to recognize them. The development of body awareness is connected to the conceptualization of the spatial orientation and therefore to the child's movement and orientation.

3   Methodology

The presented study was part of a more extensive research in developing the screening inventory for the deaf-blind students' cognitive and communicative profile and in implementing a program of intervention. The study used a qualitative research methodology and adopted an interpretative position. The aim of the inquiry was descriptive. The case study methodology was followed and data were collected using the method of direct observation.

The uniqueness of each deaf-blind child and the heterogeneity among the deaf-blind population led us to the application of the methodological approach of case study.

The research program consisted of three stages; In the first stage the student's communicative and cognitive profile was screened. In the second stage the recorded screened difficulties framed the planning and the implementation of the educational program of intervention. In the third stage the results from the implementation of the educational program of intervention were assessed.

The screening inventory and the interventional program were implemented through specially designed educational material (multisensory activities which focused on the use of the remaining senses of touch, smell and sensory cues from temperature, air blow and vibration), assistive technology (assistive devices, Braille), and augmentative and alternative expressive and receptive communication systems (tactile sign language, pictograms, objects of reference, tactile symbols).

3.1   The Cognitive and Communicative Profile

The Laboratory of Special and Curative Education (LABESPEC) at the University of Ioannina, in Greece, has implemented the research program which aimed at developing a screening inventory of the deaf-blind students' cognitive and communicative profile. The students' profile was screened through the use of alternative and adapted multisensory approaches. The recorded difficulties aimed to define and direct the planning of the individual educational interventional program.

The profile aspired to present an initial and structured framework of the deaf-blind students' potential focusing on specific sections (communication, sensory, social-emotional, motor, cognitive development and daily living skills). The gathered elements were used for the purpose of developing the interventional educational plan which would be incorporated into an adapted curriculum.

3.2   The Program of Intervention

The individual educational plan consisted of and elaborated on the elements of communication, motor, cognitive and social-emotional development. The individual educational plan aimed at encouraging the active presence and direct interaction of deafblind students in their environment. An environment, which they controlled, understood and anticipated through multisensory and accessible approaches. [23]

The educational program with deaf-blind students starts with the senses, moves to perception and ends up in cognition [19]. Deaf-blind students should be introduced to concepts which rely on their emotional experiences and sensory exploration [20]. They need not only to perceive and evaluate the sources of information, but also to develop and widen their conceptual background in order to correlate new experiences to previous ones by understanding, reasoning and interpreting the sensory inputs. Thus, an active and communicative environment is needed, which offers the child opportunities for interaction. As a result, early concept development is promoted and grounded on concrete experiences [1,2,21,22].

An educational program of intervention is presented in this article, which aimed at introducing the body schema concepts through a tactile and more concrete perspective. In planning this intervention we took into consideration the difficulties in concept formation and in the body schema perception, the limited incidental interaction with the environment and the significance of the concept development, through which the conceptual background is defined.

3.3   The Interventional Approach

The case study reported here was conducted at a school for deaf-blind students and was incorporated in the educational program for school-age students who are deaf-blind. The student was diagnosed with brain damage, congenital blindness and deafness due to viral infection, and therefore was eligible to follow a special educational program for deaf-blind students. The student communicated in others people's hands through tactile sign language. She also used objects of reference, pictograms, tactile symbols, tactile cues, Braille and facial expressions, as means of receptive and expressive communication. The student liked to share social interaction with other students and educators.

In this presentation we focused on the student's difficulty to form the concept of body schema. The program of intervention was designed according to the screened difficulties of the student. More specifically, the student could partially locate the stimulated body parts but she could not always name and match her body parts. Through this program of intervention the student was expected to name the body parts (through tactile sign language, tactile and touch cues), locate them in herself and others, and recognize their function.

The educator chose activities that encouraged the student to tactually locate the body parts and areas that are stimulated. Student's interest was stimulated and triggered using multisensory material (such as body lotion, skin cream, cloths, sticky tape, clips), while the student was asked, through tactile sign language and tactile, touch cues, to locate the body part, find and remove what has being attached to.

The activities were structured experientially, designed in sequences of movements and focusing on daily activities in which the student was actively participating (dressing, washing, role playing). At the beginning the student's hand was guided to find the named body part co-actively with the educator. Gradually, the student controlled the interaction and initiated movements and sequence of sensory activities. The student was asked, through tactile sign language, tactile and touch cues, to move the appropriate body part and match it to those of the educator or a doll. The student communicated through alternative communicative systems (tactile sign language, tactile symbols, tactile and touch cues). In parallel, objects (plastic rings, bracelets, cloth) were placed on the students' named body parts as cues to discerning them. The student seemed to enjoy participating in rhythmical activities. These activities were enriched with songs in tactile sign language and finger games involving naming and matching body parts.

Gradually, guidance was given to family members to emphasize naming the body parts as the child was actively participating in her daily routines at home (dressing, bathing, eating). The deaf-blind students, because of the restricted sensory information, benefited from exposure to multiple opportunities to generalize a concept in different contexts. [9].

The process that may at first be described as a sequence of sensory activities has been gradually developed to trigger internal representations. The student became aware of her body and controlled her body parts, while matching them with those of the educator. Associated with the body schema formation was the fact that the student became aware of having a body with a center (midline) and two sides. Thus, cognitive development was fostered through the expansion of experiences [24, 2, 1, 25]. In parallel, she expanded her body schema awareness through the activity of sorting clothes and matching them with the body part the clothes are worn to. The student succeeded in discerning her body parts, moving and isolating them purposefully.

4   Conclusions

In the case of deaf-blind students, who face a combination of visual and hearing loss, the deprivation of environmental stimuli is often profound and of primary concern for their educators. The learning capacity is often so greatly reduced that special intervention is required in the form of alternative communication and teaching techniques. The education of deaf-blind students presents a special challenge underlined by a key concept; a structured, predictable, adapted and accessible environment leads to controlled, intelligible and anticipated received information and stimuli. As a result, the communication and concept development, together with the student's autonomy should be promoted when teaching this student population.

References

1. Alsop, L.: Understanding Deafblindness. Issues, Perspectives, and Strategies. SKI-HI Institute, Utah (2002)
2. McInnes, J., Treffry, J.: Deaf-Blind Infants and Children. University of Toronto Press, Canada (1993)
3. Prickett, J., Welch, T.: Deaf-Blindness: Implications for Learning. In: Huebner, K., Prickett, J., Welch, T., Joffee, E. (eds.) Hand in Hand, pp. 25–60. AFB Press, NY (1995)
4. U.S. Dept. of Education (IDEA§300.8(C)(2)]). http://idea.ed.gov/explore/view/p/,root, regs,300,A,300%252E8
5. Mamer, L., Alsop, L.: Intervention. In: Alsop, L. Understanding Deafblindness. Issues, Perspectives and Strategies, pp. 57–94. SKI-HI Institute, Utah (2000)
6. Stavrou, L., Gibello, B., Sarris, D. : Les problèmes de symbolisation chez l'enfant déficient mental: Approche conceptuelle et étude clinique. Scientific Review of School of Education, vol. A', pp. 187–217. University of Ioannina (1997)
7. Aitken, S.: Understanding deafblindness. In: Aitken, S. (ed.) Teaching Children Who Are Deafblind, pp. 1–34. David Fulton Publishers, London (2000)
8. Stavrou, L.: Teaching Methodology in Special Education (Διδακτική Μεθοδολογία στην Ειδική Αγωγή). Anthropos, Athens (2002)
9. Hodges, L.: Effective Teaching and Learning. In: Aitken, S. (ed.) Teaching Children Who Are Deafblind, pp. 167–199. David Fulton Publishers, London (2000)
10. Stavrou, L., Sarris, D.: L'image du corps chez les infirmes moteurs cérébraux (IMC) au travers des épreuves projectives. European Journal On Mental Disability 4(16), 17–23 (1997)
11. Stavrou, L., Zgantzouri, K.A., Stavrou, P.-D. : Mechanisms involved in the formation of body and self image in individuals with psychotic symptoms. In: 6th International Psychoanalytic Symposium of Delphi, Greece, p.29 (2004)
12. Stavrou, P.-D.: Corporal Schema and Body Image (Σωματικό Σχήμα και Εικόνα Σώματος). In: Pourkos, M. (eds.) Tangible Mind. Frameworked Knowledge and Education (Ενσώματος Νους. Πλαισιοθετημένη Γνώση και Εκπαίδευση), pp. 377–402. Gutenberg, Athens (2008)
13. Stavrou, P.-D. : Mediation and guidance of containers and contents of children's thoughts : prevention and risk treatment of disharmony and early psychotic disorders / médiation et guidance des contenants et contenus des pensées enfantines : prévention et soin des risques de dysharmonies et de troubles psychotiques précoces. Dissertation, European doctorate in clinical psychopathology, Université de Picardie Jules Verne, France (Laboratoire de Psychologie Appliqué - LPA) and University of Lund: Sweden (Institute of Psychology) (2014)
14. Cash, T., Pruzinsky, T.: Body image. A handbook of theory, research, and clinical practice. The Guilford Press, London (2002)
15. Karabatzaki, Z.: Topics on Special Education (Θέματα Ειδικής Αγωγής και Εκπαίδευσης). Paralos, Athens (2010)
16. De Preester, H., Knockaert, V.: Body Image and Body Schema. Jon Benjamins B.V., Amstredam (2005)
17. Holmes, N.P., Spence, C.: The body schema and multisensory representation(s) of peripersonal space. Cognitive Process 5(2), 94–105 (2004)
18. Maravita, A., Spence, C., Driver, J.: Multisensory Integration and the Body Schema: Close to Hand and Within Reach. Current Biology 13, 531–539 (2003)

19. Walsh, S., Holzberg, R.: Understanding and Educating the Deaf-Blind Severely and Profoundly Handicapped. C.C.Thomas Publisher, USA (1981)
20. Miles, B., Riggio, M.: Remarkable Conversations. Perkins School for the Blind, Massachusetts (1999)
21. de Vignemont, F.: Body Schema and Body Image- pros and cons. Neuropsychologia **48**(3), 669–680 (2010)
22. Meng Ee, W., Chia Kok Hwee, N.: The effects of impaired visual exteroception on body schema in the drawing and 3-D plasticine modeling of human figures: A case study of three preschool children with total congenital blindness due to retinopathy of prematurity. Journal of the American Academy of Special Education Professionals, 5–24 (Winter 2010)
23. Zeza, M., Stavrou, P.-D.: Program of intervention in deafblind students: the framework of the cognitive and communicative profile of deafblind students and the application of educational plan of intervention. In: 8th ICEVI European Conference on Education and Rehabilitation of People with Visual Impairment « A Changing Future with ICF», Istanbul – Turkey (2013)
24. Recchia, S.L.: Play and Concept Development in Infants and Young Children with Severe Visual Impairments. Journal of Visual Impairments and Blindness **91**(4), 401–406 (1997)
25. van Dijk, J.: An educational curriculum for deaf-blind multi- handicapped persons. In: Ellis, D. (eds.) Sensory Impairments in Mentally Handicapped People, pp. 374–382. Croom Helm Ltd., London (1986)

Using Map-Based Interactive Interface for Understanding and Characterizing Crime Data in Cities

Zhenxiang Chen[1(✉)], Qiben Yan[2], Lei Zhang[1], Lizhi Peng[1],
and Hongbo Han[1]

[1] University of Jinan, Jinan 250022, People's Republic of China
{czx,Zhanglei,plz,nic_hanhb}@ujn.edu.cn
[2] Shape Security, Mountain View, CA 94040, USA
yanqiben@gmail.com

Abstract. Crime Data Analysis is vital to all cities and has become a major challenge. It is important to understand crime data to help law enforcement and the public in finding solutions and in making decisions. Data mining algorithms in conjunction with information system technologies and detailed public data about crimes in a given area have allowed the government and the public to better understand and characterize crimes. Furthermore, using visualization tools, the data can be represented in forms that are easy to interpret and use. This paper describes the design and implementation of a map-based interactive crime data web application that can help identify spatial temporal association rules around various types of facilities at different time, and extract important information that will be visualized on a map.

Keywords: Map-based interactive interface · Crime Data · Data Mining · Association rules

1 Introduction

A major challenge facing law enforcement agencies and the public is understanding crime patterns to predict future occurrences of offences. For example, in areas where there are repetitive offences, these offences can be spatially and/or temporally related. The banks in a local shopping center with low police coverage might be targets for robbery crimes. Traditional data mining techniques such as spatial-temporal association rules can be used to determine and predict these crime patterns. This information will help law enforcement and the public in making decisions and finding solutions, i.e. improving police coverage, avoiding certain locations at certain times, enhancing visibility, etc. A traditional data mining technique that has been used in crime data mining is the association analysis technique. It discovers frequently occurring items in a database and presents the association rules between events.

© Springer International Publishing Switzerland 2015
Y. Tan et al. (Eds.): ICSI-CCI 2015, Part III, LNCS 9142, pp. 479–490, 2015.
DOI: 10.1007/978-3-319-20469-7_51

In this study, the association rule technique will be used to determine crime patterns based on a spatial-temporal crime data. The association rules will be derived from patterns found in the dataset which will predict the occurrences of similar crimes pattern or other related crimes. The occurrences of some crimes can also be related to the occurrences of other crimes in an area.This research describes the design and implementation of an integrated data mining web application that can help identify spatial-temporal patterns and extract useful information from a crime database system. The data will be visualized on a map which can help researchers, law enforcement, and the public explore the dataset and retrieve valuable information. The components of such application consist of a database engine, a data analysis component, and a visualization Interface.

The remainder of this paper is organized as follows. Section 2 of this research discusses some related work in the field of crime data mining. Section 3 of this research discusses the proposed data analysis approach, and section 4 focuses on the detailed design of the web application. Section 5 presents the implementation of web interface. We show the association results in Section 6. Finally, we conclude this paper in section 7.

2   Related Work

There are several related work published in the field of crime data mining. Some of those publications are written on the subject of spatial and temporal pattern discovery and association rule mining techniques [1-6], while others are focusing on real-world applications [7-9]. The most related works to our study are [2, 8]. In [8], a general framework is suggested for crime data mining with some examples. The paper presents a general framework for crime data mining that draws from the experience gained with the Coplink project. The paper emphasizes using available data mining techniques to increase efficiency and reduce data analysis errors for real-world applications of law enforcement organizations. Our study is related to [8] in the sense that this general framework helps our study define different crime types and related security concerns. The suggested framework and the case study are informative and beneficial for this study.

In [2], the author discusses the methods for mining spatial association rules in a geographic information system (GIS) database. This paper demonstrates the effectiveness and efficiency of developing an association rule algorithm that can be used to provide spatial association rules based on users specific query. This paper provides methods that efficiently explore spatial association rules at multiple approximation and abstraction levels. Our study is related to [2] in the sense that a similar spatial association rule and technique can be adopted in exploiting the spatial association rules for crime data and facilities suggested in this study. Our method is different from [2] in three aspects: (1). we consider crime data applications; (2). we also consider many non-spatial objects in our rule discovery process; (3). our user query is more naïve than the complicated users query dealt with in their study.

3   Proposed Approach

Our general data analysis engine consists of three major parts: data preprocessing, ranking service and spatial association service. The data preprocessing
will import the data sets, delete incomplete (or invalid) data and list the data
in a way facilitating efficient data retrieval. In our project, we have two data
sets to deal with: the crime data sets around the CITY1 area and spatial data
sets for different facilities around the same area. The importation of data sets is
explained in section 4.3. After the data importation, the crime data are listed
in the database server, while the spatial data for facilities are also listed in the
same database server. During the data importation, we search for the invalid
crime data which are discovered to miss some important attributes for analysis, and delete them afterwards. Finally, both crime data and spatial data for
facilities are ordered and organized in a way that efficient data retrieval can be
performed.

3.1   Crime Ranking Service

Crime ranking service is the first service provided by the data analysis engine.
Rank is one important outcome which is quite meaningful to the users, and can
be visualized in various graphical manners. The users are prompted to input the
addresses that they are interested in, which can be their travelling destinations,
home addresses, workplaces, etc. Also, the users need to specify an area range
around this specific location. The users can also input the time period that they
are concerned with, and by default which is that of the whole crime data sets.

According to the users query, the crime data inside the interesting area within
the time period are retrieved. We provide two ranking results for one specific
query: (1). rank by crime types; (2). rank by TPD (time per day). To obtain the
first ranking result, we can compute the numbers of crimes for each crime type in
the retrieved data. Then, it becomes straightforward to rank different crime types
according to their calculated occurrence frequencies. This rank provides the users
a sense of which types of crimes are most likely to happen in the specified area.
For the second ranking result, we first specify different time notations in one
day, listed as:(1)Day:8am to 4pm (2)Evening:4pm to 6pm; (3)Mid-Night:12pm
to 8am.Without losing generality, other notations can also be used in this service.

3.2   Multiple-Level Spatial Association Rules Discovery

The spatial association service is another major service for data analysis. The
main technique follows the efficient algorithm for mining spatial association rules
in [2]. Each spatial association rules will be evaluated according to their support
and confidence:

1)The support of a conjunction of predicate P in a set S, denoted as $\sigma(P/S)$,
is the number of objects in S which satisfy P versus the cardinality (i.e. the total
number of objects) of S.

2)The confidence of a rule P→ Q is in S, ψ(P→ Q/S), is the ratio of σ(P∧ Q/S) versus σ(P/S), i.e. the possibility that Q is satisfied by a member of S when P is satisfied by the same member of S.

The users are mostly interested in the patterns which occur relatively frequently (or with large supports) and the rules which have strong implications (or with large confidence). The rules with the above patterns with large supports and large confidence are called strong rules. The main techniques of the association rule discovery service are illustrated as follows. Our data have the following database relations for representing spatial and nonspatial objects: (1)facility (name, type, geo);(2)crime (type, geo, time) and (3) TPD (name).

In the above relationship schemata, "geo" represents the spatial object. The facility "types" correspond to the types of facilities such as schools, hospitals, etc. "TPD" represents the time per day object, illustrated in the previous section. We consider multiple-level association rules. For example, high-level association rules are written as follows:

close_to(facility) ∧is_in(nigh time)→has(crime)

Basically, we consider simple two-level association rules, concept hierarchies are defined as follows:

-A concept hierarchy for facility:(facility(hospital, school, bar, shopping center))

-A concept hierarchy for crime:(crime(theft, robbery, homicide))

-A concept hierarchy for TPD: (TPD(daytime(morning, afternoon),nighttime (evening, late night)))

For the coarse concept level, Table 1 will be built according to [2]:

Table 1. First/coarse concept level

location	facility	crime	TPD
$around(x1, y1)$	$< close\ to, facility >$	$< has, crime >$	$< is\_in, daytime >$
$around(x2, y2)$	$< close\ to, facility >$	$< has, crime >$	$< is\_in, nighttime >$
...	...	...	...

The above location is the location of the corresponding facility. The size of this table is equal to the number of facilities considered. According to the above coarse concept level table, we can compute the counts for large k-predicate set. During the computation, we compare the number of counts to the thresholds for minimal support and confidence. The entries with lower support or confidence than the thresholds are filtered out.Then, refined computations of detailed spatial relationships at the second concept level are done based on the following Table 2:

After that, the large k-predicate set for the second concept level can also be derived. Finally, the corresponding mining rules can be discovered. The whole association rules discovery method proceeds as follows:

Step1: Prepare the spatial relationships for two different concept levels;

Table 2. Second/fine concept level

location	facility	crime	TPD
$around(x1, y1)$	$< close\ to, hospital >$	$< has\_crime, theft >$	$< is\_in, morning >$
$around(x2, y2)$	$< close\ to, school >$	$< has\_crime, robbery >$	$< is\_in, afternoon >$
...	...	...	...

Step2: Compute the counts of k-predicates for the coarse concept level;

Step3: Filter out the predicates with supports or confidence lower than the minimal thresholds at the first level;

Step4: Compute the counts for k-predicates of refined concept level;

Step5: Find the large predicate at the refined concept level and find association rules.

3.3   Overall Structure for Data Analysis

The overall structure starts from users query, as illustrated in Fig. 1. Because we only have fixed crime and facility data sets, we are not able to update the association rules. Therefore, we only show the pre-computed association rules for the users. How to efficiently generate association rules for newly input data with an online manner will be a promising future direction.

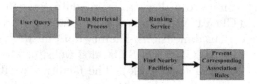

Fig. 1. Data Analysis Procedure

4   System Design

4.1   System Architecture

The system development architecture is drawn in Fig.2. The data analysis engine will deal with data preprocessing and data association,as mentioned in section 3.

4.2   Facility and Crime Data

Facility and crime data in the system architecture of Fig.2 is obtained from the CITY1 GIS Catalog website. The crime data was in XML format that was imported into the database engine. The facility data was in a Shape-file format (ESRI format). After importing the facility data into ArcMap, the facilities

Fig. 2. Data Analysis Procedure

attributes tables were exported into an excel file. The most important attributes in the facility table are the type of a facility, its geographical location (x and y coordinates,) its address, its description, etc. Since the location data (x and y) are the projected values (in meters,) it must be converted to latitude and longitude (in degrees) using geographical coordinate system conversion methods (R tools.) Another way is to geocode an address of a facility using geocoding functions part of the Google Map API V3. We are planning to use the techniques of DIV and CSS to make the web-based application look more organized and fancy.

4.3   Data Set Preprocessing

The data set to be analyzed during this project has been provided to us. And all of the data we are planning to utilize are crime data sets which are associated with the areas around CITY1. With a relational database, we are able to analyze the crime data from various dimensions. Facility data in the system architecture of Fig.2 is obtained from the CITY1 GIS Catalog website. The facility data was in a Shape-file format (ESRI format file). The facilities attributes tables were exported into an excel file.Since we have the extracted text files for each facility where we separated each attribute with tabs, we are able to move forward to process every text file with a small program written in C♯. After we obtained all the text files we are willing to analyze, we can utilize the information inside to get the actual address for each facility. With obtaining the actual address to each facility, we can plug in the address into the function of geocoding, and get the precise latitude and the longitude as the result, in addition, we will generate a facility number to every facility, and then we can insert all the data we have into the database for the facilities. This is the basic idea of our geocoding part.

While the data preprocessing didnt just stop here, there are still something to be taken care of. For instance, when we are applying the geocoding to the addresses that we got from the text files, we cannot send the request too fast, since if we send the request too fast to the Google server, the Google server will send one error result back to us: "OVER_QUERY_LIMIT", which means the server is too busy to respond all the requests we are sending it. And if such situation happened the overall database will be ruined, since when the server is sending the "OVER_QUERY_LIMIT" back, the result latitude will not be updated, which means, several facility will share one same pair of latitude and longitude as a terrible result.

4.4   Visualization Component Architecture

This is the visualization component architecture, it consists of several parts. The layout of the web-based application will be formed by the Ext Js 4.1 which is a Javascript application framework that works on all modern browsers from IE6 to the latest version of Chrome. It enables us to create the best cross-platform applications using nothing but a browser, and has a phenomenal API. With the utilization from the Ext Js 4.1 we can make our application more fancy and user friendly, for instance, with applying the APIs from libraries which are provided by the Ext Js, we can arrange our layout with windows, tabs, forms and charts that can make our web-based application visualized with a better understanding of the boring raw data output from the analysis.

The most important part of visualization is to show every facility on the Google map and show their ranking status and distribution to the users. Here we applied the functions of addMarkers (IMMarkerConfigs) and removeMarkers (IMCurMarkers) to control the markers on the Google map. The parameter of IMCurMarkers is an array of the result from the json data from upper steps. It contains the information about the latitude and the longitude, with these parameters, we are able to pinpoint the icons on the Google map. The layout of the web page is designed under the framework named Extjs, this is a javascript API library, we utilized some of its APIs to fulfill some of our web applications function and the layout format of our web page on the left hand side is named "accordion", under this layout each window can be expanded or collapsed, with this layout we can make our web page more compact.

5   System Implementation

We can firstly start from the association rules data table generation. The way we generate the association rules table is that, we utilize the two datasets that we have generated in the previous months, namely, the crime data and the facilities data. Since we have more than 120,000 crime cases, and 988 facilities contained in the facility dataset. It is much easier for us to generate an association rules table with the association rules starts from the facilities to the crime counts, for example, the order of our association rules table is like this: FID, OTYPE, LAT, LON, STOLENAUTO, ROBBERY, ARSON, THEFTFAUTO, SEXABUSE, HOMICIDE, BURGLARY, ADW, THEFT, OTHER, EVE, MID, DAY, UNK, ADD, which are the facility id number, facility type , the latitude of the facility, the longitude of the facility, number of stolen auto cases, number of robbery cases, number of arson cases, number of theft from auto cases, number of sex abuse cases, number of homicide cases, number of burglary cases, number of assault with deadly weapon cases, number of theft cases, other cases, number of case happened in the evening, midnight and daytime, the last one is the case with no specific time period is noted. This table is the heart for our research, since all the data generated here will be later visualized and displayed on our web-based application.

As a result to this approach, we obtained a table with expanded columns, from original 5 columns to finally 19 columns. Since some of the area near one specific facility was not containing any crime cases, we regarded these facilities as "safe" area, in order to make the further analysis more convenient, we also transformed this table into binary format that if the area contains no crime case, we assign binary number "0" to the crime type, otherwise, we assign binary number "1" to the crime type indicating that the type of crime was observed within the defined area around the facility. We applied the approach to obtain a better insight of the correlation between one specific facility and one type of crime in the perspective of association rule. We also take the time interval into consideration. According to the flaws in the raw data provided by the CITY1 catalog web site, all the crime data are lack of the information about what time the crime case was reported, for every case, the information in the column of reported time was all 00:00:00. The layout of our web-based application was written in Javascript mostly, and there are several API were utilized to fulfill the function of the interaction for the data and control requests. The Extjs Javascript APIs library was utilized mostly for the generation of our layout to the application. There are mainly two parts of components embedded in the layou. the control panel and the web-based interface was displayed in the Fig.3.

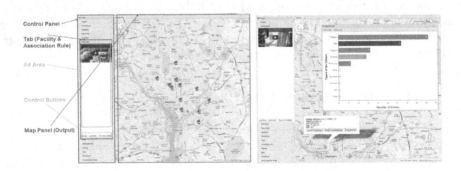

Fig. 3. Application Layout and Web-based Application Interface

In order to fulfill the function of data and command interaction, we applied two libraries to complete the functionality, the jQuery and the Extjs both applied Ajax technology. The basic idea to the data interaction is this: we formed a web service in the back stage of our website, and all the direct control commands such as the SQL queries are functioned in the web service class. And then, we applied the APIs provide by the Javascript libraries to send some web request back to the web service class formed in the back stage of the site via the Ajax technology. Later, as long as the web service get the web request it will send the direct SQL command to the database where store all the data we have generated earlier through our data analysis process. Then, the database will send all the requested data to the web service, after a brief data format transformation, the data will be sent to the front stage web application similarly via Ajax.

6   Results

We have obtained a lot of interesting association rules from crime to facility indicating which facility has high frequency of a specific crime, from crime to time per day indicating which time period has high frequency of a specific crime,in Fig.4. We have association rule from another direction, i.e. from facility to crime indicating which crime is most likely to happen in this specific facility. Also we have association rule from facility to time per day indicating which time-per-day is most likely to have crimes around this specific facility.

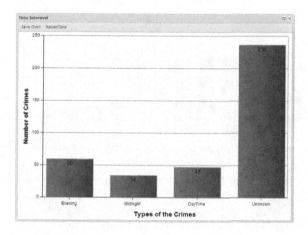

Fig. 4. Time Interval result

On the MAP interface, we only show the high frequency association results, and delete all the infrequent association results. We determine the high frequency by manually setting a threshold for support and confidence. In general, each case will generate five or six rules. We show the most useful results for our users. Homeless service area is the most dangerous place to stay. Homeless service area has highest density of StolenAuto crimes, Robbery crimes, SexAbuse crimes, Burglary, A sault with Deadly Weapon. Hotel has the highest density of Theft from Auto crime, because a lot of cars parked around hotel. Bank has highest density of Theft crimes, which is very straightforward. Criminals like to wait for the persons coming out of the banks to steal their money. One interesting point we find out is that Police station has very high density of ADW crimes, i.e. people are likely to use deadly weapons to deal with police officers. Many police stations are located in low income residential areas where crime is high, top five facilities Most crimes happen are Homeless Service,Public School,Metro Station,Hotel and Gas Station (Fig.5).

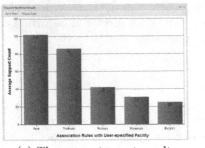 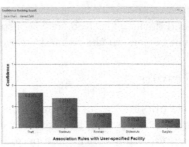

(a) The support count result   (b) The confidence result

Fig. 5. Association Rules Results from Homeless Service Facility to Crimes

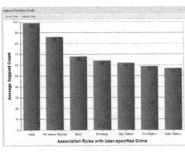 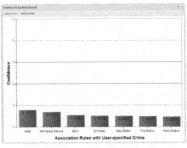

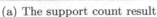 

(a) The support count result   (b) The confidence result

Fig. 6. Association Rules Results from Theft-From-Auto Crimes to Facilities

Regarding time, Bank, metro station and hotels have high density of crimes in day and evening. In midnight, Bank, Homeless service, and hotel have high density of crimes. Look at that, homeless service again is very dangerous even in mid night. For example, during the night areas around homeless shelters are usually crowded with homeless persons seeking shelters.From another association direction, we find that, around most facilities, theft and theft from auto are highest density crimes. Except in public school and police station, ADW is very high. Criminals are more likely to use deadly weapons around police station and public schools. So, stay safe if you are around these areas. Overall,around most facilities, the top five crimes are:Theft,Theft from Auto,Robbery,Stolen Auto,Burglary.(Fig.6)

Regarding time, most crimes happen during unknown time. If we do not consider unknown time. Around almost all facilities, evening has the highest crime density, midnight has the lowest , except around police station and embassy. Both of them have very high crime density during Midnight. Which probably means around police station and embassy, many officers are patrolling during midnight. So they will more likely to find crimes happening during midnight.Based on the findings of the study, it is vital for police and the public to consider such findings

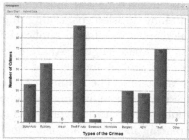
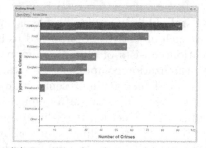

(a) The histogram for the number of crimes

(b) Ranking for the number of the crimes

Fig. 7. The Crime Count Ranking Results

when making decisions. Police may consider where and how to allocate resources, what measures are needed to prevent crimes, and what other options are available. The public may also consider these finding to determine which areas are safe and which areas they should avoid.

When it comes to crimes occurring at specific locations(Fig.7), the crime patterns inferred will motivate the police to create a list of potential measures that may prevent or reduce these crimes. For example, at locations where crimes such as theft, burglary, robbery, auto theft, etc., the police may allocate resources such as personals, cameras, signs, and other security measures to prevent or ease the crime rate at these locations.In the specific case of locations such as Homeless Services and shelters where many different types of crimes occur, the police may find ways to reduce the crimes in these areas using several methods and techniques such as allocating more resources in the area for monitoring or through community programs and services. The same can be said about other locations such as police station, banks, Hotels, etc. Decisions can also be made regarding high density crimes during the night where crimes are much higher than during the day.

In regard to the public, they become more educated and aware of the hot spot areas.Their decisions become based on facts not on myths. As an example, a person may not wonder in a hot spot area in the middle of the night, a person may become more aware and alert when visiting a bank area where crime density is high, and finally, a person may avoid hanging around location that serves the homeless population. Most importantly the public will collaborate with police to reduce crime. Crime rates will become much less and safety is much better.

7   Conclusion

In this paper, we have laid out the objective of the project, technologies used, proposed approach, and the system design. Our main objective is to analyze a given crime dataset and present the information to users in a meaningful and more understandable forms so that it can be utilized by law enforcement and the general public in solving problems and making decision concerning safety. To achieve such objective, we proposed to implement a web based application.The application consists of several components: Database engine, Analysis engine, and a visualization component. The result of the project showed that such technique is very beneficial in presenting data meaningfully.The findings of the study showed the vital rule that data mining techniques plays in helping the police and the public make decision regarding crimes and safety. An informative data will enable the police to finding solutions by creating measures against crime. The same data will also help the public understand crime data and act upon it.

Acknowledgments. This work was supported by the National Natural Science Foundation of China under Grants No.61472164 and No.61203105,the Natural Science Foundation of Shandong Province under Grants No.ZR2014JL042 and No.ZR2012FM010.

References

1. Mohan, P., Shekhar, S., Shine, J.A., Rogers, J.P.: Cascading spatio-temporal pattern discovery: A summary of results. In: Proceedings of the 10th SIAM Data Mining (SDM 2013), pp. 327–338. SIAM (2013)
2. Koperski, K., Han, J.: Discovery of spatial association rules in geographic information system. In: Egenhofer, M., Herring, J.R. (eds.) SSD 1995. LNCS, vol. 951, pp. 47–66. Springer, Heidelberg (1995)
3. Huang, Y., Shekhar, S., Xiong, H.: Discovering Colocation Patterns from Spatial Data Sets: A General Approach. IEEE Transactions on Knowledge and Data Engineering **16**(12), 1472–1485 (2004)
4. Clementini, E., Felice, P.D., Koperski, K.: Mining Multiple-level Spatial Association Rules for Objects with A Broad Boundary. Journal of Data and Knowledge Engineering **34**(3), 251–270 (2000)
5. Bogorny, V., Kuijpers, B., Alvares, L.: Reducing Uninteresting Spatial Association Rules in Geographic Databases Using Background Knowledge: A Summary of Results. International Journal of Geographical Information Science **22**(4), 361–386 (2008)
6. Bogorny, V., Kuijpers, B., Alvares, L.: Semantic-based Pruning of Redundant and Uninteresting Frequent Geographic Patterns. GeoInformatica **14**(2), 201–220 (2011)
7. Liu, X., Jian, C., Lu, C.T.: Demo paper: A spatio-temporal crime search engine. In: Proceedings of the ACM GIS 2012, pp. 528–529. ACM (2012)
8. Chen, H., Jie, J., Qin, Y., Chau, M.: Crime Data Mining: A General Framework and Some Examples. IEEE Computer **37**(4), 50–56 (2004)
9. Shah, S., Bao, F., Lu, C.T., Chen, I.R. : CROWDSAFE : Crowd sourcing of crime incidents and safe routing on mobile devices (demo paper). In: Proceedings of the ACM GIS 2014, pp. 521–524. ACM (2014)

Author Index

Printed in the United States
By Bookmasters